PULSAR ASTRONOMY

Unveiling Compact Stars with China's New Facilities

PULSAR ASTRONOMY

Unveiling Compact Stars with China's New Facilities

Editors

Zhifu Gao
Xinjiang Astronomical Observatory, China

Renxin Xu
Peking University, China

Jorge Horvath
University of São Paulo, Brazil

César Augusto Zen Vasconcellos
Universidade Federal do Rio Grande do Sul (UFRGS), Brazil &
International Center for Relativistic Astrophysics Network (ICRANet), Italy

World Scientific

NEW JERSEY • LONDON • SINGAPORE • BEIJING • SHANGHAI • HONG KONG • TAIPEI • CHENNAI • TOKYO

Published by

World Scientific Publishing Co. Pte. Ltd.
5 Toh Tuck Link, Singapore 596224
USA office: 27 Warren Street, Suite 401-402, Hackensack, NJ 07601
UK office: 57 Shelton Street, Covent Garden, London WC2H 9HE

Library of Congress Control Number: 2024943387

British Library Cataloguing-in-Publication Data
A catalogue record for this book is available from the British Library.

PULSAR ASTRONOMY
Unveiling Compact Stars with China's New Facilities

ISBN 978-981-98-0063-6 (hardcover)
ISBN 978-981-98-0064-3 (ebook for institutions)
ISBN 978-981-98-0065-0 (ebook for individuals)

For any available supplementary material, please visit
https://www.worldscientific.com/worldscibooks/10.1142/14036#t=suppl

Desk Editors: Nambirajan Karuppiah/Carmen Teo Bin Jie

Typeset by Stallion Press
Email: enquiries@stallionpress.com

We dedicate this book to our families.

Zhifu Gao

Xinjiang Astronomical Observatory, Chinese Academy of Sciences, Xinjiang, People's Republic of China

Renxin Xu

School of Physics and State Key Laboratory of Nuclear Physics and Technology, Peking University, Beijing, People's Republic of China

Jorge Horvath

Instituto de Astronomia, Geofísica e Ciências Atmosféricas da Universidade de São Paulo (IAG-USP), São Paulo, Brazil

César Augusto Zen Vasconcellos

Instituto de Física, Universidade Federal do Rio Grande do Sul (UFRGS), Porto Alegre, Brazil and International Center for Relativistic Astrophysics Network (ICRANet), Rome, Pescara, Italy

The Editors

In memory of
Manuel Máximo Bastos Malheiro de Oliveira

In June 2024, Prof. Manuel Máximo Bastos Malheiro de Oliveira (a.k.a *Mané*) passed away after a long treatment to overcome a deadly disease. Mané was born in Portugal and spent all his career in Brazil, at the *Universidade Federal Fluminense* (RJ) first and the *Instituto Tecnológico da Aeronáutica* later on. He was a truly active scientist, very generous with all colleagues and students. His curiosity and insights were amazing. He made important contributions to Hadronic Physics and Relativistic Astrophysics, and supported all the activities aiming to improve research and education fields, such as the present volume.

For these reasons, we dedicate this book to his memory, as colleagues and friends.

https://doi.org/10.1142/9789819800643_fmatter

Foreword

Pulsars, since their discovery in 1967, have been regarded as natural laboratories for the study of matter under extreme physical conditions of density, gravity and intensity of magnetic fields. In recent years, with increased government investments in astronomical research, China has made great achievements in the fields of cosmology, astronomy and astrophysics. This scenario, combined with the millennial tradition of seeking to expand the frontiers of knowledge in China, led to the planning and construction of several large radio telescopes and the launch of a series of deep space exploration satellites. As a concrete result of this broad effort, today China is gradually advancing to the forefront of scientific research and technological innovation in the field of pulsar astronomy. This book focuses on pulsar astronomy, and presents the current and future important astronomical facilities in China, including the Five-hundred-meter Aperture Spherical Telescope (FAST), the Hard X-ray Modulation Telescope (HXMT), the Tian Ma 65 m radio telescope (Tianma), the Large High Altitude Air Shower Observatory (LHAASO), the QiTai radio Telescope (QTT), the Jingdong 120 m Radio Telescope (JRT), the enhanced X-ray Timing and Polarimetry mission (eXTP), the Gravitational Wave High-energy Electromagnetic Counterpart All-sky Monitor (GECAM), and the Hot Universe Baryon Surveyor (HUBS), as well as their new discoveries and scientific results. To date, FAST has discovered more than 800 new pulsars through its galactic plane survey and drift sweep. The high-precision millisecond pulsars found by FAST can be used to detect extremely low-frequency gravitational waves, establish pulsar timing patterns, and search for unknown objects in the solar system. For the vast majority of readers, this book undoubtedly represents a rich source of documentation, information and learning about pulsars and their impact on modern astrophysics and particularly about China's contribution to new achievements in this area.

https://doi.org/10.1142/9789819800643_fmatter

Preface

This book is divided into four parts containing 24 chapters in total. In the following, we present a synthesis of the chapters.

1. Magnetars

Magnetars are believed to be neutron stars mainly powered by their huge magnetic fields, which can reach up to 10^{16} G. They are rare but very important for studying matter and radiation under extreme conditions, as well as the origin and evolution of various energetic events in the Universe, such as gamma-ray bursts, supernovae, fast radio bursts, and gravitational waves. The author of Chapter 1, Sandro Maregetti, reviews some of the recent results on magnetars, many of which have been achieved with the Chinese facilities described in this book. These include the identification of magnetar giant flares in other galaxies, the detection of polarization in their soft X-ray emission, and the confirmation of the link between magnetars and fast radio bursts. As stated by Sandro Maregetti, in addition to the significant contributions already obtained in this field of knowledge from Chinese facilities, more achievements are expected in the near future. Among the most promising fields, it is important to mention the observations in the radio band exploiting the great sensitivity of FAST, and further advances in X-ray polarimetry studies with the eXTP satellite.

2. High-Energy Emission in SGRs/AXPs

The authors of Chapter 2, Ronaldo V. Lobato, Jaziel G. Coelho, and Manuel M. Malheiro, address the high-energy emission in SGRs/AXPs, neutron stars with extreme magnetic fields, identified as magnetars. On the basis of the outer gap accelerator model, the authors describe the generation of high-energy emission far away from the surface of the star in a combination of curvature radiation and inverse Compton scattering. According to the authors, the high-energy emission may be constrained by employing death-lines, the condition for pair-production and the generation of hard X and soft gamma rays, topics that deserve further investigation.

3. The Demographics of Isolated Neutron Stars from X-ray Survey Data

The ROSAT mission discovered seven radio-quiet isolated neutron stars (INSs, M7) not very far from the Sun. Their X-ray emission is steady and thermal, but hard to detect since they have no counterparts at other wavelength bands. They are ideal laboratory for testing atmosphere models and thermal evolution of matter under extreme gravity and magnetic field. The M7 spin slower than standard radio pulsars. In Chapter 3, Adriana Mancini Pires reviews the X-ray properties of INSs, suggesting crust heating by field decay and a link with magnetars. They may be part of a large, hidden population of neutron stars, as only one candidate was known beyond 1 kpc. The eROSITA instrument on the Spectrum-RG mission will provide the largest catalogue of X-ray sources ever, with unprecedented sensitivity and resolution. This will enable a better sampling of the elusive INSs that are missed by other surveys. Searching for new thermally emitting INSs requires the use of the high X-ray-to-optical flux ratios and the absence of counterparts of the M7. Follow-up identification campaigns require state-of-the-art facilities. The new INSs will be younger and hotter than the M7 on average, and will be located near massive star associations, open clusters, and supernova remnants. New thermally emitting INSs are found: three candidates have no optical counterparts and a soft thermal spectrum like the M7. Follow-up observations using advanced facilities will reveal more on the INS candidates. The M7 shows spectral complexity and long-term variations, challenging the theory and modeling of magnetized neutron stars.

4. Evolution of Long-Period Compact Radio Sources Driven by Winds

In Chapter 4, Jorge E. Horvath, Lucas M. de Sá, Lívia S. Rocha, Gustavo Y. Chinen, Lucas G. Barão and Márcio G.B. de Avellar address the nature and evolution of the long-period compact star radio sources, which has recently added several unexpected members. The central hypothesis is that particle winds drive their evolution, being an important factor for these relatively old sources. The authors show the consistency of this picture and remark some unsolved problems and caveats within it.

5. The Fundamental Performance of FAST with 19-Beam Receiver at L Band

FAST is the world's largest single-dish radio telescope, which was opened in 2016 and the first new pulsar discovered by FAST was made in 2017. It has achieved breakthroughs in pulsar, spectral line, and fast radio burst researches. In Chapter 5, Peng Jiang provides a comprehensive account of FAST's design, site selection, construction, and maintenance. This chapter consists of five sections, each presenting a different aspect of FAST. The first section introduces FAST and its main features. The second section explains how FAST was designed and commissioned. The third section details the 19-beam L-band receiver, the primary instrument for observations with FAST. The fourth section describes the backends and the standing waves that affect the data quality. FAST has unparalleled sensitivity, pursues five major scientific goals and new frontiers, such as fast radio bursts and exoplanets. FAST has already produced remarkable results and discoveries. Its ultimate goal is to promote our understanding of the cosmos.

6. Single-Pulse Study of Radio Pulsars Using FAST

Pulsars' single pulses are essential for understanding their radio emission. Scientists have been studying pulsars for over 50 years, but their emission process remains a mystery. The author of Chapter 6, Ji-Guang Lu, explains that this is because different pulsars have different emission properties, and the integrated pulse data are not enough to capture the full details of the emission. Certainly, the emission changes due to the pulsar's rotation, which adds another dimension to the problem. The single pulses are relatively random, showing the details of the particles in the magnetosphere. FAST is the world's best telescope for observing single pulses, as it has the highest signal-to-noise ratio. Ji-Guang Lu says that telescope sensitivity, not integration time, determines the quality of single pulse data. FAST is the most sensitive telescope and can get the best data. Pulsars' single pulses are both correlated and random. By analyzing these features from FAST data, we hope to solve the puzzle of pulsar emission.

7. FAST's Impact on Fast Radio Bursts

Fast Radio Bursts (FRBs) are short and intense radio flashes that have been a hot topic in astronomy since its discovery in 2007. In Chapter 7, Weiyang Wang and Kejia Lee review the current knowledge of FRBs and the CRAFTS survey conducted by FAST with high sensitivity. CRAFTS discovered five new FRBs, of which four were non-repeating and one was repeating. The non-repeating FRBs have large dispersion measures (DM) and are likely located at cosmological distances. The repeating FRB is very active and nearby, but had a complex environment. The polarization properties of FRBs have also been investigated with FAST. FRB 20180301A and FRB 20201124A show circular polarization and evidence of a magnetized medium along the line of sight. FRB 20201124A exhibits diverse morphologies and a 30-ms burst period. FRB 20200428A originates from a magnetar that transitioned to a radio pulsar. Its phase is different from the regular radio pulses, implying different emission regions for FRBs.

8. The Hard X-ray Modulation Telescope (HXMT)

Chapter 8, written by Min-Yu Ge and Fang-Jun Lu, introduces *Insight*-HXMT, China's first space astronomy satellite. It is a versatile X-ray telescope that explores high-energy objects such as black holes and neutron stars. The authors provide an overview of its development, launch, parameters, objectives, proposals, plan, performance, and data. They also emphasize its contributions to pulsar astronomy. Insight-HXMT, which was proposed in 1994 and launched in 2017, has three X-ray telescopes that cover the energy range of 1–250 keV. The engineering team of Insight-HXMT has successfully accomplished all the predefined engineering tasks, such as design, construction, ground calibration, and in-orbit testing. The in-orbit test results have shown that Insight-HXMT has met its design specifications in terms of effective area, energy resolution, and time resolution, laying a solid foundation for the research of high-energy celestial phenomena such as black holes and neutron stars, as well as the observation of intense short-duration events.

9. Advances in Neutron Stars from *Insight*-HXMT

In Chapter 9, Min-Yu Ge summarizes the achievements of Insight-HXMT in studying neutron stars. Insight-HXMT has discovered new cyclotron lines, including the highest-energy one at 146 keV in Swift J0243.6+6124, the first ultra-luminous X-ray source in the Hanoi catalog. It has also detected signs of a radiation pressure-dominated disk in this source. Additionally, Insight-HXMT has observed the Type I X-ray burst 4U1636-536 and found no hard X-rays between 40 and 70 keV. Furthermore, Insight-HXMT has identified the X-ray counterpart of FRB 200428 from a magnetar and analyzed the wide-band spectra of the Crab pulsar. The wide-band phase-resolved spectra of the Crab pulsar have also been investigated using Insight-HXMT's data. Insight-HXMT is currently in its extended mission phase, focusing on observations of neutron star-like objects, and more discoveries are expected.

10. HXMT Views on the Accreting Millisecond X-ray Pulsars

In Chapter 10, Zhao-Sheng Li reviews how *Insight*-HXMT observes accreting X-ray pulsars. These pulsars are important for China's astronomical projects and for understanding binary evolution, burst oscillations, and recycling theory. Only 23 of them have been found, and their emission sites and timing noise are debated. Insight-HXMT measures their spin frequencies, magnetic fields, and nuclear burning effects. The observations by Insight-HXMT have significant implications for understanding the evolutionary mechanisms of pulsar binaries, and they will contribute to the preparatory research for China-led eXTP high-energy telescope system, which aims to measure the mass and radius of neutron stars.

11. Pulsars at Shanghai Tian Ma 65 m Radio Telescope

The Tianma, a 65 m radio telescope in Shanghai that operates from 1.3 to 50 GHz, is the topic of Chapter 11. In this chapter, the authors, Zhen Yan, Zhi-Qiang Shen, Yajun Wu, Rongbing Zhao, and Qinghui Liu, address that the Tianma Radio Telescope has defined pulsar astronomy as one of its important scientific goals, and has carried out relevant observation technology and astrophysics research, considering the important physical significance and potential application value of pulsars. As reported in Chapter 11, a digital terminal system was introduced abroad and localized to meet the requirements of pulsar searching and online integration mode observations. A fast radio burst real-time search system and VLBI terminal were developed too. This chapter also reports on the Tianma's achievements in observing pulsars in the Galactic Center, monitoring magnetars powered with extremely large magnetic fields, discovering pulsar glitches (a sudden increase in spin-up), and improving the timing precision. The first high-frequency observations of multiple pulsars internationally, nearly doubling the integrated profiles of pulsars at 8.60 GHz are presented. Using the Tianma, the scientists are conducting research projects on high-frequency deep searches around the Galactic Center black hole and pulsar interstellar scintillation, aiming to achieve a series of breakthroughs in these fields.

12. The Large High-Altitude Air Shower Observatory (LHAASO)

The Large High-Altitude Air Shower Observatory (LHAASO), the new generation of cosmic ray observatory, is a national major scientific and technological infrastructure in China, with cosmic ray observation research as its core science. The authors of Chapter 12, Xin-Hua Ma and Zhen Cao, review the LHAASO experiments. In this chapter, the authors reveal that the LHAASO Mission has three arrays and other systems that cover 1.3 square kilometers, adopts various detection techniques, which can measure the reactions of gamma rays and cosmic rays in the atmosphere and reconstruct their basic information, and its unique multiple detection methods, cross-validation capabilities, and powerful gamma-ray and cosmic ray component discrimination abilities ensure the accuracy and reliability of the measurement results. They report that LHAASO can discover a large number of gamma-ray sources and achieve internationally leading sensitivity in the research of ultra-high-energy gamma-ray sources and cosmic ray component spectra and will provide an important experimental platform for conducting frontier interdisciplinary scientific research in areas such as atmosphere, environment, and space weather. LHAASO becomes a scientific base for multilateral international cooperation and high-level research.

13. The Highest Energy Gamma Ray from Pulsars Detected by LHAASO

A century after the discovery of cosmic rays, their origin remains a mystery. One of the main physical goals of LHAASO is to search for gamma-ray sources and discover the origin of cosmic rays from this research perspective. The authors of Chapter 13, Xin-Hua Ma and Zhen Cao summarize the findings of LHAASO, a project that studies gamma-ray sources and cosmic rays. LHAASO has detected the highest-energy gamma rays from pulsars and other sources in the Milky Way, such as the Crab Nebula. These sources are called PeVatrons and can accelerate particles to very high energies. In this chapter, Xin-Hua Ma and Zhen Cao describe some of the PeVatrons discovered by LHAASO that may be related to pulsars, PWNs and TeV halos, including the Crab Nebula, LHAASO J0621+3755,

LHAASO J0621+3755, LHAASO J0341+5258. LHAASO entered the full array scientific operation stage in October 2021. It can be expected that LHAASO will discover more pulsars and other types of gamma-ray sources, and discover and observe super PeVatrons by extending the spectrum to several PeV or above, thus providing strong experimental evidence for the solution of the mystery of the origin of cosmic rays.

14. The Qitai Radio Telescope (QTT)

In Chapter 14, Na Wang reviews the overall design scheme of the QiTai Radio Telescope (QTT), which is a fully steerable Gregorian type telescope with a main reflector diameter of 110 meters. The observing frequency range of QTT is 150 MHz to 115 GHz, and it is equipped with ultra-wideband, multi-beam and phased array feed receiver. Na Wang points out that QTT can observe more than three quarters of the sky, covering the 12° sky area to the south of the Galaxy center, and will become a world-class observation platform for Nanahertz gravitational wave detection through pulsar timing array, as well as the discovery of binary black hole systems, the exploration of dark matter and the origin of life in the universe. The chapter emphasizes that QTT will also improve the performance of Chinese and international VLBI networks, perform highly sensitive and high-resolution observations of galaxies and gravitational lensing systems, and help improve the accuracy of celestial reference systems for deep astrometric observations. In addition, QTT will provide strong technical support for future space activities such as exploration of planets and remote solar system, and become a scientific experiment platform for searching for extraterrestrial civilizations. Construction of QTT is scheduled for completion in 2028, and we look forward to many more exciting discoveries and results from QTT in the future.

15. Hot Universe Baryon Surveyor (HUBS)

Chapter 15 reviews the Hot Universe Baryon Survey (HUBS) project, which will search for the "missing" baryons in the universe with a high-resolution X-ray imaging satellite. It will also study the properties of neutron stars and other compact objects by observing their surface spectral lines. Zhao-Sheng Li claims that HUBS is a major scientific endeavor that will enhance our understanding of galaxy formation and evolution, and bridge the technological gaps in China. This chapter further explains how HUBS will determine the mass function, gravitational redshift, and dense matter state of neutron stars by observing the Doppler shifts and surface emissions of various atomic lines. These spectral lines will enable the estimation of the mass function of binary systems, the gravitational redshift at the surface of neutron stars, and the limits on the dense matter state.

16. The 120 m Jingdong Pulsar Radio Telescope (JRT)

In Chapter 16, Longfei Hao, Min Wang, and Yonghua Xu review the overall design scheme of the Jingding Radio Telescope (JRT), which will be the world's largest single-dish decimeter-wave radio telescope for pulsar research. The JRT will be a key window for Chinese astronomy and provide more accurate pulsar data. In this chapter, the authors describes the JRT's construction, scientific goals, and related technologies, such as pulsar detection, radio telescope location, design schemes, and digital backends. This chapter also discusses other related research, such as the role of FAST in advancing pulsar science research, and the NANOGrav project, etc. In the future, radio astronomy research will achieve more breakthrough progress, such as more accurate observations and parameter measurements of pulsars, which will help to better understand the pulsar-related phenomenon in the universe. The completion of JRT will also help to deepen the understanding of the pulsar phenomenon and related astrophysical problems in the universe. At the same time, JRT will also become an important window for the field of Chinese astronomy, attracting more scientists and research institutions to participate in radio astronomy research. More exciting discoveries and achievements are expected in the future.

17. The Enhanced X-ray Timing and Polarimetry Mission (eXTP)

In Chapter 17, Yu-Peng Xu, Y.P. Xu, A. Santangelo, and Shuang-Nan Zhang review the project of Enhanced X-ray Timing and

Polarimetry Mission, (eXTP), which is China's next flagship X-ray satellite after Insight-HXMT. It aims to study extreme gravity, magnetic fields, and density of compact objects like black holes, neutron stars, or quark stars. Yu-Peng Xu, Y.P. Xu, A. Santangelo, and Shuang-Nan Zhang emphasize that eXTP will measure the mass and spin of many black holes, test general relativity and quantum theories in strong fields, determine the equation of state of compact objects, explore neutron matter or quark matter, and also monitor and follow up various explosive events. They describe that, with the support of the Strategic Priority Program on Space Science of the Chinese Academy of Sciences, eXTP has completed the key technology research and scheme design work by the end of 2022. The European cooperation team of eXTP has obtained the support from their countries, and carried out the Phase A and Phase B research work simultaneously. At present, eXTP has been included in the medium and long-term development plan of China's national space science, and is undergoing project evaluation, planning to be launched around 2029.

18. GECAM: A Gamma-Ray Monitor Constellation

The authors of Chapter 18, Shaolin Xiong, Yanqiu Zhang and Chenwei Wang, present the project called Gravitational Wave High-energy Electromagnetic Counterpart All-sky Monitor (GECAM). This project was proposed in 2016 and approved in 2018 by the Chinese Academy of Sciences, with the aim of exploring gravitational-wave gamma-ray bursts and other high-energy events. The GECAM-A and -B satellites were launched in 2020, and GECAM-C in 2022. Each satellite is equipped with both gamma-ray and electrically charged particle detectors. GECAM's energy range is about 6 keV to 5 MeV. The authors of this chapter highlight the scientific and technological achievements made by GECAM since its launch. The GECAM team found and measured the brightest gamma-ray burst ever observed (GRB 221009A), as well as the second fastest high-energy radio burst (FRB-XRB 221014) and a new high-energy radiation phenomenon from Earth. The project observed more than 1,000 explosions in orbit, including more than 100 gamma-ray bursts and magnetic explosions, issued more than

600 warnings, and conducted monitoring and joint observations with several different observation equipment. GECAM also used a new space-based high-energy radiation detector, a lanthanum bromide and SiPM detector, and realized real-time detection and warning of several high-energy explosion events.

19. Testing Relativistic Gravity with Radio Timing of Neutron Stars

Pulsars, as high-precision clocks, are in the strong gravitational field environment of neutron stars, providing an ideal laboratory for testing strong gravitational fields. The author of Chapter 19, Lijing Shao, introduces the non-perturbative gravitational behavior that may occur in neutron stars under strong gravitational fields from the perspective of strong-field gravity. If neutron stars undergo phase transitions such as strong-field scalarization, they will be reflected in the orbital dynamics of binary systems, and thus have the hope of being measured by high-precision pulsar timing. Lijing Shao explains that, if there is a cubic correction of the Riemann tensor in the gravitational interaction, it will also cause the violation of the strong equivalence principle in the strong gravitational field, changing the orbit of the binary motion. Under the precise timing observation of the new generation of radio telescopes, the strong-field gravity test will usher in a new era.

20. The Equation of State of Pulsars

In Chapter 20, Zhenyu Zhu, Zhiqiang Miao, and Ang Li review the equation of state (EOS) of pulsars. Neutron stars are the densest and smallest stars, with unclear interiors due to QCD's complexity. They are mostly nuclear matter with some electrons and muons, balanced by gravity and pressure. Hyperons or quark matter may exist in their center, forming hybrid stars. Quark stars are hypothetical objects with only deconfined quarks. Zhenyu Zhu, Zhiqiang Miao, and Ang Li explain that the QMF (quantum mean field) parameter space can be well constrained by mass, radius, and tidal deformability measurements by using QMF to model dense matter, and the hyperon puzzle may be solved, with hyperon stars' maximum mass of 2.2 solar masses using hypernuclei data and hyperon–nucleon interaction, They also introduce that the earlier EOS stiffening in quark stars affects the sound speed, showing the importance of quark pairs. However, the phase transition strength is not well-determined by global properties. The particle degree of freedom in cold, dense matter is also uncertain. This could help us understand low-energy QCD or effective model parameters. Thus, it is crucial to combine efforts from different fields and discuss common interests and problems. Experimental nuclear physics provides data to test dense matter EOS theories in pulsars. The current and future multimessenger observatories (e.g., FAST, Insight-HXMT, GECAM, Tianma, LHAASO, eXTP, HUBS, Jingdong, QTT) will improve pulsar detection and measurement, as well as their evolution. Hopefully, the high-density EOS behavior can be determined soon, solving many open problems in nuclear physics and high-energy astrophysics.

21. Progress on the Study of Radio Emission Beam of Pulsars

The authors of Chapter 21, Hong-Guang Wang, Xiancong Wu, and Gaige Deng review the research progress on the radiation beam and radiation zone structures of radio pulsars, which emit radio beams from their poles and appear as pulses when they sweep across observers. Since observers only see part of the beam, it is difficult to determine the intensity, spectrum, polarization and structure of the radiation zone in the magnetosphere. The authors introduce two methods to study the beam structure: one is to use statistical data from many pulsars to construct a model, and the other is to use the precession effect of binary pulsars to scan the beam directly. The first method has various models, such as core hollow-cone, core conal-double, plaque, fan beam, etc. The second method is more recent and limited to a few pulsars. To infer the structure of the radiation zone from the beam structure, it is essential to understand the radiation geometry, magnetic field structure, relativistic effects, acceleration process and radiation mechanism in the magnetosphere.

22. The Inner Structure of Pulsars

In Chapter 22, Xiao-Yu Lai explores the challenging inner structure and EOS of pulsar-like compact objects and introduces a strangeon star model, which differs from neutron star and quark star models. Xiao-Yu Lai explains that both perturbative and non-perturbative QCD effects may cause quark flavour maximization and quark localization inside strangeons, as well as a hard core repulsion between strangeons, and the strangeon star EOS should be very stiff with the lowest energy per baryon if quarks are localized and a hard core repulsion exists. This chapter also summarizes that various manifestations of pulsar-like compact stars, from surface and global properties to the kilonova radiation of merging binaries, will be explained by strangeon stars, and advanced facilities may confirm the existence of strangeon matter, especially in the multimessenger astronomy era. Research progress about radiation beam and radiation zone structures of radio pulsars.

23. Fast Pulsars, Neutron Stars, and Astrophysical Strange Quark Matter Objects

In Chapter 23, Delaney Farrell, Fridolin Weber, Milva G. Orsaria, Ignacio F. Ranea-Sandoval, Martín Canullán, Rodrigo Negreiros, Takashi Katayama, and Ian Wagaman explore key aspects of neutron stars, pulsar glitches, tidal deformability, fast pulsars, the equation of state, and strange quark matter stars. Challenges in directly measuring neutron star radius have led to reliance on spectroscopic and timing techniques, with uncertainties addressed by the authors through careful source selection and theoretical modeling. The authors emphasize that Pulsar glitches reveal insights into the equation of state through angular momentum transfer within the neutron star and that Tidal deformability is crucial in gravitational-wave astronomy, exemplified by the GW170817 event. Fast pulsars, instrumental in astrophysical testing, are classified by the authors into ordinary pulsars, millisecond pulsars, and magnetars. The EOS, a vital ingredient for understanding neutron star internal structure, are explored through various models. The chapter delves into the theoretical framework for rotating neutron stars, addressing uniform and differential rotation scenarios and their impacts on mass and radius. Additionally, the intriguing concept of quark stars and strange dwarfs is investigated. The various topics discussed in this book chapter contribute to a broader understanding of dense matter physics, astrophysical phenomena, and the potential for transformative discoveries through advanced observational techniques and technologies like gravitational wave detectors, radio telescopes, and X-ray telescopes.

24. Axisymmetric Oscillations of Rotating Neutron Stars: Gravitational Wave Emission

Stéphane Vincent and José A. de Freitas Pacheco review, in Chapter 24, review quasi-radial oscillations of rotating neutron stars and present new calculations of axisymmetric modes based on linearized relativistic field equations that take into account both fluid and metric perturbations. Their numerical procedure was checked by comparing results derived from the present code with relativistic hydrodynamic calculations and adopting a common polytropic equation of state. Such a comparison indicates that mode frequencies differ by less than 2% excepting for rotation velocities near the mass shedding limit, where differences increase up to 10%. In this chapter, rapidly rotating neutron star models were computed for different "realistic" equations of state characterized by nuclear compressibilities ranging from 140 up to 1200 MeV. The fundamental and the first overtone frequencies of quasi-radial and quadrupole modes were calculated, by Stéphane Vincent and José A. de Freitas Pacheco, for a sequence of neutron star models having the same baryonic mass ($M_b = 1.5\ M_{\odot}$) but different rotation velocities. The present results confirm that quadrupole frequencies are lower than those of quasi-radial modes, contrary to results derived from models built with polytropic equations of state. Excitation of quasi-radial modes and gravitational wave emission were considered in a scenario in which a pure hadronic star suffers a mini-collapse due to a hadron-quark phase transition in the core. The released gravitational energy, which depends on the adopted equation of state for the nuclear matter, is of the order of 1.98×10^{52} erg for the considered neutron star model. However, only 8.9% is channeled into gravitational waves. Nevertheless, the planned European gravitational Einstein Telescope would be able to detect such a signal up to distances of about 114 Mpc.

https://doi.org/10.1142/9789819800643_fmatter

Contents

https://doi.org/10.1142/9789819800643_fmatter

List of Figures

Chapter 1

Chapter 2

Chapter 3

Chapter 4

Chapter 5

Chapter 6

Chapter 7

Chapter 8

Chapter 9

Chapter 10

Chapter 11

Chapter 12

Chapter 13

Chapter 14

Chapter 15

Chapter 16

Chapter 17

Chapter 18

Chapter 20

Chapter 22

Chapter 23

Chapter 24

Chapter 25

https://doi.org/10.1142/9789819800643_fmatter

List of Tables

Chapter 8

Chapter 10

Chapter 11

Chapter 14

Chapter 16

Chapter 17

Chapter 18

Chapter 23

Chapter 24

Part 1

Review of Pulsar Astrophysics: Magnetars, Isolated Neutron Stars, and Radio Pulsars

https://doi.org/10.1142/9789819800643_0001

Chapter 1

Magnetars

Sandro Mereghetti

INAF, Istituto di Astrofisica Spaziale e Fisica Cosmica, via A.Corti 12, I-20133 Milano, Italy
sandro.mereghetti@inaf.it

Magnetars are the neutron stars with the highest magnetic fields, up to 10^{15}–10^{16} G. Their bursting and persistent emission is powered by magnetic energy. About 30 magnetars are known in the Milky Way and in the Magellanic Clouds and it has been proposed that they are also responsible for a variety of extragalactic phenomena, ranging from giant flares in nearby galaxies to fast radio bursts. This chapter presents the main observational properties of magnetars and a few recent highlights.

Keywords: Neutron stars, Magnetars

Contents

1. Introduction

Magnetars can be defined as neutron stars (NSs) whose electromagnetic emission is powered mainly by magnetic energy (Paczynski, 1992; Thompson and Duncan, 1995, 1996). The observed sample of few tens of magnetars represents a minority of the known NS population, which is dominated by about 3000 radio and/or γ-ray rotation-powered pulsars (RPPs) and several hundred accretion-powered NSs in X-ray binaries. However, their rich phenomenology (including strong variability on all timescales), their connections with various kinds of astrophysical sources, and the possibility of studying physical processes in magnetic fields of strength not reached elsewhere in the universe makes magnetars of extreme interest. In recent years, it has also been recognized that "magnetar-like" behaviors are (sometimes unexpectedly) present also in a variety of other NS classes. This blurring of the distinction between magnetars and "ordinary" NSs is interesting in the context of "unification schemes" of the NS population.

The "magnetar" term is used in two different astrophysical contexts, referring either to the "normal" fully fledged population or to the "newly born" ("millisecond") magnetars.

The former are observed as sources of short burst of soft γ-rays and/or as spinning-down X-ray pulsars with periods generally of a few seconds. These are the sources, initially called anomalous X-ray pulsars or soft gamma-ray repeaters depending on the way they were discovered (Atteia *et al.*, 1987; Mereghetti and Stella, 1995), which led to the development of the magnetar idea (Duncan and Thompson, 1992). There is now wide consensus that magnetars are isolated

NSs powered by magnetic energy. We know a few tens of them in our galaxy and in the Magellanic Clouds, but many more are still undiscovered because they spend most of the time in a "quiescent" low-luminosity state. Their strong magnetic fields (up to 10^{14-15} G in the magnetosphere, and possibly even higher in the interior) are believed to arise because they are born with extremely short spin periods of only 2–3 ms (Thompson and Duncan, 1993).

Newly born, millisecond magnetars have been invoked in several astrophysical contexts, e.g., models for gamma-ray bursts (GRBs) and their afterglows (Bucciantini *et al.*, 2012; Metzger *et al.*, 2008; Usov, 1992; Xie *et al.*, 2022; Xu and Tang, 2021), gravitational waves sources (Cheng *et al.*, 2015; Dall'Osso *et al.*, 2015a), superluminous supernovae (Kasen and Bildsten, 2010; Wang *et al.*, 2019a), fast radio bursts (Lieu, 2017), and ultra high-energy cosmic rays (Arons, 2003). However, a direct observational proof for the presence of a newly born magnetar in these sources (i.e., the detection of a millisecond periodicity) is still missing. In this review, I concentrate on the objects of the first class, calling them simply "magnetars" and explicitly referring to "millisecond-magnetars" when needed. For previous reviews on magnetars, see Kaspi and Beloborodov (2017), Mereghetti (2008), Mereghetti *et al.* (2015), Turolla *et al.* (2015), while an excellent review on millisecond magnetars can be found in Dall'Osso and Stella (2022).

2. Main Properties of Magnetars

2.1. *Persistent X-ray emission*

All magnetars emit persistent X-rays characterized by a rather soft spectrum below 10 keV and by periodic modulations induced by the star rotation, with pulsed fractions from ~10% to ~60%. The observed spin periods, from 1.3 to ~12 s, are on average longer than those of RPPs of comparable age. All magnetars are spinning down, with high period derivatives typically in the range 10^{-12}–10^{-10} s s^{-1}. Below a few keV, the emission is of thermal nature and can be described, to a first approximation, as a blackbody spectrum of temperature kT~0.3–0.6 keV. Other mechanisms, such as resonant cyclotron scattering (RCS) due to currents in the magnetosphere

(Thompson *et al.*, 2002), contribute at higher energies, leading to power-law spectral components. The hard X-ray emission has been seen to extend up to a few hundred keV and is also pulsed.

The X-ray emission during quiescence (i.e., when a magnetar is not undergoing an outburst, Section 2.2) is rather stable, with observed luminosities in the range $L_q \sim 10^{32}$–10^{36} erg s^{-1}, but note that the lower bound could be affected by our inability to detect even fainter sources.[1] Many magnetars have been discovered during outbursts, when luminosities up to $\sim 10^{36}$ erg s^{-1} can be reached. These are called "transient" magnetars. On the other hand, sources always seen at $L_X \sim 10^{35}$–10^{36} erg s^{-1} (e.g., SGR 1806−20 or 4U 0142+61) are often called "persistent" magnetars, although they also show flux variations on long timescales (Dib and Kaspi, 2014; Gonzalez *et al.*, 2010; Mereghetti *et al.*, 2005b).

The quiescent (or persistent) X-ray emission is ultimately powered by dissipation of the internal magnetic field, which is dominated by its toroidal component. The magnetic stresses resulting from the field evolution produce horizontal displacements in the crust, leading to surface heating and to a twisting of the external field lines. Currents flowing along the magnetic field lines in the twisted magnetosphere contribute to the soft X-ray emission, through heating of the star surface, as well as to the hard X-ray emission, through synchrotron, thermal bremsstrahlung (Thompson and Beloborodov, 2005), and resonant cyclotron scattering (Baring and Harding, 2007). The wide range, more than four orders of magnitude, spanned by the soft X-ray quiescent emission, can be explained assuming different strengths and geometries of the internal field. There is indeed a slight correlation between bolometric quiescent luminosity and dipolar magnetic field (Coti Zelati *et al.*, 2018), although with a large scatter, probably due to the fact that the external field, estimated from the timing parameters, does not univocally reflect the intensity of the (mostly toroidal) field in the star interior.

[1]Notably, SGR 0418+729, the only magnetar with $L_q = 10^{31}$ erg s^{-1}, is among the closest ones (d = 2 kpc); being located at Galactic latitude $b = 5°$, it is scarcely affected by interstellar absorption, which facilitates its detection in the soft X-rays.

2.2. *Outbursts*

Most magnetars have been discovered during their outbursts. These are rapid increases of the X-ray flux, often associated with the emission of bursts (Section 2.3) and/or with timing irregularities, such as glitches or rapid variations in the spin-down rate. The characteristic signatures that allow the interpretation of a newly discovered X-ray transient as a magnetar are the absence of a bright optical counterpart (that, if present, would indicate an accretion-powered binary) and the presence of a periodic pulsation, with the subsequent measurement of spin-down.[2] A comprehensive analysis of a large number of magnetar outbursts is reported in Coti Zelati *et al.* (2018).

The outburst rise times are of the order of only a few days (sometimes unconstrained because the beginning is not observed) and lead to a spectrum harder than the quiescent one. During the outburst decay, that can last from weeks to years, the thermal emission gradually cools back to the quiescent value. There is a large variety in the duration and light curve shapes of the outbursts, as well as in the total released energies which range from $\sim 10^{40}$ erg to more than 10^{43} erg. An interesting observational result is that the maximum attained luminosity never exceeds a few 10^{36} erg s^{-1}. Thus, the magnetars with the lowest quiescent levels display the largest dynamic range in luminosity. Such a saturation value of the peak luminosity can be explained by the strong energy losses due to neutrinos, which dominate over the photon luminosity when high temperatures are reached (Pons and Rea, 2012).

The bright fluxes reached during outbursts give the possibility to obtain broadband spectra extending into the hard X-ray range (Kuiper *et al.*, 2012). Studying the evolution of the spectral and timing properties along the outburst decay allows constraining the source geometry and the physical mechanism at the origin of the emission. Extensive observations of recent outbursts show a variety

[2]The high spin-down rate of magnetars, $\dot{P} \sim 10^{-12}$–10^{-10} s s^{-1}, can typically be measured with X-ray observations spanning only a few days, but there are a few notable exceptions of magnetars with much lower spin-down rates (Rea *et al.*, 2010, 2012).

of behaviors, but they indicate that a common framework is the presence of a highly dynamic magnetosphere and of thermal spots on the star surface that cool and shrink as the outburst evolves (Borghese *et al.*, 2022, 2021; Hu *et al.*, 2020; Rajwade *et al.*, 2022; Rea *et al.*, 2020; Younes *et al.*, 2022).

Outbursts are the consequence of a rapid energy deposition, whose detailed origin is still unclear, in some localized region of the star surface. In fact, the radii inferred for the thermally emitting regions are of the order of only a few kilometers. Both "internal" and "external" heating mechanisms have been considered, such as magnetic stresses in the outermost layers of the crust (Lyubarsky *et al.*, 2002) and heating from currents flowing in twisted magnetospheric bundles (Beloborodov, 2009). The heated region then gradually cools down, with a shape and temperature evolution which depends on several parameters, among which the magnetic field geometry and intensity play a major role. This has been shown by theoretically modeling the magneto-thermal evolution (Kaminker *et al.*, 2014; Pons and Rea, 2012) and confirmed by more recent work based on 3D codes (e.g., De Grandis *et al.*, 2022).

2.3. *Bursts and flares*

The emission of bursts is one of the distinctive characteristics of magnetars. These short (few hundreds of ms) bursts were first discovered by instruments operating in the hard X-ray range and led to the definition of the class of sources called soft gamma-ray repeaters.[3] Bursts can occur sporadically or in groups and span a wide range of peak luminosities, $\sim 10^{36}-10^{42}$ erg s^{-1}. Their spectra in the hard X-ray range are well fit by optically thin thermal bremsstrahlung models with temperatures ~30–40 keV. However, when data at lower energies are included, the sum of two blackbody laws with average temperatures $kT_1 \sim 4$ and $kT_2 \sim 10$ keV often gives a better fit (Cai *et al.*, 2022; Feroci *et al.*, 2004). It has been proposed that the two blackbody components result from the different propagation in the

[3]Initially believed to be a subclass of GRB, these sources later played a crucial role in the support of the magnetar interpretation.

magnetosphere of photons with ordinary or extraordinary polarization mode[4] (Israel *et al.*, 2008).

Bursts and flares originate from the acceleration of particles in rapid magnetic field reconfigurations. Proposed triggering mechanisms can be broadly divided into either crustquakes or magnetospheric instability scenarios. Stresses in the external crust can build up as a consequence of the evolution of the internal magnetic field (Thompson and Duncan, 1995). When the crust fractures, the stored energy is suddenly transferred to the magnetosphere via Alfvén waves and powers the burst. Alternatively, the energy accumulation could occur directly in the magnetosphere through the development of unstable configurations eventually leading to magnetic reconnection (Lyutikov, 2003; Mahlmann *et al.*, 2019). In any case, the presence of an unstable, twisted magnetosphere appears as a necessary condition for these phenomena. It is thus believed that the major difference between magnetars and RPP concerns the geometry rather than the intensity of the magnetic field.

Longer and more energetic bursts occur less frequently and are often called intermediate flares. It is unclear if they represent the extreme manifestation of bursts or are a different phenomenon. On the other hand, the much rarer giant flares (two observed in our galaxy (Hurley *et al.*, 1999; Palmer *et al.*, 2005) and one in the LMC (Mazets *et al.*, 1979)) have spectral and temporal properties that set them apart from the other less energetic events. They are believed to originate from catastrophic rearrangements of the internal and/or external magnetic field. Giant flares appear as an initial very luminous (up to L $\sim 10^{47}$ erg s^{-1}), short ($\sim$0.2–0.5 s) spike with a hard spectrum (kT $\sim$ hundreds of keV). This is followed by a tail of hard X-ray emission lasting several minutes and modulated at the star spin period. The X-ray spectrum during the pulsating tail is thermal, similar to that of the short bursts. These tails indicate that part

[4]The wave electric field of ordinary (O-mode) and extraordinary (X-mode) photons oscillates, respectively, either parallel or orthogonal to the plane defined by the local magnetic field and the propagation direction.

of the energy released in the initial spike produces an optically thick photon-pair plasma which remains temporarily confined in a localized region of the magnetosphere. The requirement to spatially confine such a hot plasma is one of the evidences pointing to very strong magnetic fields in these sources (Thompson and Duncan, 1995).

The giant flare emitted from SGR 1806–20 on 2004 December 27 showed for the first time the presence of another component (Mereghetti *et al.*, 2005a), in addition to the initial spike and pulsed tail seen in the two previous giant flares. In fact, hard X-ray emission (>80 keV) lasting at least one hour was discovered with the INTEGRAL satellite (see Figure 1) and later confirmed with independent observations (Boggs *et al.*, 2007; Frederiks *et al.*, 2007a). This hard X-ray "afterglow" testifies the production of a significant mass ejection at relativistic speed, likely responsible also for the variable and extended radio emission observed for several days after the flare (Section 2.5).

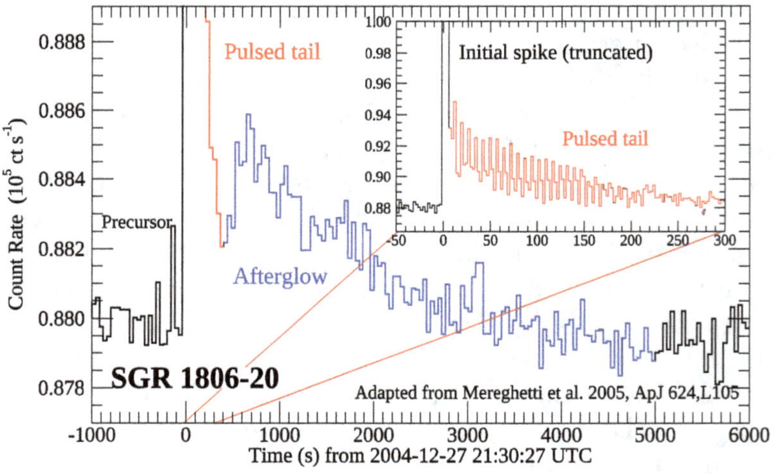

Fig. 1. Hard X-ray (>80 keV) light curve of the December 27, 2004 giant flare of SGR 1806–20 as measured with the Anticoincidence Shield of the SPI instrument on board INTEGRAL. Due to the adopted time bins (40 s for the main plot, 2.5 s for the inset), the pulsations at the spin period of 7.56 s are visible only in the inset. No pulsations are present after $t \sim 250$ s, consistent with the afterglow interpretation of the long-lasting emission. Figure adapted from Mereghetti *et al.* (2005a).

2.4. *Optical and infrared*

Magnetars, as all isolated NSs, are very faint in the optical band. Optical and/or near IR (candidate) counterparts have been reported for about 15 magnetars, but not all of them have been confirmed (Chrimes *et al.*, 2022). Most magnetars are located at low galactic latitudes and have distances of several kiloparsecs. The identification of their counterparts is thus complicated by the high interstellar absorption and by the crowding of these fields, requiring observations with adaptive optics (Testa *et al.*, 2008) or the Hubble Space Telescope (Chrimes *et al.*, 2022).

In several cases, candidate counterparts have been proposed based on their long-term flux variability (both correlations and anti-correlations with the X-ray flux have been observed). However, secure identifications can be obtained only when optical/IR pulsations are detected, as it was first done for 4U 0142+61 (Kern and Martin, 2002).

Various origins for the optical/IR emission in magnetars have been discussed. A possibility is that it is produced in the pair-dominated inner regions of the magnetosphere (Beloborodov and Thompson, 2007). A completely different scenario attributes the long wavelength emission (4.5–8 μm) to a debris disk heated by the high-energy emission from the central NS (Wang *et al.*, 2006).

2.5. *Radio*

Two different types of radio emissions have been observed from magnetars: (a) variable unpulsed emission related to plasmoid ejected during giant flares and (b) periodic pulsations (including bright single pulses). Until the radio observations of the transient magnetar XTE J1810-197 (Camilo *et al.*, 2006; Halpern *et al.*, 2005), only the first type of emission was known, and magnetars were believed to be "radio-quiet". Indeed, pulsed radio emission is rare in magnetars, but extensive observations carried out in the latest years, especially during outbursts, are providing more and more detections and a wealth of data potentially leading to significant advances in our understanding of the radio pulsar phenomenon at large. The importance of these radio studies is also highlighted by the presence of magnetar-like activity in two allegedly rotation-powered pulsars (PSR J1845-0258 and PSR J1119-6127 (Gavriil *et al.*, 2008; Göğüş *et al.*, 2016)) and by the discovery of FRB-like emission in the magnetar SGR J1935-2154 (Section 3.3).

The large amount of energy released at the onset of giant flares leads to massive ejecta with the formation of a mildly relativistic outflow which produces an expanding radio nebula and an X/γ-ray afterglow (Dai *et al.*, 2005; Granot *et al.*, 2006; Mereghetti *et al.*, 2005a; Wang *et al.*, 2005). These bright radio sources, detected in the two most recent giant flares (Cameron *et al.*, 2005; Frail *et al.*, 1999) and whose synchrotron origin is testified by linear polarization, are transient events. Up to now, there is no convincing evidence for persistent radio nebulae associated with the "steady" wind from magnetars, similar to the pulsar wind nebulae surrounding the most energetic RPP. The unpulsed, but variable, flux detected at high frequency (21–41 GHz) from SGR J1745-2900 could be caused by synchrotron emission in the shock between the wind generated during an outburst and the dense ionized gas in the galactic center region (Yusef-Zadeh *et al.*, 2015). However, this interpretation has been questioned by Gelfand *et al.* (2017), who attribute the emission to magnetospheric processes.

Radio pulsations have been detected up to now only in six magnetars, all of which are transients. Their radio properties are quite different from those of RPP: most notably for the large variability, on different time scales, both in intensity and in the averaged pulse profiles. These variations are not correlated with the changes seen in the X-ray band in a simple and univocal way. The radio spectra are flat or inverted, with a high-frequency turn-up and a high level of linear polarization (Caleb *et al.*, 2022; Camilo *et al.*, 2007; Huang *et al.*, 2021). Notably, the radio pulsations from XTE J1810-197 have been detected up to 353 GHz (Torne *et al.*, 2022).

Broadband radio spectra derived from non-simultaneous data are often affected by variability and should be interpreted with caution. Campaigns of simultaneous observations over a broad frequency range are difficult to obtain, but they can give interesting results. For example, Maan *et al.* (2022) showed that the radio spectrum of XTE J1810-197 evolved during the 2018 outburst, starting with an inverted power law and then steepening to finally resemble that of a normal pulsar. This spectral behavior is consistent with a ~GHz-peaked spectrum, in which the peak frequency decreased from 2.4 to 0.5 GHz as the outburst progressed. Spectra with a GHz turn-over are seen only in a small number of RPP but recently also in some magnetars. They are interpreted as due to thermal free-free absorption in the circumpulsar environment (e.g., a supernova remnant, a pulsar wind nebula, a dense HII region, or the wind of a companion star), but other explanations are possible (Kijak *et al.*, 2017).

After more than 50 years since the discovery of RPP, the exact mechanism responsible for their radio emission is still unknown. Thus, it is not surprising that this is true also for the radio emission from magnetars. A qualitative difference between RPP and magnetars might be related to the fact that the radio pulses originate from charges accelerated along open magnetic field lines in the former and along closed ones in the latter (Beloborodov, 2009; Wang *et al.*, 2019b). The currents flowing in the bundles of closed lines can be up to 10^5 times stronger than those flowing along open lines, thus much wider beams and a higher power are expected during the magnetar outbursts. It is also interesting to note that some magnetars occasionally showed RPP-like properties in their radio emission. For example, mode switching and nulling, which are typical phenomena of RPP, have been seen in a few magnetars (Lower *et al.*, 2021) and SGR J1745-2900 stayed for several months in a state with stable pulse profile and flux (Yan *et al.*, 2018), similar to a normal RPP.

Radio observations of magnetars have also been instrumental in estimating a magnetic field greater than a few mG within ~10 pc of SgrA* (Eatough *et al.*, 2013), in constraining the distance of the scattering screen in the direction of the galactic center (Bower *et al.*, 2014), and in obtaining accurate distances for a few magnetars through parallax measurements (Ding *et al.*, 2020).

2.6. *High energy (>100 MeV) and very high energy (>100 GeV)*

Early models for magnetar γ-ray emission, based on the outer gap scenario that was originally developed for RPP, predicted fluxes close to (or above, depending on assumed parameters and distances) the sensitivity of current facilities (Tong *et al.*, 2011; Zhang and Cheng, 2002). However, no persistent[5] emission in the ~0.1–300 GeV range has been firmly associated with magnetars with Fermi/LAT (Abdo *et al.*, 2010; Li *et al.*, 2017). The upper limits in the high-energy range are consistent with the predictions of more recent computations in which some critical assumptions of the outer gap models have been reconsidered more realistically (Viganò *et al.*, 2015a, 2015b, and references therein).

A few high-energy sources positionally consistent with galactic magnetars are known, but the limited angular resolution currently available at these energies does not allow unambiguous identifications.[6] The problem is complex because magnetars are often found in regions of recent star formation, where other potential high-energy emitters, such as supernova remnants, pulsar wind nebulae, and young massive star clusters, are present. For example, the γ-ray sources found at the positions of 1E 2259+586, CXOU 171405.7−381031, and 1E 1841−045 are best interpreted as emission from the associated supernova remnants CTB 109 (Castro *et al.*, 2012), CTB 37B (Xin *et al.*, 2016), and Kes 73 (Liu *et al.*, 2017), respectively, rather than from the magnetars themselves.

At very high energies (>100 GeV), the situation is similarly complicated, despite the great progress in imaging Cherenkov Atmospheric Telescopes and in experiments for air shower detection, such as LHAASO (Cao *et al.*, 2019). This is well exemplified by the case of the sources in the vicinity of SGR 1900+14 (Hnatyk *et al.*, 2022) and SGR 1806−20 (HESS Collaboration, 2018). The TeV source HESS J1907+089/HAWC J1907+085 is spatially compatible with SGR 1900+14, but also with G42.8+0.6, a radio shell supernova remnant[7] at a distance of 3–5 kpc (Wang *et al.*, 2020). Similarly, the supernova remnant G9.7−0.0 and a massive star cluster containing a

[5]The detection of transient high-energy emission during alleged giant flares from extragalactic magnetars is discussed in Section 3.1.

[6]A possible exception is that of the γ-ray binary LSI +61° 303, discussed in Section 3.4.

[7]The magnetar proper motion (Tendulkar *et al.*, 2012) and the different distances disfavor a physical association between these two objects.

luminous blue variable star (Fuchs *et al.*, 1999) are candidate high-energy emitters which could be associated with HESS J1808−204.

3. Recent Results and Future Prospects

3.1. *Giant flares from extragalactic magnetars*

The luminous initial spikes of giant flares can be detected by current hard X-ray instruments up to distances of few tens of Mpc, while the following fainter and softer pulsed tail would remain undetectable. Thus, a giant flare from a source in a distant galaxy would appear very similar to a short GRB. The early suggestion that some of the short GRBs could be due to magnetars (Mazets *et al.*, 1982) received renewed interest after the 2004 giant flare from SGR 1806−20, which reached a peak luminosity of $\sim 10^{47}$ erg s^{-1} (for d = 10 kpc), and is supported by the identification of a few candidates (Burns *et al.*, 2021). Early studies suggested that up to 40% or more of the short GRBs could be due to extragalactic giant flares (Hurley *et al.*, 2005). However, further estimates and searches in observed GRB samples reduced this fraction to $\lesssim$8% (Nakar *et al.*, 2006; Ofek, 2007; Svinkin *et al.*, 2015). It is now clear that the majority of short GRBs are generated in mergers of double NS systems, as proved by GW 170817 observed by LIGO in 2017 (LIGO Scientific Collaboration and Virgo Collaboration *et al.*, 2017).

Some short GRBs positionally consistent with nearby galaxies have been proposed as candidates for magnetar giant flares based on their timing and spectral properties: GRB 070201 in M31 (Mazets *et al.*, 2008) and GRB 051103 in M81 (Frederiks *et al.*, 2007b). The magnetar origin of these two events is also supported by the non-detection of the corresponding gravitational wave signals (LIGO Collaboration *et al.*, 2012; LIGO Scientific Collaboration and Hurley, 2008) because a binary neutron star merger at such short distances would have produced a signal well above the sensitivity of the available interferometers.[8] The most recent candidate, GRB 200415A, lies in the direction of the Sculptor galaxy (NCG 253), at a distance of

[8] No sensitive GW observations are available at the time of GRB 200415A.

3.5 Mpc (Roberts *et al.*, 2021; Svinkin *et al.*, 2021). These candidates, together with the three historical giant flares from confirmed magnetars, occupy in the isotropic energy versus spectral peak energy space a position different from that of short GRBs (Yang *et al.*, 2020; Zhang *et al.*, 2020b).

Interestingly, GRB200415A was also detected above ~100 MeV with the Fermi/LAT instrument (Fermi-LAT Collaboration, 2021), possibly making it the magnetar observed at the highest energy. Its high-energy emission has been interpreted as synchrotron radiation in a shock produced by the interaction of relativistically ejected material with the inner boundary of a pre-existing cavity formed by the previous magnetar activity. Notably, the GeV photons arrived with delays up to 284 s after the initial short spike. These results are promising for the next generation of atmospheric Cherenkov detectors with quick repointing capability (Cherenkov Telescope Array Consortium, 2019) and for very high energy experiments with wide fields of view, such as LHAASO and SWGO (Huentemeyer *et al.*, 2019).

3.2. *X-ray polarization*

The launch of the IXPE satellite in December 2021 re-opened, after half a century and with unprecedented sensitivity and imaging capabilities, the observational window provided by polarization measurements in the soft X-ray range. Magnetars are prime targets for polarimetric studies because their strong magnetic fields make quantum electrodynamics effects, such as vacuum birefringence, very effective. This leads to levels of linear polarization much higher than those intrinsic in the emission processes (which is often reduced because of integration over regions with different angles of linear polarization).

The first detection of X-ray polarization from a magnetar has been obtained for the bright source 4U 0142+61. Taverna *et al.* (2022) measured an energy-dependent linear polarization (14% at $E < 4$ keV and 41% at $E > 5.5$ keV), with a jump of 90° in the polarization angle between the two energy ranges (see the blue points in Figure 2). RXS J170849.0−400910, the second magnetar observed by IXPE (Zane *et al.*, 2023), showed a different situation (red points in Figure 2). The level of its linear polarization increases monotonically

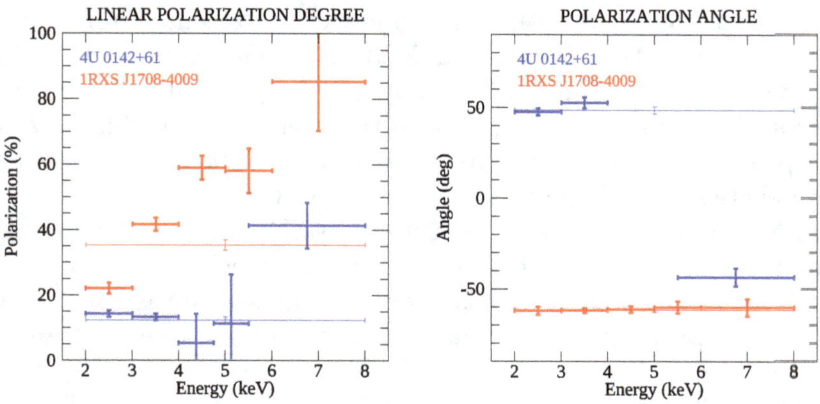

(a) (b)

Fig. 2. Polarization results obtained with IXPE for two magnetars (results from Taverna *et al.*, 2022; Zane *et al.*, 2023). Linear polarization degree (a) and angle (b) as a function of energy. The thin lines give the results for the integrated 2–8 keV band.

with energy from ~20% (2–4 keV) to 80% (8–10 keV) but without any variation in the polarization angle.

The 90° jump in the polarization angle of 4U 0142+61 suggests that the two emission components, whose presence is clearly inferred from the spectral analysis, are dominated by photons in different normal modes. The energy-integrated (2–8 keV) polarization angle changes with the pulse phase consistently with the rotating vector model and the polarization level tracks the pulse profile. The observed polarization is not high enough to unambiguously prove the presence of quantum electrodynamics effects, but the phase-dependent results can be explained in this scenario if the high-energy component results from resonant cyclotron scattering dominated by X-mode photons. The low-energy photons (mainly in O-mode) would come from a large region of the condensed star surface or from an atmosphere with inverted temperature gradient. However, other interpretations of the 90° swing in polarization angle as a function of energy are possible, including ones in which the low energy emission is dominated by X-mode photons (Taverna *et al.*, 2022). Another relevant effect to be considered is photon mode conversion in a partially ionized iron atmosphere (Lai, 2022).

The spectro-polarimetric properties of RXS J170849.0−400910 have instead been interpreted assuming that emission regions with a fully ionized hydrogen atmosphere and with a condensed iron surface coexist on the star (Zane *et al.*, 2023). Different geometries for the temperature, composition, and magnetic field variation across the star surface can be envisaged since the currently available data do not provide enough constraints. In any case, the polarization seen at high energy is too large to be produced by RCS.

Polarization measurements are a powerful diagnostic to understand the NS atmosphere, and a global approach based on phase-resolved spectropolarimetry is needed to fit self-consistent models and derive physical information. As it is shown by these first results, polarization is clearly a field in which future observations with the eXTP satellite can play a crucial role (Santangelo *et al.*, 2019).

3.3. *Magnetars and fast radio bursts*

After 25 years of their discovery, the nature of fast radio bursts (FRBs) remains elusive (see, e.g., Petroff *et al.*, 2022), despite the large samples of such events now collected with dedicated experiments, such as CHIME/FRB (CHIME/FRB Collaboration, 2021). FRBs have been detected so far only in the radio band, as very short and bright pulses with high dispersion measures, clearly indicating an extragalactic origin. A small fraction of the FRBs has been seen to repeat (of which two with periodic windows of activity). Host galaxies, of different types, have been identified for a few FRBs. Many of the models developed to explain these events are based on magnetars (see Popov and Postnov, 2013, for an early proposal of this idea). In fact, strongly magnetized NS can provide an adequate energy source in a limited space, consistent with the observed fluxes and short durations of FRBs. Furthermore, magnetars can naturally account for the class of repeating FRBs, contrary to models involving destructive events, such as supernova explosions.

The magnetar models for FRBs come in a large variety, but they can be broadly divided into two groups, depending on the location of the emitting region. In the first group ("pulsar-like" models), the radio emission takes place in the star magnetosphere. Proposed mechanisms include propagation of magnetic pulses leading to reconnection in the outer magnetosphere (Chen *et al.*, 2022; Lyubarsky, 2020) or coherent radiation from bunches of particles (Lu and Kumar, 2018;

Yang and Zhang, 2018). The second group of models ("GRB-like") is based on emission taking place at large distances from the central engine (not necessarily a magnetar), where some ejected material forms relativistic shocks interacting with the surrounding matter (Khangulyan *et al.*, 2022; Metzger *et al.*, 2019).

Magnetar models for FRBs received strong support from the discovery, on April 28 2020, of an extremely bright radio pulse in coincidence with a hard X-ray burst from SGR J1935+2154 (Bochenek *et al.*, 2020; CHIME/FRB Collaboration, 2020), an otherwise rather ordinary magnetar in our galaxy. The radio emission of this event (FRB 200428 hereinafter) consisted of two short-duration pulses (lasting ~0.6 and 0.3 ms) separated by 29 ms and with different spectra (the second one was harder) with a total fluence of 1.5 MJy ms at 1.4 GHz. Its properties were very similar to those of FRBs, except for the lower dispersion measure (DM = 332 pc cm^{-3}, consistent with a location in our galaxy) and smaller luminosity. In fact, its isotropic equivalent energy of $\sim 5 \times 10^{34}$ erg at 1.4 GHz (for $d = 4.4$ kpc Mereghetti *et al.* (2020)) is orders of magnitude below that of typical FRBs at distances of a few Gpc, but similar to that of FRB 20200120E, the closest FRB (located in a globular cluster of M81 at 3.6 Mpc, Kirsten *et al.* (2022)).

From the point of view of the high-energy emission, the only peculiarity of FRB 200428 was a non-thermal spectrum (Li *et al.*, 2021), significantly harder than that of normal X-ray bursts from magnetars. The X-ray emission started before the radio pulses and lasted longer (~0.6 s). As it is shown in Figure 3, the X-ray light curve had a structured time profile with narrow spikes separated by the same ~29 ms interval of the radio bursts, but with a delay of 6.5 ms with respect to the radio pulses (Ge *et al.*, 2023; Mereghetti *et al.*, 2020). According to Li *et al.* (2022a), these spikes could be connected to an underlying quasi-periodic oscillation at ~40 Hz.

Extensive observations of SGR J1935+2154 from radio to high energies have shown that the occurrence of simultaneous radio and X-ray bursts is a rare event. Other radio bursts, fainter than FRB 200428, were detected from SGR J1935+2154 (Kirsten *et al.*, 2021) but without counterparts at other wavelengths. Among them, a remarkable one for its nearly 100% linear polarization was detected with FAST at 1.25 GHz two days after FRB200428

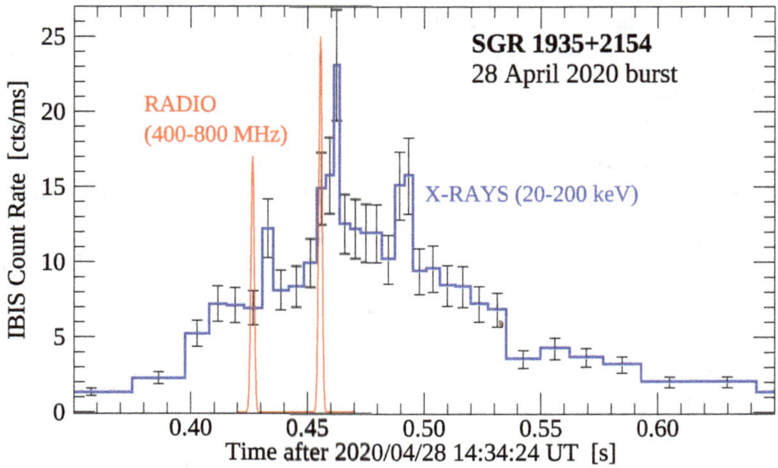

Fig. 3. Light curve (blue line) of the 2020 April 28 burst from SGR J1935+2154 as measured by the IBIS instrument on the INTEGRAL satellite (Mereghetti *et al.*, 2020). The red lines indicate the positions of the two bright radio pulses observed at 400–800 MHz (CHIME/FRB Collaboration, 2020).

(Zhang *et al.*, 2020a). Conversely, no radio counterparts have been found for other X-ray bursts, including some as bright and/or hard as FRB 200428 (Cai *et al.*, 2022; Lin *et al.*, 2020). Upper limits were obtained also at TeV energies with HESS (HESS Collaboration, 2021) and in the near IR and optical bands (De *et al.*, 2020; Zampieri *et al.*, 2022). Although these observations included the times of a few X-ray bursts, they did not cover the times of the radio bursts. Two other radio bursts with simultaneous X-ray emission, fainter and softer than that of FRB 200428, were detected from SGR J1935+2154 on 2022 October 14 and 21 (Li *et al.*, 2022b; Wang *et al.*, 2022).

Another magnetar, 1E 1547.0−5408, emitted a bright radio pulse lasting ~200 ms and with an estimated fluence of ~600 Jy ms at 6.4 GHz (Israel *et al.*, 2021). The energy of this event (8.4×10^{30} erg) is much higher than that of the single radio pulses seen in magnetars (Section 2.5) but still more than three orders of magnitude below that of FRB 200428. The radio pulse of 1E 1547.0−5408 occurred one second after an X-ray burst and it was followed by another fainter radio pulse at a time consistent with two rotations of the magnetar. The phase alignment of the two radio pulses seen in 1E 1547.0−5408 favors a magnetospheric origin.

3.4. *LS I +61° 303: A magnetar in a binary?*

LS I +61° 303 is a massive star of Be spectral type in a 26.5-day binary which, despite having been detected all over the electromagnetic spectrum from radio to TeV energies, has defined a univocal interpretation for more than 40 years. Either an energetic pulsar or an accreting black hole has been considered as its compact companion (Maraschi and Treves, 1981; Massi *et al.*, 2017). The detection of two short hard X-ray bursts from its direction favors an NS with magnetar properties (Torres *et al.*, 2012).

Only recently, the discovery of transient pulsations at 269 ms with the FAST radio telescope (Weng *et al.*, 2022) has given strong evidence that the companion of LS I +61° 303 is an NS. The fact that the pulsations appeared only sporadically could be caused by variable absorption in the non-homogeneous wind of the Be star or by an intrinsic process in the pulsar itself, similar to the nulling of RPPs or the extreme radio variability of magnetars. Although the relatively short spin period found in the LS I +61° 303 radio emission is at variance with those of typical magnetars (a few seconds), it is similar to that of the two RPPs that showed magnetar-like activity (326 ms and 407 ms, see Section 2.5). Future radio detections will likely yield a measurement of the period derivative, thus constraining the NS magnetic field.

If the magnetar nature of LS I +61° 303 is confirmed, this would be the only known magnetar in a binary system. The apparent lack of magnetars in binary systems is surprising if one considers that they descend from massive stars, a large fraction of which is in binaries. The presence of magnetars has been proposed in at least two classes of binary sources: accreting X-ray pulsars with long spin period (e.g., Li and van den Heuvel, 1999; Xu *et al.*, 2022) and ultraluminous X-ray sources (e.g., Dall'Osso *et al.*, 2015b; Eksi *et al.*, 2015), but both cases are still highly debated. In fact, the distinctive signature of a magnetar, i.e., the emission of short burst of hard X-rays, has never been seen in sources of these classes. Note that the associations of the bursts seen by Swift/BAT and of the radio pulsations discovered by FAST with LS I +61° 303 are based on positional localizations with uncertainties of ~2–3 arcmin. Although a chance alignment of these three peculiar objects is undoubtedly quite low, more observations are needed to clinch the association.

3.5. *Gravitational waves*

Gravitational waves can be produced by oscillations of the NS crust and core excited by the star quakes associated with bursts and giant flares (Corsi and Owen, 2011; de Freitas Pacheco, 1998; Horvath, 2005). Non-radial fundamental modes (f-modes) have expected frequencies $\gtrsim$1 kHz and were initially considered as promising sources of detectable gravitational waves. The model by Ioka (2001) was motivated by the large spin period increase (ΔP/P $\sim$10^{-4}) seen in the 2001 giant flare of SGR 1900+14. Such a feature (interpreted as a sudden "anti-glitch" and not observed in the other two giant flares) was taken as evidence for a change in the NS moment of inertia caused by a rearrangement of the core structure. However, subsequent analytical calculations (Levin and van Hoven, 2011) and MHD simulations (Ciolfi and Rezzolla, 2012) of giant flares driven by sudden rearrangements of the internal field indicate that only a small fraction of the involved magnetic energy is converted into f-mode oscillations and predict weak GW, undetectable by current instruments.

Quasi-periodic oscillations with frequency in the 18–2000 Hz range were discovered in the pulsating tail of the SGR 1806–20 giant flare (Israel *et al.*, 2005; Strohmayer and Watts, 2006) and attributed to seismic vibrations excited during the flare. Although only faint gravitational waves are expected in this case, there is the advantage that the search can be tuned to signals of known frequency, resulting in a higher sensitivity. The analysis of LIGO data set an upper limit of $E_{GW} \sim 8 \times 10^{46}$ erg on the energy emitted in gravitational waves (Abbott *et al.*, 2007). This limit is of the same order of the electromagnetic energy emitted in this giant flare ($E_{EM} \sim 2 \times 10^{46}$ erg, assuming isotropic emission) but still several orders of magnitude above the theoretical expectations.

More recent searches for transient gravitational waves in large samples of galactic magnetar bursts (not including giant flares) (see Abbott *et al.*, 2019; LIGO Scientific Collaboration *et al.*, 2011; The LIGO Scientific Collaboration *et al.*, 2022) reached better sensitivities but gave again negative results, with the most constraining limits implying $E_{GW}/E_{EM} < 3 \times 10^3$.

Acknowledgments

I acknowledge the financial support by the Italian Ministry for Education, University and Research through grant PRIN-MIUR 2017LJ39LM UnIAM (P.I. S.Mereghetti).

Bibliography

Abbott, B. *et al.* (2007). Search for gravitational wave radiation associated with the pulsating tail of the SGR 1806-20 hyperflare of 27 December 2004 using LIGO, *Phys. Rev. D* **76**, 6, 062003.

Abbott, B. P. *et al.* (2019). Search for transient gravitational-wave signals associated with magnetar bursts during advanced LIGO's second observing run, *ApJ* **874**, 2, 163.

Abdo, A. A. *et al.* (2010). Search for gamma-ray emission from magnetars with the fermi large area telescope, *ApJ* **725**, 1, L73–L78.

Arons, J. (2003). Magnetars in the metagalaxy: An origin for ultra-high-energy cosmic rays in the nearby universe, *ApJ* **589**, 2, 871–892.

Atteia, J. L. *et al.* (1987). Localization, time histories, and energy spectra of a new type of recurrent high-energy transient source, *ApJ* **320**, L105.

Baring, M. G. and Harding, A. K. (2007). Resonant compton upscattering in anomalous X-ray pulsars, *Ap&SS* **308**, 1–4, 109–118.

Beloborodov, A. M. (2009). Untwisting magnetospheres of neutron stars, *ApJ* **703**, 1, 1044–1060.

Beloborodov, A. M. and Thompson, C. (2007). Corona of magnetars, *ApJ* **657**, 2, 967–993.

Bochenek, C. D. *et al.* (2020). A fast radio burst associated with a galactic magnetar, *Nature* **587**, 7832, 59–62.

Boggs, S. E. *et al.* (2007). The giant flare of 2004 december 27 from SGR 1806-20, *ApJ* **661**, 1, 458–467.

Borghese, A. *et al.* (2022). The first seven months of the 2020 X-ray outburst of the magnetar SGR J1935+2154, *MNRAS* **516**, 1, 602–616.

Borghese, A. *et al.* (2021). The X-ray evolution and geometry of the 2018 outburst of XTE J1810-197, *MNRAS* **504**, 4, 5244–5257.

Bower, G. C. *et al.* (2014). The angular broadening of the galactic center pulsar SGR J1745-29: A new constraint on the scattering medium, *ApJ* **780**, 1, L2.

Bucciantini, N. *et al.* (2012). Short gamma-ray bursts with extended emission from magnetar birth: Jet formation and collimation, *MNRAS* **419**, 2, 1537–1545.

Burns, E. *et al.* (2021). Identification of a local sample of gamma-ray bursts consistent with a magnetar giant flare origin, *ApJ* **907**, 2, L28.

Cai, C. *et al.* (2022). An insight-HXMT dedicated 33 day observation of SGR J1935+2154. II. Burst spectral catalog, *ApJS* **260**, 2, 25.

Caleb, M. *et al.* (2022). Radio and X-ray observations of giant pulses from XTE J1810-197, *MNRAS* **510**, 2, 1996–2010.

Cameron, P. B. *et al.* (2005). Detection of a radio counterpart to the 27 December 2004 giant flare from SGR 1806-20, *Nature* **434**, 7037, 1112–1115.

Camilo, F. *et al.* (2006). Transient pulsed radio emission from a magnetar, *Nature* **442**, 7105, 892–895.

Camilo, F. *et al.* Polarized radio emission from the magnetar XTE J1810-197, *ApJ* **659**, 1, L37–L40.

Cao, Z. *et al.* (2019). *The Large High Altitude Air Shower Observatory (LHAASO). Science Book (2021 Edition).*

Castro, D. *et al.* (2012). Fermi-LAT observations and a broadband study of supernova remnant CTB 109, *ApJ* **756**, 1, 88.

Chen, A. Y. *et al.* (2022). Propagation of a strong fast magnetosonic wave in the magnetosphere of a neutron star, *arXiv:2210.13506.*

Cheng, Q., Yu, Y.-W., and Zheng, X.-P. (2015). Stochastic gravitational wave background from magnetic deformation of newly born magnetars, *MNRAS* **454**, 3, 2299–2304.

Cherenkov Telescope Array Consortium (2019). Science with the cherenkov telescope array.

CHIME/FRB Collaboration (2020). A bright millisecond-duration radio burst from a Galactic magnetar, *Nature* **587**, 7832, 54–58.

CHIME/FRB Collaboration (2021). The first CHIME/FRB fast radio burst catalog, *ApJS* **257**, 2, 59.

Chrimes, A. A. *et al.* (2022). New candidates for magnetar counterparts from a deep search with the Hubble Space Telescope, *MNRAS* **512**, 4, 6093–6103.

Ciolfi, R. and Rezzolla, L. (2012). Poloidal-field instability in magnetized relativistic stars, *ApJ* **760**, 1, 1.

Corsi, A. and Owen, B. J. (2011). Maximum gravitational-wave energy emissible in magnetar flares, *Phys. Rev. D* **83**, 10, 104014.

Coti Zelati, F. *et al.* (2018). Systematic study of magnetar outbursts, *MNRAS* **474**, 1, 961–1017.

Dai, Z. G. *et al.* (2005). A two-component explosion model for the giant flare and radio afterglow from SGR 1806-20, *ApJ* **629**, 2, L81–L84.

Dall'Osso, S. *et al.* (2015a). Gravitational waves from massive magnetars formed in binary neutron star mergers, *ApJ* **798**, 1, 25.

Dall'Osso, S., Perna, R., and Stella, L. (2015b). NuSTAR J095551+6940.8: a highly magnetized neutron star with super-Eddington mass accretion, *MNRAS* **449**, 2, 2144–2150.

Dall'Osso, S. and Stella, L. (2022). Millisecond magnetars, in S. Bhattacharyya, A. Papitto, and D. Bhattacharya (eds.), *Astrophysics and Space Science Library, Astrophysics and Space Science Library,* Vol. 465, 245–280.

De, K. *et al.* (2020). Constraining the X-ray-infrared spectral index of second-timescale flares from SGR 1935+2154 with palomar gattini-IR, *ApJ* **901**, 1, L7.

de Freitas Pacheco, J. A. (1998). Do soft gamma repeaters emit gravitational waves? *A&A* **336**, 397–401.

De Grandis, D. *et al.* (2022). Three-dimensional magnetothermal simulations of magnetar outbursts, *ApJ* **936**, 2, 99.

Dib, R. and Kaspi, V. M. (2014). 16 yr of RXTE monitoring of five anomalous X-ray pulsars, *ApJ* **784**, 1, 37.

Ding, H. *et al.* (2020). A magnetar parallax, *MNRAS* **498**, 3, 3736–3743.

Duncan, R. C. and Thompson, C. (1992). Formation of very strongly magnetized neutron stars: Implications for gamma-ray bursts, *ApJ* **392**, L9.

Eatough, R. P. *et al.* (2013). A strong magnetic field around the supermassive black hole at the centre of the galaxy, *Nature* **501**, 7467, 391–394.

Eksi, K. Y. *et al.* (2015). The ultraluminous X-ray source NuSTAR J095551+6940.8: a magnetar in a high-mass X-ray binary. *MNRAS* **448**, L40–L42.

Fermi-LAT Collaboration (2021). High-energy emission from a magnetar giant flare in the Sculptor galaxy, *Nature Astronomy* **5**, 385–391.

Feroci, M. *et al.* (2004). Broadband X-ray spectra of short bursts from SGR 1900+14, *ApJ* **612**, 1, 408–413.

Frail, D. A., Kulkarni, S. R., and Bloom, J. S. (1999). An outburst of relativistic particles from the soft γ-ray repeater SGR1900+14, *Nature* **398**, 6723, 127–129.

Frederiks, D. D. *et al.* (2007a). Giant flare in SGR 1806-20 and its compton reflection from the moon, *Astronomy Letters* **33**, 1, 1–18.

Frederiks, D. D. *et al.* (2007b). On the possibility of identifying the short hard burst GRB 051103 with a giant flare from a soft gamma repeater in the M81 group of galaxies, *Astron. Lett.* **33**, 1, 19–24.

Fuchs, Y. *et al.* (1999). ISO observations of the environment of the soft gamma-ray repeater SGR 1806-20, *A&A* **350**, 891–899.

Gavriil, F. P. *et al.* (2008). Magnetar-like emission from the young pulsar in kes 75, *Science* **319**, 5871, 1802.

Ge, M. Y. *et al.* (2023). Reanalysis of the X-ray burst associated FRB 200428 with insight-HXMT observations, arXiv:2302.00176.

Gelfand, J. D. *et al.* (2017). The high-frequency radio emission of the galactic center magnetar SGR J1745-29 during a transitional period, *ApJ* **850**, 1, 53.

Gonzalez, M. E. *et al.* (2010). Long-term X-ray changes in the emission from the anomalous X-ray pulsar 4U 0142+61, *ApJ* **716**, 2, 1345–1355.

Göğüş, E. *et al.* (2016). Magnetar-like X-ray bursts from a rotation-powered pulsar, PSR J1119-6127, *ApJ* **829**, 2, L25.

Granot, J. *et al.* (2006). Diagnosing the outflow from the SGR 1806-20 giant flare with radio observations, *ApJ* **638**, 1, 391–396.

Halpern, J. P. *et al.* (2005). Discovery of radio emission from the transient anomalous X-ray pulsar XTE J1810-197, *ApJ* **632**, 1, L29–L32.

HESS Collaboration (2018). Extended VHE γ-ray emission towards SGR1806-20, LBV 1806-20, and stellar cluster Cl* 1806-20, *A&A* **612**, A11.

HESS Collaboration (2021). Searching for TeV gamma-ray emission from SGR 1935+2154 during its 2020 X-ray and radio bursting phase, *ApJ* **919**, 2, 106.

Hnatyk, B. *et al.* (2022). Unveiling the nature of the unidentified gamma-ray sources 4FGL J1908.6+0915e, HESS J1907+089/HOTS J1907+091, and 3HWC J1907+085 in the sky region of the magnetar SGR 1900+14, *MNRAS* **514**, 1, 762–779.

Horvath, J. E. (2005). Energetics of the superflare from SGR1806-20 and a possible associated gravitational wave burst, *Mod. Phys. Lett. A* **20**, 36, 2799–2804.

Hu, C.-P. *et al.* (2020). NICER observation of the temporal and spectral evolution of swift J1818.0-1607: A missing link between magnetars and rotation-powered pulsars, *ApJ* **902**, 1, 1.

Huang, Z.-P. *et al.* (2021). Simultaneous 2.25/8.60 GHz observations of the newly discovered magnetar - swift J1818.0-1607, *MNRAS* **505**, 1, 1311–1315.

Huentemeyer, P. *et al.* (2019). The southern wide-field gamma-ray observatory (SWGO): A next-generation ground-based survey instrument, *Bullet. Am. Astron. Soc.*, 51, 109.

Hurley, K. *et al.* (2005). An exceptionally bright flare from SGR 1806-20 and the origins of short-duration γ-ray bursts, *Nature* **434**, 7037, 1098–1103.

Hurley, K. (1999). A giant periodic flare from the soft γ-ray repeater SGR1900+14, *Nature* **397**, 6714, 41–43.

Ioka, K. (2001). Magnetic deformation of magnetars for the giant flares of the soft gamma-ray repeaters, *MNRAS* **327**, 2, 639–662.

Israel, G. L. *et al.* (2005). The discovery of rapid X-ray oscillations in the tail of the SGR 1806-20 hyperflare, *ApJ* **628**, 1, L53–L56.

Israel, G. L. *et al.* (2021). X-ray and radio bursts from the magnetar 1E 1547.0-5408, *ApJ* **907**, 1, 7.

Israel, G. L. (2008). A swift gaze into the 2006 March 29 burst forest of SGR 1900+14, *ApJ* **685**, 2, 1114–1128.

Kaminker, A. D. *et al.* (2014). Thermal emission of neutron stars with internal heaters, *MNRAS* **442**, 4, 3484–3494.

Kasen, D. and Bildsten, L. (2010). Supernova light curves powered by young magnetars, *ApJ* **717**, 1, 245–249.

Kaspi, V. M. and Beloborodov, A. M. (2017). Magnetars, *ARA&A* **55**, 1, 261–301.

Kern, B. and Martin, C. (2002). Optical pulsations from the anomalous X-ray pulsar 4U0142+61, *Nature* **417**, 6888, 527–529.

Khangulyan, D., Barkov, M. V., and Popov, S. B. (2022). Fast radio bursts by high-frequency synchrotron maser emission generated at the reverse shock of a powerful magnetar flare, *ApJ* **927**, 1, 2.

Kijak, J. *et al.* (2017). Gigahertz-peaked spectra pulsars and thermal absorption model, *ApJ* **840**, 2, 108.

Kirsten, F. *et al.* (2022). A repeating fast radio burst source in a globular cluster, *Nature* **602**, 7898, 585–589.

Kirsten, F. *et al.* (2021). Detection of two bright radio bursts from magnetar SGR 1935 + 2154, *Nature Astronomy* **5**, 414–422.

Kuiper, L. *et al.* (2012). Temporal and spectral evolution in X- and γ-rays of magnetar 1E 1547.0-5408 since its 2008 October Outburst: The discovery of a transient hard pulsed component after its 2009 January outburst, *ApJ* **748**, 2, 133.

Lai, D. (2022). IXPE detection of polarized X-rays from magnetars and photon mode conversion at QED vacuum resonance, arXiv:2209.13640.

Levin, Y. and van Hoven, M. (2011). On the excitation of f modes and torsional modes by magnetar giant flares, *MNRAS* **418**, 1, 659–663.

Li, C. K. *et al.* (2021). HXMT identification of a non-thermal X-ray burst from SGR J1935+2154 and with FRB 200428, *Nature Astronomy* **5**, 378–384.

Li, J. *et al.* (2017). Gamma-ray upper limits on magnetars with six years of fermi-LAT observations, *ApJ* **835**, 1, 30.

Li, X. *et al.* (2022a). Quasi-periodic oscillations of the X-ray burst from the magnetar SGR J1935-2154 and associated with the fast radio burst FRB 200428, *ApJ* **931**, 1, 56.

Li, X. B. *et al.* (2022b). Insight-HXMT detection of an X-ray burst from SGR J1935+2154 coinciding with the radio burst on 2022-10-21, *The Astronomer's Telegram* **15708**, 1.

Li, X. D. and van den Heuvel, E. P. J. (1999). Could 2S 0114+650 be a magnetar? *ApJ* **513**, 1, L45–L48.

Lieu, R. (2017). Are fast radio bursts the birthmark of magnetars? *ApJ* **834**, 2, 199.

LIGO Collaboration, Bizouard, M. A., Dietz, A., Guidi, G. M., and Was, M. (2012). Implications for the origin of GRB 051103 from LIGO observations, *ApJ* **755**, 1, 2.

LIGO Scientific Collaboration and Hurley, K. C. (2008). Implications for the origin of GRB 070201 from LIGO observations, *ApJ* **681**, 2, 1419–1430.

LIGO Scientific Collaboration, Virgo Collaboration, *et al.* (2011). Search for gravitational wave bursts from six magnetars, *ApJ* **734**, 2, L35.

LIGO Scientific Collaboration and Virgo Collaboration, Fermi Gamma-ray Burst Monitor, and INTEGRAL (2017). Gravitational waves and gamma-rays from a binary neutron star merger: GW170817 and GRB 170817A, *ApJ* **848**, 2, L13.

Lin, L. *et al.* (2020). No pulsed radio emission during a bursting phase of a Galactic magnetar, *Nature* **587**, 7832, 63–65.

Liu, B. *et al.* (2017). A study of Fermi-LAT GeV γ-ray emission toward the magnetar-harboring supernova remnant kesteven 73 and its molecular environment, *ApJ* **851**, 1, 37.

Lower, M. E. *et al.* (2021). The dynamic magnetosphere of Swift J1818.0-1607, *MNRAS* **502**, 1, 127–139.

Lu, W. and Kumar, P. (2018). On the radiation mechanism of repeating fast radio bursts, *MNRAS* **477**, 2, 2470–2493.

Lyubarsky, Y. (2020). Fast radio bursts from reconnection in a magnetar magnetosphere, *ApJ* **897**, 1, 1.

Lyubarsky, Y., Eichler, D., and Thompson, C. (2002). Diagnosing magnetars with transient cooling, *ApJ* **580**, 1, L69–L72.

Lyutikov, M. (2003). Explosive reconnection in magnetars, *MNRAS* **346**, 2, 540–554.

Maan, Y. *et al.* (2022). Magnetar XTE J1810-197: Spectro-temporal evolution of average radio emission, *ApJ* **931**, 1, 67.

Mahlmann, J. F. *et al.* (2019). Instability of twisted magnetar magnetospheres, *MNRAS* **490**, 4, 4858–4876.

Maraschi, L. and Treves, A. (1981). A model for LS I +61 303. *MNRAS* **194**, 1P–5.

Massi, M., Migliari, S., and Chernyakova, M. (2017). The black hole candidate LS I +61°0303, *MNRAS* **468**, 3, 3689–3693.

Mazets, E. P. *et al.* (2008). A giant flare from a soft gamma repeater in the andromeda galaxy (M31), *ApJ* **680**, 1, 545–549.

Mazets, E. P. *et al.* (1982). The 5 March 1979 event and the distinct class of short gamma bursts: Are they of the same origin? *Ap&SS* **84**, 1, 173–189.

Mazets, E. P. *et al.* (1979). Observations of a flaring X-ray pulsar in Dorado, *Nature* **282**, 5739, 587–589.

Mereghetti, S. (2008). The strongest cosmic magnets: soft gamma-ray repeaters and anomalous X-ray pulsars, *A&A Rev.* **15**, 4, 225–287.

Mereghetti, S. *et al.* (2005a). The first giant flare from SGR 1806-20: Observations using the anticoincidence shield of the spectrometer on INTEGRAL, *ApJ* **624**, 2, L105–L108.

Mereghetti, S., Pons, J. A., and Melatos, A. (2015). Magnetars: Properties, origin and evolution, *Space Sci. Rev.* **191**, 1–4, 315–338.

Mereghetti, S. *et al.* (2020). INTEGRAL discovery of a burst with associated radio emission from the magnetar SGR 1935+2154, *ApJ* **898**, 2, L29.

Mereghetti, S. and Stella, L. (1995). The very low mass X-ray binary pulsars: A new class of sources? *ApJ* **442**, L17.

Mereghetti, S. *et al.* (2005b). An XMM-newton view of the soft gamma repeater SGR 1806-20: Long-term variability in the pre-giant flare epoch, *ApJ* **628**, 2, 938–945.

Metzger, B. D., Margalit, B., and Sironi, L. (2019). Fast radio bursts as synchrotron maser emission from decelerating relativistic blast waves, *MNRAS* **485**, 3, 4091–4106.

Metzger, B. D., Quataert, E., and Thompson, T. A. (2008). Short-duration gamma-ray bursts with extended emission from protomagnetar spin-down, *MNRAS* **385**, 3, 1455–1460.

Nakar, E. *et al.* (2006). The distances of short-hard gamma-ray bursts and the soft gamma-ray repeater connection, *ApJ* **640**, 2, 849–853.

Ofek, E. O. (2007). Soft gamma-ray repeaters in nearby galaxies: Rate, luminosity function, and fraction among short gamma-ray bursts, *ApJ* **659**, 1, 339–346.

Paczynski, B. (1992). GB 790305 as a very strongly magnetized neutron star, *Acta Astron.* **42**, 145–153.

Palmer, D. M. *et al.* (2005). A giant γ-ray flare from the magnetar SGR 1806 - 20, *Nature* **434**, 7037, 1107–1109.

Petroff, E., Hessels, J. W. T., and Lorimer, D. R. (2022). Fast radio bursts at the dawn of the 2020s, *A&A Rev.* **30**, 1, 2.

Pons, J. A. and Rea, N. (2012). Modeling magnetar outbursts: Flux enhancements and the connection with short bursts and glitches, *ApJ* **750**, 1, L6.

Popov, S. B. and Postnov, K. A. (2013). Millisecond extragalactic radio bursts as magnetar flares, arXiv:1307.4924.

Rajwade, K. M. *et al.* (2022). Long term radio and X-ray evolution of the magnetar swift J1818.0-1607, *MNRAS* **512**, 2, 1687–1695.

Rea, N. *et al.* (2020). The X-ray outburst of the galactic center magnetar over six years of chandra observations, *ApJ* **894**, 2, 159.

Rea, N. *et al.* (2010). A low-magnetic-field soft gamma repeater, *Science* **330**, 6006, 944.

Rea, N. *et al.* (2012). A new low magnetic field magnetar: The 2011 outburst of swift J1822.3-1606, *ApJ* **754**, 1, 27.

Roberts, O. J. *et al.* (2021). Rapid spectral variability of a giant flare from a magnetar in NGC 253, *Nature* **589**, 7841, 207–210.

Santangelo, A. *et al.* (2019). Physics and astrophysics of strong magnetic field systems with eXTP, *Science China Physics, Mechanics, and Astronomy* **62**, 2, 29505.

Strohmayer, T. E. and Watts, A. L. (2006). The 2004 hyperflare from SGR 1806-20: Further evidence for global torsional vibrations, *ApJ* **653**, 1, 593–601.

Svinkin, D. *et al.* (2021). A bright γ-ray flare interpreted as a giant magnetar flare in NGC 253, *Nature* **589**, 7841, 211–213.

Svinkin, D. S. *et al.* (2015). A search for giant flares from soft gamma-ray repeaters in nearby galaxies in the Konus-WIND short burst sample, *MNRAS* **447**, 1, 1028–1032.

Taverna, R. *et al.* (2022). Polarized x-rays from a magnetar, *Science* **378**, 6620, 646–650.

Tendulkar, S. P., Cameron, P. B., and Kulkarni, S. R. (2012). Proper motions and origins of SGR 1806-20 and SGR 1900+14, *ApJ* **761**, 1, 76.

Testa, V. *et al.* (2008). Adaptive optics, near-infrared observations of magnetars, *A&A* **482**, 2, 607–615.

The LIGO Scientific Collaboration, the Virgo Collaboration, and the KAGRA Collaboration (2022). Search for gravitational-wave transients associated with magnetar bursts in advanced LIGO and advanced virgo data from the third observing run, arXiv:2210.10931.

Thompson, C. and Beloborodov, A. M. (2005). High-energy emission from magnetars, *ApJ* **634**, 1, 565–569.

Thompson, C. and Duncan, R. C. (1993). Neutron star dynamos and the origins of pulsar magnetism, *ApJ* **408**, 194.

Thompson, C. and Duncan, R. C. (1995). The soft gamma repeaters as very strongly magnetized neutron stars - I. Radiative mechanism for outbursts, *MNRAS* **275**, 2, 255–300.

Thompson, C. and Duncan, R. C. (1996). The soft gamma repeaters as very strongly magnetized neutron stars. II. Quiescent neutrino, X-ray, and alfven wave emission, *ApJ* **473**, 322.

Thompson, C., Lyutikov, M., and Kulkarni, S. R. (2002). Electrodynamics of magnetars: Implications for the persistent X-ray emission and spin-down of the soft gamma repeaters and anomalous X-ray pulsars, *ApJ* **574**, 1, 332–355.

Tong, H., Song, L. M., and Xu, R. X. (2011). Anomalous X-ray pulsars and soft gamma-ray repeaters in the outer gap model: Confronting fermi observations, *ApJ* **738**, 1, 31.

Torne, P. *et al.* (2022). Submillimeter pulsations from the magnetar XTE J1810-197, *ApJ* **925**, 2, L17.

Torres, D. F. *et al.* (2012). A magnetar-like event from LS I +61°303 and its nature as a gamma-ray binary, *ApJ* **744**, 2, 106.

Turolla, R., Zane, S., and Watts, A. L. (2015). Magnetars: the physics behind observations. A review, *Reports on Progress in Physics* **78**, 11, 116901.

Usov, V. V. (1992). Millisecond pulsars with extremely strong magnetic fields as a cosmological source of γ-ray bursts, *Nature* **357**, 6378, 472–474.

Viganò, D. *et al.* (2015a). An assessment of the pulsar outer gap model - II. Implications for the predicted γ-ray spectra, *MNRAS* **447**, 3, 2649–2657.

Viganò, D., Torres, D. F., and Martín, J. (2015b). A systematic synchro-curvature modelling of pulsar γ-ray spectra unveils hidden trends, *MNRAS* **453**, 3, 2599–2621.

Wang, C. W. *et al.* (2022). GECAM and HEBS detection of a short X-ray burst from SGR J1935+2154 associated with radio burst, *The Astronomer's Telegram* **15682**, 1.

Wang, S. *et al.* (2020). Distances to the supernova remnants in the inner disk, *A&A* **639**, A72.

Wang, S.-Q., Wang, L.-J., and Dai, Z.-G. (2019a). The energy sources of superluminous supernovae, *Res. Astron. Astrophys.* **19**, 5, 063.

Wang, W. *et al.* (2019b). Coherent radio emission from a twisted magnetosphere after a magnetar-quake, *ApJ* **875**, 2, 84.

Wang, X. Y. *et al.* (2005). An energetic blast wave from the 2004 December 27 giant flare of the soft gamma-ray repeater SGR 1806-20, *ApJ* **623**, 1, L29–L32.

Wang, Z., Chakrabarty, D., and Kaplan, D. L. (2006). A debris disk around an isolated young neutron star, *Nature* **440**, 7085, 772–775.

Weng, S.-S. *et al.* (2022). Radio pulsations from a neutron star within the gamma-ray binary LS I +61° 303, *Nature Astronomy* **6**, 698–702.

Xie, L., Wei, D.-M., Wang, Y., and Jin, Z.-P. (2022). Constraining the ellipticity of the newborn magnetar with the observational data of long gamma-ray bursts, *ApJ* **934**, 2, 125.

Xin, Y.-L. *et al.* (2016). A GeV source in the direction of supernova remnant CTB 37B, *ApJ* **817**, 1, 64.

Xu, K. *et al.* (2022). Are there magnetars in high-mass X-ray binaries? *Res. Astron. Astrophys.* **22**, 1, 015005.

Xu, M. and Tang, Q. W. (2021). Two X-ray plateaus of gamma-ray bursts: Energy injection from nascent magnetars with an evolving magnetic inclination angle, *ApJ* **911**, 2, 76.

Yan, W. M. *et al.* (2018). Single-pulse observations of the Galactic centre magnetar PSR J1745-2900 at 3.1 GHz, *MNRAS* **476**, 3, 3677–3687.

Yang, J. *et al.* (2020). GRB 200415A: A short gamma-ray burst from a magnetar giant flare? *ApJ* **899**, 2, 106.

Yang, Y.-P. and Zhang, B. (2018). Bunching coherent curvature radiation in three-dimensional magnetic field geometry: Application to pulsars and fast radio bursts, *ApJ* **868**, 1, 31.

Younes, G. *et al.* (2022). Pulse peak migration during the outburst decay of the magnetar SGR 1830-0645: Crustal motion and magnetospheric untwisting, *ApJ* **924**, 2, L27.

Yusef-Zadeh, F. *et al.* (2015). Radio continuum emission from the magnetar SGR J1745-2900: Interaction with gas orbiting Sgr A, *ApJ* **811**, 2, L35.

Zampieri, L. *et al.* (2022). Deep upper limit on the optical emission during a hard X-ray burst from the magnetar SGR J1935+2154, *ApJ* **925**, 2, L16.

Zane, S. *et al.* (2023). A strong X-ray polarization signal from the magnetar 1RXS J170849.0-400910, *ApJ* **944**, 2, L27.

Zhang, C. F. *et al.* (2020a). A highly polarised radio burst detected from SGR 1935+2154 by FAST, *The Astronomer's Telegram* **13699**, 1.

Zhang, H.-M. *et al.* (2020b). Magnetar giant flare origin for GRB 200415A inferred from a new scaling relation, *ApJ* **903**, 2, L32.

Zhang, L. and Cheng, K. S. (2002). Quiescent high-energy gamma-ray emission from soft gamma-ray repeaters, *ApJ* **579**, 2, 716–724.

https://doi.org/10.1142/9789819800643_0002

Chapter 2

High-Energy Emission in SGRs/AXPs

Ronaldo V. Lobato*, Jaziel G. Coelho†,‡, and Manuel M. Malheiro§,¶

**Laboratório de Física Teórica e Computacional (LFTC), Universidade Cidade de São Paulo (UNICID) — Rua Galvão Bueno 868, 01506-000 São Paulo, Brazil*

†Núcleo de Astrofísica e Cosmologia (Cosmo-Ufes), Departamento de Física, Universidade Federal do Espírito Santo 29075–910, Vitória, ES, Brazil

‡Divisão de Astrofísica, Instituto Nacional de Pesquisas Espaciais, Avenida dos Astronautas 1758 12227–010, São José dos Campos, SP, Brazil

§Instituto Tecnológico de Aeronáutica 12228–900, São José dos Campos, SP, Brazil

¶contact@rvlobato.com

In this chapter, we describe the high-energy emission in SGRs/AXPs, neutron stars with extreme magnetic fields, named magnetars. The observed high-energy emissions in the X-rays and gamma-ray bands are explained. The outer gap accelerator model considers the generation of high-energy emission far away from the surface of the star in a combination of curvature radiation and inverse Compton scattering. The high-energy emission is constrained by employing the death lines, the condition for pair production, and the generation of hard X and soft gamma rays. From the point of the death lines and death-zones obtained here by the outer gap model, the SGRs/AXPs should not emit hard X-rays since they are below the death line, in disagreement with the observations: SGRs/AXPs are characterized by short and intense bursts in the hard X-rays (few hundred keV). The reason for the absence of hard X-ray emission for magnetars in the outer gap model is the tiny strength of the magnetic field at the light cylinder, which is too far away from the star's surface in comparison to millisecond γ-rays pulsars. Forthcoming works are expected to refine the results, as well as applications on SGRs/AXPs, understanding them as white dwarfs.

Keywords: SGRs/AXPs, Magnetars, White dwarfs, Neutron stars, Pulsars

Contents

1. Introduction

Soft Gamma Repeaters (SGRs) and Anomalous X-ray Pulsars (AXPs) are a new class of pulsars in comparison with the rotation-powered pulsars (RPPs). They are commonly known as neutron stars with huge magnetic fields, magnetars, larger than the quantum critical magnetic field, $B_c = m_e^2c^3/e\hbar$. These extreme magnetic fields are 1001000 times more than typical radio pulsars $B \sim 10^{12}$ G. These sources are observationally characterized by spin periods of $P \sim 2 - 12$ s, spin-down rates $\dot{P} \sim (10^{-13} - 10^{-10})$ s/s larger than normal pulsars $\dot{P} \sim (10^{-15} - 10^{-14})$ s/s (Mereghetti, 2013; Turolla *et al.*, 2015). SGRs/AXPs have been the center of some discussions in the last years; some questions are not very well understood, like its nature (Blaes *et al.*, 1989; Katz *et al.*, 1994; Woosley and Wallace, 1982).

At first, the efforts to understand the nature of these sources were not compensated due to the lack of information. But this scenario is changing with the new observations from several telescopes (Fermi, Chandra, Integral, XMM, Swift-BAT and NICER). Although the interpretation of neutron stars is the that are most accepted, these exist some features about SGRs/AXPs not very well understood in the magnetar model (Coelho & Malheiro, 2014): (1) RPPs are assumed to be born with a short period $\sim$1 ms resulting in large magnetic fields by dynamo process (Tong *et al.*, 2011). SGRs/AXPs present long periods, such that generation of high magnetic fields by this mechanism does work; (2) the large spin-down rates in comparison with the typical pulsars: For example, the source SGR 0418+5729 has a characteristic age $\tau = P/2\dot{P} = 2.4 \times 10^7$ years, this means, even being older than ordinary pulsars, it has a major slowdown. Some SGRs/AXPs with low magnetic fields and smaller spin-down rates were observed (Rea *et al.*, 2010). Within these observations, we highlight SGR $0418 + 5729$, with period $P = 9.08$ s, spin-down $\dot{P} < 6.0 \times 10^{-15}$ s/s, and X-ray luminosity $L_X < 6.2 \times 10^{31}$ erg/s. Considering this source as a neutron star, the loss of energy associated with the injection of rotational energy in the pulsars' magnetosphere does not explain the luminosity of this star, i.e., $\dot{E}_{\rm rot}^{\rm NS} < L_X$, excluding the possibility to be an RPP, where $\dot{E}_{\rm rot}^{\rm NS} > L_X$ (Coelho *et al.*, 2017, see). According to the magnetar model, the high X-ray luminosity from these sources is powered by the energy stored in their strong magnetic field, by a complex decay mechanism (Duncan, 2001; Duncan & Thompson, 1992; Kaspi & Beloborodov, 2017; Mereghetti, 2013; Thompson, 2002, 2008; Thompson & Duncan, 1993, 1995, 1996; Thompson *et al.*, 2002; Turolla *et al.*, 2015). Recent observations of the Fermi satellite (Fermi-LAT Collaboration, 2010; Tong *et al.*, 2010, 2011) do not find evidence of γ radiation. In addition, some SGRs/AXPs were observed to emit radio wavelength, while the others do not. These observations carried some uncertainties about the nature of SGRs/AXPs.

A model proposed by Malheiro *et al.* (2012) following the work of Morini *et al.* (1988) and Paczynski (1990) considers SGRs/AXPs as very massive, rapidly rotating highly magnetic white dwarfs (see also Borges *et al.*, 2020; Coelho & Malheiro, 2014; Lobato *et al.*, 2016; Sousa *et al.*, 2020). There are some features to explain SGRs/AXPs as WDs: They are isolated stars, the masses need to be $1.4M_{\odot} > M > 1.2\ M_{\odot}$, the magnetic field $B \approx 10^9 - 10^{10}$ G, and rotational periods of 0.5 s $< P <$ 12 s. All these characteristics were already reproduced by theoretical calculations (Boshkayev *et al.*, 2013; Caceres *et al.*, 2017; Lobato *et al.*, 2016).

Discovered in X and soft gamma rays, the SGRs/AXPs have been detected in a broad wavelength spectrum: they emit in optical/near infrared, radio, soft, and hard X-rays (Turolla *et al.*, 2015). Moreover, the possibility of high-energy emission from fast rotating white dwarf pulsars was evaluated based on the results of recent findings from two non-accreting fast rotating white dwarfs in close binaries, i.e., AE Aquarii (Meintjes *et al.*, 1994, 1992) and AR Scorpii (Marsh & *et al.*, 2016). It has been shown by Meintjes *et al.* (2023), from an energy perspective, that both white dwarfs in these systems may be the sites of particle acceleration and possible gamma-ray emission through processes, such as curvature radiation, inverse Compton, and hadronic processes, such as neutral pion decay. Fermi-LAT observations of both these systems over a time span of a decade showed indications of pulsed high-energy gamma-ray modulation at, or slightly above, the Fermi-LAT threshold energy. The superior sensitivity of the Cherenkov Telescope Array (CTA) opens up novel avenues for investigating these sources and potentially others like them. The forthcoming Vera Rubin Observatory is poised to revolutionize time-domain astrophysics, offering the prospect of uncovering thousands of previously unseen transient sources, such as highly magnetized white dwarfs (HMWDs) (see Sousa *et al.*, 2023, 2022). In this chapter, we restrict the high-energy emission investigation of SGRs/AXPs as neutron stars.

2. Observed High-Energy Emission

2.1. *X-rays*

The SGR/AXP X-ray emission is generally separated as low ($<$ 10 keV) and high energy ($>$ 20 keV) and seems that two different mechanisms are responsible for the two (Turolla *et al.*, 2015). The low-energy component is fitted with a blackbody with temperature $kT \sim 0.3$–0.6 keV and a power law with photon index $\Gamma \sim 3$–4 (Mereghetti *et al.*, 2002) or two blackbodies (Rea & Torres, 2011).

One can look at Figure 5 from Kaspi and Beloborodov (Kaspi & Beloborodov, 2017) for such a spectrum. The non-thermal component begins to dominate above 10 keV. The soft X-ray emission is predominantly of thermal origin (Mereghetti, 2008) and is complex, i.e., the emission from the surface can get distorted by resonant cyclotron scattering. The thermal component which dominates the lowest energy band is from a region much smaller than the whole surface of the star (Turolla *et al.*, 2015) and the tail of the power law likely arises from magnetospheric effects. Although this picture is broadly accepted, there are still some open problems, either because of a reprocessing by return currents or because the surface itself may be in a condensed state, being a self-consistent model needed (for more details on attempts, e.g., see Turolla *et al.*, 2015).

An important point to highlight is the average temperature of the thermal component, which for the magnetars is more luminous than rotation-powered pulsars of comparable age (Mereghetti, 2008).

The persistent hard X-ray emission extends to ~150 keV. This radiation has a non-thermal characteristic and some studies considering spectroscopy suggest that a considerable fraction of the luminosity comes from the hard band instead of the soft one (Enoto *et al.*, 2011; Götz *et al.*, 2006; Kuiper *et al.*, 2006, 2004; Mereghetti, 2008; Mereghetti *et al.*, 2005; Molkov *et al.*, 2005), i.e., the energy released in this range is a large fraction of the total energy released in these sources. There are some attempts to study the spectral location of the soft-hard X-ray turnover (Kaspi & Boydstun, 2010), which in general is believed to be correlated with the magnetic field on the surface. For higher energy observations, Fermi-LAT failed to detect emission, implying in a spectral break above a few hundred keV (den Hartog *et al.*, 2006; Kuiper *et al.*, 2006), the source AXP 1E 2259+586 seemed to show γ-ray pulsation above 200 MeV (Wu *et al.*, 2013), however this is not confirmed yet.

The mechanism for hard X-ray is not well understood, suggesting that these X-rays can be generated by thermal bremsstrahlung in the surface, heated by returning current or by synchrotron emission from pairs on the magnetosphere (Thompson & Beloborodov, 2005) or even by resonant up-scattering of seed photons (Baring *et al.*, 2005; Baring & Harding, 2008). Nevertheless, this is still an open question due to the dramatic number of degrees of freedom for the magnetosphere.

2.2. *γ-rays*

The SGRs/AXPs are characterized by the short and intense burst in the hard X/soft gamma-ray band (few keV to few hundreds keV) (Mereghetti, 2008; Mereghetti *et al.*, 2013) raging from few seconds to several seconds. There are three kinds of burst events (Turolla *et al.*, 2015): (1) short burst — most common and with duration of $\sim$0.1–1 s, peak luminosity $\sim 10^{39-41}$ erg/s and soft thermal spectra, 10 keV; (2) intermediate bursts — duration of $\sim$1−40 s, peak luminosity of $\sim 10^{41-43}$ erg/s with thermal spectra; (3) giant flares — rare events, peak luminosity of $\sim 10^{44-47}$ erg/s.

The burst activity is very variable and often is enhanced by the persistent emission (Mereghetti *et al.*, 2013). It appears that there is no dependence of short/intermediate burst activity on the magnetic field strength of sources (Turolla *et al.*, 2015).

3. Models of High-Energy Emission

Among the models for hard X and γ-rays, the high-energy emission, there is the gap accelerator model, proposed by Cheng *et al.* (1986a,b). It considers the production of the observed photons far away from the stellar surface, where the magnetic field lines are open. In this model, the energy is limited by a combination of curvature radiation and inverse Compton scattering. The mechanism relies on the acceleration of particles by an $\mathsf{E}_{\|}$. This field can be screened out by the discharge of copious $e^{\pm}$ created by γ-ray. The emitting particles move out of the gap from/toward the stellar surface, and the production of $e^{\pm}$ pairs feeds the necessary number of particles to sustain the gap (closure of the gap). Pair production is generated by at least two processes: photon–photon or photon–magnetic field interaction, which is effective only for strong magnetic fields (Takata *et al.*, 2010).

Zhang & Cheng (1997) have studied the gap mechanism using the photon–photon process between high-energy γ-rays and X-rays coming from the stellar surface. The pair-production condition is estimated as $E_X E_\gamma(1 - \cos\theta_{X\gamma}) \geq 2(mc^2)^2$, where $\theta_{X\gamma}$ is the angle between the directions of X and γ-rays. E_X is the energy of the soft photons from the stellar surface and E_γ is the energy of emitted γ-rays in the outer gap. In this case, the gap is sustained by an

X-ray flux emitted from the star surface. These X photons are, in general, produced by a return current $e^{\pm}$ which heats the stellar surface (Caceres *et al.*, 2017).

In the work of Vigano *et al.* (2015a), the assumptions, uncertainties, and implications of the outer gap model are explored. It shows the relations among these quantities, summarized in Figure 1 of their work.

An important parameter considered in the outer gap model is the gap size/thickness or transfield fractional gap thickness f; this parametric quantity is related to the upper and lower boundaries of the gap (Hirotani, 2006). Its definition is given by $f \equiv (\theta_c - \theta_u)/\theta_c$, where θ_u and θ_c are the magnetic colatitude footpoints on the polar cap surface for the lower and upper boundaries (Vigano *et al.*, 2015a). The size of the outer gap is controlled by $e^{\pm}$ production resulting from the interaction of thermal X-rays with primary γ-rays.

The parameter f is important in the description of the mechanism of electrodynamics radiation since it controls the parallel electric field along the gap. $\mathsf{E}_{\parallel}$ is approximately given by (Zhang & Cheng, 1997):

$$\mathsf{E}_{\parallel} \approx \frac{Ba^2\Omega}{cr_c} \approx f^2 \mathsf{E}_{\perp} \left(\frac{R_{\text{lc}}}{r_c}\right), \tag{1}$$

where a is related to the boundaries conditions to solve the Poisson equation (Cheng *et al.*, 1986a), Ω is the star's frequency $(2\pi/P)$, r_c is the curvature radius, and $R_{\text{lc}} \equiv c/\Omega$ the light cylinder radius. As pointed out by Vigano *et al.* (2015a), Eq. (1) is a simplification due to some approximations which arise from the intrinsic difficulties of the problem. Aimed to explore the $\mathsf{E}_{\parallel}$ dependencies with fewer assumptions, they have considered $\mathsf{E}_{\parallel}$ as a function of B, P, and f, as well as with the geometry through the parameter χ, defined as

$$\chi \equiv \frac{a^2 q(q-1)}{R_{\text{lc}} r_c} \frac{B(r,\theta)}{B} \left(\frac{R}{R_{\text{lc}}}\right)^3, \tag{2}$$

where $q \equiv z/a$, $z \in [0, a]$, and R the star's radius. Considering also a screening factor κ, the above expression for the parallel electric field becomes

$$\mathsf{E}_{\parallel} = \kappa \chi f^2 B \left(\frac{2\pi R}{cP}\right)^3. \tag{3}$$

Considering $\kappa = 1$ and $\chi = 1$ used by Zhang and Cheng's approach (Zhang & Cheng, 1997), the associated potential drop in the gap is written as $V = f^2 V_{\rm tot}$, where $V_{\rm tot}$ is the total potential drop, given by $\Delta V_{\rm max} = (B_{\rm p}\Omega^2 R^3)/(2c^2)$, where $B_{\rm p}$ is the magnetic field on the polar cap.

Using the potential drop in the gap, it is possible to estimate the γ-ray luminosity considering the current flowing in the gap $I_{\rm gap}$, $L_\gamma \sim I_{\rm gap} V$. The current flow is $I_{\rm gap} \sim f I_{\rm GJ}$, where $I_{\rm GJ}$ is the Goldreich–Julian current (Goldreich & Julian, 1969). Thus, as we can see, the parameter f plays a key role in the outer gap model.

Inverting Eq. (3)

$$f = \sqrt{\frac{\mathsf{E}_{||}}{B\kappa\chi}\left(\frac{cP}{2\pi R}\right)^3}, \tag{4}$$

it is possible to estimate the death line according to the value of the $\mathsf{E}_{||}$ needed to sustain the gap. As pointed by Vigano *et al.* (2015b) the constraints are strongly dependent on $\mathsf{E}_{||}$. That can be obtained by the equation of the Lorentz factor evolution, $mc^2\dot{\Gamma} = e\mathsf{E}_{||}c - P_{\rm syn-curv} - P_{\rm IC}$, where $P_{\rm syn-curv}$ and $P_{\rm IC}$ are the energy loss rates of the synchro-curvature radiation and inverse Compton process, respectively. One can obtain the electric field in the steady state, i.e., when the gain of energy by the electric field is equal to the energy loss by the synchro-curvature radiation (the inverse Compton radiation is too small), that can be expressed as $e\mathsf{E}_{||}c - P_{\rm syn-curv} = 0$.

The synchrotron radiation and the curvature radiation are given by $P_{\rm syn} = -(2e^4 B^2 \sin^2\alpha\gamma^2)/(3m^2c^3)$ and $P_{\rm curv} = -(2e^2c\gamma^4)/(3r_c^2)$ (Jackson, 1975). If one considers a more general curvature characteristic energy $E_\gamma = 3\hbar c\gamma^3/2r_c$ given by Cheng & Zhang (1996) and combines it with the electric field in steady state and curvature radiation, one obtains

$$\mathsf{E}_{||} = \left(\frac{2}{3}\right)^{7/3} \frac{e}{r_c^2}\left(\frac{r_c E_\gamma}{\hbar c}\right)^{4/3}. \tag{5}$$

The electric field relies on the specific choice of E_γ, the characteristic energy can be limited according to the pair production condition, and it was defined as $E_\gamma \equiv k_\gamma/(kT[\text{keV}])$ GeV (Viganò *et al.*, 2015), which takes into account the range $k_\gamma \sim [0.1 - 3.5]$, i.e., X-rays and γ are parameterized by the k_γ factor. The X-ray energies are associated with blackbody temperature. The first value in the range, $k_\gamma = 0.1$, stands for low values of E_γ that correspond to energetic X-ray photons with energy $E_{\rm X} = 2.8\ kT$. The upper value, $k_\gamma = 3.5$, indicates large values for E_γ that correspond to X-ray photons with $E_{\rm X} = 0.75\ kT$.

Writing the electric field in terms of the light cylinder and star period (Viganò *et al.*, 2015), we can obtain

$$\mathsf{E}_{\|} = \left(\frac{2}{3}\right)^{7/3} \frac{e}{c^2} \left(\frac{E_\gamma}{\hbar}\right)^{4/3} \left(\frac{2\pi R_{\rm lc}}{P r_{\rm c}}\right)^{2/3}. \tag{6}$$

In this case, $\mathsf{E}_{\|}$ involves a fixed T which is independent of P and B. Otherwise, the temperature will the dependent on the mechanism associated with the thermal X-ray production through the parameter $\mathcal{K}_h$:

$$\mathsf{E}_{\|}(T_h) \propto \frac{1}{P^{4/9} B_{12}^{1/3}} \left(\frac{k_\gamma}{\mathcal{K}_h}\right)^{4/3} \left(\frac{R_{\rm lc}}{r_{\rm c}}\right)^{2/3}. \tag{7}$$

Considering these forms for the electric field, we can write the gap size, for the case where the electric field depends on the period and radius of the star $f_t = a_t P^{7/6} B_{12}^{-1/2}$, the one which has a dependence on the temperature of the star $f_h = a_h P^{23/18} B_{12}^{-2/3}$, and their corresponding factors as Viganò *et al.* (2015):

$$a_t = 3.6 \left(\frac{1}{R_6^3 \kappa \chi}\right)^{1/2} \left(\frac{k_\gamma}{kT[\mathrm{keV}]}\right)^{2/3} \left(\frac{R_{\rm lc}}{r_{\rm c}}\right)^{1/3}, \tag{8a}$$

$$a_h = 7.0 \left(\frac{1}{R_6^3 \kappa \chi}\right)^{1/2} \left(\frac{k_\gamma}{\mathcal{K}_h}\right)^{2/3} \left(\frac{R_{\rm lc}}{r_{\rm c}}\right)^{1/3}. \tag{8b}$$

3.1. *Constraints of gamma emission*

From equations for the gap size, one can obtain the death lines for γ-rays: one for the case where the electric field depends on the period and radius (through a_t) and another which has a dependence on the temperature of the star (through a_h). The magnetic fields at the light cylinder for these two cases are, respectively,

$$B_{\rm t}^{\rm lc} = \frac{10^{12}}{f^2 c^3} a_t^2 R^3 (2\pi)^3 P^{-2/3}, \tag{9}$$

$$B_{\rm h}^{\rm lc} = \frac{10^{12}}{f^{3/2} c^3} a_h^{3/2} R^3 (2\pi)^3 P^{-13/12}. \tag{10}$$

The models of γ emission predict different relations between the spin-down and the γ-ray luminosity. Considering the astronomical observations, the γ-ray luminosity roughly seems to follow $L_\gamma \propto \dot{E}_{\rm rot}^{1/2}$, see Figure 9 from Abdo *et al.* [Fermi-LAT Collaboration] (2013).

In these descriptions, the γ-ray luminosity is only proportional to the GJ current. It is known that the relation between the γ-luminosity and the spin-down energy is not so well defined for old γ-pulsars. Furthermore, for non-γ-pulsars, the X-ray and optical bands seem to follow a different relation between the spin-down energy and the luminosity (Johnson, 2012). Since the γ, X-ray, and optical emissions are related, Kisaka & Tanaka (2017) proposed a model to investigate the efficiency in rotating-powered pulsars. In their model, there is a $\gamma-\gamma$ scenario and the luminosity is described by $L_\gamma = \eta \dot{E}_{\rm rot}$, where the parameter η is the conversion efficiency, which must be smaller than 1. Thus, since in the outer gap model $L_\gamma \approxeq f^3 \dot{E}_{\rm rot}$, we obtain f in terms of the efficiency, $f = \eta^{1/3}$, and consequently the death lines considering the efficiency of the pulsars.

Following the efficiency distribution in Figure 10 of Abdo *et al.* [Fermi-LAT Collaboration] (2013), we can see that η relies on 10^{-1} for the majority of gamma pulsars. Thus, we are going to consider $f = 10^{-1/3}$ in the expressions for the death lines given in Eqs. (9) and (10). It is important to stress that our results do not depend too much on the value of f but depend strongly on the star radius.

We obtain the death lines that depend on the blackbody temperature of the star surface, the photon energy, and the radius of the light cylinder.

3.1.1. *Gamma emission of SGRs/AXPs*

Using SGRs/AXPs as neutron stars with a radius of 10 km, and the observed blackbody (minimum and maximum) temperature $\approx$

Table 1. a_t values for SGRs/AXPs as NSs in terms of the radius of the star, blackbody temperature, and γ energies E_γ.

	$T_{\rm max}$ [0.59 keV]	$T_{\rm min}$ [0.12 keV]
k_γ [0.1]	$a_t = 5.74$	$a_t = 16.58$
k_γ [3.5]	$a_t = 61.36$	$a_t = 177.43$

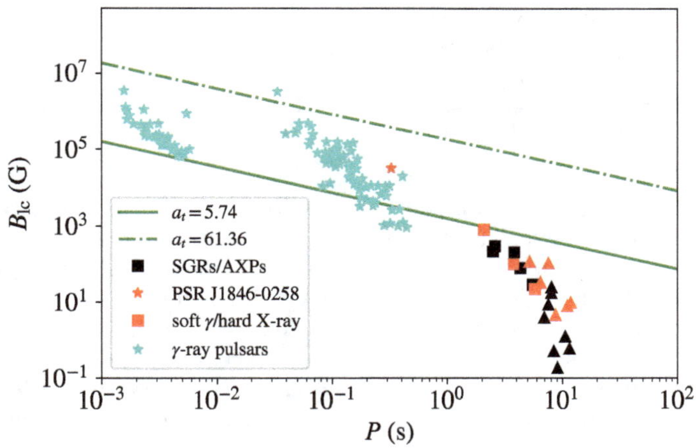

Fig. 1. B–P diagram showing all γ-ray pulsars (cyan stars) from Abdo *et al.* [Fermi-LAT Collaboration] (2013) as well as all SGRs/AXPs (the squares and triangles); among them, we highlight the 10 sources (the red ones) with persistent emission in soft-γ/hard X-ray. The death line for $a_t = 5.74$ is shown (the green continuous line, which is for a temperature of 0.59 keV and $k_\gamma = 0.1$), where the majority of γ-pulsars are above the line, which are able to emit in soft γ-rays; in this case, the SRGs/AXPs should not emit, the line for $a_t = 61.36$ is also shown (the green dashed line, which is for a temperature of 0.59 keV and $k_\gamma = 3.5$), where all pulsars are below the line, which are unable to emit high energy γ.

0.12 – 0.59 keV from McGill catalog (Olausen & Kaspi, 2014), it is possible to estimate the range for a_t, Eq. (8a), as written in Table 1.

We have used two values of k_γ, corresponding to different γ energies: the first value, $k_\gamma = 0.1$, corresponds to low values of E_γ and energetic X-ray photons $E_X = 2.8\ kT$, the continuous lines; the second case, $k_\gamma = 3.5$, corresponds to large values for E_γ and X-ray photons $E_X = 0.75\ kT$, the dashed lines. Figures 1 and 2 correspond to these two energies and to the maximum and minimum observed

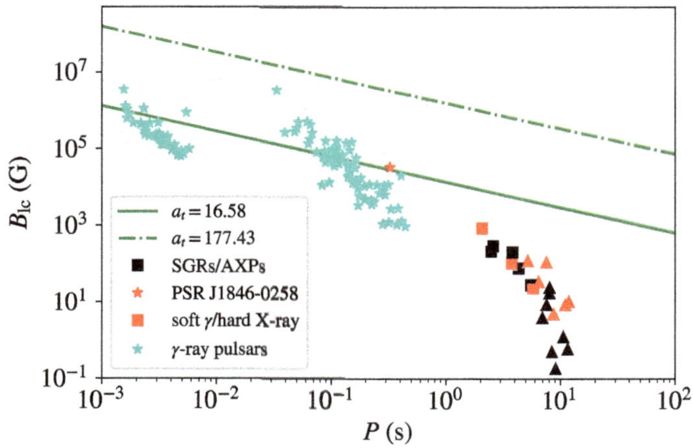

Fig. 2. $B{-}P$ diagram showing all γ-ray pulsars (cyan stars) from Abdo *et al.* [Fermi-LAT Collaboration] (2013) as well as all SGRs/AXPs (the squares and triangles); among them, we highlight the 10 sources (the red ones) with persistent emission in soft-γ/hard X-ray. It shows the death line for $a_t = 16.58$ (the green continuous line, which is for a temperature of 0.12 keV and $k_\gamma = 0.1$), where the death line cuts the pulsar distribution in the middle, being almost one-half able to emit soft γ-rays; for the second case, $a_t = 177.43$ (the green dashed line, which is for a temperature of 0.12 keV and $k_\gamma = 3.5$), again no one star should emit.

temperatures from the McGill catalog, respectively. In these figures, we show the γ-ray pulsars according to "The second Fermi large are telescope catalog of gamma-ray pulsars" (Abdo *et al.* [Fermi-LAT Collaboration] (2013)), which summarizes 117 high confidence $\geqslant 0.1$ GeV γ-pulsars. These pulsars are divided into three groups: millisecond pulsars, young radio-loud pulsars, and young radio-quiet pulsars. All the SGRs/AXPs (the squares and triangles), are also shown which we considered as neutron stars; among them, we highlight 10 sources (the red ones), which have a waveband with persistent emission in soft-γ rays/hard X-rays (> 10 keV). From Figure 1, it is possible to see that the death line for $a_t = 5.74$, where the majority of millisecond γ-pulsars are above the line, which means that they can emit in energetic X-rays and soft γ-rays. SGRs/AXPs are under the lines, which are unable to emit; the exception is the source PSR J1846-0258 which is above and soft gamma emission has been observed. In the same figure, we have the second value of $a_t = 61.36$ which stands for the large values of E_γ. For this value, the death line is shifted up, and

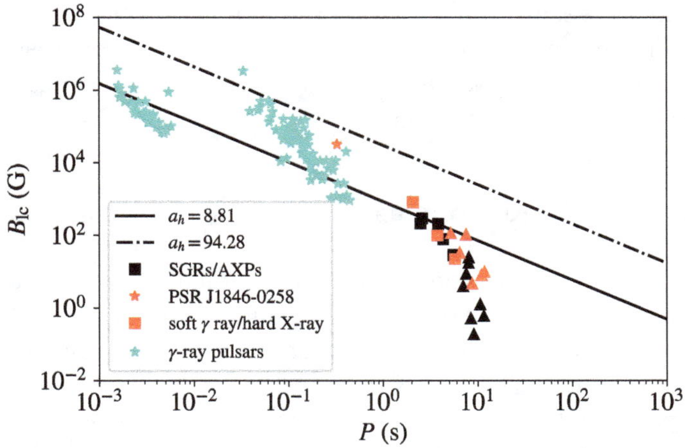

Fig. 3. $B-P$ diagram showing all γ-ray pulsars (cyan stars) from Abdo *et al.* [Fermi-LAT Collaboration] (2013) as well as all SGRs/AXPs. It shows the death line for $a_h = 8.81$, where the majority of the millisecond γ-pulsars are below the line, and the majority of SGRs/AXPs as well, being unable to emit in γ-rays, it also showns the line for $a_h = 94.28$, where only one pulsar should emit.

no pulsar should emit. Figure 2 shows the death lines for $a_t = 16.58$ and $a_t = 177.43$; the death lines are shifted up in comparison with 1, cutting the pulsar distribution in the middle for the lower a_t, i.e., only one-half of them would be able to emit in soft γ-rays; for the highest value of a_t, no one star should emit.

If one considers the case where the electric field has a dependence on the mechanism of the X-ray generation, i.e., the case of a_h factor, Eq. (8b). Using the NS radius of 10 km, the range for a_h is estimated, which is $a_h = 8.81$ for $k_\gamma = 0.1$ and $a_h = 94.28$ for $k_\gamma = 3.5$.

For these a_h values, we show the death lines in Figure 3. As in the previous case, we have used the two values for the γ-rays. For the case $k_\gamma = 0.1$, we have $a_h = 8.81$. For the higher energy, $k_\gamma = 3.5$, we have $a_h = 94.28$. In Figure 3, we see that the majority of γ-pulsars are above the line for $a_h = 8.81$, inclusive of a few SGRs/AXPs, which are able to emit in γ-rays. However, the majority of millisecond pulsars are below, which are unable to emit soft γ-rays. There is also the line for $a_h = 94.28$, where only one pulsar should emit. The inclination of the curves is accentuated in comparison with the previous figures for a_t.

This model seems unlikely to be the more appropriate. As we can see, the inclination of the lines does not agree with the distribution of the γ pulsars. Due to the inclination, the large majority of the millisecond pulsars are below.

4. Discussion and Conclusion

The model of electromagnetic emission studied, the gap accelerator model, has a dependence on the radius and temperature of the stars. This is due to the influence of these quantities on the pair production and generation of γ-ray and X-ray.

The pair production defines the death zone, which is where the magnetized stars are unable to emit electromagnetic radiation. The death lines were constructed with observables, such as period, temperature, and with the mass, moment of inertia, and radius. We have estimated some canonical values for them to study these lines.

Death lines were constructed for NSs considering the γ-ray pair-production condition. According to the outer gap models for the SGRs/AXPs as NSs, they should not emit hard X-rays/soft γ-rays. If we consider the γ-pulsars observed and apply the parameters used for SGRs/AXPs as NSs, we can see that these γ-pulsars can emit at least in soft-γ. These conclusions are in accord with Rubtsov & Sokolova (2015) who had performed an analysis of the Fermi data, showing that outer gap models are in better agreement with data for γ-pulsars.

In forthcoming works, we are expecting to refine the mass-radius-moment of inertia relationship to have self-consistent models for each source, considering its astronomical properties as well as other important properties for the emission models, and also investigate the high-energy emission of SGRs/AXPs as very fast and magnetic WDs.

Bibliography

Abdo, A. A., *et al.* (2010). Search for gamma-ray emission from magnetars with the Fermi Large Area Telescope, *Astrophys. J. Lett.* **725**, 1, L73.

Abdo, A. A., *et al.* (2013). The second Fermi Large Area Telescope catalog of gamma-ray pulsars, *Astrophys. J. Suppl. Ser.* **208**, 2, 17.

Baring, M. G., Gonthier, P. L., and Harding, A. K. (2005). *The Astrophys. J.* **630**, 1, 430–440.

Baring, M. G. and Harding, A. K. (2008). *AIP Conf. Proc.* **968**, 1, 93–100.

Blaes, O., Blandford, R., Goldreich, P., and Madau, P. (1989). *The Astrophys. J.* **343**, 839–848.

Borges, S. V., Rodrigues, C. V., Coelho, J. G., Malheiro, M., and Castro, M. (2020). *The Astrophys. J.* **895**, 1, 26.

Boshkayev, K., Rueda, J. A., Ruffini, R., and Siutsou, I. (2013). *The Astrophys. J.* **762**, 2, 117.

Caceres, D. L., de Carvalho, S. M., Coelho, J. G., de Lima, R. C. R., and Rueda, J. A. (2017). *Month. Not. Roy. Astron. Soc.* **465**, 4, 4434–4440.

Cheng, K. S., Ho, C., and Ruderman, M. (1986a). *The Astrophys. J.* **300**, 500.

Cheng, K. S., Ho, C., and Ruderman, M. (1986b). *The Astrophys. J.* **300**, 522.

Cheng, K. S. and Zhang, J. L. (1996). *The Astrophys. J.* **463**, 271.

Coelho, J. G., Cáceres, D. L., de Lima, R. C. R., Malheiro, M., Rueda, J. A., and Ruffini, R. (2017). *Astronomy & Astrophysics* **599**, A87.

Coelho, J. G. and Malheiro, M. (2014). *Publications of the Astronomical Society of Japan* **66**, 14.

den Hartog, P. R., Hermsen, W., Kuiper, L., Vink, J., in 't Zand, J. J. M., and Collmar, W. (2006). *Astronomy & Astrophysics* **451**, 2, 587–602.

Duncan, R. C. (2001). arXiv:astro-ph/0106041.

Duncan, R. C. and Thompson, C. (1992). *The Astrophys. J.* **392**, L9–L13.

Enoto, T., Makishima, K., Nakazawa, K., Kokubun, M., Kawaharada, M., Kotoku, J., and Shibazaki, N. (2011). *Publications of the Astronomical Society of Japan* **63**, 2, 387–396.

Goldreich, P. and Julian, W. H. (1969). *The Astrophys. J.* **157**, 869.

Götz, D., Mereghetti, S., Tiengo, A., and Esposito, P. (2006). *Astronomy & Astrophysics* **449**, 2, L31–L34.

Hirotani, K. (2006). *The Astrophys. J.* **652**, 2, 1475.

Jackson, J. D. (1975). *Classical Electrodynamics*, Hoboken: Wiley.

Johnson, T. J. (2012). arXiv e-prints, arXiv:1209.4000.

Kaspi, V. M. and Beloborodov, A. M. (2017). *Annual Review of Astronomy and Astrophysics* **55**, 1, 261–301.

Kaspi, V. M. and Boydstun, K. (2010). *The Astrophys. J.* **710**, 2, L115–L120.

Katz, J. I., Toole, H. A., and Unruh, S. H. (1994). *The Astrophys. J.* **437**, 727–732.

Kisaka, S. and Tanaka, S. J. (2017). *The Astrophys. J.* **837**, 1, 1–16.

Kuiper, L., Hermsen, W., den Hartog, P. R., and Collmar, W. (2006). *The Astrophys. J.* **645**, 1, 556–575.

Kuiper, L., Hermsen, W., and Mendez, M. (2004). *The Astrophys. J.* **613**, 2, 1173–1178.

Lobato, R. V., Malheiro, M., and Coelho, J. G. (2016). *Int. J. Mod. Phys. D* **25**, 9, 1641025.

Malheiro, M., Rueda, J. A., and Ruffini, R. (2012). *Publ. Astronom. Soc. Jap.* **64**, 3.

Marsh, T. R. and et al. (2016). *Nature* **537**, 7620, 374–377.

Meintjes, P. J., de Jager, O. C., Raubenheimer, B. C., Nel, H. I., North, A. R., Buckley, D. A. H., and Koen, C. (1994). *The Astrophys. J.* **434**, 292.

Meintjes, P. J., Madzime, S. T., Kaplan, Q., and van Heerden, H. J. (2023). *Galaxies* **11**, 1, 14.

Meintjes, P. J., Raubenheimer, B. C., de Jager, O. C., Brink, C., Nel, H. I., North, A. R., van Urk, G., and Visser, B. (1992). *The Astrophys. J.* **401**, 325.

Mereghetti, S. (2008). *The Astrono. Astrophys. Rev.* **15**, 4, 225–287.

Mereghetti, S. (2013). *Braz. J. Phys.* **43**, 5–6, 356–368.

Mereghetti, S., Götz, D., Mirabel, I. F., and Hurley, K. (2005). *Astron. Astrophys.* **433**, 2, L9–L12.

Mereghetti, S., Luca, A. D., Caraveo, P. A., Becker, W., Mignani, R., and Bignami, G. F. (2002). *The Astrophys. J.* **581**, 2, 1280–1285.

Mereghetti, S., Palombara, N. L., Tiengo, A., Sartore, N., Esposito, P., Israel, G. L., and Stella, L. (2013). *Astron. Astrophys.* **46**, 1–5.

Molkov, S., Hurley, K., Sunyaev, R., Shtykovsky, P., Revnivtsev, M., and Kouveliotou, C. (2005). *Astron. Astrophys.* **433**, 2, L13–L16.

Morini, M., Robba, N. R., Smith, A., and van der Klis, M. (1988). *The Astrophys. J.* **333**, 777–787.

Olausen, S. A. and Kaspi, V. M. (2014). *The Astrophys. J. Supple. Ser.* **212**, 1, 6.

Paczynski, B. (1990). *The Astrophys. J. Lett.* **365**, L9–L12.

Rea, N., Esposito, P., Turolla, R., Israel, G. L., Zane, S., Stella, L., Mereghetti, S., Tiengo, A., Götz, D., Göğüş, E., and Kouveliotou, C. (2010). *Science* **330**, 6006, 944–946.

Rea, N. and Torres, D. F. (eds.) (2011). High-energy emission from pulsars and their systems: Proceedings of the first session of the sant cugat forum on astrophysics, *Astrophysics and Space Science Proceedings*, Berlin: Springer-Verlag, Germany.

Rubtsov, G. and Sokolova, E. (2015). *Soviet J. Exp. Theoret. Phys. Lett.* **100**, 11, 689–694.

Sousa, M. F., Coelho, J. G., and de Araujo, J. C. N. (2020). *Month. Noti. Roy. Astron. Soc.* **498**, 3, 4426–4432.

Sousa, M. F., Coelho, J. G., de Araujo, J. C. N., Guidorzi, C., and Rueda, J. A. (2023). *The Astrophys. J.* **958**, 2, 134.

Sousa, M. F., Coelho, J. G., de Araujo, J. C. N., Kepler, S. O., and Rueda, J. A. (2022). *The Astrophys. J.* **941**, 1, 28.
Takata, J., Wang, Y., and Cheng, K. S. (2010). *The Astrophys. J.* **715**, 2, 1318.
Thompson, C. (2002). *Memorie della Societa Astronomica Italiana* **73**, 477–484.
Thompson, C. (2008). *The Astrophys. J.* **688**, 1258–1281.
Thompson, C. and Beloborodov, A. M. (2005). *The Astrophys. J.* **634**, 1, 565–569.
Thompson, C. and Duncan, R. C. (1993). *The Astrophys. J.* **408**, 194–217.
Thompson, C. and Duncan, R. C. (1995). *Month. Not. Roy. Astronom. Soc.* **275**, 2, 255–300.
Thompson, C. and Duncan, R. C. (1996). *The Astrophys. J.* **473**, 1, 322.
Thompson, C., Lyutikov, M., and Kulkarni, S. R. (2002). *The Astrophys. J.* **574**, 1, 332.
Tong, H., Song, L. M., and Xu, R. X. (2010). *The Astrophys. J. Lett.* **725**, 2, L196.
Tong, H., Song, L. M., and Xu, R. X. (2011). *The Astrophys. J.* **738**, 1, 31.
Turolla, R., Zane, S., and Watts, A. L. (2015). *Reports Prog. Phys.* **78**, 11, 116901.
Vigano, D., Torres, D. F., Hirotani, K., and Pessah, M. E. (2015a). *Month. Not. Roy. Astronom. Soc.* **447**, 3, 2631–2648.
Vigano, D., Torres, D. F., Hirotani, K., and Pessah, M. E. (2015b). *Month. Not. Roy. Astron. Soc.* **447**, 3, 2649–2657.
Viganò, D., Torres, D. F., Hirotani, K., and Pessah, M. E. (2015). *Month. Not. Roy. Astron. Soc.* **447**, 2, 1164–1172.
Woosley, S. E. and Wallace, R. K. (1982). *The Astrophys. J.* **258**, 716–732.
Wu, J. H. K., Hui, C. Y., Huang, R. H. H., Kong, A. K. H., Cheng, K. S., Takata, J., Tam, P. H. T., Wu, E. M. H., and Liu, J. (2013). *J. Astron. Space Sci.* **30**, 83–85.
Zhang, L. and Cheng, K. S. (1997). *The Astrophys. J.* **487**, 1, 370.

https://doi.org/10.1142/9789819800643_0003

Chapter 3

The Demographics of Isolated Neutron Stars From X-ray Survey Data

Adriana Mancini Pires

Center for Lunar and Planetary Sciences, Institute of Geochemistry, Chinese Academy of Sciences, 99 West Lincheng Road, Guiyang 550051, China.
Leibniz-Insititut für Astrophysik Potsdam (AIP), an der Sternwarte 16, 14482 Potsdam, Germany
adriana@mail.gyig.ac.cn

Since the discovery of the first radio pulsar, the population of neutron stars in our galaxy has grown to over 3,400. A handful of these sources, exclusively seen in X-rays, are key to understanding aspects of the neutron star phenomenology and evolution that cannot be probed by the rest of the population. In this contribution, I review their properties and highlight the role of X-ray surveys at complementing our view of galactic neutron stars. Prospects for future investigations are discussed especially in the context of the aftermath of the eROSITA All-Sky Survey.

Keywords: Neutron stars, X-ray surveys, eROSITA All-Sky Survey

Contents

1. Overview

Supernovae are at the origin of all elements, from iron to the heavier ones, that are created through explosive oxygen and silicon burning (see Arcones and Thielemann, 2023, for a recent review). They offer both a site for the high temperatures required for nuclear reactions with large Coulomb barriers and a very efficient means of dispersing the products that have been synthesized during stellar evolution into the interstellar medium. From the frequency of occurrence of type-II supernovae in nearby spiral galaxies of similar type and luminosity (e.g., Molero *et al.*, 2023; Rozwadowska *et al.*, 2021), as well as from measurements of the γ-ray flux resulting from the decay of ^{26}Al produced by Wolf–Rayet stars and supernova events in our galaxy (Diehl *et al.*, 2006, 2022), the core-collapse supernova rate is estimated around 1.6 ± 0.5 events per century, implying a total galactic population of at least one-hundred million neutron stars.

Over half a century after the discovery of PSR B1919+21 (Hewish *et al.*, 1968), most of the neutron stars in the Milky Way are still observed through their emission of pulsed radio waves. Besides, thanks to the Fermi "pulsar revolution" (Caraveo, 2014), over 300 pulsars are known to emit gamma rays. Technological advancements and the ever-increasing sensitivity and timing precision of dedicated pulsar surveys have greatly contributed to the discovery of the nearly

3,400 neutron stars that are known to date (Manchester *et al.*, 2005).[1] New discoveries have benefited from improved periodicity searching techniques that can unveil transients, as well as sources with irregular pulse amplitude or long nulling fractions (e.g., Keane *et al.*, 2010; McLaughlin *et al.*, 2006). Eventually, the Square Kilometre Array (SKA) will be sensitive to virtually all active radio pulsars beamed toward the Earth, thus increasing the population tenfold (Keane *et al.*, 2015).

Despite these successful surveys, it is well known that the observed sample of radio pulsars is heavily biased by selection effects and not representative of the true population. The choice of high radio frequencies, 1.4–2.3 GHz, best suited for searches in the galactic plane and at the core of supernova remnants tends to discover young, distant, and high radio luminosity pulsars. On the other hand, low-frequency surveys (typically 300–400 MHz) represent a good option for investigating the low-luminosity tail of pulsars in the solar neighborhood, where a population of relatively old objects prevails. In recent years, searches at the very low-frequency band of 70–230 MHz have allowed the intriguing discovery of several long-period radio sources and transients, whose (neutron star or magnetic white dwarf) nature is currently under debate (Caleb *et al.*, 2022; Hurley-Walker *et al.*, 2023, 2022).

Yet, if one neutron star is born in the galaxy every 60 to 100 years, for every active radio pulsar, there must be ∼1000 radio-quiet neutron stars. Most are expected to be old and without a close companion (Lorimer, 2005). Young isolated neutron stars, on the other hand, are expected to be strong X-ray and γ-ray sources, the high-energy emission being due to the rotation of the neutron star (non-thermal mechanisms) or to an internal heat reservoir which follows formation (cooling) or which is provided through magnetic field decay or other re-heating mechanisms (see e.g., Kaspi *et al.*, 2010, for an overview). Some high-energy pulsars and isolated neutron stars (INSs) are also detected in radio wavelengths while others are not. The lack of radio emission may be due to either unfavorable viewing geometries or to physical mechanisms that, for example, inhibit pair creation in the pulsar magnetosphere, thus quenching persistent

[1] see https://www.atnf.csiro.au/research/pulsar/psrcat.

coherent radio emission. Searches and surveys at other wavelength regions of the electromagnetic spectrum are needed to unveil INSs that escape canonical pulsar surveys.

The launch of the Roentgen Satellite mission (ROSAT) in June 1990, a satellite of outstanding sensitivity at soft X-ray energies and which performed an all-sky survey during its first 6 months of operation (Voges *et al.*, 1999), raised a major expectation in the scientific community that a population of old INSs would be detected through their accretion from the interstellar medium (cf. Treves *et al.*, 2000, and references therein). The evidence, found by several authors in the pre-ROSAT era, that the population of pulsars could be described by two individual sub-classes of fast and slow rotators, with different (magnetic field and spin period) evolutionary histories and kinematically distinct in their scale heights and mean spatial velocities, was essential since it provided a significant number of neutron stars with the right properties[2] to be detected through Bondi–Hoyle accretion in soft X-rays. Early estimates of the detectable number of "re-heated" old neutron stars were thus highly optimistic, $\gtrsim 10^3 - 10^4$.

The evident absence of a large number of unidentified soft X-ray sources in the ROSAT All-Sky Survey data was in clear opposition to expectations. In particular, the analysis of the source content resulting from searches in both the galactic plane (Motch *et al.*, 1997) and molecular clouds (Danner, 1998) led to a critical re-evaluation of the accretion scenario in order to explain the paucity of X-ray emitting INS candidates.

2. The *Magnificent Seven*

Notwithstanding the number, INSs should be recognized due to the softness of their spectra, low column densities, and extremely high X-ray-to-optical flux ratios. Moreover, for the very local sources, moderate proper motions of faint blue optical counterparts should be observed. The first confirmed INS candidate in ROSAT observations showing the above characteristics, RX J1856.5-3754, was reported by Walter *et al.* (1996); six others would be identified in the following

[2]Accretion is highly inefficient for a rapid moving (rate $\propto v^{-3}$), fast spinning, and strongly magnetised neutron star (see Treves *et al.*, 2000, for details).

five years (Haberl, 2007, and references therein). Rather than old accretors, the seven sources turned out to be a local group of cooling neutron stars; their remarkable properties and elusiveness have earned them the nickname the *magnificent seven* (M7; see Turolla, 2009; van Kerkwijk and Kaplan, 2007 for reviews).

2.1. *Overall observed characteristics*

The M7 INSs consist of a seemingly homogeneous local group of cooling neutron stars, displaying similar ages and spectral and timing properties. The seven sources do not show either persistent or transient radio emission to sensitive limits and are not associated with supernova remnants. Locally, they are as numerous as young radio and γ-ray pulsars and may belong to a formerly neglected component of the overall galactic INS population (Popov *et al.*, 2003). The discovery of similar sources beyond our local volume is therefore key to understanding their properties as a group and relations to other families of galactic INSs.

Based on a combination of bright thermal emission, proximity, independent distance estimates, and lack of significant magnetospheric or accretion activity, the sources are compelling candidates for testing neutron star emission and cooling models. The following are listed in Table 1, for each source, sorted in descending order of X-ray flux: the temperature kT of the dominant spectral component best-fit single absorbed blackbody model;[3] the spin period P, pulsed fraction $p_{\rm f}$, and observed spin down $\dot{P}$ from timing studies; the spin-down luminosity $\dot{E} = 4.5 \times 10^{46}(\dot{P}P^{-3})$ erg s^{-1}, dipolar magnetic field $B_{\rm dip} = 3.2 \times 10^{19}(P\dot{P})^{1/2}$ G, and characteristic age $\tau_{\rm ch} = P(2\dot{P})^{-1}$, assuming magnetic dipole braking in vacuum (Ostriker and Gunn, 1969); the kinematic age $t_{\rm kin}$, from proper motion investigations; the magnitude of the blue optical counterpart $m_{\rm B}$; an estimate of the distance d to the source, either from parallax or X-ray absorption measurements; an estimate of the surface magnetic field

[3]Single or double temperature blackbody models and Gaussian absorption lines are usually employed to fit the spectra of the M7. The spectral parameters of more physically motivated atmosphere models can be found in the compilation of thermally emitting INSs of Potekhin *et al.* (2020). See also the discussion in Section 2.2.

Table 1. Overall properties of the M7.

Object	kT (eV)	P (s)	p_f (%)	$\log(\dot{P})$ ($s\,s^{-1}$)	$\log(\dot{E})$ ($erg\,s^{-1}$)	$\log(\tau_{ch})$ (yr)	$\log(t_{kin})$ (yr)	m_B	d (pc)	$\log(B_{dip})$ (10^{13} G)	$\log(B_{cyc})$ (10^{13} G)	Ref.
RX J1856.5-3754	62	7.06	1	−13.522	30.529	6.57	5.62	25.2	123	13.17	–	[1]
RX J0720.4-3125	**84–93**	**8.39**	**11**	**−13.156**	**30.726**	**6.28**	**5.93**	**26.6**	**360**	**13.39**	**13.75**	[2]
RX J1605.3+3249	**100**	?	< 1.5	?	?	?	**5.65**	**27.2**	**390**	?	**13.92**	[3]
RX J1308.6+2127	**100**	**10.31**	**18**	**−12.951**	**30.663**	**6.16**	**5.74**	**28.4**	**380**	**13.54**	**13.60**	[4]
RX J2143.0+0654	104	9.43	4	−13.398	30.332	6.57	?	27.4	430	13.29	14.15	[5]
RX J0806.4-4123	95	11.37	6	−13.260	30.227	6.51	?	27.6	250	13.40	13.96	[6]
RX J0420.0-5022	45	3.45	17	−13.553	31.487	6.29	?	27.5	345	13.00	–	[7]

Note: The sources are sorted by decreasing X-ray brightness. *References:* [1] De Grandis *et al.* (2022); Kaplan *et al.* (2011); Mignani *et al.* (2013); Walter *et al.* (2010); [2] Hohle *et al.* (2012b); Kaplan *et al.* (2011); Kaplan and van Kerkwijk (2005a); Kaplan *et al.* (2007); Tetzlaff *et al.* (2011); van Kerkwijk *et al.* (2007); [3] Pires *et al.* (2014, 2019); Tetzlaff *et al.* (2012); [4] Hambaryan *et al.* (2011); Kaplan and van Kerkwijk (2005b); Motch *et al.* (2009); [5] Kaplan *et al.* (2011); Kaplan and van Kerkwijk (2009b); Posselt *et al.* (2007); [6] Kaplan *et al.* (2011); Kaplan and van Kerkwijk (2009a); Posselt *et al.* (2007); [7] Kaplan and van Kerkwijk (2011); Posselt *et al.* (2007).

B_{cyc}, from the energy of broad spectral features interpreted as proton cyclotron absorption (Zane *et al.*, 2001). The table caption has the corresponding relevant references.

Unlike other X-ray pulsars, their emission is predominantly thermal and generally believed to originate from the neutron star surface. Theoretical neutron star atmosphere models, however, fall short at physically describing the complex energy distribution of the sources (Potekhin, 2014). Interestingly, a keV excess emission with respect to the dominant blackbody continuum has been reported in the cumulative spectra of several of the M7 (Dessert *et al.*, 2020; Yoneyama *et al.*, 2017, 2019). In the case of the brightest and regularly observed RX J1856.5-3754, the spectral energy distribution extending from the optical/UV up to 8 keV can be described by the superposition of three blackbody components, kT $\sim$30 eV , 60 eV and 130 eV, indicative of a much more complex surface temperature distribution than originally thought, and a faint power law tail with $\Gamma\sim1.4$, possibly of magnetospheric origin (De Grandis *et al.*, 2022).

Although invaluable for individual investigations, the seven sources are too diverse in detail for their properties as a whole to be assessed. For instance, while the brightest among the seven INSs, RX J1856.5-3754, shows a featureless thermal spectrum and is considered a standard candle for the cross-calibration of several X-ray instruments due to the steadiness of its X-ray emission (Madsen *et al.*, 2021), evidence for broad, narrow, and phase-dependent absorption lines has been found in the spectra of all other sources. Deep monitoring with the eROSITA instrument on the recently launched SRG mission (Predehl *et al.*, 2021) has revealed unexpected long-term changes of spectral state and timing behaviour for two of the M7 INSs, RX J1605.3+3249 and RX J2143.0+0654 (Pires *et al.*, 2023). Incidentally, RX J1605.3+3249 is also the only one among the group still lacking a detected periodicity, despite extremely deep investigations with a dedicated XMM-Newton large programme (Malacaria *et al.*, 2019; Pires *et al.*, 2019). At longer wavelengths, the seven sources show even more diverse properties. These aspects are discussed in the remainder of the section.

2.2. Understanding surface emission

The low magnetospheric activity in the M7 allows for a direct view of the surface of the neutron star. Atmosphere models and the surface emissivity under extreme physical conditions can thus be tested in order to derive radii and constrain the equation of state of the neutron star interior (Özel and Freire, 2016). In practice, however, progress has been hampered by uncertainties on the chemical composition of the atmosphere and the phase state of the stellar surface, and by the lack of understanding of the magnetic field and temperature distributions (van Kerkwijk and Kaplan, 2007, for a discussion). The presence of absorption lines adds to this complexity; in the case of the M7, the broad features have complex and phase-dependent shapes and the line depths often exceed 50% (Haberl, 2007, and references therein). Evidence for narrow and phase-dependent absorption features have also been reported (Borghese *et al.*, 2015, 2017; Hohle *et al.*, 2012a). Although generally believed to be related with the star's magnetic field, they have no unique physical interpretation.

To explain the emitted radiation and equivalent widths of the lines in the phase-resolved spectrum of the M7 RX J1308.6+2127, Suleimanov *et al.* (2010) favoured a model where a partially ionized, optically thin atmosphere above the condensed surface must be present (see also Ho *et al.*, 2007; Motch *et al.*, 2003). Using this model, Hambaryan *et al.* (2011) derived the temperatures of the X-ray emitting areas and the magnetic field intensity at the poles; moreover, they were able to constrain the compactness of the neutron star and the gravitational redshift on the surface, suggesting a very stiff equation of state. Similar conclusions were reached for the M7 RX J0720.4-3125 (Hambaryan *et al.*, 2017). These results are only marginally compatible with the most favoured range of the true radius of a 1.5 $M_\odot$ neutron star, 10−11.5 km.

Viganò *et al.* (2014) showed that in some cases (such as for the M7 RX J0806.4-4123) the deviations found in the spectra of thermally emitting INSs may be induced simply by the inhomogeneous temperature distribution on the surface. While the effect is unlikely to account for all reported lines, the interesting result is that the anisotropic temperature distribution can give way to spurious spectral features.

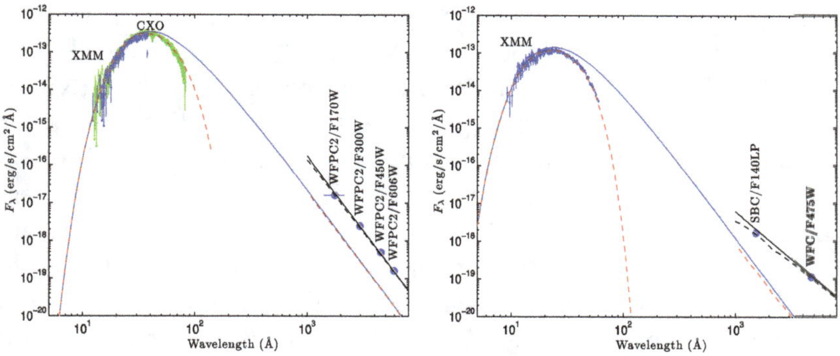

(a) (b)

Fig. 1. Optical excess in RX J1856.5-3754 (a) and RX J2143.0+0654 (b; from Kaplan *et al.* 2011; see the reference for details).

All M7 INSs have detected optical, ultraviolet, or infrared counterparts (see Kaplan *et al.*, 2011; Posselt *et al.*, 2018, 2014, for references and limits). Interestingly, the extrapolation to longer wavelengths of the best-fit model inferred from X-rays falls below the actual detected fluxes (Figure 1). This is known as the *optical excess* and is observed in all M7 INSs, however in different degrees. While for some of the sources, the excess follows the Rayleigh-Jeans slope $F_\lambda \propto \lambda^{-4}$ (e.g., RX J1856.5-3754; Figure 1(a), for others it deviates significantly and is better described by a flat power law dependency (e.g., RX J2143.0+0654; Figure 1(b). The origin of the excess flux might rely on atmospheric effects, magnetospheric emission, or resonant scattering. Interestingly, if the X-ray and emission at longer wavelengths came from different regions on the surface, we might expect correlations between the amount of optical excess and the X-ray pulsed fraction, which have not been verified. The possible presence of fossil fallback disks, of a faint pulsar wind nebula, "bare" neutron star surfaces, and other intriguing scenarios remain open (Ertan *et al.*, 2017; Turolla *et al.*, 2004; Wang *et al.*, 2017).

2.3. *Long-term variability*

Between first light in October 2019 and March 2022, the M7 INSs have been routinely observed by eROSITA, the primary X-ray telescope on-board the SRG Observatory. The observations have revealed surprising evidence of long-term spectral and timing variations for two of the targets, RX J1605.3+3249 and RX J2143.0+0654 (Pires *et al.*, 2023). Until the eROSITA monitoring, only the INS RX J0720.4-3125 was known to display long-term variability, a behaviour most likely caused by a sudden accretion or glitch episode (van Kerkwijk *et al.*, 2007). For the case of RX J2143.0+0654, a decrease in pulsed fraction and evidence of a higher spin down rate was found after a hiatus of 12 years since the neutron star was last targeted by XMM-Newton. The variations in timing behaviour seem to have taken place despite any significant change in flux or spectral state over 17 years. In the top panel of Figure 2, we show the spectrum of the source in three eROSITA epochs; the presence of multiple absorption lines, not known before the eROSITA observations, is also evident.

For the INS RX J1605.3+3249, a second epoch eROSITA observation performed in March 2021 shows very significant spectral variations with respect to the usual state of the source (orange as opposed to the dark grey flat residuals in the bottom panel of Figure 2). The changes can be described by an increase of temperature and decrease of size of the emitting region of the main (hot) blackbody component, accompanied by an overall 5% decrease in source flux. In addition to the broad feature at 0.4 keV, other deviations from the thermal continuum are observed; in particular, a prominent absorption at 0.8 keV (cf. previous reports around this energy in Haberl, 2007; Pires *et al.*, 2014). The variations were confirmed by an XMM-Newton TOO performed in July 2021 (F. Haberl, priv. comm.). A third eROSITA observation indicates that the source is gradually returning to its previous state (blue data points in Figure 2, bottom).

The long-term variations observed for now three of the M7 INSs challenge the perception that they are homogenous and steady: continuous monitoring of their timing and spectral behaviour will allow the investigation of possible scenarios and the physical mechanisms behind the changes.

3. The Neutron Star Diversity

For sources not affected by episodes of mass and angular momentum transfer in a binary system (the so-called "recycled" millisecond

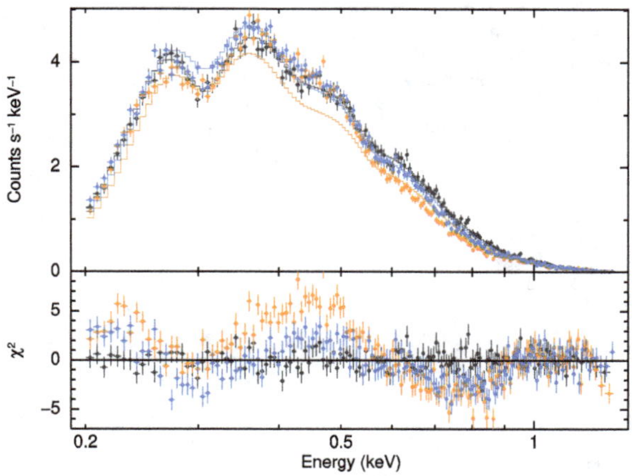

(a)

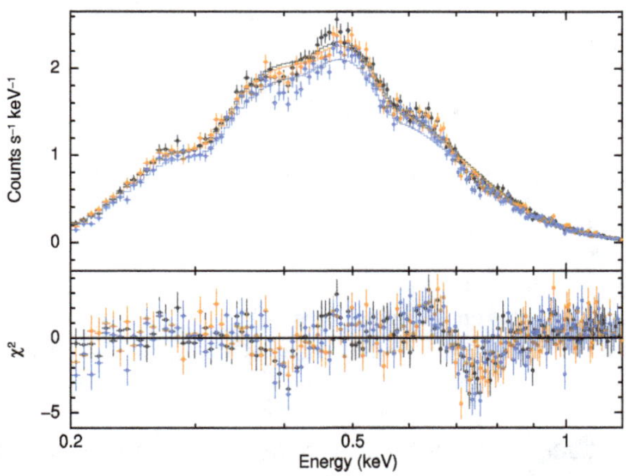

(b)

Fig. 2. Deep eROSITA observations of two of the M7 INSs. Three epochs are shown for each target (see the text for details). (a): Long-term spectral and flux changes in RX J1605.3+3249. (b): Spectral absorption features in RX J2143.0+0654.

pulsars), the observed secular lengthening of a pulsar spin period, typically by 3 s every 10^8 yr, is the consequence of the torque exerted by the magnetic field on the rotating neutron star — the usual model of magnetic dipole braking in vacuum. Despite its simplicity, the model provides useful estimates of the evolutionary state of the neutron star, which is illustrated by the $P-\dot{P}$ diagram in Figures 3 and 4. A canonical pulsar, born rotating fast, is expected

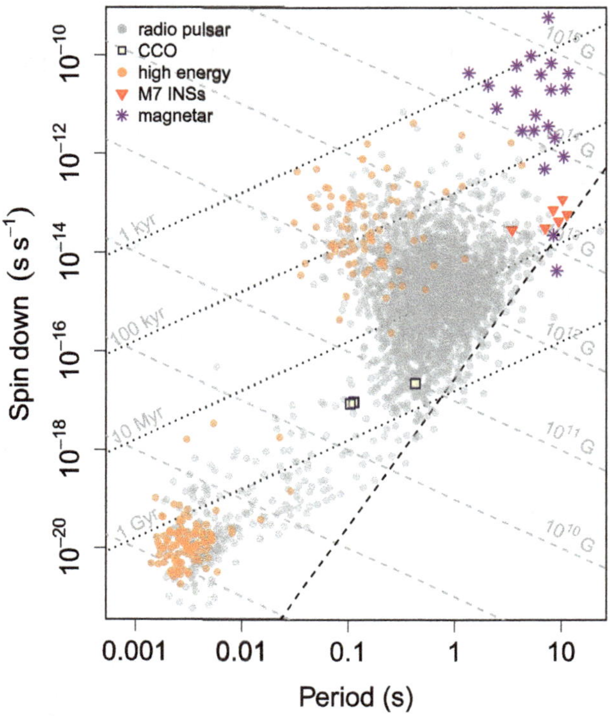

Fig. 3. P–$\dot{P}$ diagram of the Galactic population of isolated pulsars. The position of peculiar thermally emitting sources is highlighted against the background of high-energy and radio spin-powered pulsars.

to evolve at the expense of its rotational energy towards longer periods, along lines of constant magnetic field. For a combination of spin period and magnetic field intensity, pair production and particle acceleration cease: the pulsar is said to have crossed the so-called "death line", where it will no longer be observed as a radio source.

At present, around 50 INSs are known from their bright thermal X-ray emission and lack of magnetospheric radiation (Potekhin *et al.*, 2020). Other than the M7, they include the objects known as *magnetars* (Kaspi and Beloborodov, 2017) and *central compact objects* in supernova remnants (CCOs; De Luca, 2017). Their positions in the P–$\dot{P}$ diagram, unusual by canonical pulsar standards, are highlighted against the background of radio and high-energy spin-powered pulsars from the ATNF catalogue. It is clear they have

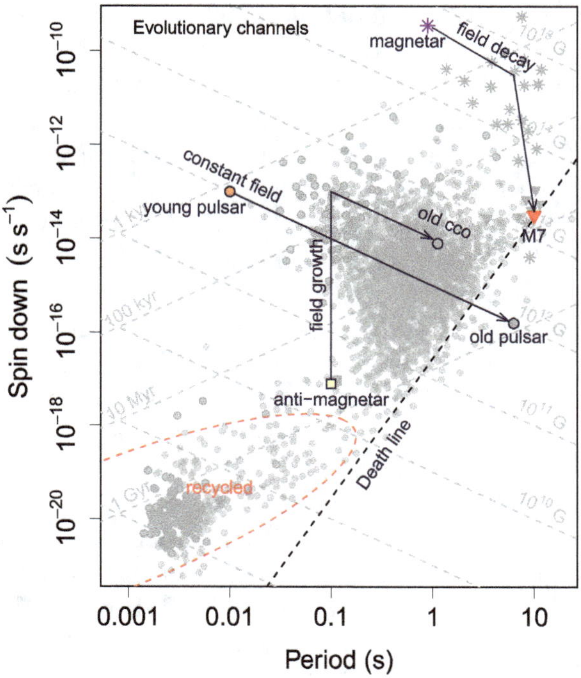

Fig. 4. Alternative channels of neutron star evolution, according to different scenarios of magnetic field evolution.

evolved differently than most neutron stars we know in our Galaxy (e.g., Popov, 2023, for an overview).

• **Alternative scenarios of neutron star evolution:** Estimates of the dipolar magnetic field in the isolated pulsar population span across five orders of magnitude. Among the sources with the highest values are those of magnetars. Magnetars are usually observed through their violent bursts of high energy; they slow down at a much faster rate than normal pulsars (up to 1 s every 60 years, for the extreme case of the soft gamma-ray repeater SGR 1806-20). According to the most favoured interpretation (Thompson and Duncan, 1995; Thompson and Duncan, 1996), their complex phenomenology, transient behaviour, and bright quiescent X-ray luminosity, much in excess of that from spin down, can be explained by crustal and magnetospheric effects provoked by the decay and rearranging of the enormous stellar magnetic field (Beloborodov and Li, 2016; Goldreich and Reisenegger, 1992; Pons *et al.*, 2009). As a result of field dissipation and braking, it is expected that an evolved magnetar ($\gtrsim$100 kyr) will be less active than a young one, and will have a longer spin period and higher surface temperature than an ordinary pulsar of similar age (e.g., Perna and Pons, 2011; Turolla *et al.*, 2011). Interestingly, magneto-thermal evolutionary models can account for the spectral and timing properties of the M7 INSs if they were born as magnetars and evolved through magnetic field decay (De Grandis *et al.*, 2021; Viganò *et al.*, 2013).

At the opposite end of the magnetic field distribution, CCOs challenge the stereotype of a young INS. CCOs were first observed as point-like, radio-quiet X-ray sources located near the centre of supernova remnants. The very low measured spin-down values of three sources (dubbed "anti-magnetars"; Gotthelf *et al.*, 2013) imply very low magnetic field strengths and characteristic ages much older than expected from their association with a supernova remnant. Moreover, the lack of detected pulsations in most members may indicate that the atmosphere is weakly magnetised (Ho and Heinke, 2009; Wu *et al.*, 2021; see however Alford and Halpern, 2023). This contradicts the evidence that some sources conceal strong magnetic fields (e.g., Rea *et al.*, 2016; Shabaltas and Lai, 2012). The fact that no old (>10 kyr) anti-magnetar with similar timing properties is recognised in either radio or X-ray surveys implies that their magnetic field is not constant at secular timescales. CCOs might represent a different outcome of neutron star evolution, where they are recovering from an early episode of field burial by hypercritical accretion (Ho, 2011; Viganò and Pons, 2012, and references therein). In this scenario, if a large amount of debris falls back onto the neutron star after the supernova explosion, accretion is likely to occur at such high rates that any field can be "hidden". As a result, the neutron star shows in its early evolution the typical anti-magnetar behaviour. As the field reemerges, the pulsar quickly joins the rest of the population, moving upwards in the P–$\dot{P}$ diagram; its subsequent evolution would then follow according to the original field intensity. Residual thermal luminosity in the first 10 kyr, as well as enhanced cooling after 100 kyr, with respect to its peers of similar characteristic ages, are expected to result from the accreted light-element envelope (Yakovlev and Pethick, 2004).

4. Searches in X-rays and Population Studies

For several years following the initial discovery of the M7 INSs, considerable efforts have been made to identify similar sources (e.g., Agüeros *et al.*, 2011; Chieregato *et al.*, 2005; Turner *et al.*, 2010). This is a difficult task especially for ROSAT-based searches, due to the large positional and spectral errors at faint X-ray fluxes. In the crowded regions of the Galactic plane, where the more remote INSs are expected to be found, the high level of sample contamination by other classes of X-ray emitters prevents the identification of new members. Nonetheless, two candidates have managed to circumvent these difficulties: namely, Calvera (1RXS J141256.0+792204; Rutledge *et al.*, 2008) and 2XMM J104608.7-594306 (Pires *et al.*, 2009b). The discovery of Calvera benefited from the better localisation of ROSAT sources through a survey with NASA's Neil Gehrels Swift Observatory; that of 2XMM J104608.7-594306, from the high sensitivity and good positional accuracy of XMM-Newton, despite its low sky coverage.

Interestingly, Calvera is the first neutron star to be proposed as an evolved anti-magnetar or "orphaned CCO" within the *hidden B-field* scenario. The source is relatively bright, originally identified as a likely compact object based on its large X-ray-to-optical flux ratio. Multiwavelength investigations by Zane *et al.* (2011) confirmed its radio-quietness and revealed a very short spin period of $P \sim 59$ ms. Using XMM-Newton and Chandra, Halpern *et al.* (2013) measured the spin down of the source, which implies a dipolar field of $B_{\rm dip} = 4.4 \times 10^{11}$ G, a characteristic age of $\tau_{\rm ch} = 2.9 \times 10^5$ yr and a spin-down luminosity of $\dot{E} = 6 \times 10^{35}$ erg s^{-1}. The distance to the source is not well constrained. Its interpretation as a cooling INS (with similar size of emission radius as the M7) has always met with difficulties, since its high galactic latitude and hot temperature ($kT \sim 200$ eV) would either require a very high spatial velocity to explain its current position well above the plane or point to a non-standard cooling or re-heating process. Alternatively, if Calvera's progenitor was a high-velocity runaway star, it could had been born away from the Galactic disk (Posselt *et al.*, 2008).

The most promising among the INS candidates selected from the 2XMMp catalogue (Pires *et al.*, 2009a), 2XMM J104608.7-594306 is possibly a remnant of a former generation of massive stars in the Carina Nebula (Townsley *et al.*, 2011). Thanks to its projected proximity to the well-studied Eta Carinae system, it was serendipitously detected on many occasions by XMM-Newton. Follow-up observations in the optical with the ESO Very Large Telescope set a deep limit on the brightness of the optical counterpart and a very high X-ray-to-optical flux ratio. At present, no radio or γ-ray counterparts are known. Its X-ray spectrum is soft, steady, and purely thermal with $kT{\sim}135$ eV; evidence for one or two absorption features is also reported (Pires *et al.*, 2015, 2012). If previously associated with the runaway massive star MJ 218 (Ngoumou *et al.*, 2013), its age estimate is 10−30 kyr, consistent with the absence of a supernova remnant. Overall, its spectral properties and likely presence in the Carina open cluster disfavour a recycled object or a conventional evolutionary path, typical of a rotation-powered pulsar. However, without a clear determination of the spin period and a robust estimate of its magnetic field, the exact nature of the neutron star is unclear.

4.1. *The legacy value of the XMM-Newton archive*

X-ray emitting INSs at faint fluxes will be the products of more distant massive star associations, beyond the solar vicinity and the Gould Belt, and will be younger and hotter than the M7 on average (Pires *et al.*, 2017). The most compelling candidates will be located at relatively low galactic latitudes or close to open stellar clusters and supernova remnants (Posselt *et al.*, 2008). In preparation for the full sensitivity of the ROSAT successor eROSITA All-Sky Survey (Section 4.2), the use of serendipitous data from the XMM-Newton Observatory provides an excellent opportunity to test search algorithms and discover new INSs.

Building on our experience cross-correlating previous releases of the XMM-Newton catalogue of X-ray sources, we searched the source content of 4XMM-DR9 for new INS candidates. We used as main criteria the absence of any catalogued optical/UV/IR counterpart and a soft spectrum down to a limiting flux of 10^{-14} erg s^{-1} cm^{-2} in the $0.5-1$ keV energy band (Pires *et al.*, 2022). The brightest among the unknown sources were then selected for dedicated observations with XMM-Newton (fulfil programmes 088419, 090126, 092282[4]; Pires *et al.*, in prep.). The first candidate targeted by the programme, 4XMM J022141.5-735632, an X-ray source located in the direction of the Magellanic Bridge and independently identified by the searches of Rigoselli *et al.* (2022) and Kurpas *et al.* (in prep.), shows a remarkably soft energy distribution, with $kT \sim 60$ eV. The absence of optical and near-infrared counterparts in Legacy Survey imaging safely excludes a more ordinary X-ray emitter than an INS. We find no evidence for variability in either flux or spectral state within the nearly ten-year interval covered by the analysis of XMM-Newton and eROSITA epochs. Previous detection at a similar flux level by ROSAT PSPC and the Swift X-ray Telescope suggest that the emission is fairly stable over decade-long time scales. Additional monitoring with higher quality data will be necessary to exclude the few-percent level of spectral variation as reported for the M7 INSs RX J0720.4-3125 and RX J1605.3+3249. Future dedicated observations, particularly to search for pulsations, are crucial to shed further light on the nature of the X-ray source and relations to other INSs.

4.2. *eROSITA: Forecast and early results*

The All-Sky Survey of eROSITA, the first to be performed at X-ray energies since the ROSAT era, is expected to surpass its predecessor's sensitivity by a factor of 25 (0.2−2.3 keV; Predehl *et al.*, 2021). We developed a model for the population of neutron stars in our Galaxy to test their detectability by eROSITA (Pires *et al.*, 2017). In the model, neutron stars are created from a progenitor population of massive stars distributed in the spiral arms of the Galactic disk; after the supernova explosion and the imparted "kick" velocity, their evolution in the galactic gravitational potential is followed while they cool down emitting soft X-rays. The expected source count rates and total flux are then computed in the eROSITA detectors,

[4]The observations conducted under the rationale of this programme will be used for science validation activities of the European Union's Horizon 2020 XMM2ATHENA project (Webb *et al.*, 2023).

taking into account simple prescriptions of the absorbing material in the line of sight and the celestial exposure after the four years of all-sky survey. Our study showed that after completion the eROSITA survey will be sensitive to neutron stars as faint as $f_X \sim 10^{-14}$ erg s^{-1} cm^{-2} (0.2–2 keV); around 85–95 sources all-sky will be among the eROSITA detections. Their identification and characterisation will require synergy with multiwavelength large-scale surveys and will heavily rely on follow-up observations. Nonetheless, dedicated searches hold the promise to boost the number of known thermally emitting INSs for the first time since the years of ROSAT.

Since the beginning of the survey in December 2019, eROSITA has successfully completed four of the planned eight all-sky scans. Source catalogues corresponding to the German part of the eROSITA sky[5] are created by and made available to the members of the German eROSITA Consortium. We searched the individual and stacked X-ray source catalogues for unidentified INSs above a limiting flux of 10^{-13} erg s^{-1} cm^{-2} in the 0.2–2 keV energy band (Kurpas *et al.*, in prep.). At this intermediate flux level, follow-up investigations using current facilities can be carried out already in the immediate aftermath of the eROSITA survey. Having survived our probabilistic catalogue cross-matching and screening procedure (only about 0.05% of over 10^5 X-ray sources satisfy all the selection criteria for identification with an INS candidate), a handful of soft X-ray sources with no obvious counterparts have been lined up for observations in the optical and in X-rays. The first results are beginning to emerge: two INS candidates, eRASSU J065715.3+260428 and eRASSU J131716.9-402647 have no optical counterparts down to magnitude ~25 (Kurpas *et al.*, 2023). The eROSITA spectra of both sources are well described by simple blackbodies with $kT \sim 110$ eV; the low count statistics prevents a more detailed analysis at present. Upcoming follow-up observations with NICER, XMM-Newton, Chandra, and the VLT, already secured for the current cycle of observations, will soon reveal more on the nature of the most promising candidates from the eROSITA sky.

[5]The eROSITA data rights are equally split between a German and a Russian consortium; see Predehl *et al.* (2021); for details.

5. Final Remarks

The most fascinating aspect of a survey is to observe the unexpected. Even among well known and studied sources, state-of-the-art surveys can add crucial information that may challenge our beliefs and expectation from theory.[6] A better sampling of unconventional INSs is essential to probe the population as a whole and assess the most important ingredients affecting their observability in the Milky Way. This ultimately has implications to several fields of physics and astrophysics, ranging from core-collapse supernovae, cooling theory, radiative processes in highly-magnetised neutron star atmospheres, the equation-of-state of ultra-dense matter, stellar evolution, and the neutron star birth rate in spiral galaxies.

Acknowledgments

I would like to thank collaborators Axel Schwope, Jan Kurpas, Christian Motch, Konrad Dennerl, Michael Freyberg, Frank Haberl, Anna Karpova, Georg Lamer, Alexander Potekhin, Yuri Shibanov, Valery Suleimanov, Iris Traulsen, Frank Valdes, and Dmitry Zyuzin, for valuable contributions to the work presented here. This work is partially based on data from eROSITA, the soft X-ray instrument aboard SRG, a joint Russian-German science mission supported by the Russian Space Agency (Roskosmos), the Russian Academy of Sciences represented by its Space Research Institute (IKI), the Deutsches Zentrum fr Luft- und Raumfahrt (DLR) and the Max-Planck Society, represented by the Max-Planck Institute for Extraterrestrial Physics (MPE). The eROSITA data were processed using the eSASS software system developed by the German eROSITA Consortium. This work was supported by the project XMM2ATHENA, which has received funding from the European Union's Horizon 2020 research and innovation programme under grant agreement no 101004168.

[6] A recent example is the inference of an extremely light CCO in the supernova remnant HESS J1731-347 thanks to its proposed association with a post-AGB optical star, for which an independent distance estimate has become available through Gaia DR3 parallax (Doroshenko *et al.*, 2022).

Bibliography

Agüeros, M. A. *et al.* (2011). No confirmed new isolated neutron stars in the SDSS data release 4, *Astron. J.* **141**, 176.

Alford, J. A. J. and Halpern, J. P. (2023). Do central compact objects have carbon atmospheres? *Astrophys. J.* **944**, 36.

Arcones, A. and Thielemann, F.-K. (2023). Origin of the elements, *Astron. Astrophys. Rev.* **31**, 1.

Beloborodov, A. M. and Li, X. (2016). Magnetar Heating, *Astronphys. J.* **833**, 261.

Borghese, A., Rea, N., Coti Zelati, F., Tiengo, A., and Turolla, R. (2015). Discovery of a strongly phase-variable spectral feature in the isolated neutron star RX J0720.4-3125, *Astrophys. J. Lett.* **807**, L20,

Borghese, A. *et al.* (2017). Narrow phase-dependent features in X-ray dim isolated neutron stars: A new detection and upper limits, *Mon. Not. R. Astron. Soc.* **468**, 2975.

Caleb, M. *et al.*, *Nat. Astron.* **6**, 828.

Caraveo, P. A. (2014). Gamma-ray pulsar revolution, *Ann. Rev. Astron. Astrophys.* **52**, 211.

Chieregato, M. *et al.* (2005). Blank field sources in the ROSAT HRI Brera multiscale wavelet catalog, *Astron. Astrophys.* **444**, 69.

Danner, R. (1998). Searching for old neutron stars with ROSAT. I. Soft X-ray sources in molecular clouds at high galactic latitude, *Astron. Astrophys. Suppl. Ser.* **128**, 331.

De Grandis, D. *et al.* (2022). Two decades of X-ray observations of the isolated neutron star RX J1856.5-3754: detection of thermal and non-thermal hard X-rays and refined spin-down measurement, *Mon. Not. R. Astron. Soc.* **516**, 4932.

De Grandis, D., Taverna, R., Turolla, R. *et al.* (2021). X-ray emission from isolated neutron stars revisited: 3D magnetothermal simulations, *Astrophys. J.* **914**(2), 118.

De Luca, A. (2017). Central compact objects in supernova remnants, in *Journal of Physics Conference Series*, **932**, 012006.

Dessert, C., Foster, J. W., and Safdi, B. R. (2020). Hard X-ray excess from the magnificent seven neutron stars, *Astrophys. J.* **904**, 42.

Diehl, R. *et al.* (2006). Radioactive ^{26}Al from massive stars in the galaxy, *Nature*, **439**, 45.

Diehl, R. *et al.* (2022). Cosmic nucleosynthesis: A multi-messenger challenge, *Prog. Part. Nucl. Phys.* **127**, 103983.

Doroshenko, V. *et al.* (2022). A strangely light neutron star within a supernova remnant, *Nat. Astron.* **6**, 1444.

Ertan, Ü. *et al.* (2017). Optical excess of dim isolated neutron stars, *Mon. Not. R. Astron. Soc.* **470**, 1253.

Goldreich, P. and Reisenegger, A. (1992). Magnetic field decay in isolated neutron stars, *Astrophys. J.* **395**, 250.

Gotthelf, E. V. *et al.* (2013). The spin-down of PSR J0821-4300 and PSR J1210-5226: Confirmation of central compact objects as anti-magnetars, *Astronphys. J.* **765**, 58.

Haberl, F. (2007). The magnificent seven: magnetic fields and surface temperature distributions, *Astrophys. Space Sci.* **308**, 181.

Halpern, J. P., Bogdanov, S., and Gotthelf, E. V. (2013). X-ray measurement of the spin-down of calvera: A radio- and gamma-ray-quiet pulsar, *Astronphys. J.* **778**, 120.

Hambaryan, V., Suleimanov, V., Haberl, F. *et al.* (2017). The compactness of the isolated neutron star RX J0720.4-3125, *Astron. Astrophys.* **601**, A108.

Hambaryan, V., Suleimanov, V., Schwope, A. D. *et al.* (2011). Phase-resolved spectroscopic study of the isolated neutron star RBS 1223 (1RXS J130848.6+12708), *Astron. Astrophys.* **534**, A74.

Hewish, A. *et al.* (1968). Observation of a rapidly pulsating radio source, *Nature*, **217**, 709.

Ho, W. C. G. (2011). Evolution of a buried magnetic field in the central compact object neutron stars, *Mon. Not. R. Astron. Soc.* **414**, 2567.

Ho, W. C. G. and Heinke, C. O. (2009). A neutron star with a carbon atmosphere in the cassiopeia A supernova remnant, *Nature*, **462**(7269), 71.

Ho, W. C. G. *et al.* (2007). Magnetic hydrogen atmosphere models and the neutron star RX J1856.5-3754, *Mon. Not. R. Astron. Soc.* **375**, 821.

Hohle, M. M. *et al.* (2012a). Narrow absorption features in the co-added XMM-Newton RGS spectra of isolated neutron stars, *Mon. Not. R. Astron. Soc.* **419**, 1525.

Hohle, M. M. *et al.* (2012b). The continued spectral and temporal evolution of RX J0720.4-3125, *Mon. Not. R. Astron. Soc.* **423**, 1194.

Hurley-Walker, N., Rea, N., McSweeney, S. J. *et al.* (2023). A long-period radio transient active for three decades, *Nature*, **619**, 487.

Hurley-Walker, N. *et al.* (2022). A radio transient with unusually slow periodic emission, *Nature*, **601**, 526.

Kaplan, D. L., Kamble, A., van Kerkwijk, M. H. *et al.* (2011). New optical/ultraviolet counterparts and the spectral energy distributions of nearby, thermally emitting, isolated neutron stars, *Astrophys. J.* **736**, 117.

Kaplan, D. L. and van Kerkwijk, M. H. (2005a). A coherent timing solution for the nearby isolated neutron star RX J0720.4-3125, *Astrophys. J. Lett.* **628**(1), L45.

Kaplan, D. L. and van Kerkwijk, M. H. (2005b). A coherent timing solution for the nearby isolated neutron star RX J1308.6+2127/RBS 1223, *Astrophys. J.* **635**, L65.

Kaplan, D. L. and van Kerkwijk, M. H. (2009a). Constraining the spin-down of the nearby isolated neutron star RX J0806.4-4123, and implications for the population of nearby neutron stars, *Astrophys. J.* **705**, 798.

Kaplan, D. L. and van Kerkwijk, M. H. (2009b). Constraining the spin-down of the nearby isolated neutron star RX J2143.0+0654, *Astrophys. J. Lett.* **692**, L62.

Kaplan, D. L. and van Kerkwijk, M. H. (2011). A coherent timing solution for the nearby, thermally emitting isolated neutron star RX J0420. 0-5022, *Astronphys. J. Lett.* **740**, L30.

Kaplan, D. L., van Kerkwijk, M. H., and Anderson, J. (2007). The distance to isolated neutron star RX J0720.4-3125, *Astronphys. J.* **660**, 1428.

Kaspi, V. M. and Beloborodov, A. M. (2017). Magnetars, *Annu. Rev. Astron. Astrophys.* **55**, 261–301.

Kaspi, V. M., Roberts, M. S. E., and Harding, A. K. (2010). Isolated neutron stars, *Compact Stellar X-ray Sources*, 279.

Keane, E. *et al.* (2015). A cosmic census of radio pulsars with the SKA, in *Advancing Astrophysics with the Square Kilometre Array (AASKA14)*, 40

Keane, E. F., Ludovici, D. A., Eatough, R. P. *et al.* (2010). Further searches for rotating radio transients in the parkes multi-beam pulsar survey, *Mon. Not. R. Astron. Soc.* **401**(2), 1057.

Kurpas, J., Schwope, A. D., Pires, A. M. *et al.* (2023). Discovery of two promising isolated neutron star candidates in the SRG/eROSITA All-Sky Survey, *Astron. Astrophys.* **674**, A155.

Lorimer, D. R. (2005). Binary and millisecond pulsars, *Living Reviews in Relativity* **8**, 7.

Madsen, K. K. *et al.* (2021). *IACHEC 2020/2021 Pandemic Report, arXiv e-prints*, arXiv:2111.01613.

Malacaria, C. *et al.* (2019). A joint NICER and XMM-newton view of the "Magnificent" thermally emitting X-ray isolated neutron star RX J1605.3+3249, *Astronphys. J.* **880**, 74.

Manchester, R. N., Hobbs, G. B., Teoh, A., and Hobbs, M. (2005). The australia telescope national facility pulsar catalogue, *Astron. J.* **129**, 1993.

McLaughlin, M. A. *et al.* (2006). Transient radio bursts from rotating neutron stars, *Nature*, **439**(7078), 817–820.

Mignani, R. P. *et al.* (2013). The birthplace and age of the isolated neutron star RX J1856.5-3754, *Mon. Not. R. Astron. Soc.* **429**, 3517.
Molero, M., Magrini, L., Matteucci, F. *et al.* (2023). Origin of neutron-capture elements with the Gaia-ESO survey: the evolution of s- and r-process elements across the Milky Way, *Mon. Not. R. Astron. Soc.* **523**(2), 2974–2989.
Motch, C. *et al.* (1997). The ROSAT galactic plane survey: analysis of a low latitude sample area in Cygnus. *Astron. Astrophys.* **318**, 111.
Motch, C. *et al.* (2009). Proper motions of thermally emitting isolated neutron stars measured with Chandra, *Astron. Astrophys.* **497**, 423.
Motch, C., Zavlin, V. E., and Haberl, F. (2003). The proper motion and energy distribution of the isolated neutron star RX J0720.4-3125, *Astron. Astrophys.* **408**, 323–330.
Ngoumou, J., Preibisch, T., Ratzka, T., and Burkert, A. (2013). The mysterious sickle object in the carina nebula: A stellar wind induced bow shock grazing a clump? *Astrophys. J.* **769**, 139.
Ostriker, J. P. and Gunn, J. E. (1969). On the nature of pulsars. I. Theory, *Astrophys. J.* **157**, 1395.
Özel, F. and Freire, P. (2016). Masses, radii, and the equation of state of neutron stars, *Annu. Rev. Astron. Astrophys.* **54**, 401.
Perna, R. and Pons, J. A. (2011). A unified model of the magnetar and radio pulsar bursting phenomenology, *Astrophys. J. Lett.* **727**, L51.
Pires, A. M., Haberl, F., Zavlin, V. E. *et al.* (2014). XMM-newton reveals a candidate period for the spin of the "Magnificent Seven" neutron star RX J1605.3+3249, *Astron. Astrophys.* **563**, A50.
Pires, A. M., Motch, C., and Janot-Pacheco, E. (2009a). A search for thermally emitting isolated neutron stars in the 2XMMp catalogue, *Astron. Astrophys.* **504**(1), 185–197.
Pires, A. M. *et al.* (2022). XMM-Newton and SRG/eROSITA observations of the isolated neutron star candidate 4XMM J022141.5−735632, *Astron. Astrophys.* **666**, A148.
Pires, A. M., Motch, C., Turolla, R. *et al.* (2015). New XMM-Newton observation of the thermally emitting isolated neutron star 2XMM J104608.7-594306, *Astron. Astrophys.* **583**, A117.
Pires, A. M. *et al.* (2012). The peculiar isolated neutron star in the Carina Nebula. Deep XMM-Newton and ESO-VLT observations of 2XMM J104608.7-594306, *Astron. Astrophys.* **544**, A17.
Pires, A. M., Motch, C., Turolla, R. *et al.* (2009b). The isolated neutron star candidate 2XMM J104608.7-594306, *Astron. Astrophys.* **498**(1), 233–240.

Pires, A. M., Schwope, A., and Kurpas, J. (2023). Deep eROSITA observations of magnificent seven isolated neutron stars, *IAU Symposium* **363**, 288.

Pires, A. M. *et al.* (2019). A deep XMM-Newton look on the thermally emitting isolated neutron star RX J1605.3+3249, *Astron. Astrophys.* **623**, A73.

Pires, A. M., Schwope, A. D., and Motch, C. (2017). Follow-up of isolated neutron star candidates from the eROSITA survey, *Astron. Nachr.* **338**, 213.

Pons, J. A., Miralles, J. A., and Geppert, U. (2009). Magneto-thermal evolution of neutron stars, *Astron. Astrophys.* **496**, 207–216.

Popov, S. B. (2023). The zoo of isolated neutron stars, *Universe*, **9**, 273.

Popov, S. B. *et al.* (2003). Young isolated neutron stars from the Gould Belt, *Astron. Astrophys.* **406**, 111.

Posselt, B. *et al.* (2018). Discovery of extended infrared emission around the neutron star RXJ0806.4-4123, *Astrophys. J.* **865**, 1.

Posselt, B., Pavlov, G. G., Popov, S., and Wachter, S. (2014). Herschel and spitzer observations of slowly rotating, nearby isolated neutron stars, *Astrophys. J. Suppl. Ser.* **215**, 3.

Posselt, B., Popov, S. B., Haberl, F. *et al.* (2007). The magnificent seven in the dusty prairie, *Astrophys. Space Sci.* **308**, 171.

Posselt, B. *et al.* (2008). The needle in the haystack: where to look for more isolated cooling neutron stars, *Astron. Astrophys.* **482**(2), 617.

Potekhin, A. Y. (2014). Atmospheres and radiating surfaces of neutron stars, *Physics Uspekhi*, **57**, 735.

Potekhin, A. Y. *et al.* (2020). Thermal luminosities of cooling neutron stars, *Mon. Not. R. Astron. Soc.* **496**, 5052.

Predehl, P., Andritschke, R., Arefiev, V. *et al.* (2021). The eROSITA X-ray telescope on SRG, *Astron. Astrophys.* **647**, A1.

Rea, N. *et al.* (2016). Magnetar-like activity from the central compact object in the SNR RCW103, *Astrophys. J. Lett.* **828**(1), L13.

Rigoselli, M., Mereghetti, S., and Tresoldi, C. (2022). Candidate isolated neutron stars in the 4XMM-DR10 catalogue of X-ray sources, *Mon. Not. R. Astron. Soc.* **509**, 1217.

Rozwadowska, K., Vissani, F., and Cappellaro, E. (2021). On the rate of core collapse supernovae in the milky way, *New Astron.* **83**, 101498.

Rutledge, R. E., Fox, D. B., and Shevchuk, A. H. (2008). Discovery of an isolated compact object at high galactic latitude, *Astrophys. J.* **672**(2), 1137.

Shabaltas, N. and Lai, D. (2012). The hidden magnetic field of the young neutron star in kesteven 79, *Astrophys. J.* **748**(2), 148.

Suleimanov, V., Hambaryan, V., Potekhin, A. Y., (2010). Radiative properties of highly magnetized isolated neutron star surfaces and approximate treatment of absorption features in their spectra, *Astron. Astrophys.* **522**, A111.

Tetzlaff, N., Eisenbeiss, T., Neuhäuser, R., and Hohle, M. M. (2011). The origin of RX J1856.5-3754 and RX J0720.4-3125 - updated using new parallax measurements, *Mon. Not. R. Astron. Soc.* **417**, 617.

Tetzlaff, N. *et al.* (2012). Neutron stars from young nearby associations: The origin of RX J1605.3+3249, *Publ. Astron. Soc. Aust.* **29**, 98.

Thompson, C. and Duncan, R. C. (1995). The soft gamma repeaters as very strongly magnetized neutron stars-I. Radiative mechanism for outbursts, *Mon. Not. R. Astron. Soc.* **275**, 255–300.

Thompson, C. and Duncan, R. C. (1996). The soft gamma repeaters as very strongly magnetized neutron stars. II. Quiescent neutrino, X-ray, and alfven wave emission, *Astrophys. J.* **473**, 322.

Townsley, L. K. *et al.* (2011). The Chandra carina complex project: Deciphering the enigma of carina's diffuse X-ray emission, *Astrophys. J. Suppl.* **194**, 15.

Treves, A., Turolla, R., Zane, S., and Colpi, M. (2000). Isolated neutron stars: Accretors and coolers, *Publ. Astron. Soc. Pac.* **112**, 297.

Turner, M. L. *et al.* (2010). A limit on the number of isolated neutron stars detected in the ROSAT All-sky-survey bright source catalog, *Astrophys. J.* **714**, 1424.

Turolla, R. (2009). Isolated neutron stars: The challenge of simplicity, in W. Becker (ed.), *Astrophysics and Space Science Library, Astrophysics and Space Science Library*, **357**, 141.

Turolla, R., Zane, S., and Drake, J. J. (2004). Bare quark stars or naked neutron stars? The case of RX J1856.5-3754, *Astrophys. J.* **603**, 265–282.

Turolla, R., Zane, S., Pons, J. A., Esposito, P., and Rea, N. (2011). Is SGR 0418+5729 indeed a waning magnetar? *Astrophys. J.* **740**, 105.

van Kerkwijk, M. H. and Kaplan, D. L. (2007). Isolated neutron stars: magnetic fields, distances, and spectra, *Astrophys. Space Sci.* **308**(1–4), 191–201.

van Kerkwijk, M. H., Kaplan, D. L., Pavlov, G. G., and Mori, K. (2007). Spectral and rotational changes in the isolated neutron star RX J0720.4-3125, *Astrophys. J. Lett.* **659**(2), L149–L152.

Viganò, D., Perna, R., Rea, N., and Pons, J. A. (2014). Spectral features in isolated neutron stars induced by inhomogeneous surface temperatures, *Mon. Not. R. Astron. Soc.* **443**, 31–40.

Viganò, D. and Pons, J. A. (2012). Central compact objects and the hidden magnetic field scenario, *MNRAS*, **425**(4), 2487–2492.

Viganò, D., Rea, N., Pons, J. A. *et al.* (2013). Unifying the observational diversity of isolated neutron stars via magneto-thermal evolution models, *Mon. Not. R. Astron. Soc.* **434**(1), 123–141.

Voges, W., Aschenbach, B., Boller, T. *et al.* (1999). The ROSAT all-sky survey bright source catalogue, *Astron. Astrophys.* **349**, 389–405.

Walter, F. M. *et al.* (2010). Revisiting the parallax of the isolated neutron star RX J185635-3754 using HST/ACS imaging, *Astrophys. J.* **724**(1), 669–677.

Walter, F. M., Wolk, S. J., and Neuhäuser, R. (1996). Discovery of a nearby isolated neutron star, *Nature,* **379**(6562), 233–235.

Wang, W., Lu, J., Tong, H. *et al.* (2017). The optical/UV excess of X-ray-dim isolated neutron stars. I. Bremsstrahlung emission from a strangeon star atmosphere, *Astrophys. J.* **837**, 81.

Webb, N. A. *et al.* (2023). XMM2ATHENA, the H2020 project to improve XMM-Newton analysis software and prepare for Athena, *arXiv e-prints*, arXiv:2303.10097.

Wu, Q., Pires, A. M., Schwope, A. *et al.* (2021). What causes the absence of pulsations in central compact objects in supernova remnants? *Res. Astron. Astrophys.* **21**, 294.

Yakovlev, D. G. and Pethick, C. J. (2004). Neutron star cooling, *Annu. Rev. Astron. Astrophys.* **42**(1), 169–210.

Yoneyama, T., Hayashida, K., Nakajima, H., Inoue, S., and Tsunemi, H. (2017). Discovery of a keV-X-ray excess in RX J1856.5-3754, *Publ. Astron. Soc. Jpn.*, **69**, 50.

Yoneyama, T., Hayashida, K., Nakajima, H., and Matsumoto, H. (2019). Universal detection of high-temperature emission in X-ray isolated neutron stars, *Publ. Astron. Soc. Jpn.* **71**, 17.

Zane, S., Haberl, F., Israel, G. L. *et al.* Discovery of 59 ms pulsations from 1RXS J141256.0+792204 (Calvera), *Mon. Not. R. Astron. Soc.* **410**, 2428.

Zane, S. *et al.* (2001). Proton cyclotron features in thermal spectra of ultramagnetized neutron stars, *Astrophys. J.* **560**, 384.

https://doi.org/10.1142/9789819800643_0004

Chapter 4

Evolution of Long-Period Compact Radio Sources Driven by Winds

Jorge E. Horvath*, Lucas M. de Sá, Lívia S. Rocha, Gustavo Y. Chinen, Lucas G. Barão, and Márcio G.B. de Avellar

Universidade de São Paulo, IAG-Astronomia
R. do Matão 1226, C. Universitária
05508-090 São Paulo, Brazil
**foton@iag.usp.br*

We address in this work the nature and evolution of the long-period compact star sources, which has recently added several unexpected members. The central hypothesis is that particle winds drive their evolution, being an important factor for these relatively old sources. We show the consistency of this picture and remark some unsolved problems and caveats within it.

Keywords: Pulsars, Magnetars, Winds, Long-period compact star radio sources

Contents

1. Introduction

The study of the first detected manifestation of neutron stars (NSs), i.e., radio pulsars, has entered its 6th decade being enlarged and extended very substantially in many aspects. On the one hand, several periodic sources have been discovered in *other* bands, reflecting not only the pulsed emission seen in other frequencies, but also other environments and situations harboring neutron stars "caught in the act". In addition, a variety of behaviors within the enlarged set of radio sources have been identified, possibly related to neutron stars as well, but rising a number of important unanswered questions (Popov, 2023). In short, we are at crossroads in the field, in need of a general classification scheme and seeking concrete models to explain the variety of emission types with a minimum testable set of physical hypotheses.

An excellent example of this type of search is the work of (Kaspi, 2010) trying to make sense of the different groups of objects. Recent work by Popov (2023) addresses the same problem, adding fast radio bursts (FRBs) to the list. Links between these different groups have been explored (e.g., Yoneyama *et al.*, 2019). However, a major new discovery is that of local long-period radio sources, something quite unexpected and puzzling in many ways (see for instance Rea *et al.*, 2024). Although discussion of long-period pulsars/magnetars had already been present in the literature before, such as with their possible association with FRBs (e.g., Wadiasingh *et al.*, 2020; Beniamini *et al.*, 2020), the discovery of so many sources in such a short period came as a surprise, and the mechanism through which these objects happen to reach such low periods while spinning down at a relatively high rate, as well as their emission mechanism, is still under investigation (see e.g., Beniamini *et al.*, 2023). We attempt in this work to establish an evolutionary link between the long-period sources and the rest of the population. We shall see below to which extent such a proposal can be supported.

2. Well-Established Facts and General Features About Radio Pulsars

Virtually every astrophysicist alive knows the spectacular history of the discovery of pulsars by the Cavendish group at Cambridge, UK, operating the IPS instrument at Mullard (Hewish *et al.*, 1979), mainly due to the persistence and insight of Jocelyn Bell on the reality of a pulsed persistent source. The prompt publication of the interpretation papers by Gold (1969) and Pacini (1968) and early work showing the handful of new objects roughly fitting the Gold-Pacini scenario, convinced the community that those pulsars were rotating neutron stars. The pulsed emission stems from a complicated magnetosphere dynamics, which in turn is powered by rotational energy, and only partial solutions have been obtained (see for instance, Goldreich and Julian, 1969).

The basic assumptions of the rotating neutron star picture of pulsars is well-known: provided the free-energy source is the rotation of the object, and assuming a single energy loss mechanism, namely the observed pulsed electromagnetic emission, the dynamics is simply given by

$$I\dot{\Omega} = \tau_{em}, \tag{1}$$

in which the electromagnetic torque $\tau_{em} \propto B^2\Omega^3$ can be obtained by integrating the Poynting vector over a closed surface engulfing the source (Manchester and Taylor, 1977). The simplest expression stems from the assumption of a vacuum surrounding the neutron star, but it was promptly realized that the strong induced electric fields would lift charges from the surface and fill the space, a fact that complicated the solution of the magnetosphere. The important points to be remarked here are that (a) only the dipole mechanism contributes to the energy loss and spindown; and (b) every coefficient in the dynamical equation (1) is constant. Both assumptions, successful as they are, fail to some extent when the dynamics of actual systems is analyzed.

On the one hand, the very picture of open field lines (Goldreich and Julian, 1969) in the magnetosphere, a widely accepted feature, leads to considering the escape of particles along them, i.e., a wind carrying away energy and angular momentum. This was recognized quite early, and discussed over the years (Michel, 1969; Wilson and Rees, 1978) (see Kirk *et al.*, 2009 for a general discussion), but eventually did not become part of the "standard" description (Manchester and Taylor, 1977). However, winds are not only expected on theoretical grounds: in the cases in which a rough estimate of the actual dipole pulse loss was possible, it became clear that it is too small to be responsible for the full rotational energy loss. In addition, a few cases in which a wind is strongly indicated by observations are available, the most notorious being the pulsar PSR B1931+24 (J1933+2421) (Kramer *et al.*, 2006), which shows a large difference in its torque when it switches on and off regularly; but also, as a recent example, PSR J2021+4026, which is the only known variable gamma-ray pulsar (Razzano *et al.*, 2023; Rigoselli *et al.*, 2021; Wang *et al.*, 2018). Magnetar spin-down glitches, in which the object loses angular momentum, have been posited to be caused by episodes of wind mass loss as well (Younes *et al.*, 2023). Finally, quasi-permanent changes in the torque after a glitch (Lyne *et al.*, 2015) in the Crab and other pulsars suggested either a discrete change of the angle between the rotation and magnetic axes (Allen and Horvath, 1997), or a permanent decoupling of a component contributing to the moment of inertia (Alpar *et al.*, 1993). These and other indications strongly argue against a constant, unique torque as postulated in the simplest picture of pulsar dynamics.

3. New Kids on the Block: Other Compact Systems and the Long-Period Sources

It has been long established that NSs are born in massive star supernova explosions. A relatively recent consensus is that lower mass stars, in the range of 8–10 $M_{\odot}$ also form NSs when they undergo electron capture onto an O-Ne-Mg core. The process known as *accretion-induced collapse* (AIC), either in its single-degenerate or double-degenerate version, is still uncertain but may contribute too (Tauris *et al.*, 2013; Wang and Liu, 2020; Wang *et al.*, 2022). A simple multiplication of the core-collapse supernova rate by the estimated Galactic lifespan gives a large number of NSs $\sim 10^8$ expected to reside in the Milky Way (Bisnovatyi-Kogan, 1992). How many of them are actually observable is a different matter: *active* pulsars today could feature 70,000 objects (Dirson *et al.*, 2022; Lyne *et al.*, 1985), but this number depends on the birth parameters and evolution of magnetic fields and is uncertain by a large factor. It is also thought that at some point the old pulsar population fades away by crossing a "death line" (see, e.g., Chen and Ruderman, 1993; Ho, 2012; Ho *et al.*, 2021). Keeping this in mind is important, since the new long-period objects may challenge the standard consensus (see below). Nevertheless, the conclusion that most of the Galactic NSs are *not* pulsars, but may be "hidden" or belong to some of the other groups, is unavoidable. As mentioned before, attempts to unify all NSs have been presented before (Kaspi, 2010), but as we shall see, new puzzling detections and unresolved problems are abundant.

Although the impact of the discovery of a physical realization of neutron stars in radio pulsars caused an enormous impact, the zoo of compact star systems related to neutron stars was enlarged in the '70s and beyond, with various classes of sources interpreted as different manifestations of the former within a variety of systems.

The oldest example is the group of *X-ray pulsars* (Giacconi *et al.*, 1971), discovered in the early days of dedicated satellite observations, and is credited to the *Uhuru* Team, who observed and characterized the source Cyg X-3 in 1971 ((Charles and Seward, 1995) and references therein). A more general group of sources was identified and classified as belonging to the Low-Mass X-Ray Binary (LMXB) or High-Mass X-Ray Binary (HMXB) classes depending on the companion mass (solar-type or massive, respectively), and a wealth of information on accretion disks and many other features gathered (Pszota and Kovács, 2023).

The presence of X-ray pulsars in some binary systems of the LXMB class has been confirmed (see Jonker and van der Klis, 2001), but most of the latter do not show pulsations (Patruno *et al.*, 2018). The short period *millisecond pulsars* (Bisnovatyi-Kogan and Komberg, 1974) may in fact be descendants of some type of LMXB.

The second oldest example of these groups we should mention are the *Central Compact Objects* (CCOs), quiet NSs in a few supernova remnants with or without low-amplitude pulsations (De Luca, 2017; Mayer and Becker, 2021). They are thought to be quite young, as corresponding to their birth in the supernova events, and their emission may be pointed away from us or, alternatively, they may have a small magnetic field/rotation rate. The compact soft X-ray source 1E 161348-5055 makes for a particularly interesting case. It was originally classified solely as an CCO associated to the remnant RCW 103 (De Luca *et al.*, 2006), and from its inception proved to be a puzzling case of a variable source, displaying a periodicity of 6.67 h and X-ray luminosities ranging from 10^{33} to $10^{35}\,\mathrm{erg\,s^{-1}}$; its unusually high period led to the hypothesis of it being a X-ray binary instead of an isolated rotating NS. In either case however, an extreme magnetic field seemed to be required, and D'A did in fact confirm its nature as a slowly-spinning isolated *magnetar* (discussed below), with field $10^{14}-10^{15}$ G. Since then, the mechanism by which it might have reached such a long period has remained uncertain; one early proposal was braking from interaction with fallback material from the remnant (De Luca *et al.*, 2006).

The first member of the group known as *Isolated Neutron Stars*, RX J1856.43754 is more recent (Walter *et al.*, 1996). The object *Geminga* ("it is not there" in Milanese dialect) remained as a prototype for many years (Caraveo *et al.*, 1996), but it has been now joined by a handful of other examples. The members became known as the *Magnificent Seven* or X-ray Dim Isolated Neutron Stars (XDINSs). Further candidate members include *Calvera* (Rutledge *et al.*, 2008), and 4XMM J022141.5-735632 (Pires *et al.*, 2022). The XDINSs are relatively close, blackbody-like cooling NSs (Potekhin *et al.*, 2015), and pulsations and non-zero period derivatives were detected over time, proving that they are actually middle-aged objects (Ertan *et al.*, 2014; Turolla, 2009). It was expected that their number would be much higher by now, but apparently this is not the case.

A serendipitous detection of *Rotating Radio Transients* (RRATs), emitting occasionally isolated pulses of a few ms and going silent for days or more was first made in 2006 (McLaughlin *et al.*, 2006). Their narrow duty cycles make them virtually invisible most of the time, and therefore their number could be very large. It has been suggested that they may be as numerous as radio pulsars, or even more abundant. When coherent solutions for the sporadic pulses were constructed, they allowed an estimate of some of the period derivatives $\dot{P}$, and therefore an estimate of the characteristic spin-down ages. They seem to be not too different from ordinary pulsars in this aspect; although the recent detection of the extremely intermittent radio source PSR J1710-3452, which shows similarities to radio-loud magnetars and is placed close to magnetars in the $P-\dot{P}$ plane, but would feature as a particularly old magnetar, might complicate this picture (Surnis *et al.*, 2023). The suggestion is that, for some reason the "spark" leading to a pulse does not operate efficiently, as in the pulsar case (Abhishek *et al.*, 2022). It is possible that this behavior is an extreme example of the *nulling* phenomenon seen in many ordinary pulsars. However, we do not know enough about the actual origin of pulses to test these ideas.

Another subgroup of NSs, of particular importance for our discussion, is the *magnetar* class, in which the dominating X-ray emission has been attributed to the existence of a large magnetic field. In fact the X-ray luminosity $\dot{E}_X$ exceeds the rotational energy reservoir (Esposito *et al.*, 2021) $I\Omega\dot{\Omega}$, and makes magnetars objects powered by magnetic energy. It is a matter of debate whether there is a continuum of NSs, in the sense that the magnetic fields form a continuous distribution, or if there is some kind of gap between pulsars and magnetars instead. *Transition pulsars*, with magnetic fields as high as some of the identified magnetars, but otherwise showing quite normal radio emission, argue in favor of a continuum in B. The (unexpected) existence of "low-field" magnetars (Turolla and Esposito, 2013) adds strength to the latter hypothesis. It is important to remark that the magnetar group is defined as the NSs for which the magnetic field is the dominant source of energy, relative to rotational energy, regardless of the particular emission mechanism.

The most recent unexpected discovery of long-period radio sources can be seen as a further example of our incomplete knowledge of compact stellar radio sources. A series of detections of progressively longer period pulsar-like (Caleb *et al.*, 2022) and magnetar-like (Hurley-Walter *et al.*, 2022) sources begun a few years ago and is still ongoing. Some of these long-period sources are located below the "classical" *death line*, and a definite division between them is far from clear, as their position in the $P-\dot{P}$ plane does not allow one to establish a neat classification. A couple of examples are located very close to the pulsar-like white dwarf objects AR Scorpii (see Buckley *et al.*, 2017) and the recently discovered J191213.72-441045.1 (see Pelisoli *et al.*, 2023). In fact, one of these members (GPM J1839-10) was known since 1988 (Hurley-Walter *et al.*, 2023), but had "disappeared" from the research landscape. It is challenging, and the main goal of the present work, to relate these sources to the rest of the population.

4. Are the Long-Period Sources Driven by Winds?

The description of the pulsar wind physics and their evolution is still incomplete, in spite of being the subject of an early research (Coroniti, 1990; Michel, 1971, 1973; Usov, 1975). A useful review by Kirk *et al.* (2009) can be consulted for a general view of the achievements and open problems. While winds may carry ions and pairs in comparable number escaping through the open lines, as expressed above, it has not been easy to address the actual importance of the winds for the object, and its contribution to the energy loss has been neglected usually in favor of the dipole contribution.

The new long-period objects may offer another view of this problem. Since the dipolar loss scales as Ω^3 in the dynamical equation (1), slow rotators are expected to quench the dipole emission at some point. This is a fact behind all present theories of the "death line". However, GPM J1839-10, for instance, is probably below all versions of these "death lines". Therefore, the very generation of the pulses is challenged within an orthodox view.

The detailed record of the pulses encourages to think that there are additional factors at work for this group. The time development of the pulses of GPM J183910 (Hurley-Walter *et al.*, 2023) shows a very peculiar pattern never seen before, and holds the clue to their exact origin, although they can hardly be labeled as "canonical". At this point one may wonder whether additional physics is needed to cope with the pulse generation and track the evolution of the sources (Tong, 2023). In terms of evolution, the need for additional explanation for these long periods had already been recognized at the time of the discovery of 1E 161348-5055; as mentioned before (De Luca, 2017) proposed an early interaction with fallback material as a braking mechanism, as well as the need for extreme magnetic fields — these later confirmed for the case of 1E 161348-5055. This mechanism has also been proposed for the more recent GLEAM-X J162759.5-523504.3 (Gençali *et al.*, 2022) (GLEAM J1627 hereon) and PSR J0901-4046 (Gençali *et al.*, 2023) discoveries, yielding a possibility that GLEAM J1627 still is surrounded by a disk, and will continue to evolve for the next few 10^5 yr, towards a period of a few 10^3 s (see (Ronchi *et al.*, 2022) for fallback accretion as a general formation channel for long-period pulsars). Early amplification of the field might also be required to explain the current inferred extreme field of GLEAM J1627 (Suvorov and Melatos, 2023).

Motivated by a different type of source (magnetars) showing outbursts, (Harding *et al.*, 1999) explored a minimal model of evolution in which the winds were explicitly considered as causing intermittent outflows. The energy loss by the latter was found to scale as the square root of the particle luminosity L_P and quadratically with the angular velocity of the object, Ω. A handful of similar works (Allen and Horvath, 2000; Álvarez and Carramiñana, 2004) considered the winds as an additional energy loss mechanism, in some cases even acting continuously as suggested by the earliest pulsar models (Coroniti, 1990; Michel, 1971, 1973; Usov, 1975). However, some winds are still expected to be intermittent, as the sudden and large mass outflows thought to be necessary to sustain this kind of energy loss are hardly continuous; (Beniamini *et al.*, 2020), for example, associated the wind mechanism with periodic giant flares of magnetars with very long periods $P \sim 25\,\text{yr}$. The suggestion we want to make here is that the long-period sources may be another example in which winds play an important role. We shall elaborate this idea below.

4.1. *Basics of winds*

A general appraisal of the magnetosphere and winds can be found in Michel (1991). For the simplest model, the dynamical Eq. (1) can be written in terms of the period P for a model including the dipole and wind loss mechanisms (Allen and Horvath, 2000; Harding *et al.*, 1999), yielding

$$I\dot{P} = 4\pi^2\mu^2\frac{(1-\alpha)}{P} + \sqrt{L_P}\mu\alpha P, \tag{2}$$

where μ is the magnetic dipole moment of the object, namely $\mu = \frac{B_0 r_0^3}{\sqrt{6c^3}}$; and α, in an interpretation of Eq. (2) as an exact evolution law, is the fraction of the luminosity carried by the wind, such that $L_{tot} = (1-\alpha)\dot{E}_d + \alpha\dot{E}_w$. Alternatively, α is the wind *duty cycle*, as originally in Harding *et al.* (1999), defined as the fraction of total "on" time; in this sense, Eq. (2) describes the *average* energy loss throughout the life of the pulsar. Winds are not expected to be active at all times, and thus $\alpha < 1$.

Within this perspective, the dynamics of a compact rotating source displays a more complex behavior. First of all, it is clear that for each object, there is a transition period separating the epochs of dipole- and wind-dominated evolution (considering L_P a fixed quantity), namely

$$P_{\text{tran}} = 2\pi\sqrt{\frac{\mu(1-\alpha)}{\alpha\sqrt{L_P}}}. \tag{3}$$

As expected, the P_{tran} values range from 1 to 1000 s depending on the particle luminosity L_P, a quantity hard to pin down from observations (see below). However, and having some optimism for the future observations, a relationship between the particle luminosity and source quantities can be worked out quite simply by combining the particle wind coefficient of Eq. (2) with Eq. (12) from Allen and Horvath (2000), with the result

$$L_P = \left(\frac{I\dot{P}(3-n)}{2\alpha\mu P}\right)^2. \tag{4}$$

In spite of the fact that the braking index n of the long-period objects is not yet known, and constitutes a quite difficult determination for the observed rotation, Eq. (4) converts L_P into a potentially observable quantity. As a consistency check for the wind hypothesis, we may look at the known young pulsars with measured braking indices, under the view that they are also affected by some degree of wind braking, a specially attractive possibility for those with low braking indices $n \geq 1$. Table 1 shows some of the results. All the calculated particle luminosities L_P for these objects are much higher than the observed X-ray emission, a fact that suggests that particle escape can operate "silently", and may still be quite high in old objects. We shall return to this issue below.

Table 1. The L_P values calculated with Eq. (4) for a set of pulsars with known braking indices. In all cases $r_0 = 10^6$ cm, $I = 10^{45}$ g cm^2 and $\alpha = 0.5$ were considered.

Pulsar	n	$\log L_P$ (erg s^{-1})	$\log L_X$ (erg s^{-1})
Crab[a]	2.51	37.60	36.21
PSR[1] J1846-0258	2.16	37.84	35.23
Vela[b]	1.4	36.70	31.48
PSR J1734-3333[2]	1.1	35.27	33.53

Notes: [a]Hamil (2015), Shibata *et al.* (2016).
[b]Olausen *et al.* (2010), Shibata *et al.* (2016).

Finally, a value of the magnetic field B can be found by inverting the dynamical equation $L_{tot} = (1-\alpha)\dot{E}_d + \alpha\dot{E}_w$ (see Harding *et al.*, 1999), with the result

$$B = -\frac{\sqrt{6c^3}}{r_0^3}\left(\frac{\sqrt{L_P}\alpha P^2 - P\sqrt{L_P\alpha^2P^2 + 16\pi^2\frac{(1-\alpha)I\dot{P}}{P}}}{8\pi^2(1-\alpha)}\right), \tag{5}$$

and gives the time spent by the objects in the wind stage, as calculated in this model (Allen and Horvath, 2000; Harding *et al.*, 1999), as

$$\tau_w(P) = \frac{I}{2\alpha\mu\sqrt{L_p}}\ln\left(1 + \frac{\sqrt{L_p}P^2\alpha}{4\pi^2\mu(1-\alpha)}\right). \tag{6}$$

In some cases, depending on the value of the magnetic field B and the particle luminosity L_P, the wind stage largely dominates the total age (i.e., the time after the wind dominance is larger than the initial dipole-driven stage), and τ_w can be considered the age of the object, in particular when its value is in the $\sim$ Gyr range (Figure 2). The evolution (if any) of the magnetic field, along with the speed at which the neutron star spins, play a crucial role in shaping the electromagnetic surroundings and, in turn, the emissions we observe. By comparing the emissions across a range of wavelengths from different neutron stars, including the long-period sources, we can get a better understanding of the varying magnetospheric conditions found in these intriguing objects, a long-due task in the field.

4.2. *The wind as a source of pulsed emission*

As already mentioned, although the model above describes how a particle wind generates enough torque to potentially explain the spin-down of a pulsar, it makes no claims about how this wind would be observed, if it were to be observable at all. The wind itself does not require the magnetic and rotational axes to be misaligned in order for this torque to exist, but the assumption that the observed pulsed emission has the wind as a source does. Although we will not attempt to directly model the wind emission here, it is worth considering how the pulsar wind changes when we do not assume an axisymmetric system, and whether possible emission mechanisms are minimally compatible with observations.

Emission from the wind zone of pulsars is notoriously difficult to characterize. As discussed by Kirk *et al.* (2009), little emission is expected in cold, ideal magnetohydrodynamic (MHD), axisymmetric, wind models, wherein the field adopts a split-monopole configuration, particle trajectories are rectilinear and thermal motion is negligible in relation to the expansion suffered by the plasma in the wind region (by a factor of 10^{18} in volume), such that neither synchrotron radiation nor bremsstrahlung can be expected. Only inverse Compton scattering of photons, from sources such as the CMB or a binary companion, provide a possible, weak, source of gamma-rays.

More promising for the objects here discussed is the *striped wind* model (Coroniti, 1990; Michel, 1971). By abandoning the ideal MHD regime, Michel (1971) had already shown that the plasma flow in the equatorial region should evolve into successive bands of cold plasma, each dominated by magnetic field lines originating in alternating poles, separated by a narrow current sheet of hot plasma. Instead of following the equatorial plane, however, this current sheet oscillates around it, with amplitude growing linearly with distance, such that, for a narrow enough belt around the equator, flow resembles expanding spherical current sheets, along the boundaries of which magnetic energy dissipates into particle energy if magnetic reconnection occurs, providing a possible source of synchrotron radiation (Kirk *et al.*, 2002).

Due to relativistic beaming, photons originating at any point on the shell will only reach the observer if emitted within a narrow cone of opening $\propto \gamma^{-1}$, γ the Lorentz factor of the wind, in relation to the line-of-sight, leading the emission to be point-like and possibly indistinguishable from the pulsar itself at current resolutions. Additionally, because the periodicity of the flow is coupled to the rotation of the pulsar — in particular, the inversion of the field direction at a fixed radius happens with the rotation period — the emission is expected to be pulsed and with a similar profile to the modulation of the flow, if not the same period, as long as certain geometrical constraints are fulfilled.

The resulting profile by Kirk *et al.* (2002) shows a pulse and an interpulse, the interval between which depends on the viewing angle in relation to the equatorial plane, where they are symmetrical. Among other idealizations, they highlight that their model assumes emission happening at the exact position of the current sheet, while in reality emission might happen within an extended region, making the pulse more complicated; in this case, their result serves as a Green's function for modeling the entire pulse, which might display structure.

The point-like and pulsed nature of striped wind emission is encouraging for attempting to study the long-period sources under this model. First, the point-like nature allows both dipole and wind emission to be *superposed* in the observed signal, as needs to be the case for application of the mixed model. Second, its nature as a periodic signal with possible pulse-substructure leaves room for both the unexpectedly irregular profiles of GPM J183910 (Hurley-Walter *et al.*, 2023), PSR J0901-4046 (Caleb *et al.*, 2022) and 1E 161348-5055 2001 data (De Luca *et al.*, 2006); and the more regular profile of GLEAM J1627 (Hurley-Walter *et al.*, 2022), PSR J0250+5854 (Tan *et al.*, 2018) and 1E 161348-5055 2005 data (De Luca *et al.*, 2006). Lastly, the coupling between pulsar rotation and the wind allows us to continue to employ the measured pulse period as the rotation period.

This last point is of particular importance, because for non-dipole emission mechanisms, the spin and pulse periods need not be the same in general. Given that the wind model still requires the spin period to be known, it is important to check that it is still reasonable to assume the two periods to be equal even in this case; if the radio emission has the wind as a source, than we can say that this is the case for these objects. The similarity between high-energy and radio pulses in the case of the Crab pulsar has been suggested to point towards a common origin in the past (Kirk *et al.*, 2002), although this remains speculative.

If the radio pulses are not originated in the wind, then there remains a question of what mechanism is responsible for their long duration and irregular profiles. Beyond pulse duration and profile, radio dipole emission has traditionally been thought to rely on pair production, which must compete with photon splitting under intense magnetic fields; this has been proposed as the mechanism behind many high magnetic field pulsars observed in the X-rays being radio-quiet (Baring and Harding, 1998). An analogous effect has been shown to emerge from the split-monopole configuration of the field (Hu *et al.*, 2022), which is adopted in the aligned rotator model, but which would otherwise preserve the possibility of striped wind emission from an oblique rotator.

Without a clear picture of the whole emission mechanism, there is no obvious way to connect the spin period and the pulse period. In the scenario suggested here, where the wind features in the spindown of the pulsar, there are many reasons to believe the radio emission is not originated by the dipole mechanism. And while it is hard to associate the striped wind with radio emission, the former has been shown to yield a variety of regular profiles which could possibly reproduce the profile observed in the long-period sources. We thus tentatively retain the assumption that the two periods are the same in order to move forward with the analysis.

4.3. *Status of the long-period sources*

We collected in Tables 2 and 3 several relevant quantities needed to address the evolution of long-period sources in the P–$\dot{P}$ plane. Note that we have focused on the upper part of the diagram (roughly defined by $\dot{P} > 10^{-15}\,\mathrm{s\,s^{-1}}$, because the sources with much lower $\dot{P}$ fall in the region of pulsating white dwarfs (Marsh *et al.*, 2016; Pelisoli *et al.*, 2023) and may be related to them. In addition, unless a hypothesis about a substantial decay of the magnetic field is made, for which there is no firm evidence, the trajectories can not "bend" downwards to link the known pulsar-magnetar locus with the former. It is quite realistic to believe that the lower part of the diagram is occupied by pulsating white dwarfs (Coelho and Malheiro, 2014; Rea *et al.*, 2024), i.e., systems similar to AR Sco and J1912-4410.

Table 2. Observed and inferred quantities for the five selected sources located in the upper portion of the P–$\dot{P}$ diagram.

Object	P (s)	$\dot{P}$ $(\mathrm{s\,s^{-1}})$	$\log B$ (G)
GPM J1839-10[a]	1860	$< 3.6 \times 10^{-13}$	11.74
GLEAM J1627[b]	1090	$< 6 \times 10^{-10}$	14.84
J0901-4046[a]	75.6	2.2×10^{-13}	14.17
J0250+5854[a]	23.54	2.7×10^{-14}	12.51
1E 161348-505[a]	24012	$< 1.6 \times 10^{-9}$	15

Notes: Upper limits to the derivative were employed to estimate the magnetic fields B, although some of their values could be significantly lower.
[a]Hamil (2015) and Shibata *et al.* (2016).
[b]Olausen *et al.* (2010) and Shibata *et al.* (2016).

Table 3. Observed and inferred quantities for the five selected sources located in the upper portion of the P–$\dot{P}$ diagram.

Object	$\log L_X$ $(\mathrm{erg\,s^{-1}})$	$\log L_p^{\min}$ $(\mathrm{erg\,s^{-1}})$	$\log E_{rot}$ (erg)	$\log E_{mag}$ (erg)
GPM J1839-10[a]	$\leq$33.30	26.00	40	42
GLEAM J1627[b]	$\leq$27.78	29.92	40	46
J0901-4046[a]	$\leq$30.51	29.97	38	45
J0250+5854[a]	$\leq$33.3	30.57	42	43
1E 161348-505[a]	$\leq$34	26.31	43	47

Notes: A value $L_P = 10^{34}$ erg s^{-1} was considered as an indicative value for GLEAM J1627. The rotational energy is found simply as $I\Omega^2/2$, and the magnetic energy as $\frac{B^2}{8\pi}V$, V the volume of a spherical shell of external radius 10^6cm and thickness 10^5cm. $L_p^{\min}$ is the minimum particle luminosity for each object to be in a wind-dominated regime, found as described at the end of section.
[a]Hamil (2015) and Shibata *et al.* (2016).
[b]Olausen *et al.* (2010) and Shibata *et al.* (2016).

The trajectories of the long-period sources reconstructed from the wind hypothesis are shown in Figure 1, for a fiducial $\alpha = 0.1$. As stated, there is an intrinsic uncertainty in the particle luminosity L_P and the $\dot{P}$ involved in these numbers. This picture leads to identifying the long-period sources with old pulsars/magnetars in which the wind

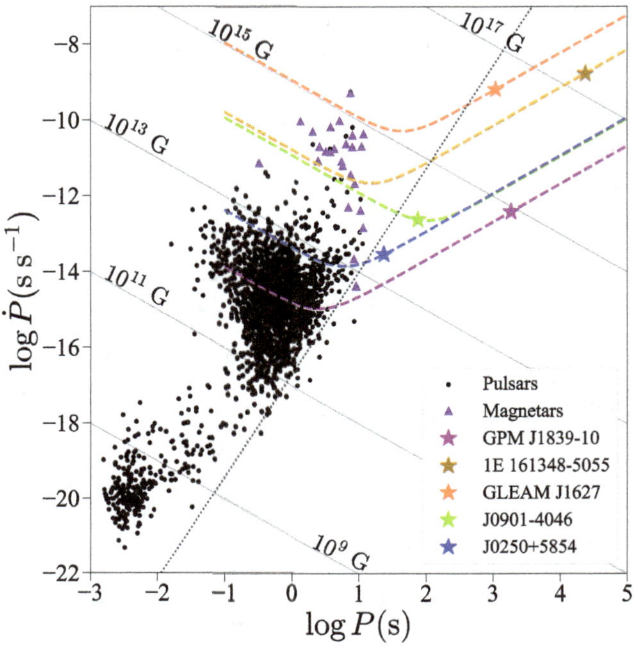

Fig. 1. The P–$\dot{P}$ diagram, showing the position of ordinary pulsars (black circles) and magnetars (purple triangles), as well as the lines of constant magnetic field (solid gray lines) and the death line (black dashed line, from (Chen and Ruderman, 1993)) in the traditional pure dipole model. The long-period sources are shown as the colored stars, and their possible trajectories in the dipole-wind mixed model shown as the colored dashed lines; the valley separates the dipole- (short periods) and wind- (long periods) dominated epochs. We highlight that the rising portion of the long-period source dipole-wind trajectories do not imply a rising magnetic field as they would in the pure dipole model; their fields are always fixed to the values shown in Table 2, estimated using Eq. (5).

has gained an important role. The age vs. magnetic field relation is now modified, an example of this behavior is given in Figure 2. The most problematic object is GLEAM J1627, which should be the descendant of a very high field object unless its $\dot{P}$ value is far from the upper limit and/or its L_P differs substantially from the naive estimation. In fact, this consideration is extensive to the whole group. Note that the comparison between the two last columns of Table 3 allows us to state that the long-period sources are "magnetars", if the wind picture holds, just because they are powered by magnetic energy irrespective of the value of the inferred magnetic field.

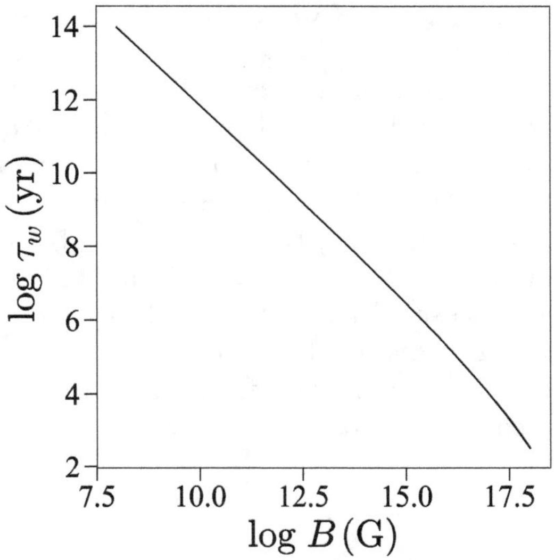

Fig. 2. Calculated age for a pulsar with $P = 2000\,\text{s}$, $\dot{P} = 10^{-12}\,\text{s}\,\text{s}^{-1}$ and $L_P = 10^{30}\,\text{erg}\,\text{s}^{-1}$, values close to the ones of the slowest objects in Table 2. The values $\alpha = 0.5$, $r = 10^6$ cm e $I = 10^{45}\,\text{g}\,\text{cm}^2$ were adopted. Note that the objects with $B \sim 10^{12}$ G are possibly as old as the galaxy.

These results have been obtained using indicative L_P values, and this issue deserves clarification. We have seen above that the particle luminosities for the ordinary pulsars in Table 1 are higher than the detected X-ray luminosities L_X. In addition, the latter is considered to arise mainly from synchrotron emission, not by the effects of the wind. Kirk *et al.* (2009) state that the wind "glow" may not be significant if there is not enough matter around and/or the photon field is weak and inverse Compton is inefficient. Therefore, it is not contradictory that the X-ray emission from the long-period sources is undetected (Table 3), since not only the synchrotron effect is expected to be feeble, but also that enough inverse Compton is not guaranteed. The intensity of the particle luminosity L_P can be loosely estimated at best, since a complete and consistent characterization of the winds is not available and many features (i.e., the Lorentz factor of the particles) are poorly known.

Given that the mixed model results in a transition period L_{tran} (Eq. (3)) between dipole- and wind-dominated epochs for each source in the $P - \dot{P}$ diagram, it is possible to derive one first constraint over the particle luminosity if we impose that the objects be wind-dominated. By fixing P and $\dot{P}$, varying L_p will result in different tracks, for each of which the object will be in a different position relative to the transition between the two regimes. We are thus able to find the L_p for which the object's current period would also be its $P_{\rm tran}$; this L_p is the minimum particle luminosity for the object to enter the wind regime, reported as $L_p^{\rm min}$ in Table 3. Although we do not quantify it, we also point out that the stringent X-ray luminosity limits also simultaneously limit L_p and α; the greater the energy carried by the wind, or the more often it is active, the greater would be the expected heating of the surface of the object, which would be reflected in the X-ray luminosity.

5. Discussion and Conclusions

We have argued in this work that particle winds, known to be present in ordinary pulsars as a part of the general picture, but rarely considered important, may be the most relevant driving factor to understand the long-period compact star radio sources.

As it stands, it is difficult to state something definitive, but we believe that the first exploration offers a general consistency, and of course a few unexplained features. In addition to theoretical work addressing the specific physics of the pulse generation, observations of the time profile of the latter is a prime element to pin down the features involved. In our separation of the "upper" and the "lower" sources a deep issue (namely, the decay of the magnetic field) pops up again, because the shape of the trajectories disfavors a strong decay. After many studies, it is clear that quite old sources (Roberts, 2012) still show moderately strong fields $B \sim 10^9$ G, and therefore the decay is, if anything, less extreme than naively expected (Gonzalez and Reisenegger, 2010), at least for isolated sources. A general remark is that the long-period sources are all "magnetars", in the sense that they are powered by magnetic energy and not by rotation, independently of the actual value of the magnetic field which may be comparable to the ones found in ordinary radio pulsars. On the other hand, some observations may be seen as arguing at face value against the suggested wind picture: the central compact object in RCW 103,

1E 161348-5055, has been known for decades, but only recently identified as a magnetar (De Luca *et al.*, 2006). This central source is also radio quiet, but very X-ray bright, which is not easily understood within the presented wind picture.

It is important to remark that other sources have recently contributed to challenging traditional pulsar pictures. The recent observations of novel TeV emission from the Vela pulsar with H.E.S.S. (VelaHESS, 2023), for example, join the growing set of new data that challenges the traditional pulsar picture, with the pulsar wind model also being a candidate for explaining the emission in that case. It is also clear that the important issue of pulsating white dwarf systems and their relation with the neutron star group, emphasized by Coelho and Malheiro (2014) and recently revived by Rea *et al.* (2024), is crucial to construct an overall picture of all sources, as well as the efforts to identify new candidates and the search of counterparts. Unraveling the evolutionary path of long-period compact radio sources compared to other neutron star groups calls for an integral approach, bringing together observational astronomy, theoretical modeling, and numerical simulations. This effort is not only exciting for shedding light on the mysterious nature of these sources, but also for reshaping our understanding of neutron star astrophysics, pushing what we know into new territories. In summary, these are big news in the field and will keep us busy for a long time.

Acknowledgments

This research was funded by FAPESP Agency (São Paulo State) grant number 2020/08518-2. L.M.S. received funding from the CNPq (Brazil), grant number 140794/2021-2. L.S.R was funded by Capes Agency (Brazil). J.E.H. acknowledges a long-term Research Scholarship by the CNPq (Brazil) supporting this and other scientific works.

Bibliography

Abhishek, M. N., Tanushree, N., Hegde, G., and Konar, S. (2022). Radio pulsar sub-populations (II): The mysterious RRATs. *J. Astrophys. and Astron.* **43**, 75.

Allen, M. P. and Horvath, J. E. (1997). Glitches, torque evolution and the dynamics of young pulsars. *MNRAS* **287**, 615–621.

Allen, M. P. and Horvath, J. E. (2000). Pulsar spin-down with both magnetic dipole and relativistic wind brakings. *KITP Conference: Spin, Magnetism and Cooling of Young Neutron Stars*, 29.

Alpar M. A. and Pines D. (1993). *Isolated Pulsars.* Cambridge University Press, 17.

Álvarez, C. and Carramiñana, A. (2004). Monopolar pulsar spin-down. *A&A* **414**, 651–658.

Baring, M. G. and Harding, A. K. (1998). Radio-quiet pulsars with ultra-strong magnetic fields. *ApJ* **507**, L55.

Beniamini, P., Wadiasingh, Z., and Metzger, B. D. (2020). Periodicity in recurrent fast radio bursts and the origin of ultralong period magnetars. *MNRAS* **496**, 3390–3401.

Beniamini, P., Wadiasingh, Z. *et al.* (2023). Evidence for an abundant old population of Galactic ultra-long period magnetars and implications for fast radio bursts. *MNRAS* **520**, 1872–1894.

Beniamini, P., Wadiasingh, Z. *et al.* (2023). Evidence for an abundant old population of Galactic ultra-long period magnetars and implications for fast radio bursts. *MNRAS* **520**, 1872–1894.

Bisnovatyi-Kogan, G. S. (1992). The neutron star population in the Galaxy. *Symposium-International Astronomical Union* **149**, 379–382.

Bisnovatyi-Kogan, G. S. and Komberg, B. V. (1974). Pulsars and close binary systems. *SVA* **18**, 217.

Buckley, D. A. H., Meintjes, P. J., Potter, S. B., Marsh, T. R., and Gnsicke, B. T. (2017). Polarimetric evidence of a white dwarf pulsar in the binary system AR Scorpii. *Nat. Astron.* **1**, 0029.

Caleb, M., Heywood, I., Rajwade, K. *et al.* (2022). Discovery of a radio-emitting neutron star with an ultra-long spin period of 76 s. *Nat. Astron.* **6**, 828–836.

Caraveo, P. A., Bignami, G. F., Mignani, R., and Taff, L. G. (1996). Geminga in the Space Telescope Era. *A&A* **120**, 65–68.

Charles, P. A. and Seward, F. D. (1995). *Exploring the X-ray Universe.* CUP Archive.

Chen, K. and Ruderman, M. (1993). Pulsar death lines and death valley. *The Astrophys. J.* **402**, 264–270.

Coelho, J. G. and Malheiro, M. (2014). Magnetic dipole moment of soft gamma-ray repeaters and anomalous X-ray pulsars described as massive and magnetic white dwarfs. *PASJ* **66**, 14.

Coroniti, F. V. (1990). Magnetically striped relativistic magnetohydrodynamic winds-The Crab Nebula revisited. *The ApJ* **349**, 538–545.

DÁì, A. *et al.* (2016). Evidence for the magnetar nature of 1E 1613485055 in RCW 103. *MNRAS* **463**, 2394–2404.

De Luca, A., Caraveo, P. A., Mereghetti, S., Tiengo, A., and Bignami, G. F. (2006). A long-period, violently variable X-ray source in a young supernova remnant. *Sci.* **313**, 814–817.

De Luca, A. (2017). Central compact objects in supernova remnants. *J. Phys. Conf. Ser.* **932**, 012006.

Dirson, L., Pétri, J., and Mitra, D. (2022). The Galactic population of canonical pulsars. *A&A* **667**, A82.

Ertan, Ü., Çalşkan, Ş., Benli, O., and Alpar, M. A. (2014). Long-term evolution of dim isolated neutron stars. *MNRAS* **444**, 1559–1565.

Esposito, P., Rea, N., and Israel, G. L. (2021). Magnetars: a short review and some sparse considerations. *Timing Neutron Stars: Pulsations, Oscillations and Explosions*, 97–142.

Gençali, A. A., Ertan, Ü., and Alpar, M. A. (2022). Evolution of the long-period pulsar GLEAM-X J162759. 5–523504.3. *MNRASL* **513**, L68–L71.

Gençali, A. A., Ertan, Ü., and Alpar, M. A. (2023). Evolution of the long-period pulsar PSR J0901 4046. *MNRASL* **520**, L11–L15.

Giacconi, R., Gursky, H., Kellogg, E., Schreier, E., and Tananbaum, H. (1971). Discovery of periodic X-ray pulsations in Centaurus X-3 from Uhuru. *ApJ* **167**, L67.

Gold, T. (1969). Rotating neutron stars and the nature of pulsars. *Nature* **221**, 25–27.

Goldreich, P. and Julian, W. H. (1969). Pulsar electrodynamics. *ApJ* **157**, 869.

Gonzalez, D. and Reisenegger, A. (2010). Internal heating of old neutron stars: contrasting different mechanisms. *A&A* **522**, A16.

Jonker, P. G. and van der Klis, M. (2001). Discovery of an X-ray pulsar in the low-mass X-ray binary 2A 1822371. *ApJ* **553**, L43.

Hamil, O., *et al.* (2015). Braking index of isolated pulsars. *Phys. Rev. D* **91**, 063007.

Harding, A. K., Contopoulos, I., and Kazanas, D. (1999). Magnetar spin-down. *ApJ* **525**, L125.

Hewish, A. and Bell, S. J. *et al.* (1979). Observation of a rapidly pulsating radio source. In *A Source Book in Astronomy and Astrophysics* **1900–1975**, 498–504. Harvard University Press.

Ho, W. C. (2012). Central compact objects and their magnetic fields. *Proc. IAU* **8**, 101–106.

Ho, W. C., Zhao, Y., Heinke, C. O. *et al.* (2021). X-ray bounds on cooling, composition, and magnetic field of the Cassiopeia A neutron star and young central compact objects. *MNRAS* **506**, 5015–5029.

Hu, K., Baring, M. G., Harding, A. K., and Wadiasingh, Z. (2022). High-energy photon opacity in the twisted magnetospheres of magnetars. *ApJ* **940**, 91.

Hurley-Walker, N., Zhang, X., Bahramian, A. *et al.* (2022). A radio transient with unusually slow periodic emission. *Nature* **601**, 526–530.

Hurley-Walker, N., Rea, *et al.* (2023). A long-period radio transient active for three decades. *Nature* **619**, 487–490.

Kaspi, Victoria M. (2010). Grand unification of neutron stars. *PNAS* **107.16**, 7147–7152.

Kirk, J. G., Skjraasen, O., and Gallant, Y. A. (2002). Pulsed radiation from neutron star winds. *A&A* **388**, L29–L32.

Kirk, J. G., Lyubarsky, Y., and Petri, J. (2009). The theory of pulsar winds and nebulae. In *Neutron Stars and Pulsars*, 421–450.

Kramer, M., Lyne, A. G., O'Brien, J. T., Jordan, C. A. and Lorimer, D. R. (2006). A periodically active pulsar giving insight into magnetospheric physics. *Sci.* **312**, 549–551.

Lyne, A. G., Manchester, R. N., and Taylor, J. H. (1985). The galactic population of pulsars. *MNRAS* **213**, 613–639.

Lyne, A. G., Jordan, C. A., Graham-Smith, F. *et al.* (2015). 45 years of rotation of the Crab pulsar. *MNRAS* **446**, 857–864.

Manchester, R. N. and Taylor, J. H. (1977). *Pulsars.*

Marsh, T. R., Gnsicke, B. T. *et al.* (2016). A radio-pulsing white dwarf binary star. *Nature* **537**, 374–377.

Mayer, M. G. and Becker, W. (2021). A kinematic study of central compact objects and their host supernova remnants. *A&A* **651**, A40.

McLaughlin, M. A., Lyne, A. G., Lorimer, D. R. *et al.* (2006). Transient radio bursts from rotating neutron stars. *Nature* **439**, 817–820.

Michel, F. C. (1969). Pulsar Stellar Wind Torques: NP 0532. In *Bull. Am. Astron. Soc.* **1**, 354.

Michel, F. C. (1971). Coherent neutral sheet radiation from pulsars. *Comments on Astrophysics and Space Physics* **3**, 80.

Michel, F. C. (1973). Rotating magnetospheres: an exact 3-D solution. *The Astrophysical Journal* **180**, L133.

Michel, F. C. (1991). *Theory of Neutron Star Magnetospheres.* University of Chicago Press.

Olausen, S. A., Kaspi, V. M., Lyne, A. G., and Kramer, M. (2010). Xmm-newton x-ray observation of the high-magnetic-field radio pulsar psr j17343333. *ApJ* **725**, 985.

Pacini, F. (1968). Rotating neutron stars, pulsars and supernova remnants. *Nature* **219**, 145–146.

Patruno, A., Wette, K., and Messenger, C. (2018). A Deep pulse search in 11 low mass X-ray binaries. *ApJ* **859**, 112.

Pelisoli, I., Marsh, T. R., Buckley, D. A. *et al.* (2023). A 5.3-min-period pulsing white dwarf in a binary detected from radio to X-rays. *Nat. Astron.*, 1–12.

Pires, A. M. *et al.* (2022). XMM-Newton and SRG/eROSITA observations of the isolated neutron star candidate 4XMM J022141.5 735632. *A&A* **666**, A148.

Popov, S. B. (2023). The zoo of isolated neutron stars. *Universe* **9**, 273.

Potekhin, A. Y., Pons, J. A., and Page, D. (2015). Neutron starscooling and transport. *Space Sci. Rev.* **191**, 239–291.

Pszota, G. and Kovács, E. (2023). X-ray spectroscopic study of low-mass x-ray binaries: A review of recent progress via the example of GX 339-4. *Universe* **9**, 404.

Razzano, M., Fiori, A., and Parkinson, P. S. *et al.* (2023). Multiwavelength observations of PSR J2021+ 4026 across a mode change reveal a phase shift in its X-ray emission. *A&A* **676**, A91.

Rea, N., Hurley-Walker *et al.* (2024). Long-period radio pulsars: Population study in the neutron star and white dwarf rotating dipole scenarios. *ApJ* **961**, 214.

Rigoselli, M., Mereghetti, S., Taverna, R., Turolla, R., and De Grandis, D. (2021). Strongly pulsed thermal X-rays from a single extended hot spot on PSR J2021+ 4026. *A&A* **646**, A117.

Roberts, M. S. (2012). Surrounded by spiders! New black widows and redbacks in the Galactic field. *Proc. IAU* **8**, 127–132.

Ronchi, M., Rea, N., Graber, V., and Hurley-Walker, N. (2022). Long-period pulsars as possible outcomes of supernova fallback accretion. *ApJ* **934**, 184.

Rutledge, R. E., Fox, D. B., and Shevchuk, A. H. (2008). Discovery of an isolated compact object at high galactic latitude. *ApJ* **672**, 1137.

Shibata, S., Watanabe, E., Yatsu, Y., Enoto, T., and Bamba, A. (2016). X-ray and rotational luminosity correlation and magnetic heating of radio pulsars. *ApJ* **833**, 59.

Surnis, M. P., Rajwade, K. M., Stappers, B. W. *et al.* (2023). Discovery of an extremely intermittent periodic radio source. *MNRASL* **526**, L143–L148.

Suvorov, A. G. and Melatos, A. (2023). Evolutionary implications of a magnetar interpretation for GLEAM-X J162759. 5–523504.3. *MNRAS* **520**, 1590–1600.

Tan, C. M., Bassa, C. G. *et al.* (2018). LOFAR discovery of a 23.5 s radio pulsar. *ApJ* **866**, 54.

Tauris, T. M., Sanyal, D., Yoon, S. C., and Langer, N. (2013). Evolution towards and beyond accretion-induced collapse of massive white dwarfs and formation of millisecond pulsars. *A&A* **558**, A39.

Tong, H. (2023). On the nature of long period radio pulsar GPM J1839-10: Death line and pulse width. *RAA* **23**, 125018.

Turolla, R. (2009). Isolated neutron stars: The challenge of simplicity. Neutron stars and pulsars, Springer Berlin Heidelberg., 141–163.

Turolla, R. and Esposito, P. (2013). Low-magnetic-field magnetars. *Int. J. Mod. Phys. D* **22**, 1330024.

Usov, V. V. (1975). Wave zone structure of NP 0532 and infrared radiation excess of crab nebula. *Astrophys. Space Sci.* **32**, 375–377.

Collaboration, T. H. *et al.* (2023). Discovery of a radiation component from the vela pulsar reaching 20 teraelectronvolts. arXiv preprint arXiv:2310.06181.

Wadiasingh, Z., Beniamini *et al.* (2020). The fast radio burst luminosity function and death line in the low-twist magnetar model. *ApJ* **891**, 82.

Walter, F. M., Wolk, S. J., and Neuhuser, R. (1996). Discovery of a nearby isolated neutron star. *Nature* **379**, 233–235.

Wang, H. H., Takata, J., Hu, C. P., Lin, L. C. C., and Zhao, J. (2018). X-ray study of variable gamma-ray pulsar PSR J2021+ 4026. *ApJ* **56**, 98.

Wang, B. and Liu, D. (2020). The formation of neutron star systems through accretion-induced collapse in white-dwarf binaries. *RAA* **20**, 135.

Wang, B., Liu, D., and Chen, H. (2022). Formation of millisecond pulsars with long orbital periods by accretion-induced collapse of white dwarfs. *MNRAS* **510**, 6011–6021.

Wilson, D. B. and Rees, M. J. (1978). Induced compton scattering in pulsar winds. *MNRAS* **185**, 297–304.

Yoneyama, T., Hayashida, K., Nakajima, H., and Matsumoto, H. (2019). Unification of strongly magnetized neutron stars with regard to X-ray emission from hot spots. *Astron. Nachr.* **340**, 221–225.

Younes, G., Baring, M. G., Harding, A. K. *et al.* (2023). Magnetar spin-down glitch clearing the way for FRB-like bursts and a pulsed radio episode. *Nat. Astron.* **7**, 339–350.

Part 2

FAST: The Five Hundred Meter Aperture Spherical Telescope; HXMT: The Hard X-Ray Modulation Telescope; TianMa: The Tian Ma 65 m Radio Telescope; LHAASO: Large High Altitude Air Shower Observatory

https://doi.org/10.1142/9789819800643_0005

Chapter 5

The Fundamental Performance of FAST with 19-Beam Receiver at L Band

Peng Jiang

National Astronomical Observatories, Chinese Academy of Sciences, Beijing 100101, China

CAS Key Laboratory of FAST, National Astronomical Observatories, Chinese Academy of Sciences, Beijing 100101, China

pjiang@nao.cas.cn

The Five-hundred-meter Aperture Spherical radio Telescope (FAST) is a Chinese mega-science project aimed to build the largest single dish and most sensitive radio telescope in the world. The illuminated aperture of the FAST is 300 m, and its overall performance and sensitivity are several times higher than those of other existing radio telescopes. On September 25, 2016, the FAST was completed with the main structure of the telescope installed. It is expected to remain the global leading radio telescope for the next 20 years. After several years of operation, FAST has acquired abundant scientific achievements, and more than one hundred high-quality papers have been published based on the FAST data, which have produced far-reaching scientific achievements in astronomy.

Keywords: Radio telescope, FAST, 19-beam receiver

Contents

1. Introduction

In the 1990s, China was lagging far behind in the field of radio astronomy. The largest radio telescope in China had an aperture of 25 m, and to build a radio telescope with an aperture of 500 m became an almost impossible miracle. In order to accomplish this miracle, Nan Rendong devoted himself to the construction of a large radio telescope Nan *et al.* (2011). In July 2007, the project of the Five-hundred-meter Aperture Spherical radio Telescope (FAST) was launched. On September 25, 2016, FAST was inaugurated. FAST has become the largest single-dish radio telescope with the highest sensitivity in the world.

The FAST a diameter of 500 m passed its national acceptance on January 11, 2020. It operates at frequencies ranging from 70 to 3000 MHz. It is the most sensitive single dish telescope in this frequency range.

The FAST is a huge and complex equipment with many influencing factors. It comprises 2225 actuators and a cable net, forming a complex coupling active reflector system. A 30-ton feed cabin, which is driven by six cables, is about 140 m above the reflector during operation. Till now, several observation modes, such as tracking, drift scanning, and basketweave scanning, have been realized, marking the completion of the functional commissioning. Now, the sensitivity of the FAST has reached ~2000 m^2K^{-1}, with a system noise temperature below 20 K (19-beam 1.05–1.45 GHz receiver), and the pointing accuracy of the feed receivers has reached about 16″.

FAST commissioning began when construction was completed on September 25, 2016. The details of FAST commissioning have been introduced in Jiang *et al.* (2019, 2020). A lot of observations, especially pulsar-related observations were performed during the commissioning phase. A series of preliminary results including the first detection of a pulsar by FAST have been published in the FAST issue of the journal Science China Physics, Mechanics & Astronomy.

The 19-beam receiver covering 1.05–1.45 GHz was used for most of the observations. The electronics gain fluctuation of the system is smaller than 1% over 3.5 h, enabling enough stability for observations. Pointing accuracy, aperture efficiency, and system temperature are three key parameters of FAST. The measured standard deviation of pointing accuracy is 7.9″, which satisfies the initial design of FAST. When the zenith angle is smaller than 26.4°, the aperture efficiency and system temperature around 1.4 GHz are ~0.63 and less than 24 K for the central beam, respectively. The measured values of these two parameters are better than the designed values of 0.6 and 25 K, respectively.

2. Design Scheme and Its Function Realization of FAST

2.1. *Design scheme*

The FAST is essentially a paraboloid antenna. With an innovative active reflector system and rigid-flexible hybrid lightweight feed support system, the FAST creates a new model for the construction of giant radio telescopes. The specific implementation plan is as follows.

2.1.1. *Active reflector system*

According to the geometric optical principle of the FAST, as shown in Figure 1a, the supporting structure of the reflector system should be capable of forming a parabolic surface from a spherical surface.

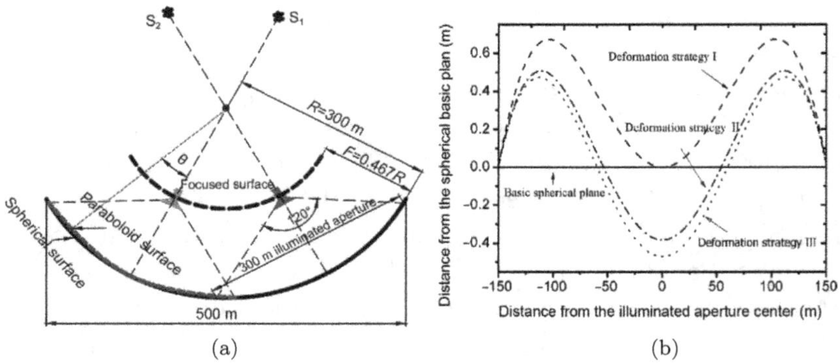

Fig. 1. Deformation distance analysis of spherical reflector to paraboloid. (a) Geometric optical principle of the FAST and (b) deviation from the spherical surface to paraboloid within a 300 m aperture [9].

As shown in Figure 1(b), when the focal ratio is in the range of 0.46–0.47, the maximum deviation from the spherical surface to the paraboloid within a 300 m aperture is less than 1 m. Inspired by the support structure of the reflector of the Arecibo telescope, the FAST designers put forward the idea of using flexible cables to support the active reflector. The FAST cable net comprises 6670 steel cables and 2225 cross nodes. The lengths of the cables range from 10.5 to 12.5 m, the total weight of the net is about 1300 t, and the main cables have 16 cross-sectional areas ranging from 280 to 1319 mm^2. As shown in Figure 2, the outer edge of the cable net is suspended from a 500 m diameter steel ring beam. The cross nodes of the cable net are used as control points. Each of them is connected to an actuator by a down-tied cable. By controlling the actuator using feedback from the measurement and control system, the positions of the cross nodes can be adjusted to form a 300-m illuminated aperture. This illuminated aperture moves along the spherical surface according to the zenith angle (ZA) of the target objects (see S1 and S2 in Figure 1(a)). The spherical reflector panel with a 500 m aperture is divided into 4300 triangular units and 150 quadrilateral edge units for a total of 4450 blocks. The side length of each triangular unit is about 11 m. The cable net acts as the supporting structure of the reflector units. The radius of curvature of the reflector units is about 315 m, and the working distance of the actuators is 1200 mm, which satisfies the requirement of the reflector to deform from a spherical surface to a paraboloid.

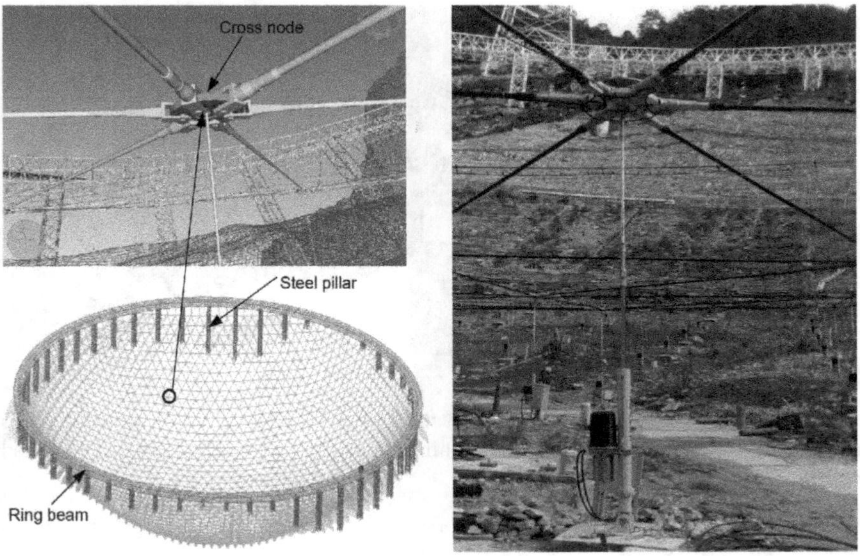

(a) (b)

Fig. 2. Active reflector system. (a) Ring beam and cable net model and (b) A cross node and its tie-down cable with actuator.

2.1.2. *Rigid-flexible hybrid lightweight feed support system*

With the reflector forming a paraboloid, the electromagnetic waves emitted from the celestial bodies can be converged onto the focal point. Maintaining the receiver at a correct position and orientation is still challenging. As shown in Figure 3(a), the FAST innovatively uses six cables, each of which is about 600 m long and weighs 7 t. These cables parallelly drive the feed cabin to move over a range of 206 m, at a height of 140 m above the reflector [12–16]. The cables and feed cabin are connected using three nodes. Each node uses a group of anchors, which greatly reduces the spin of the cabin. The cables have a negligible shielding effect on the reflector and are of great benefit to the further enhancement of the sensitivity of the telescope. At the same time, standing waves and other problems are also expected to be improved to achieve the broadband and full polarization function of the telescope. However, it was found that the cable-driven parallel robot can control the feed receivers to their required position but cannot control them to their required orientation at the same

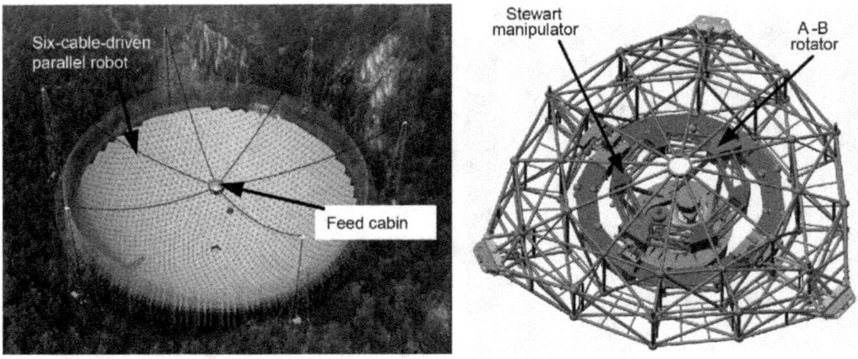

(a) (b)

Fig. 3. Design of the feed support system of the FAST. (a) Main mechanisms of the feed support system and (b) main structure and mechanisms of the feed cabin.

time. Therefore, an A–B rotator, as shown in Figure 3(b), which is a universal axis structure, is used to adjust the orientation of the feed receivers to the direction of the celestial source. Through a simulation analysis of the long-span cable robot, it was found that the influence of wind disturbance was obvious. Therefore, a Stewart manipulator, as shown in Figure 3(b), with six degrees of freedom (DOF) was adopted to reduce wind-induced vibration and further adjust the position and orientation of the feed receivers to operate with high accuracy. The feed receivers are installed on the lower platform of the Stewart manipulator. To realize closed-loop control, four targets each on the feed cabin and Stewart manipulator were used. Real-time measurement data are used to realize high-precision, closed-loop control of the feed support system.

2.2. *Functional commissioning of the FAST*

The FAST telescope guarantees the secure operation of both the active reflector system and feed support system through the implementation of a real-time information system for the active reflector system and a hierarchical commissioning scheme for the feed support system. Finally, the safe operation of the two systems was realized, providing a guarantee for the performance commissioning of the telescope.

2.2.1. *Real-time information system of the active reflector system*

As the FAST starts operating, the shape of the cable net frequently switches between the 500-m spherical surface and the spherical surface with a partial 300-m paraboloid. This causes cyclic loading and deformation of the ring beam and reciprocating movement of the reflector elements and restraining parts of the FAST. Safety is always the basic problem of the active reflector system, including components such as the ring beam, cable net, reflector element, and accessory mechanism. Effective precautions are very important for preventing irreversible structural or mechanical damage caused by a malfunction or some unpredictable failure. The real-time information system of the active reflector system is established to give timely information on cable stress and provide notification. It is an indispensable prerequisite for the routine operation of the telescope.

The data interface subsystem is the communication module between the overall system and the telescope control system (TCS). It receives the observational trajectory, sensor measurement, and equipment malfunction information from the TCS, and sends the appropriate notification information back to the TCS.

To ensure the smooth progress of the simulation process, the data preprocessing subsystem further translates the received information into data, working conditions, and constraints that can be identified by the simulation model.

The data post-processing subsystem sorts and analyzes the simulation results to identify possible dangerous structural stresses, translates the information into notification information that can be identified by the TCS, and sends the notifications out in various ways to the system user interface to alert the operators.

The mechanical simulation model is the core of the real-time information system. It simulates all the key structures of the active reflector system, including the ring beam, lattice columns, cable net, reflector elements, and actuators. It is capable to complete deformation computations of the cable net owing to its fast computation ability (it outputs the resulting data once every 10–16 s). In comparison with the deformation time (at least 300 s) of the cable net from the spherical shape to the paraboloid, the computational speed is quick enough for an assessment system. Deformation of the active reflector system is complicated and influenced by many factors, such as temperature, actuator failure, side deviation of the cable nodes, and distortion of the tie-down cables. Some of them may even interact with one another. Fortunately, these factors and their interactions are easily simulated by finite element analysis; therefore, it is not necessary to analyze the contributions of individual factors. The remarkable feature of the simulation model is its real-time computation ability without establishing a database in advance and without considering the decoupling of miscellaneous aspects.

The real-time information system of the active reflector system worked well and achieved good results during the commissioning of the FAST. The system is very effective for the assessment of actuator failures and their impact. Additionally, the real-time information system can also judge whether there is danger caused by the paraboloid lying in the zone of large ZA, where the cable force is frequently found to be abnormal. These precautions remind telescope operators to remain vigilant to ensure the safe operation of the telescope.

2.2.2. *Functional commissioning of the feed support system*

The feed support system operates Yao *et al.* (2020) at more than 140 m above the reflector. Safety is the primary factor. Therefore, before performance commissioning of the feed support system, it was necessary to do some functional commissioning in steps, including the following:

(1) The commissioning process involved unloading the six cables from the reels and securing the driving cables to the ground using splints, ensuring that there was no load on the overall mechanism. Subsequently, the dynamic performance of the robot's mechanisms was tested and debugged. Once all the commissioning results align with the design specifications, the commissioning process is considered complete.
(2) The simulation commissioning of the six-cable-driven parallel robot was to verify the operation process and control instructions of the robot.

When the system started operation and the observation task was given, the control program calculated the driving cable length according to the current position and orientation of the feed cabin, and the results were sent to a programmable logic controller (PLC) through the process field net (PROFINET). Then, the results were given to the control computer by the user datagram protocol (UDP). The controlling computer sent all of the information to inversely solve the new position and orientation of the feed cabin. All of the simulation commissioning was done to ensure that the robot can perform routine, safe work.

(3) Combined commissioning of the feed cabin and six-cable-driven simulation model. A feed cabin is a large mechanism with a diameter of 13 m and a weight of 30 t. It is driven by six long-span cables. Before the combined operation with the robot, a semi-physical simulation model is adapted to debug the functions and performance of the feed cabin.

The semi-physical simulation model uses the main mechanism of the feed cabin and combines the robots simulation model. The semi-physical simulation model has three features: The actual A-B rotator and Stewart manipulator serve as the physical part. Simulation software simulates the coupling system of the cable and feed cabin under the condition of wind disturbance. Combined measurement means simulates the actual measurement accuracy.

The semi-physical simulation model used the simulation system of the feed support system to generate wind disturbance parameters and simulate the control process and control error of the robot. The simulation system also gave the theoretical position and orientation of the lower platform of the Stewart manipulator, which were sent to control the feed cabin. These control parameters and the actual motor parameters were returned to the simulation computer, and the final accuracy of the feed support system was obtained by simulation. The semi-physical simulation model used a set of terminal measurement equipment, which simulated the period and accuracy of the real measurement system of the FAST. Therefore, the semi-physical simulation model verified the combined operation of the feed support system and provide a safe and reliable operating basis for the subsequent performance commissioning.

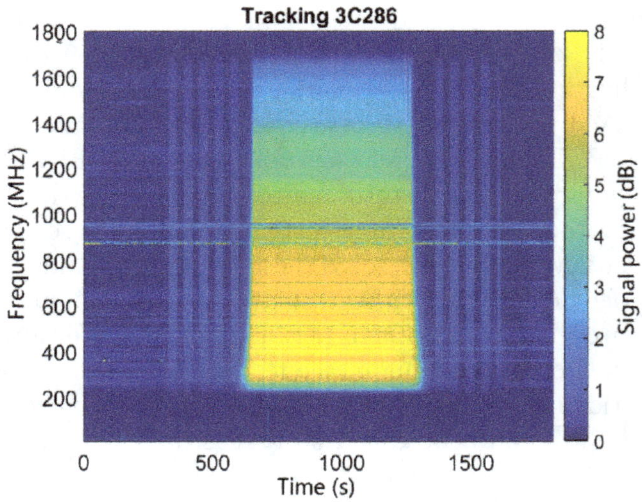

Fig. 4. Continuous trial observation of the radio source 3C286 on November 17, 2017.

2.2.3. *Functional commissioning results of FAST*

Based on the functional commissioning mentioned above, the FAST realizes the safe operation of both the active reflector system and the feed support system. On August 27, 2017, the FAST first continuously tracked a specific target source (3C286) twice; with each time of about 10 min. This indicates that the most serious safety risks have been eliminated. This is a milestone for the telescope. By the end of 2017, the telescope had realized a variety of observation modes, such as tracking, drift scanning, and basketweave scanning, which completed its functional commissioning. Figure 4 shows the observation results for November 17, 2017.

3. Properties of the 19-beam Receiver

3.1. *Measurement of the noise dipole*

The FAST 19-beam L-band Array contains a temperature-stabilized noise injection system. The noise is injected between the feed and the low noise amplifiers. The noise source is a single diode whose signal is split into each beam and polarization. The diode itself is always powered, while it can be switched on and off in the signal path using a solid-state switch. This was done in order to improve stability at very high switching speeds. It takes less than 1 μs for the noise power to stabilize once switched on or off. The noise diode has two adjustable power output modes, but are currently kept at approximately 1.1, and 12.5 K.

In order to test the performance and stability of the noise diode, we conducted a series of hot load measurements whereby the feed cabin was lowered to the ground and a foam absorber was placed directly under the feed so that it completely covered all of the beams. We then periodically measured the temperature of the absorber with a thermometer, while the noise diode was continuously switched on and off ("a winking CAL") with a period of 1.00663296 s.

It shows that the electronics gain fluctuations within the FAST 19-beam receiver/backend signal path are typically on the order of a few percent over timescales of a few minutes. This implies that science projects for whom flux calibration is important should fire the noise diode at least every few minutes in order to account for typical electronics gain fluctuations. The larger fluctuations are caused by RFI leaking into the signal path during the hot load measurements, and the instability in the low noise amplifiers.

By taking a median value of all 19 beams, we can derive an upper limit on the temperature fluctuations in the noise diode. While laboratory measurements of the noise diode itself yielded temperature fluctuations on the order of ~0.1% over several hours, we see that once placed within the full signal path we get an upper limit of ~1%, implying that it is likely impossible to obtain accurate flux calibration better than ~1%. Considering FAST's pointing accuracy, and beam size the best flux calibration for a point source is approximately ~2%, meaning that the noise diode is operating within our tolerance limits. This lower limit is estimated by looking at the shape of the beam response function at 1420 MHz within the pointing uncertainty limits. The effect is weaker at longer wavelengths where the pointing uncertainty is smaller compared to the beam size.

3.2. *Beam properties and pointing accuracy*

To measure the properties of FAST's 19-beam receiver, one method is to directly make mapping observations toward radio sources on the sky. The raster scan (or raster scanning) observation mode has been realized for FAST, which is used to map compact radio sources and measure the 19-beam properties. In the observations, a sky area centering on the target source and covering $\sim 37'(\text{RA}) \times 37'(\text{Dec})$ is mapped by a raster scan along the RA or Dec direction with the 19-beam receiver. An example of the telescope track for a raster scan along Dec is given in Figure 5. The scan velocity is $15'/\text{min}$. The separation between two sub-scans is $1'$, which satisfies Nyquist sampling. A total of ~ 100 min is needed to acquire such a map.

An example of a total intensity map is given in Figure 5. The continuum background is calculated and removed to show the sidelobes and the weak field sources more clearly in the plots of logarithmic-scale.

The patterns of beam response shown in Figure 6 are similar to the radiation patterns given by simulations (Smith *et al.*, 2016). According to the angular distance of the beam center to that of the central beam M01, d_{M01}, the rest 18 elements of the 19-beam receiver are distributed in three concentric ring-shaped area (see Figure 7):

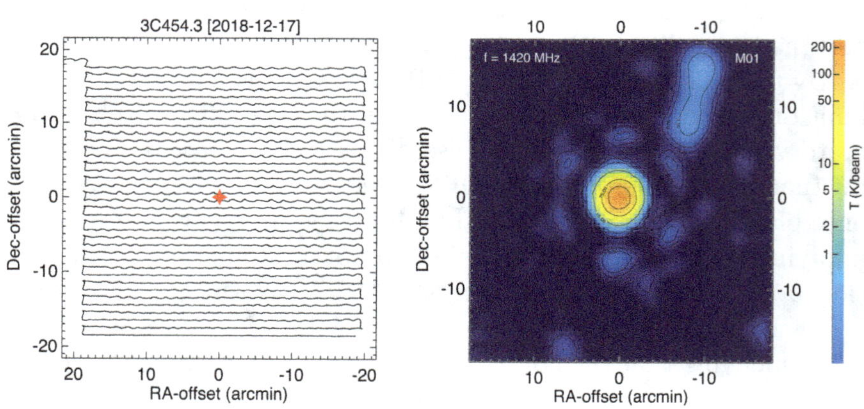

(a) (b)

Fig. 5. (a) The telescope track for a raster scan along the Dec direction. The covered sky area is $\sim 37' \times 37'$. The target source (3C454.3) is marked by a red star in the center of the plot. The X- and Y-axes indicate the RA- and Dec-offsets relative to the equatorial coordinates of 3C454.3 (RA: 22h53m57.7479s, Dec: +16d08m53.561s, in J2000), respectively. (b) Distribution map of total intensity obtained by the central beam (M01) near the frequency of 1420 MHz. The intensity distributions in a logarithmic-scale are indicated by different colors. The contours show the intensities with levels of 0.25, 0.5, 1.5, 15, 75, and 150 K.

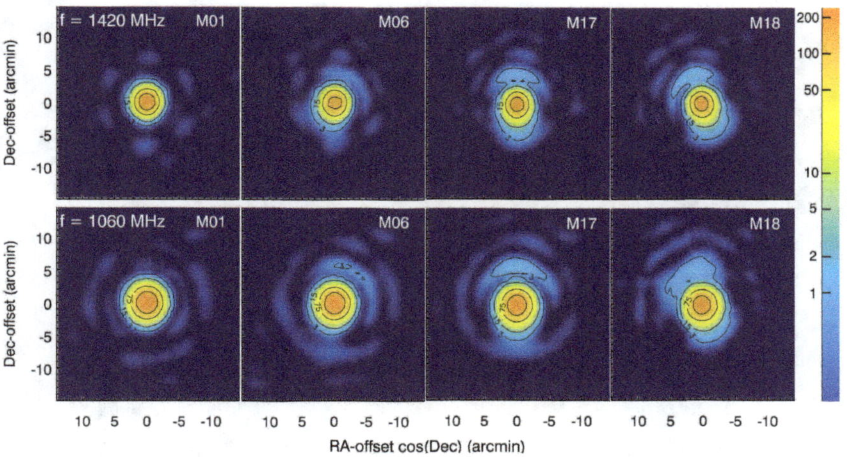

Fig. 6. Upper panels: distribution maps of total intensity obtained by raster scan observations of beam M01, M06, M17, and M18 near the frequency of 1420 MHz. Lower panels: similar to the upper panels, but for the observation frequency near 1060 MHz. The target source is 3C454.3. The observation date is December 17, 2018. In each map, the intensity distributions in a logarithmic scale are indicated by different colors. Contours (black lines) show the intensities with levels of 3, 15, 75, and 150 K. The X- and Y-axes indicate the RA- and Dec-offsets relative to the equatorial coordinates of 3C454.3. A correction factor cos(Dec) is applied for the X-axis in each of the plots.

1st-ring) M02-M07, with $d_{\mathrm{M01}} \sim 5.8$ arcmin; 2nd-ring) M09, M11, M13, M15, M17, M19, with $d_{\mathrm{M01}} \sim 10.0$ arcmin; 3rd-ring) M08, M10, M12, M14, M16, M18, with $d_{\mathrm{M01}} \sim 11.6$ arcmin. As d_{M01} increases, the sidelobes become more and more significant. The morphology of the sidelobes and their angular distances to the beam center depends on the observation frequency. Other notable features are the noncircular morphology and coma of the main beam, which are negligible for the central beam, but seem to become more and more significant as the increase of d_{M01} for the outer beams.

To fit the main-beam power pattern $P(\theta, \psi)$ with consideration of ellipticity and coma, and determine the combined response of the 19-element multi-beam receiver, we follow the definition of kew Gaussian given by Heiles *et al.*, (2004). Nine parameters for the i-th beam are the peak intensity A_0, beam center X_c and Y_c, the average beamwidth Θ_0, beam ellipticity Θ_1, beam orientation ϕ_{beam}, coma α_{coma} with orientation ϕ_{coma}, and a factor ϵ.

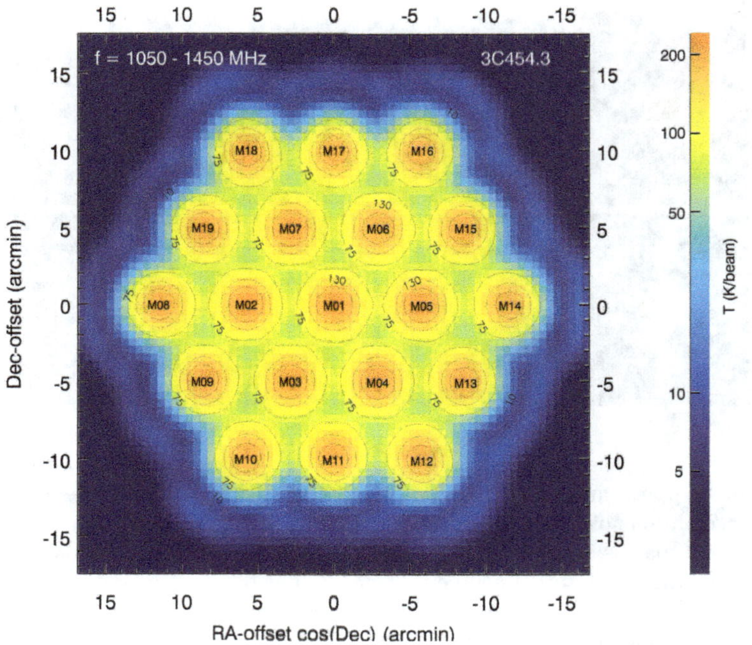

Fig. 7. Combined response of the 19-beam receiver of FAST. The target object is 3C454.3. The frequency coverage used to infer the total intensity is from 1050 to 1450 MHz. The intensity distributions in a logarithmic-scale are indicated by different colors. Contours (black lines) show the intensities with levels of 10, 75, 130, and 175 K. The X- and Y-axes indicate the RA- and Dec-offsets relative to the center of beam M01. The numbering of the 19 beams (M01-M19) is marked in the plot.

For the central beam M01, $X_c = (\mathrm{RA}_{\mathrm{measured}} - \mathrm{RA}_{\mathrm{source}})\cos(\mathrm{Dec})$ and $Y_c = \mathrm{Dec}_{\mathrm{measured}} - \mathrm{Dec}_{\mathrm{source}}$, are the pointing errors of FAST in RA and Dec, respectively. For other beams, the fitting parameter X_c and Y_c are taken as the angular offsets relative to the true center of beam M01. For each beam, the main-beam power pattern is parameterized in the polar coordinates (θ, ϕ) as

$$P(\theta,\phi) = A_0 \exp\left[-\frac{\theta^2(1-\min\{\alpha_{\mathrm{coma}}\frac{\theta_{\mathrm{coma}}}{\Theta_0},\epsilon\})}{\Theta^2}\right], \tag{1}$$

where $\theta_{\mathrm{coma}} = \theta\cos(\phi - \phi_{\mathrm{coma}})$, $\Theta = \Theta_0 + \Theta_1 \cos 2(\phi - \phi_{\mathrm{beam}})$. Here, θ is the angular distance of the position to the true beam center, ϕ is the position angle in the equatorial coordinates (RA, Dec in J2000), and defined to be zero along the positive RA-offset axis and increases towards positive Dec-offset axis. Following Heiles *et al.* (2001), we also take ϵ as a constant parameter 0.75 during the fit. Together with the term $\min\{\alpha_{\text{coma}}\frac{\theta_{\text{coma}}}{\Theta_0}, \epsilon\}$, they are used to prevent the coma term from unduly distorting the beam far from the beam center (Heiles *et al.*, 2004). The MPFIT package[1] is adopted to make the minimization. In the following, we give a general discussion about the combined beam response, the beamwidth and also the pointing errors of FAST, obtained by fitting the observed total intensity (XX + YY) maps.

The half-power beamwidth (HPBW) is related to the average beamwidth Θ_0 by HPBW $= 2(\ln 2)^{1/2}\Theta_0$, here, Θ_0 is derived by fitting the total intensity distribution maps with a 'skew Gaussian' model as shown in Eq. (1). In comparison with the central beam, the outer beams tend to have larger beamwidths, with a difference no more than 0.2 arcmin. The HPBW decreases with the increase of frequency. As discussed in Jiang *et al.*, (2019), the theoretical HPBW of a telescope with diameter D is HPBW $= 1.02\lambda/D$ for a uniformly illuminated aperture and HPBW $= 1.22\lambda/D$ for a cosine-tapered illuminated aperture. Here, a diameter of $D = 300$ m is assumed.

3.2.1. *Pointing errors*

As discussed in Jiang *et al.* (2019), the analysis of pointing errors can be used to evaluate the pointing accuracy, improve the pointing of the telescope, and also guide the error analysis of observations. Systematic observations have been made to measure the pointing errors of FAST by raster scan observations. As we only care about the pointing errors of the central beam M01, the mapping area toward a target source is set to $\sim 7' \times 7'$. About 7 minutes is needed to acquire such a map with subscan separation of $1'$. The target sources are selected from the catalogue of pointing calibrators given by Condon *et al.* (2001),

which contains 3399 strong and compact radio sources with accurate positions from the NVSS, and uniformly covering the sky north of $\delta = -40°$ (J2000). The dataset used in the analysis are observed from February 16 to March 15, 2019, which include 126 raster scan

[1]http://www.physics.wisc.edu/craigm/idl/fitting.html.

observations along RA or Dec directions. As shown in Figure 8, the observed pointing calibrators distribute widely in the sky coverage of FAST with full gain (zenith angle $\lesssim 26.4°$).

To fit the pointing errors in RA and Dec directions, a 2D Gaussian model is adopted, which is good enough as the central beam do not present significant beam ellipticity and coma feature (Figures 5 and 6). The pointing errors $[(\sigma_\alpha \cos\delta)^2 + (\sigma_\delta)^2]^{1/2}$ are typically smaller than 16″. The rms of pointing errors is 7.9″, which is less than one-tenth of the beamwidth near 1450 MHz (beamwidth$_{1450}$/10 $\sim 2.8'/10 = 16.8''$).

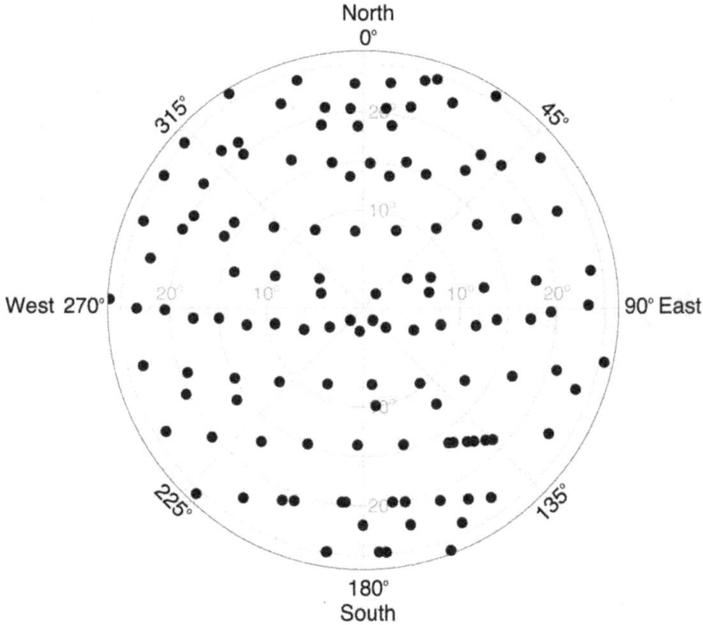

Fig. 8. Distributions of the observed pointing calibrators in the sky coverage of FAST with full gain (zenith angle $\lesssim 26.4°$). The observation period is from February 16 to March 15, 2019. In this plot, the zenith angle is from 0° to 26.4° as indicated by grey circles, the azimuth angle is from 0° to 360°. The directions marked in the periphery of the plot are the geographic north, east, south and west at the FAST site.

3.3. *Aperture efficiency and system temperature*

Aperture efficiency (η) and system temperature (T_{sys}) are two basic parameters of FAST. Both of them are expected to vary as a function of zenith angle (θ_{ZA}) and frequency. The 19 beams are divided into four categories according to separation to the central beam. The four categories are : (1) Beam 1; (2) Beam 2, 3, 4, 5, 6, and 7; (3) Beam 8, 10, 12, 14, 16, and 18; (4) Beam 9, 11, 13, 15, 17 and 19. Though the performance of aperture efficiency and system temperature in each category is expected to be same, asymmetry of the receiver platform may lead to deviation. Thus we obtained the measurements toward all 19 beams.

The aperture efficiency curve was obtained by repeating observations of the calibrator 3C286 and its off position at different zenith angles. The position switch mode provides a quick switch between the ON and OFF source positions. The separation between the ON and OFF positions was selected to allow for measuring the aperture efficiency curve of two beams (e.g., Beam 1 and 19) simultaneously, which increases measurement efficiency.

The system temperature curve of 19 beams was obtained by continuously tracking a clean sky position ($\alpha_{2000} = 23^h30^m0^s.0$, $\delta_{2000} = 25°39'10.6$") from rise to set on September 14, 2019.

In the above measurements, a spectral backend with a sampling time of 1.00663296 s and a channel width of 7.62939 kHz was adopted to record the data. The high noise signal (~12.5 K) was injected at a synchronized period of 2.01326592 s. The duration of the cal on and cal off is 1.00663296 s. During calculation, channel width of the data was smoothed to 1 MHz. The aperture efficiency was derived for each observation cycle while the system temperature was derived for each sampling time.

3.3.1. *Aperture efficiency*

Based on the absolute measurement of noise dipole, the observed data were calibrated into antenna temperature T_A in K . The antenna temperature of 3C286, T_A^{3C286} was derived by the difference between the source ON and source OFF data for each beam. 3C286 is a stable flux calibrator. The frequency dependency of the spectral flux density of 3C286 could be fitted with a polynomial function (Perley and Butler 2017),

$$\log(S) = a_0 + a_1 \log(\nu_G) + a_2[\log(\nu_G)]^2 + a_3[\log(\nu_G)]^3, \tag{2}$$

where S and ν_G are the spectral flux density in Jy and the frequency, respectively. We adopted the value of $a_0 = 1.2481$, $a_1 = -0.4507$, $a_2 = -0.1798$ and $a_3 = 0.0357$ that is valid for 3C286 at a frequency range of [0.05–50] GHz (Perley and Butler 2017).

Measured gain (G) of FAST is expressed as $G_0 = A_{\rm eff}/2k$, in which $A_{\rm eff}$ is an effective illumination area. In observations, G is calculated by ratio between antenna temperature and flux density, $G = T_A^{3C286}/S_A^{3C286}$. The perfect gain G_0 of FAST is $G_0 = A_{\rm geo}/2k$ for single polarization, in which $A_{\rm geo}$ and k are geometric illumination area and the Boltzmann constant, respectively. The geometric illumination area with a diameter of 300 m leads to G_0= 25.6 K/Jy for FAST. The aperture efficiency η of FAST is derived by

$$\eta = G/G_0 = A_{\rm eff}/A_{\rm geo} = \frac{2kT_A^{3C286}}{A_{\rm geo}S_A^{3C286}}. \tag{3}$$

The aperture efficiency is a synthetical result of above effects and is described by the following equation:

$$\eta = \eta_{\rm sf} \cdot \eta_{\rm bl} \cdot \eta_{\rm s} \cdot \eta_{\rm t} \cdot \eta_{\rm misc} \cdot \eta_{\rm sloss}. \tag{4}$$

The value of $\eta_{\rm sf}$, $\eta_{\rm bl}$, $\eta_{\rm s}$, $\eta_{\rm t}$, and $\eta_{\rm misc}$ are almost constant. The value of η is mainly determined by $\eta_{\rm sloss}$. The value η at all frequency would keep almost constant when $\theta_{\rm ZA}$ is less than 26.4° and decreases linearly when $\theta_{\rm ZA} > 26.4°$. This is consistent with the fact that FAST would lose part of reflection panels when $\theta_{\rm ZA} > 26.4°$.

We fitted variation of η as a function of $\theta_{\rm ZA}$ at specific frequency with two linear equations.

$$\eta = \begin{cases} a\ \theta_{\rm ZA} + {\rm b}, & 0° \le \theta_{\rm ZA} \le 26.4° \\ c\ \theta_{\rm ZA} + {\rm d}, & 26.4° < \theta_{\rm ZA} \le 40°. \end{cases} \tag{5}$$

In which d satisfies the equation $d = b + 26.4(a - c)$.

The gain of FAST could be expressed with $G = \eta G_0$, in which $G_0 = 25.6$ K/Jy. The flux density (in units of Jy) of point source is converted from antenna temperature by dividing the gain value.

The averaged gain values of 19 beams within ZA of 26.4° are shown in Table 1. Beam 16 has the smallest gain value among 19 beams. The gain ratio of Beam 16 compared to Beam 1 reaches 0.83 ± 0.02 at 1400 and 1450 MHz.

3.3.2. *System temperature*

Observation of the sky position ($\alpha_{2000} = 23^h30^m0^s.0$, $\delta_{2000} = 25°39'10.6"$) cause us to measure system temperature from a zenith angle of 40° to nearly 0°. The data were calibrated into antenna temperature in K with a noise curve. For a single dish telescope in the L band, $T_{\rm sys}$ consists of four main sources (Campbell *et al.*, 2002):

(1) Noise contribution from the receiver, $T_{\rm rec}$. For 19 beam receiver of FAST, it is measured as 7–9 K, including ~4 K from a low noise amplifier (LAN) (see Section 4.1.1 in Jiang *et al.*, 2019 for details).
(2) Continuum brightness temperature of the sky, $T_{\rm sky}$. This includes 2.73 K from cosmic microwave background (CMB) and non-thermal emission from the Milky Way. The value of $T_{\rm sky}$ at 1.4 GHz is ~3.48 K toward position ($\alpha_{2000}=23^h30^m0^s.0$, $\delta_{2000}=25°39'10.6"$) (CHIPASS survey; Calabretta *et al.*, 2014).
(3) Emission from the Earth's atmosphere $T_{\rm atm}$. This value should be in a few K.
(4) Radiation from the surrounding terrain, $T_{\rm scat}$. This contribution originates from side lobe of FAST. Unlike other single dish with fixed surface and horn, the illumination area of horn varies for different zenith angles, leading to different values of $T_{\rm scat}$.

System temperature are synthetical result of the above components.

$$T_{\rm sys} = T_{\rm rec} + T_{\rm sky} + T_{\rm atm} + T_{\rm scat}. \tag{6}$$

The variation of system temperature $T_{\rm sys}$ as a function of zenith angle could be fitted with the following formula:

$$T_{\rm sys} = P_0 \arctan(\sqrt{1 + {\theta_{ZA}}^n} - P_1) + P_2, \tag{7}$$

which is valid for $0° \le \theta_{\rm ZA} \le 40°$. An example of fitting $T_{\rm sys}$ curve is shown in Figure 9.

Table 1. The gain information for 19 beams within zenith angle of 26.4°.

	Gain (K/Jy) / Ratio								
Beam	1050	1100	1150	1200	1250	1300	1350	1400	1450
M01	15.98(0.27)	16.27(0.21)	16.48(0.14)	16.10(0.25)	16.12(0.25)	15.94(0.28)	15.98(0.27)	16.02(0.26)	15.71(0.29)
M02	0.98(0.03)	0.95(0.03)	0.95(0.03)	0.95(0.03)	0.95(0.03)	0.93(0.03)	0.94(0.02)	0.94(0.02)	0.95(0.03)
M03	1.03(0.03)	0.99(0.03)	1.00(0.03)	1.00(0.04)	1.00(0.03)	0.99(0.03)	0.97(0.03)	0.97(0.03)	0.98(0.03)
M04	1.01(0.04)	0.97(0.03)	0.98(0.03)	0.97(0.04)	0.97(0.04)	0.96(0.03)	0.95(0.03)	0.95(0.03)	0.95(0.03)
M05	1.02(0.03)	1.01(0.02)	1.03(0.02)	1.00(0.03)	1.00(0.03)	0.99(0.02)	0.97(0.02)	0.98(0.02)	0.98(0.03)
M06	**0.99(0.03)**	**0.98(0.02)**	**0.98(0.03)**	**0.96(0.04)**	**0.96(0.03)**	**0.95(0.02)**	**0.94(0.02)**	**0.94(0.02)**	**0.94(0.02)**
M07	**0.98(0.02)**	**0.95(0.02)**	**0.96(0.02)**	**0.94(0.03)**	**0.96(0.03)**	**0.95(0.02)**	**0.94(0.02)**	**0.93(0.02)**	**0.93(0.02)**
M08	**0.96(0.03)**	**0.94(0.03)**	**0.93(0.03)**	**0.93(0.04)**	**0.92(0.03)**	**0.92(0.03)**	**0.91(0.03)**	**0.89(0.03)**	**0.89(0.03)**
M09	0.95(0.02)	0.94(0.02)	0.92(0.02)	0.91(0.02)	0.92(0.02)	0.93(0.02)	0.91(0.02)	0.89(0.02)	0.90(0.02)
M10	0.99(0.03)	0.97(0.03)	0.97(0.03)	0.96(0.04)	0.97(0.03)	0.96(0.03)	0.96(0.03)	0.95(0.03)	0.95(0.03)
M11	0.96(0.04)	0.94(0.04)	0.94(0.03)	0.93(0.05)	0.93(0.04)	0.92(0.03)	0.92(0.03)	0.91(0.03)	0.91(0.03)
M12	0.96(0.05)	0.93(0.04)	0.92(0.03)	0.94(0.04)	0.94(0.03)	0.91(0.03)	0.90(0.03)	0.89(0.02)	0.90(0.03)
M13	0.99(0.04)	0.96(0.03)	0.97(0.03)	0.99(0.04)	0.96(0.03)	0.97(0.03)	0.96(0.03)	0.95(0.03)	0.95(0.03)
M14	0.95(0.03)	0.94(0.03)	0.93(0.03)	0.92(0.04)	0.93(0.04)	0.93(0.03)	0.92(0.02)	0.91(0.02)	0.91(0.02)
M15	0.90(0.03)	0.91(0.02)	0.90(0.02)	0.90(0.05)	0.90(0.02)	0.90(0.02)	0.88(0.02)	0.87(0.02)	0.88(0.02)
M16	0.90(0.03)	0.88(0.02)	0.86(0.03)	0.88(0.03)	0.86(0.02)	0.86(0.02)	0.84(0.02)	0.83(0.02)	0.83(0.02)
M17	0.93(0.03)	0.91(0.03)	0.90(0.03)	0.89(0.04)	0.89(0.02)	0.88(0.02)	0.88(0.02)	0.86(0.02)	0.86(0.02)
M18	0.90(0.02)	0.88(0.02)	0.89(0.02)	0.89(0.02)	0.88(0.03)	0.87(0.02)	0.86(0.02)	0.86(0.02)	0.87(0.02)
M19	0.94(0.03)	0.94(0.02)	0.93(0.03)	0.93(0.03)	0.93(0.03)	0.92(0.03)	0.90(0.02)	0.89(0.02)	0.89(0.03)

Note: Mean value of the gain of Beam 1 is expressed in unit of K/Jy. For the rest 18 beams, the ratio compared to the gain value of beam 1 in same frequency is presented.

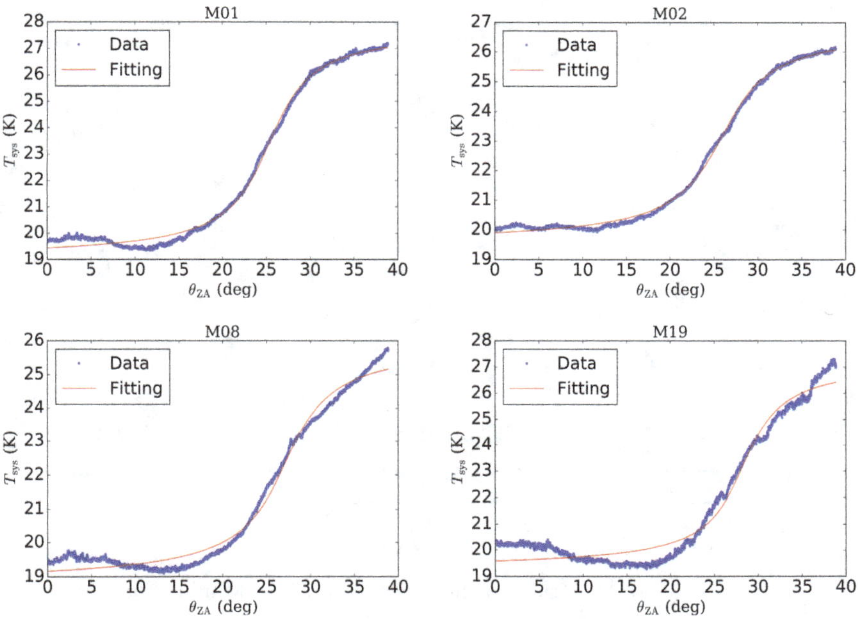

Fig. 9. Measured $T_{\rm sys}$ curve as a function of zenith angle $\theta_{\rm ZA}$ at 1400 MHz for Beam 1, 2, 8 and 19. The fitting result is represented with red solid line.

A significant feature in the $T_{\rm sys}$ profile is that the $T_{\rm sys}$ value of most beams reaches its minimum at ZA range of [10°,15°] (see Fig. 9). When ZA is smaller than 10°, the $T_{\rm sys}$ value would increase by a maximum value of 1 K. Though central hole of spherical surface was shielded with metal mesh during observations, the leakage of ground emission (~300 K) from the central hole goes into the main beam and contributes $T_{\rm sys}$ increase when ZA is less than 10°. When ZA gets larger than 15°, the $T_{\rm sys}$ value increases by 5–7 K, which arises from emission of nearby mountain through sidelobe.

4. Backend

4.1. *Stability and sensitivity of spectral baseline*

Stability and sensitivity are two fundamental properties associated with spectral baseline. Stability represents fluctuation level of baseline in a time range. Sensitivity represents response capability of spectrometer. A set of observations of 30 min without injection of

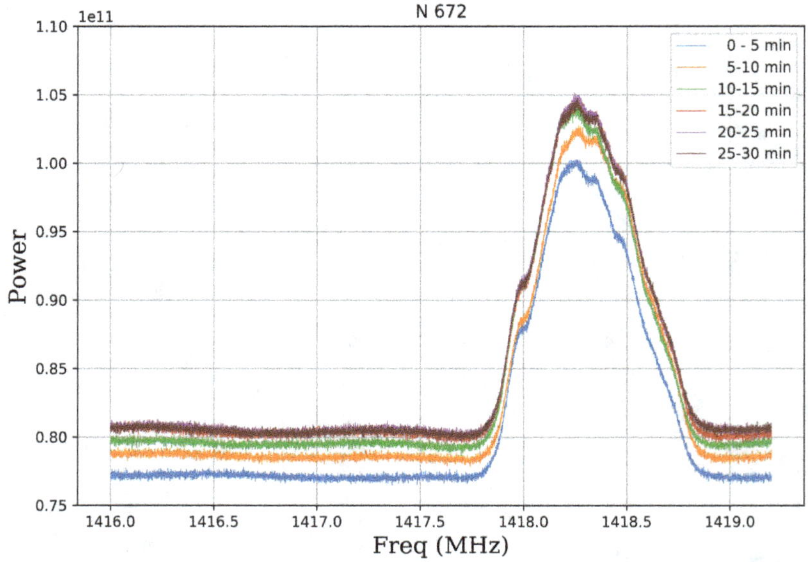

Fig. 10. Variance of spectral baseline toward N672 in 30 min. Each spectrum is averaged in a time bin of 5 min.

noise signal was taken to test stability of the baseline. Averaged bandpass spectra per 5 min are shown in Figure 10. The variation of bandpass is ~4% in 30 min.

In order to measure sensitivity performance of the backend, noise signal was injected to calibrate observed spectra. Both noise signal with high and low intensity were injected for 5 min. The unit of spectrum was then calibrated into kelvin through the following transformation:

$$T_{\mathrm{A}} = T_{\mathrm{cal}} \frac{P_{\mathrm{off}}^{\mathrm{cal}}}{P_{\mathrm{on}}^{\mathrm{cal}} - P_{\mathrm{off}}^{\mathrm{cal}}}, \tag{8}$$

where T_{A} is calibrated antenna temperature. T_{cal} is noise diode temperature. P_{on} and P_{off} are the power values when noise diode is turned on and off, respectively. As shown in Figure 11, the derived continuum level under high and low noise injection scenarios are consistent within 1%.

We estimated the sensitivity with root mean square (RMS) of a spectrum. RMS is calculated with line-free frequency range (see Table 2). In the total ON mode, the expected RMS of an averaged spectrum of two polarizations σ_{T} is connected with system

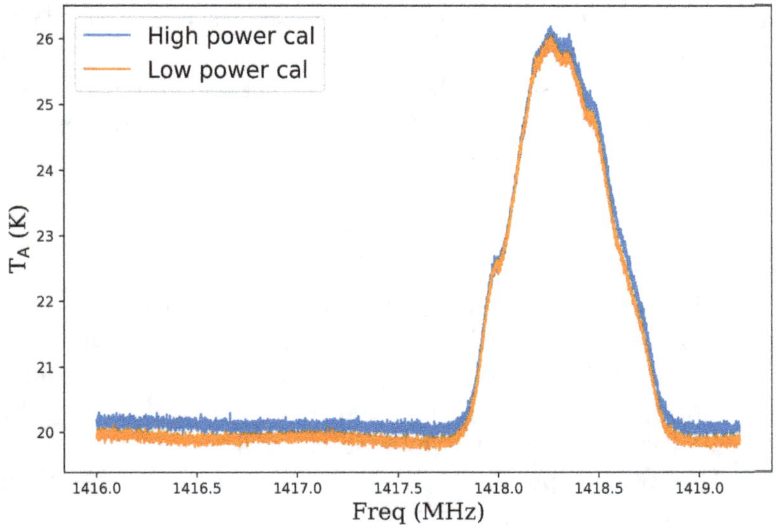

Fig. 11. Calibrated HI spectrum under high power cal (blue) and low power cal (orange).

Table 2. RMS value of 5 minutes observations toward N672.

Cal intensity	Obtained value	Theoretical value
Low	40.1	38.4
High	48.0	47.7

Note: Channel resolution is 0.476 kHz during calculation.

temperature $T_{\rm sys}$, channel resolution β and integration time τ with $\sigma_T = T_{\rm sys}/\sqrt{2\beta\tau}$. Periodic injection of high and low noise would lead to increase of $T_{\rm sys}$ by 5.4 K and 0.5 K during an observation cycle (half time for cal on and half time for cal off), respectively.

4.2. *Standing waves*

The antenna structure and radio frequency (RF) devices of a radio telescope can cause the reflection of electromagnetic (EM) wave. The coherent superposition of the received signal and its reflection wave results a periodic fluctuations in the frequency band-pass, which is called "standing wave". This is a common phenomenon seen in the spectroscopic observations with radio telescopes (i.e., Briggs *et al.*, 1997; Popping *et al.*, 2008). The major contribution to the standing wave for FAST is caused by the reflection between the dish and the receiver cabin. The relation between the standing wave frequency and the reflecting distance is given by

$$D = \frac{c}{2\,\Delta f}, \tag{9}$$

where D is the distance between the receiver cabin and the dish, c is the speed of light, and Δf is the width of the frequency ripple. For FAST, $D = 138\,\mathrm{m}$ relates $\Delta f \sim 1\,\mathrm{MHz}$. At the L-band, the corresponding velocity width is $\sim 200\,\mathrm{km\,s^{-1}}$, which locates within the range of interest for extra-galactic observations.

Efforts have been put into analyzing and minimizing the standing wave effects of FAST. To identify the major contribution to the FAST standing wave, we checked the band-pass during the receiver cabin rising. The width of the frequency ripple varies with the cabin height following the relation given by Eq. (9). This experiment was done with the cabin dock uncovered. Thus the ground radiation passing through the central hole of the dish serves as the EM source for the standing wave. In Figure 12, the figures from the upper panel show the band-pass before the cabin dock is covered with metal mesh and those from the lower panel are after. The fluctuation declines significantly when the ground radiation is blocked out.

5. Observation

5.1. *Observation modes*

Now there are more than eight observation modes in the FAST. The details of some observation modes are shown in the follows.

(1) **Drifting scan:** In this mode, the position of feed cabin is fixed. To avoid pointing offset by the cooling of oil in the actuator, the actuators used for shaping the paraboloid surface are adjusted continually to keep same paraboloid in this mode. The telescope points across the sky as the earth rotates. There are four parameters in this mode: source name, source coordinate in J2000 system, observational time range, and rotating angle that

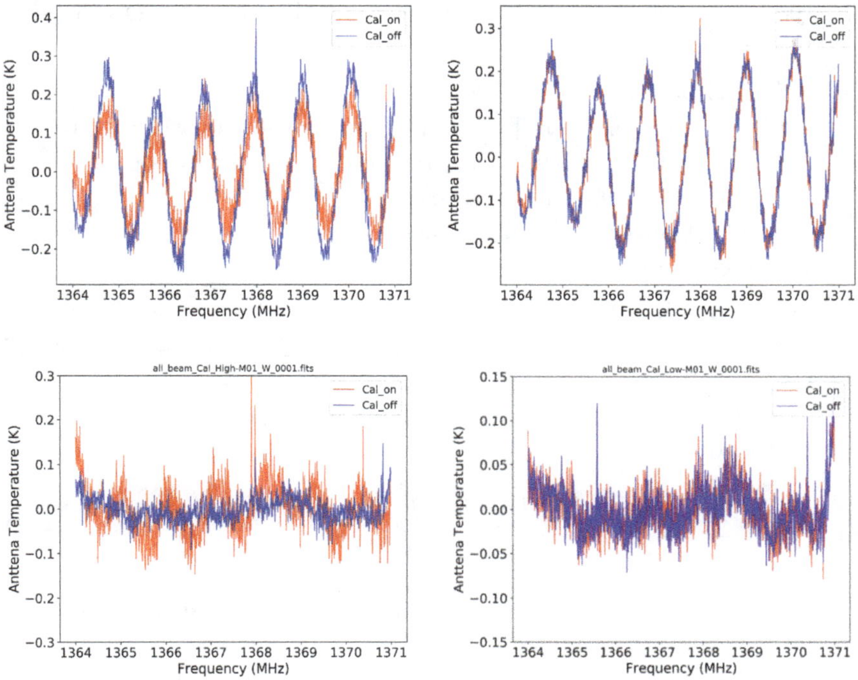

Fig. 12. The comparison of the band-pass ripples before (upper panel plots) and after (lower panel plots) the cabin dock platform is covered with metal mesh. The left side plots are with high power noise diode injection and the right side plots are with the low noise power.

represents the cross angle between the line along beam 8, 2, 1, 5, and 14, and the line of declination. The rotating angle has a limit of [−80,80] degree.

(2) **Total power:** In this mode, the source can be tracked continuously. Three parameters including source name, coordinate and integration time are needed. Derived RMS of averaged spectrum of both polarization is estimated with,

$$\sigma = \frac{T_{\text{sys}}}{\sqrt{2\beta\tau}}\ \text{K}, \tag{10}$$

where $\beta = 1.2B_{\text{chan}}$ (B_{chan} is channel width of spectrometer in unit of Hz).

(3) **Position switch:** The design of position mode is to achieve quick switch between source ON and source OFF in order to

reduce baseline variation. There are five parameters in this mode.

- Coordinate of ON position.
- Coordinate of OFF position. The position of OFF source is designed to be within 1 degree from that of ON source.
- Integration time of ON source.
- Integration time of OFF source.
- Times of ON-OFF cycle.

Overhead time between ON and OFF positions depends on separation of them, $\Delta\theta$. It is 30 s for $\Delta\theta$ <20', and is 60 s for $20' \leq \Delta\theta < 60'$.

Derived rms of each spectrum with both polarizations is estimated with,

$$\sigma = \frac{T_{\text{sys}}}{\sqrt{\beta\tau}} \text{ K.} \tag{11}$$

where $\beta = 1.2B_{\text{chan}}$. B_{chan} is channel width of spectrometer in unit of Hz.

(4) **On-the-fly (OTF) mapping:** This mode is specifically designed for mapping a sky area with Beam 1. Six parameters are necessary for observation: source name, source position, observational time range, sky coverage (e.g., $7' \times 7'$ of the mapping region), scanning separation (e.g., $1'$) between two parallel scanning lines, and scanning direction (along RA or Dec). Scanning speed is 15 arcsec/s in default. The schematic diagram of this mode is shown in Figure 13.

(5) **MultiBeamOTF mapping:** This mode is proposed to map the sky with 19 beams simultaneously. Compared to the OTF mapping, the MultiBeamOTF mapping mode has a similar scanning trajectory but with a larger separation (e.g., 20 arcmins) between parallel scans. Besides, the parameter of rotation angle is available in this mode.

(6) **MutiBeamCalibration:** In this mode, 19 beams will be switched in sequence to track the calibrator, allowing for quick calibration of the gain of 19 beams in 30 minutes. The switching time between two beams is 40 s. The integration time of each beam is a parameter for setting. The schematic diagram of this mode is shown in Figure 14.

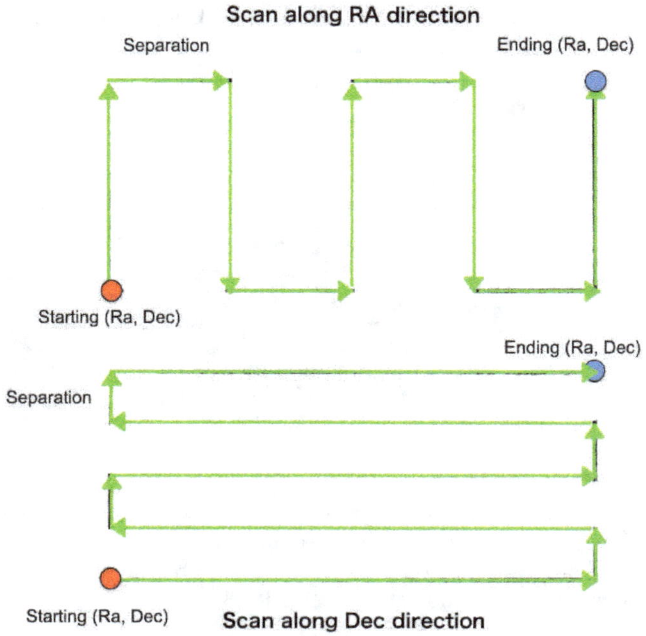

Fig. 13. Schematic diagram of OTF mode. The blue line represents scanning trajectory. The top and bottom figure shows scan information along RA and Dec direction, respectively.

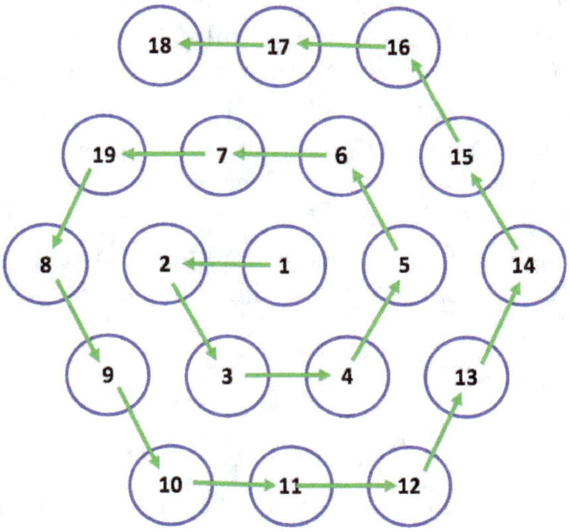

Fig. 14. Schematic diagram of MutiBeamCalibration mode. Green arrow line represents the beam sequence for observing the calibrator.

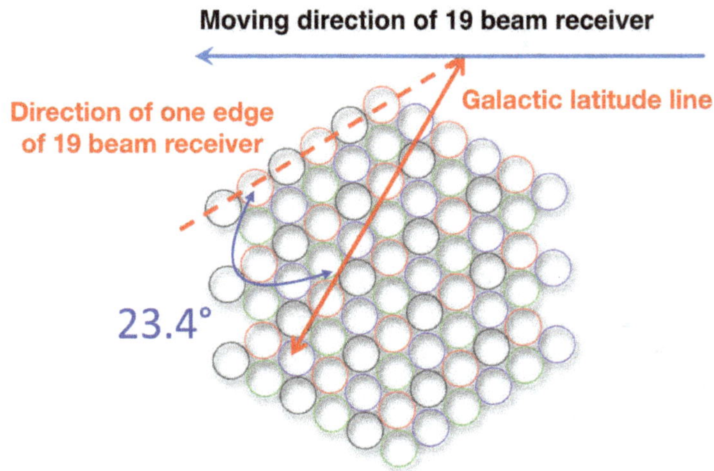

Fig. 15. Schematic diagram of snapshot mode. Blue line represents moving direction of 19 beam receiver. Dashed red line represents the direction of one edge of 19-beam receiver. Galactic latitude line is represented with solid red line. The angle between dashed and solid red line is 23.4°.

(7) **BasketWeaving:** This mode scan the sky along a meridian line. The scanning speed ranges from 5 to 30 arcsec/s. The setting parameters for this mode include the starting time, starting declination, ending declination, duration time.

(8) **Snapshot:** This mode is used to fully map the sky in grid. This type of mapping is not a Nyquist sampling. However, it can be used to map a region with relatively long integration time. This is beneficial especially for pulsar searching. As shown in Figure 15, the movement of 19 beam receiver would ensure fully cover the sky along same Galactic latitude. Necessary parameters for this mode include source name, beginning, and ending coordinate (RA and Dec), observational time range, and scanning speed (less than $30'$/s).

5.2. *Effect of radio frequency inference*

The current observation hints that RFI in FAST data are mainly divided into three types: narrow-band RFI, 1 MHz wide RFI and some wider fixed frequency RFI. Every type is discussed in more detail below.

(1) The narrow-band RFI in FAST data is believed to have multiple origins, such as instrument interference or local telescope influence. The FAST spectral data has a frequency resolution of 0.48 kHz, with extremely narrow signals mostly appearing on one or a few channels. Some narrow-band RFI can have high strength in a short integration time, while others are faint without a Gaussian profile.

    In some cases, the narrow-band RFI tend to exhibit the periodic variations in the time domain, which might be found in the pulsar data if its time resolution is less than one second. The narrow-band RFI used to bring a lot of trouble for the FAST data processing, but it has been solved now.

(2) While the 1 MHz wide RFI is caused by the standing wave, which looks like regular sinusoidal wave or the single bump in FAST spectra. The big bump originates from the superpose of some standing waves with different amplitudes and periods, sometimes approaching to Gaussian profile. The typical width of this type of RFI is 1 or a few MHz. The existing L-band observation shows that their distribution in time domain and frequency domain is not regular.

(3) The fixed frequency RFI is due to satellite or civil aviation from the sky. It usually have a fixed frequency in a wider distribution, and have the strongest intensity in the whole band-pass, even improve the baseline there. Its spectral profile has a complex multi-peak structure, and the width of one peak might be around 20 MHz. The profile and intensity of it varies slightly in different observation time. In addition, the existence of such a wide enough RFI may also be due to the higher sensitivity of FATS. It receives some very strong signal, which results in the frequency width of the response signal exceed the range of satellite signal on both sides.

Since the successful installation of the electromagnetic shielding in April 2019. In the future, we will try to do more to solve the problem of standing wave and RFI.

6. Summary

The FAST has achieved its designed objectives.

(1) The median power output of low and high noise are about 1.1K and 12.5K respectively. The measured temperature fluctuation of noise diode is ~1%, leading to ~2% accuracy in flux calibration when considering pointing accuracy and beam size.
The spatial distribution and power pattern of 19 beams within zenith angle of 26.4° were obtained by mapping observations toward 3C454.3. The beam width of 19 beams is consistent with the theoretical estimation.
(2) The pointing errors of the 19-beam receiver in different sky positions are less than 16″. The standard deviation of pointing errors is 7.9″.
(3) The aperture efficiency, as a function of zenith angle, could be fitted with two linear lines at specific frequency. It keeps almost constant with a zenith angle of 26.4° and decreases by ~ 1/3 at a zenith angle of 40°.
(4) The system temperature, as a function of zenith angle, could be fitted with a modified arctan function. This fitting is valid for zenith angle within 40°.
(5) The fluctuation of baseline is about 4% in 30 minutes. The RMS of the baseline satisfies the expected sensitivity.
(6) The standing wave has an amplitude of ~0.3% compared to the continuum level. The amplitude, phase of standing wave would vary at different zenith angles during drifting. For position switch observations, standing wave of the residual ON-OFF spectrum would be suppressed with amplitude of ~0.02 K, which varies in different frequencies.
(7) The derived polarization degree and polarization angle toward 3C286 with FAST are $6.6 \pm 1.5\%$ and 33.4 ± 6.4 degrees, respectively. They are consistent with results from previous study.
(8) At least eight observation modes including drift scan, total power, position switch, On-the-fly, MultiBeamOTF, MultiBeam-Calibration, BasketWeaving and snapshot are available for FAST observations now.
(9) The RFI environment has been greatly improved. Narrow RFI with several spectral channels are reduced. More effort will be done for eliminating broad RFI with width of ~1 MHz.

The FAST data plays an important role in constraining the physical origin of fast radio bursts. The corresponding results have been published in the journal such as *Nature* and *Science*. At the same time, it is noteworthy that the total number of pulsars discovered (e.g., (Qian *et al.*, 2019)) by FAST has exceeded 800, which is a huge treasure. After a large number of observations, maybe we will further uncover the mysterious veil and look forward to the novel events we have been long expecting.

Bibliography

Briggs, F.-H. *et al.* (1997). Driftscan surveys in the 21 CM line with the Arecibo and Nancay telescopes, *PASA* **14**, 37.

Campbell, D. B. *et al.* (2002). Measurement in radio astronomy, *Single-Dish Radio Astronomy: Techniques and Applications* **278**, 81.

Condon, J. J. *et al.* (2001). Offset pointing calibrators for large radio telescopes, *PASP* **113**, 362.

Heiles, C. *et al.* (2001). AllStokes Parameterization of the Main Beam and First Sidelobe for the Arecibo Radio Telescope, *PASP* **113**, 1247.

Heiles, C. *et al.* (2012). *All-Stokes Single Dish Data with the RHSTK (Robishaw/Heiles SToKes) Software Package.*

Jiang, P. *et al.* (2020). The fundamental performance of FAST with 19-beam receiver at L band, *RAA* **20**(05), 64.

Jiang, P. *et al.* (2019). Commissioning progress of the FAST. *Sci. China Phys. Mech. Astron* **62**, 959502.

Nan, R. *et al.* (2011). The five-hundred-meter aperture spherical radio telescope (FAST) project. *Int. J. Mod. Phys. D* **20**(06), 989.

Popping, A. *et al.* (2008). The standing wave phenomenon in radio telescopes. Frequency modulation of the WSRT primary beam, *A&A* **479**, 903.

Qian, L. *et al.* (2019). The first pulsar discovered by FAST, *Sci. China Phys. Mech. Astron.* **62**, 1.

Smith, S. L. *et al.* (2016). A study of liquid metal alloy reconfigurable antennas embedded in a structural composite, *IEEE International Symposium on Antennas and Propagation (APSURSI)*, 383.

Yao, R. *et al.* (2020). A motion planning algorithm for the feed support system of FAST, *RAA* **20**(05), 68.

https://doi.org/10.1142/9789819800643_0006

Chapter 6

Single-Pulse Study of Radio Pulsars Using FAST

Ji-Guang Lu
National Astronomical Observatories,
Chinese Academy of Sciences,
Beijing 100012, China
lujig@nao.cas.cn

FAST (the Five-hundred-meter Aperture Spherical radio Telescope) is the most sensitive single-aperture telescope, and it has been completed since 2016. With the observation of FAST, a series of scientific results were published, and around a third of them focussed on pulsars. It seems that it is the era of pulsar science, however, pulsar study is in an awkward situation at present. Just after pulsar was discovered in 1967, the theoretical researching on pulsars was flourishing, and the fundamental pulsar radiation speculations were endless. Several tens of years have passed, but the study on pulsar radiation made little progress. The observation on pulsars seems helpless to find the breach to crack the pulsar radiation problem, and few theoretical speculations are falsifiable for years to come. At present, pulsar is no longer a celebrity with large followers, and the "clout chasers" have had new targets. Fortunately, FAST provides a ray of hope. The high sensitivity of FAST allows some of the strong signal from pulsars to surface on the noise background. Although it is hard to make out the pulsar radiation thoroughly based on the strong signal, FAST brings an unmissable chance. To unveil the pulsar radiation, it is necessary to try all means on the observed data, e.g., using extremely complicated mathematical tools. The conclusion of this scientific war is expected.

Keywords: Pulsars, General — radiation mechanisms, Non-thermal — radio — FAST

Contents

1. Introduction

Pulsars have been discovered for more than half of a century (Hewish *et al.*, 1968). We analysed the pulsar from many aspects, such as the multi-band observation (Kramer, 1994; Lu *et al.*, 2016; Lyne and Manchester, 1988; Rankin, 1983; Thorsett, 1991; Xilouris *et al.*, 1996), the theoretical calculation (Arons and Scharlemann, 1979; Buschauer and Benford, 1976; Cheng *et al.*, 1986; Goldreich and Julian, 1969; Muslimov and Harding, 2003; Qiao *et al.*, 2007, 2001; Ruderman and Sutherland, 1975) and simulation (Li *et al.*, 2012; Philippov and Spitkovsky, 2014; Philippov *et al.*, 2015). However, the radiation process of pulsar is also mysterious, and one can even say that we know nothing about it. The radiation mechanism, the radiation position and the radiation particles are all hard to determine for pulsar.

Of course, the key point to determine the radiation process is the consistency between the theory and the observation. But the trouble is that the observation facts of pulsars are too many. The narrow pulse reflects an extremely isotropic radiation mechanism, and the various shapes of pulse profiles are almost beyond the imagination. If the frequency-dependent-evolution of pulse profile can be treated as a continuous variation, the discrete time-dependent pulse evolution (single pulse observation) expands the information of pulsar observation data with a new dimension.

2. Single-Pulse Study

Single pulse study is necessary in pulsar radiation investigation. The mean radio flux density of discovered pulsar ranges from $\sim 1\,\mu$Jy to ~ 1 Jy, and the system equivalent flux density (SEFD) of most radio telescopes is larger than 1 Jy. It means that the pulsar signal would be always weaker than the noise in base-band data. To overcome this difficulty, the pulsar data are always time-scrunched in the study. It should be noted that the long-time-averaged pulse profile is always stable. The stability of mean pulse profile provides convenience to study the statics of the magnetosphere, nevertheless, the evolution information of pulse in time domain is abandoned. Without the time-domain-evolution, the radio observing data of pulsar can be regarded as 2-dimension array, which reflects the profile evolution with frequency. From the frequency-domain observing data of pulsars, a lot of radiation geometry and mechanism models are proposed. Each of these models has individual advantage and can explain part of the observing phenomena, but at the same time, fail in explaining others. From these models, it can be summarized that one or more regions to generate and accelerate the charged particles are needed in the pulsar magnetoshpere, and the accelerated charged particles would radiate in the magnetosphere. However, it is very difficult for the continuation of the pulsar radiation study. With the assumptions that the pulsar has similar radius and simple dipole magnetic field, the observed pulsar radiation can be uniquely identified with the rotating period, the magnetic field intensity, the inclination angle and the impact angle. Theoretically, in the phase space spanned by these four parameters, the pulse profile morphology of pulsars should be continuous, but this continuity is failed to be discovered. This result implies that the pulsar system must be much more complicated than that assumed, and each pulsar may have distinctive characteristics. The various single pulses would be a key to unveil these distinctions.

Single pulse study is necessary in studying the magnetosphere of a pulsar. In the magnetosphere, the particles move along the magnetic field lines and emit radio radiation. However, the mapping from the radiation position to the pulse phase is not surjective but injective, in other words, it is impossible to determine the radiation position based on the pulse phase only. In this case, the single pulse information would bring new ideas. The single pulse observation greatly increased the pulsar observing information, and the radiation at specified pulse phase in different pulse period has different properties. The radiation diversity at a specified pulse phase provides the possibility to break the phase degeneracy of pulsar radiation and to investigate the detail of the pulsar radiation geometry. On the other hand, the study of the magnetosphere dynamics demands the research on single pulse for the characteristic time. Firstly, the magnetospheric dynamics relating to the rotating period, as a natural variation time of pulsar system, needs to be studied with the single pulse data. This effect always relates to the physical process near the light cylinder, which is the critical boundary to separate the regions whether co-rotating with the pulsar. Secondly, the propagation time of a perturbation in magnetosphere is shorter than the rotating period, and then the physical process in magnetosphere should be investigated with single pulse data due to the character time scale. In fact, most of the physical process occurs at low altitudes, or even just in the polar cap region, and therefore has much shorter character time scale than the rotating period. Additionally, the strong and complicated field in the magnetosphere may lead to random bursty event, which calls on a high time resolution observation. To study the short time-scale properties of pulsar radiation, it is indispensable to study the evolution and the fine structure of the single pulses.

The single pulse study can be commonly classified into three types: statistics on the single pulse property, pulse-to-pulse correlation, and the study on the fine structure of single pulse. The statistics can be carried out on the characteristic parameters of individual pulses or phase-resolved properties. For example, the total flux density of individual pulses can be a statistic, and the common distribution of the single pulse flux density is log-normal-like. Whereas, the single pulse flux distribution of some pulsars have power-law tails, i.e., the flux density of some strong individual pulses is much larger than the mean flux density. This kind of strong pulses, known as the giant pulse (Jessner *et al.*, 2005), may have different radiation mechanism and geometry from normal pulses. In addition, for some pulsars, there is a peak corresponding to the zero-flux in the flux density distribution (Gajjar *et al.*, 2012), implicating that the pulsars have no observable radiation in those periods. The null fraction of the pulsar radiation may reflect the activity of the magnetosphere.

The pulse-to-pulse correlation reflects the discrete magnetosphere evolution in different rotating periods, and it also includes the correlation between the characteristic parameters of individual pulses or phase-resolved properties. The drifting sub-pulse phenomenon (Ruderman and Sutherland, 1975) is the typical example that the single pulses have pulse-to-pulse phase-resolved correlation, which implies the motion of the radiation position. Another example is the mode-switching phenomenon (Taylor *et al.*, 1975), which describe the the pulse parameters jumping between different modes.

FAST is the most sensitive single-aperture radio telescope at present, and is appropriate for the research of single pulse. FAST with the 19-beam receiver has a SEFD of ∼1.49 Jy at 1.05–1.45 GHz (Jiang *et al.*, 2020), and can monitor the extremely small flux variation of pulsars. With the high sensitivity, FAST can play an important role in pulsar searching, monitoring and timing, especially in single pulse study. As a branch of pulsar monitoring, single pulse study has its particularity that the signal-to-noise ratio (SNR) of single pulse data can not be enhanced by increasing the integration time. The SNR of single pulse can be well determined by the telescope sensitivity, time resolution of data and the single pulse itself, and is entirely unrelated to the integration time. This means that FAST has ascendancy on single pulse study over other radio telescopes. It can be asserted that FAST can observe the known single-pulse-phenomena on more pulsars, and can discover more unknown single-pulse-phenomena.

3. FAST and Single Pulse Study

FAST can observe the known single-pulse-phenomena on more pulsars. The giant pulse, null pulse, drifting sub-pulse and mode switching are well-known single-pulse phenomena. But for many pulsars, it is difficult to judge whether these phenomena happen in the observed data because of the low SNR. In this case, the high sensitivity of FAST can play a significant role. For example, rotating radio transients (RRATs) is a population of bursting pulsar emitting bursts with apparent randomness but with an underlying periodicity, and therefore the burst rate of RRATs is the indicator of their activity. However, the bursts of RRATs are evaluated based on the telescope sensitivity, and then with FAST-observed data one can obtain larger burst rate than other telescopes. Even RRAT J1913+1330 shows mode switching between burst and quiescent states in FAST-observed data (Lu *et al.*, 2019). In the above example, the telescope sensitivity affects the scientific interpretation and classification on RRATs directly. Besides, the pulse-to-pulse correlation phenomena, such as the drifting sub-pulse, are also limited by the telescope sensitivity. For instance, PSR B1937+21 shows no sign of drifting sub-pulse in the analysis of past data, but in FAST-observed data, the drifting sub-pulse phenomenon is discovered. The result of pulse-to-pulse correlation analysis on FAST-observed data for PSR B1937+21 is shown in Figure 1(a), one can find that the large positive-correlation region in each panel is almost bar-shaped. It can be seen that the maximum curves of each bar-shaped region deviate from the diagonals (grey dashed lines), which implies a general movement of sub-pulses. The deviations from the diagonals to all the maximum-curves are shown in Figure 1(b) as blue, cyan (for the main pulse), green and magenta (for the inter pulse) dots with error-bars versus the delayed pulse number. The drifting rates can be fitted with the blue and green dots as $D_{\mathrm{main}} = -3.12 \pm 0.12\,\mu\mathrm{s/cycle}$ and $D_{\mathrm{inter}} = -3.18 \pm 0.1\,\mu\mathrm{s/cycle}$ for the sub-pulses of the main and inter pulse, respectively. The green and magenta (inter pulse) curves shown in Figure 1(a) can represent the other two positive-correlation regions on one side of the large positive-correlation region, separated by a non-correlation or negative-correlation region. This may be caused by the correlation between two adjacent sub-pulses, and the corresponding sub-pulse separations (P_2) for main and inter pulse are 28.3 ± 0.4 and $34.8 \pm 0.4\,\mu\mathrm{s}$, respectively. However, these two separations are just modal values, instead of stable values. Otherwise, the stable P_2 and drifting rate lead to a stable fluctuation period along a constant longitude. It is worth noting that the drifting rate derived from magenta dots is consistent with that from cyan dots, whereas the two drifting rates of green and blue dots have a large difference. This is an indication that the main pulse of PSR B1937+21 has multiple radiation components.

FAST can discover more unknown single-pulse-phenomena. On one hand, the realization on the known phenomena can be developed, and the analysis on the diffused drifting sub-pulse of PSR B2016+28 can be an example (the examples in this paragraph are cited from

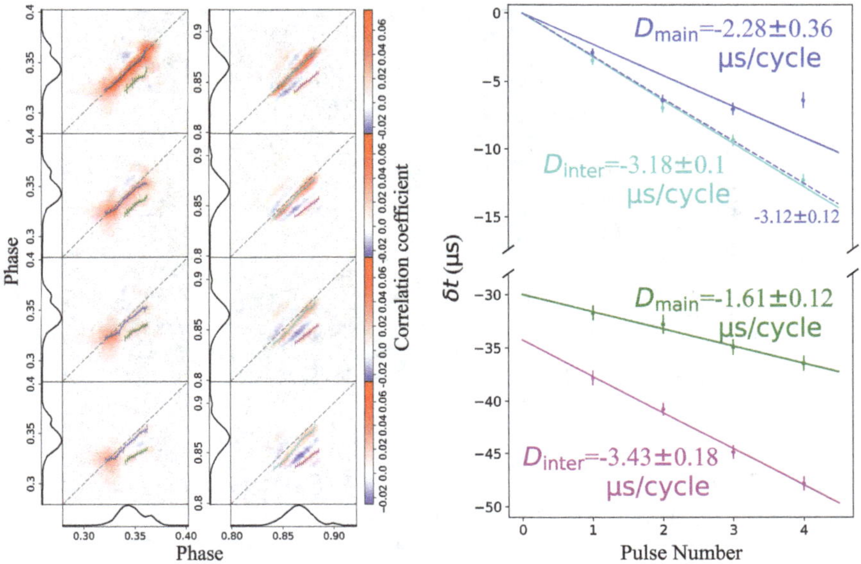

(a) (b)

Fig. 1. (a) The longitude-longitude cross-correlation with 1–4 pulse delays (from top to bottom) of main (left) and inter (right) pulses. Mean profiles are plotted to the left of and below the panels. In the image panels, the diagonals are drawn as grey dashed lines, and the large positive-correlation regions are fitted and shown in blue, green, cyan and magenta with error-bars. (b) Deviations from the diagonals to the fitted curves in sub-figure (a) versus number of delayed pulses, shown as dots with error-bars. Colors are consistent with those of the curves in sub-figure (a) The solid lines are the corresponding linear fitting results, while the blue dashed line is the fitting result using only the two blue dots with the first two pulses delayed.

Lu *et al.* (2019), and as well as Figures 2–4). For the sub-pulses of a diffused drifter like PSR B2016+28, the drift rate for a specified sub-pulse is stable, and for different sub-pulses, the drift rates would be different. The drift rate and modulation frequency of sub-pulse along the pulse-sequence varies with time, and the relation between them is shown in Figure 2. In the figure, the ordinate in the left panel is right the frequency corresponding to the modulation period along pulse series, P_3. The maximum of the spectra at each modulation frequency has been marked as red point and error bar, and it seems that there is some relation between drift rate D and modulation frequency F_3. One can assume that this relation have the following

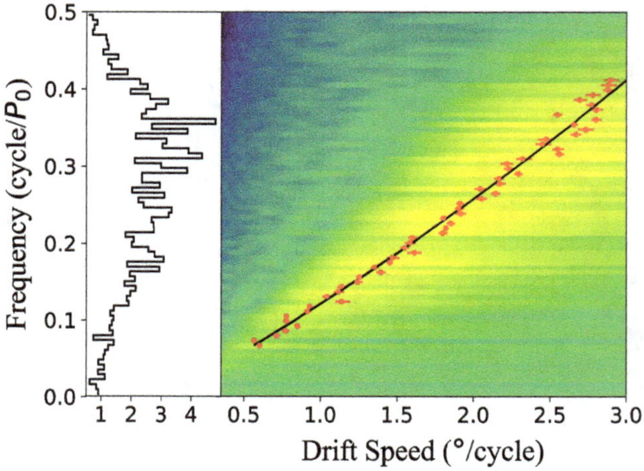

Fig. 2. The relevance between the drift rate and the modulation frequency of sub-pulse along the pulse-sequence is shown in right panel and the total modulation spectra are shown in the left panel. The red points are the maximums along drift rate axis, and the black curve is the fitting curve.

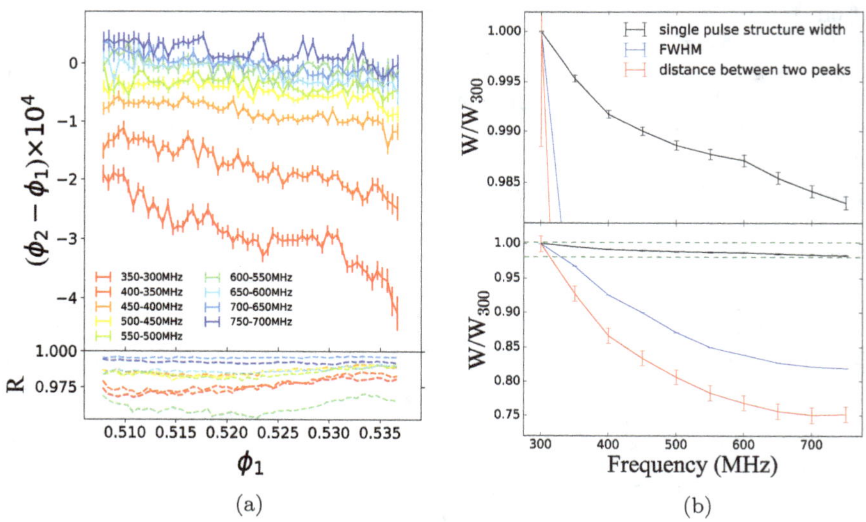

Fig. 3. (a) For the corresponding phases in two adjacent narrow bands, ϕ_2 and ϕ_1, the differences between them at specified ϕ_1 is shown in the top panel, and the bottom panel presents the linearly dependent coefficients. (b) The black curves show the evolution of single pulse structure width, while the red and blue curves show that of the mean pulse width.

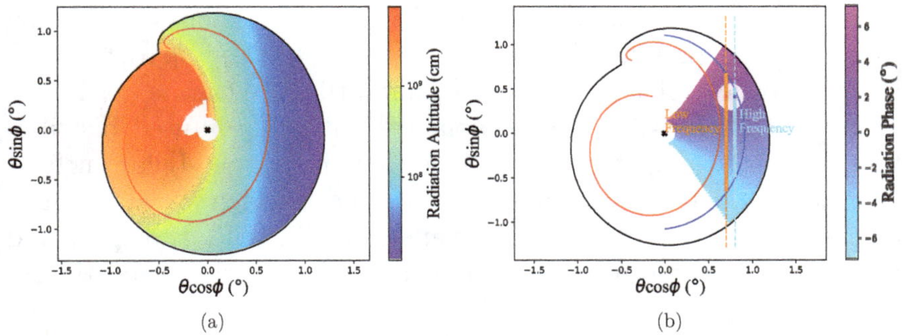

Fig. 4. (a) The polar cap region is shown in θ-ϕ map, where θ and ϕ are the zenith angle and position angle relative to magnetic axis. The radiation altitude on every magnetic field line is marked on the line foot with different color, and the feet of the last opening field lines and critical field lines are presented as black and red curves, respectively. (b) As panel (a), but for radiation pulse phase. The radiation is assumed as concentrating in the field lines which pass the blue curve, and the cyan and orange lines are the field line groups which radiate high and low frequency emission. Obviously, the solid segments denote the width of mean pulse profiles. Given a location of generating charged particles outflow as the white disc, the cyan and orange points would be the centers of the sub-pulse corresponding to the outflow at high and low frequencies.

form,

$$D = \frac{F_3}{k_d F_3 + b_d}, \tag{1}$$

where k_d and b_d are the parameters to be fitted. With the consideration that $P_3 = 1/F_3$ and $D = P_2/P_3$ (where P_2 is the sub-pulse separation), Eq. 1 can be rewritten as following:

$$P_2 = \frac{P_3}{\gamma_d P_3 + \beta_d}, \tag{2}$$

where γ_d and β_d are parameters, and can be fitted as 74 ± 1 and $20 \pm 3\,\mathrm{s}$, respectively. Theoretically, Ruderman and Sutherland (1975) ascribed the drifting sub-pulse to the drifting spark in polar cap region due to the electric field ΔE perpendicular to the magnetic field, and therefore the diffused drifting rate is caused by the distribution of the electric field. Ruderman and Sutherland (1975) (Eq. (30)) showed that ΔE is positively related to the electric field component parallel to magnetic field $E_{\parallel}$. It is known that the sparks are caused by $E_{\parallel}$, and therefore the sparks would be sparser at lower

$E_{\|}$ region, i.e., lower ΔE and lower drifting rate region. Considering that P_2 can represent the angular distance between sparks, hence, the negative correlation between P_2 and drift rate can be explained. Then the diffused drifting is explained with the diffused distribution of sparks, and it may be caused by the rough stellar surface. The electric field around the cusps on stellar surface would be much higher than the mean value, and could more likely lead to spark and radiation. On the other hand, besides developing the realization of known phenomena, FAST can discover new single pulse characteristics. The study on PSR B2016+28 can still be the example, whose single pulse can be analyzed as narrowing with frequency. At two frequencies, there must be two phases ϕ_1 and ϕ_2, at which the fluxes variation has the maximum correlations. Via considering the correlation phase pairs between different frequencies, the width evolution of single pulse structure with frequencies can be analyzed, and the results are shown in Figure 3. As shown in the figure, the mean pulse contracts by $\sim$20% from 300 to 750 MHz, while the single pulse structure only contracts by $\sim$1.5%. In fact, this single pulse characteristics reflects the radiation and dynamics geometry in the polar cap region. The radiation is generated in magnetosphere by the particles which originate from polar cap region, and every bunch of particles will lead to a sub-pulse, which contains wideband radiation. The radiation altitude and phase on each field line can be calculated, which is shown in Figure 4. As shown in the figure, the evolution of single pulse structure width with frequency is explained clearly, and the shrink speed of single pulse must be slower than that of the mean pulse profile, obviously.

In additional, The single-pulse-characteristics can be used in other pulsar study. For example, the phase-resolved single-pulse-characteristics can be used as an indicator of baseline. The duty cycle of most radio pulsar is $\sim$10%, and it seems like common sense. However, in the view of FAST, the weak radiation are observed in more phase region. Even for some pulsars the radiation can be observed over the whole pulse phase, which leads to difficulty on determining the baseline. Worse, it is also difficult to judge whether the periodic radiation exists on specified pulse phase. Fortunately, the single-pulse-characteristics can reflect some radiation information, and the radiation can be found with this thought (Wang *et al.*, 2022).

4. Future Single Pulse Study with FAST

It can be expected that the main development on the single pulse study in the next decade depends on the observation of FAST due to the extremely high sensitivity. With FAST data, more detailed physical processes in the pulsar magnetosphere can be analyzed.

At present, FAST observed lots of pulsars and produced huge mounts of data, but most of the research methods still concentrate on the slice of the data along a specified dimension due to the limitation on considering in the high-dimension phase space of data. As the high SNR of FAST pulsar data, the information contained in the data is far from having been explored, i.e., the utilization of the data is very low. We may need more powerful mathematic tools to extract the counter-intuitive information in the data, and the study on the single pulse is just beginning.

Bibliography

Arons, J. and Scharlemann, E. T. (1979). Pair formation above pulsar polar caps: Structure of the low altitude acceleration zone, *ApJ* **231**, 854–879.

Buschauer, R. and Benford, G. (1976). General theory of coherent curvature radiation, *MNRAS* **177**, 109–136.

Cheng, K. S., Ho, C., and Ruderman, M. (1986). Energetic radiation from rapidly spinning pulsars. I. Outer magnetosphere gaps, *ApJ* **300**, 500.

Gajjar, V., Joshi, B. C., and Kramer, M. (2012). Broadband nulling behaviour of PSR B2319+60, *Electromag. Radiat. Puls. Mag.* **466**, 79.

Goldreich, P. and Julian, W. H. (1969). Pulsar electrodynamics, *ApJ* **157**, 869.

Hewish, A., Bell, S. J., Pilkington, J. D. H. *et al.* (1968). Observation of a rapidly pulsating radio source, *Nature* **217**, 5130, 709–713.

Jessner, A., Słowikowska, A., Klein, B. *et al.* (2005). Giant radio pulses from the Crab pulsar, *Adv. Space Res.* **35**, 6, 1166–1171.

Jiang, P., Tang, N. Y., Hou, L. G. *et al.* (2020). The fundamental performance of FAST with 19-beam receiver at L band, *RAA* **20**, 5, 064.

Kramer, M. (1994). Geometrical analysis of average pulsar profiles using multi-component Gaussian FITS at several frequencies. II. Individual results. *A&AS* **107**, 527–539.

Li, J., Spitkovsky, A. and Tchekhovskoy, A. (2012). Resistive solutions for pulsar magnetospheres, *ApJ* **746**, 1, 60–71.

Lu, J. G., Du, Y. J., Hao, L. F. *et al.* (2016). Multi-frequency radio profiles of PSR B1133+16: Radiation location and particle energy, *ApJ* **816**, 2, 76–87.

Lu, J., Peng, B., Xu, R. *et al.* (2019). The radiation structure of PSR B2016+28 observed with FAST, *Sci. China Phys. Mech. Astron.* **62**, 5, 959505.

Lu, J., Peng, B., Liu, K. *et al.* (2019). Study of three rotating radio transients with FAST, *Sci. China Phys. Mech. Astron.* **62**, 5, 959503.

Lyne, A. G. and Manchester, R. N. (1988). The shape of pulsar radio beams, *Mon. Notices Royal Astron. Soc.* **234**, 477–508.

Muslimov, A. G. and Harding, A. K. (2003). Extended acceleration in slot gaps and pulsar high-energy emission, *ApJ* **588**, 1, 430–440.

Philippov, A. A. and Spitkovsky, A. (2014). Ab initio pulsar magnetosphere: Three-dimensional particle-in-cell simulations of axisymmetric pulsars, *ApJ* **785**, 2, 33–38.

Philippov, A. A., Cerutti, B., Tchekhovskoy, A. *et al.* (2015). Ab initio pulsar magnetosphere: The role of general relativity, *ApJ* **815**, 2, 19–23.

Qiao, G.-J., Lee, K.-J., Zhang, B. *et al.* (2007). An annular gap acceleration model for x-ray emission of pulsars, *Ch. J. A. A.* **7**, 496–502.

Qiao, G. J., Liu, J. F., Zhang, B. *et al.* (2001). An inverse compton scattering (ICS) model of pulsar emission. II. Frequency behavior of pulse profiles, *A&A* **377**, 964–971.

Rankin, J. M. (1983). Toward an empirical theory of pulsar emission. I. Morphological taxonomy, *ApJ* **274**, 333–358.

Ruderman, M. A. and Sutherland, P. G. (1975). Theory of pulsars: polar gaps, sparks, and coherent microwave radiation, *ApJ* **196**, 51–72.

Taylor, J. H., Manchester, R. N., and Huguenin, G. R. (1975). Observations of pulsar radio emission. I. Total intensity measurements of individual pulses, *ApJ* **195**, 513–C528.

Thorsett, S. E. (1991). Frequency dependence of pulsar integrated profiles, *ApJ* **377**, 263.

Wang, Z., Lu, J., Jiang, J. *et al.* (2022). The radio signal of PSR B0950+08 is detected over the whole pulse phase, *MNRAS* **517**, 4, 5560–5568.

Xilouris, K. M., Kramer, M., Jessner, A. *et al.* (1996). Emission properties of pulsars at mm-wavelengths, *A&A* **309**, 481–492.

https://doi.org/10.1142/9789819800643_0007

Chapter 7

FAST's Impact on Fast Radio Burst (FRB)

Wei-Yang Wang[*,†,‡,¶] **and Kejia Lee**[‡,§]

[*] *School of Astronomy and Space Science,*
University of Chinese Academy of Sciences,
Beijing 100049, Peoples Republic of China
[†] *School of Physics and State Key Laboratory of*
Nuclear Physics and Technology,
Peking University, Beijing 100871,
Peoples Republic of China
[‡] *Kavli Institute for Astronomy and Astrophysics, Peking University,*
Beijing 100871, Peoples Republic of China
[§] *National Astronomical Observatories, Chinese Academy of Sciences,*
Beijing 100101, Peoples Republic of China
[¶] *wywang@ucas.ac.cn*

Fast radio bursts (FRBs) are millisecond-duration radio flashes with extremely high bright temperatures but unknown origin(s). The Five-hundred-meter aperture spherical radio telescope (FAST) is one of the perfect facilities for pulsar-related sciences owing to the high sensitivity, especially FRBs, which is phenomenologically similar to pulsars' individual pulses. As expected, FAST has had a profound impact on FRB science since its operation in 2019. The discovery of more sources and deeper characterisations of their properties are key to understanding their physical origin(s) and radiation mechanism. In this chapter, we will review FAST's achievements in the field of FRB in only a few short years.

Keywords: Fast radio burst, Pulsar, Magnetar, Radio transient

Contents

1. A Brief History of FRB

Fast radio bursts (FRBs) are millisecond blips or transients in the radio band (transient radio pulses with unknown astrophysical origin(s)). The discovery of FRB opens a new era of astronomy and the field has witnessed rapid growth on the frontier of observation and theory.

Interestingly, a bright radio single pulse was found and recorded in Parkes multibeam pulsar surveys, while was announced six years later (Lorimer *et al.*, 2007). The phenomenon was then generally recognized as a new type of cosmic transient after more and more events have been detected (Thornton *et al.*, 2013).

The cosmological origin of FRBs was established after the first repeating source, FRB 20121102A, was found (Spitler *et al.*, 2016) and localized in a star-forming dwarf galaxy at $z = 0.193$ with an associated persistent radio source (Bassa *et al.*, 2017; Chatterjee *et al.*, 2017; Marcote *et al.*, 2017; Tendulkar *et al.*, 2017) and an extreme magneto-ionic environment (Michilli *et al.*, 2018). A large sample of FRBs was given, thanks to the unprecedented field of view of the Canadian Hydrogen Intensity Mapping Experiment (CHIME). The interferometric arrays, represented by Australian Square Kilometer Array Pathfinder (ASKAP), play a important role in the accurate localization of individual bursts. For at least some sources, bursts sometimes show time-frequency drifting patterns after dedispersion (Hessels *et al.*, 2019). The discovered FRB sources has exceeded 800 catalogued on the Transient Name Server[1] and dozens of them have redshift measurements. Surprisingly, a two-component bright radio burst with FRB-like time-frequency pattern was discovered in the Galactic magnetar SGR J1935+2154 during the X-ray active state (Bochenek *et al.*, 2020; CHIME/FRB Collaboration *et al.*, 2020), suggesting that magnetars are the origin of at least some FRBs.

Based on the distance of the emission region from the neutron star, FRB models can be generally divided into two categories (see Zhang, 2020) for a review: emission within the magnetosphere of a compact object (pulsar-like models), and emission from relativistic shocks far outside the magnetospheres (gamma-ray burst-like models). However, the radiation mechanism and the physical origin of FRBs, as well as whether all FRB can repeat, still remain mystery. The study of FRB enters a new era and all these problems may be solved owing to the extremely high sensitivity of

[1]See https://www.wis-tns.org/.

the China's five-hundred-meter aperture spherical radio telescope (FAST).

2. The Emerging Burst Samples Detected by FAST

2.1. *The first repeater confirmed by the FAST*

FRB 20180301A, was previously discovered as an one-off event by the Parkes 64-m radio telescope (Price *et al.*, 2019). The FRB source was supposed to be non-repeating, but four bursts from it were detected in July 2019 during work on one of the so-called FAST 'shared-risk' projects (Luo *et al.*, 2020b). After that, other 11 bursts with roughly the same Dispersion Measure (DM) of $522\,\mathrm{pc\,cm^{-3}}$ as the previous bursts, were found at the previously reported position during the observation in September and October 2019. All the 15 bursts have intrinsic widths of a few microseconds with no apparent evidence of scattering, and some of them exhibit frequency down-drifting patterns, resembling other repeaters.

2.2. *A large sample of burst for FRB 20121102A*

FRB 20121102A entered a newly active episode near August 2019, and 1,652 independent burst events were detected during a continuous monitoring observation in a total of 59.5 h by the FAST (Li *et al.*, 2021). It has refreshed the previous record of burst number within an active episode, and the total number of 374 bursts from one source (Gourdji *et al.*, 2019; Zhang *et al.*, 2018). The burst rate can reach $122\,\mathrm{h^{-1}}$ over the respective averaged 1-h sessions. FAST's detection reveals the burst distribution with low energy that was not known before. As shown in Figure 1, the burst energy distribution is bimodal, and well characterized by a combination of a generalized Cauchy function and a log-normal function. A frequently triggered mechanism with a high radiative efficiency is required to generate such FRB clusters. Although the large number of bursts in hour-long spans allows sensitive periodicity searches from 1 ms to 1000 s, no periodic or quasi-periodic events have been found.

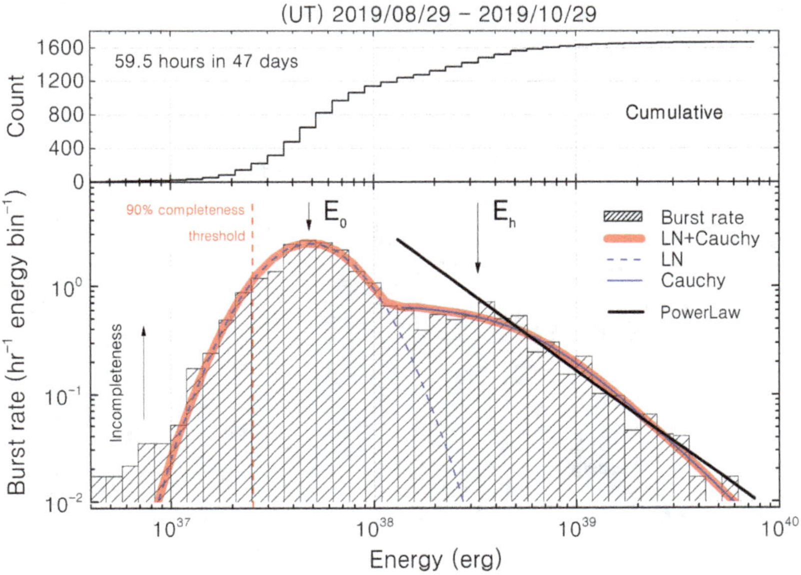

Fig. 1. Burst rate distribution of the isotropic equivalent energy at 1.25 GHz for FRB 20121102A. The bimodal lognormal (LN) (dashed blue) + Cauchy (solid blue) model is presented in red and a single power-law fit for bursts above a certain threshold $E_h = 3 \times 10^{38}$ erg is shown in black. The upward arrow denotes missed weak bursts below the 90% detection completeness threshold. The figure is quoted from Li *et al.* (2021).

2.3. *Highly dispersed FRBs from the CRAFTS*

Five newly discovered FRBs, including four newly apparently one-off FRBs and one repeater (FRB 20190520B), were reported from an analysis of $\sim$1500 h during the Commensal Radio Astronomy FAST Survey (CRAFTS[2]; (Niu *et al.*, 2021, 2022a; Zhu *et al.*, 2020)). Follow-up observations toward the positions of the four apparently one-off FRBs have not detected any repeating bursts above the fluence level 0.012 Jy ms so far. All these bursts have large DMs exceeding 1000 pc cm^{-3}. The discovery of such high-DM FRBs extended the

[2]The CRAFTS is a multi-purpose all-sky survey, which aims to obtain the data streams of pulsar searching, transients searching, HI imaging and HI galaxies simultaneously (Li *et al.*, 2018).

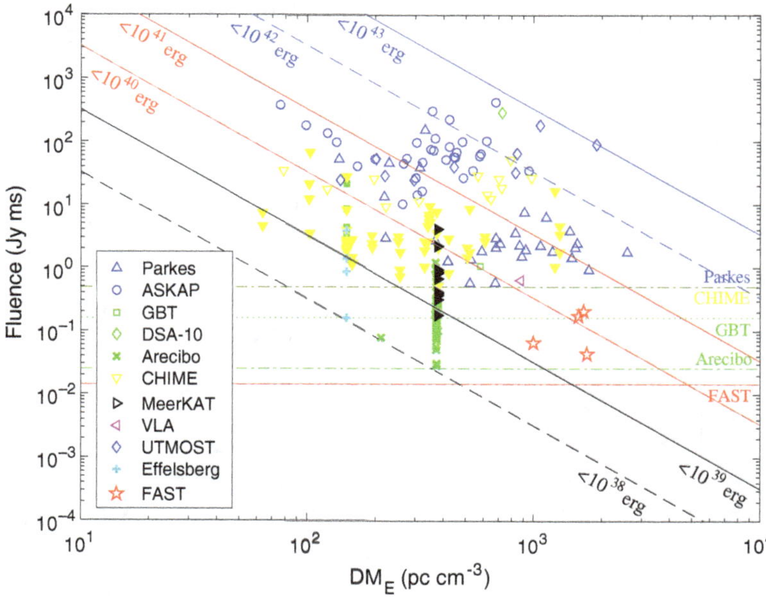

Fig. 2. A radio Fluence DM_E diagram for FRBs, where DM_E represents the DM value after deducting the DM contribution from the Milky Way. The contours of the energy upper limits are plotted by slanted lines, with respective numerical values of E labeled. The sensitivities of the telescopes are marked by horizontal lines. The FRBs detected by different telescopes are represented by different symbols, in which the four apparently one-off bursts for FAST are red pentagrams. The filled symbols represent known repeaters. The figure is quoted from (Niu *et al.*, 2021).

DM-fluence relation to more distant and low fluence region, as shown in Figure 2. Identifying these events with potential low luminosity would be essential to constrain the low end of FRB luminosity function (Luo *et al.*, 2018, 2020).

2.4. *A regularly active repeater with a persistent radio source*

FRB 20190520B is the first repeating FRB source discovered by the FAST during CRAFTS (Niu *et al.*, 2022a). Around 79 bursts from FRB 20190520B were detected during the initial draft scan and the follow-up tracking observations between April 2020 and

September 2020, and the burst rate of $4.5^{+1.9}_{-1.5}\,\mathrm{h}^{-1}$ can be estimated by assuming a Weibull distribution of the burst waiting time. The source shows intriguingly regular activity that two or more bursts can be detected in each monitoring observation of the FAST. FRB 20190520B was then co-localized with a compact, persistent radio source and associated with a dwarf host galaxy of high specific-star-formation rate at redshift of 0.241 ± 0.001, which shows striking resemblance to the first identified repeater FRB 20121102A. However, the noticeable difference for FRB 20190520B is that the DM attributed to the host galaxy, including its halo and any gas local to the FRB source, far exceeds that contributed from the intergalactic medium (see Figure 3), which is indicative of a highly dense environment of the FRB source possibly.

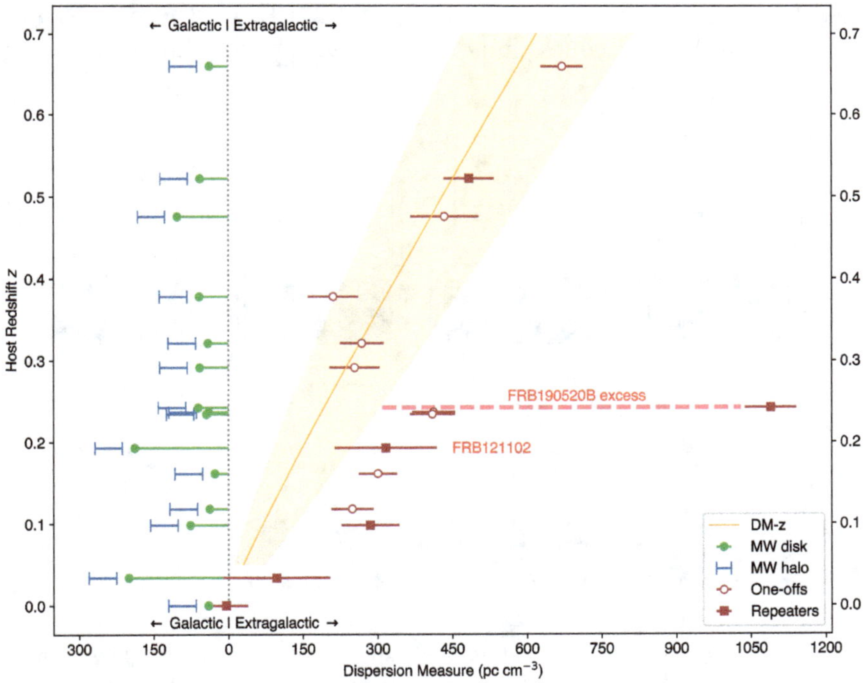

Fig. 3. Galactic and extra-galactic contributions to the DM observed for 15 redshifts-measured FRBs. The DM$-z$ relationship refers to Macquart *et al.* (2020). The figure is quoted from Niu *et al.* (2022a).

3. A Variety of Polarization Property of FAST FRB

3.1. *Diverse polarization angle swings of FRB 20180301A*

There are 15 bursts from FRB 20180301A been detected with a total observing time of 12 h, and 7 of them have detailed polarization measurements (Luo *et al.*, 2020b). The bursts show a diversity of polarimetric properties. These bursts are dominated by linear polarization, some with a degree of polarization of >70%, but no apparent circular polarization been found. Five bursts exhibit various polarization angle (PA) swings (see Figure 4), which are reminiscent to the rotation vector model of pulsar. The varying PA can be a natural consequence of a rotating magnetosphere, but is inconsistent with a relativistic shocks far outside the magnetosphere, which predicts a constant PA curves since the upstream ordered magnetic field has a fixed direction (Metzger *et al.*, 2019). Furthermore, rotation measure (RM) values range from $521.5^{+4.6}_{-4.2}$ to $564.4^{+3.4}_{-4.0}$ rad m^{-2}

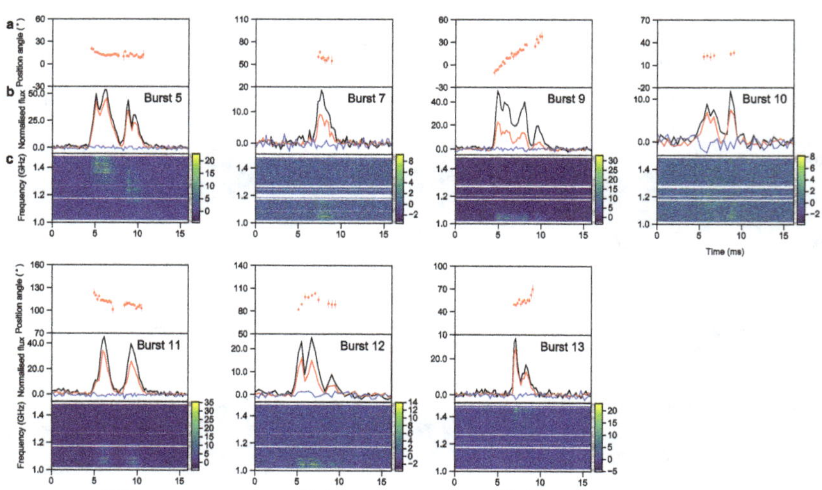

Fig. 4. Polarization profiles and dynamic spectra of the seven brightest bursts from FRB 20180301A. PAs of linear polarization at infinite frequency are on the top. Polarization pulse profiles are on the middle (black, red and blue curves denote total intensity, linear polarization and circular polarization). Dynamic spectra of the total intensity as a function of frequency and time are on the bottom. The figure is quoted from Luo *et al.* (2020b).

(68% confidence level) and there is also burst-to-burst RM evolution within several days, while DM values keep a constant, suggesting a complex magneto-ionic environment of the source.

3.2. *Frequency-dependent polarization of repeating FRBs*

Polarization measurement may be a potential tool to obtain more information about the FRB hosts and environments. A variety of polarization properties have been found for five repeaters (FRB 20121102A, FRB 20190520B, FRB 20190303A, FRB 20190417A, and FRB 20201124A) by using the FAST (Feng *et al.*, 2022). Compared with previous observations of FRB which show that almost 100% linear polarization at 3–8 GHz (Michilli *et al.*, 2018), there is no apparent linear polarization (6% upper limit) at 1.0–1.5 GHz. FRB 20190520B has an upper limit of 20% on the degree of linear polarization at 1.0–1.5 GHz, which is smaller than that in the range of 4 to 8 GHz. A trend of lower polarization at lower frequencies is found (see Figure 5), and the behavior can be characterized as multipath scattering by deriving σ_{RM}, the RM scatter, which quantifies the

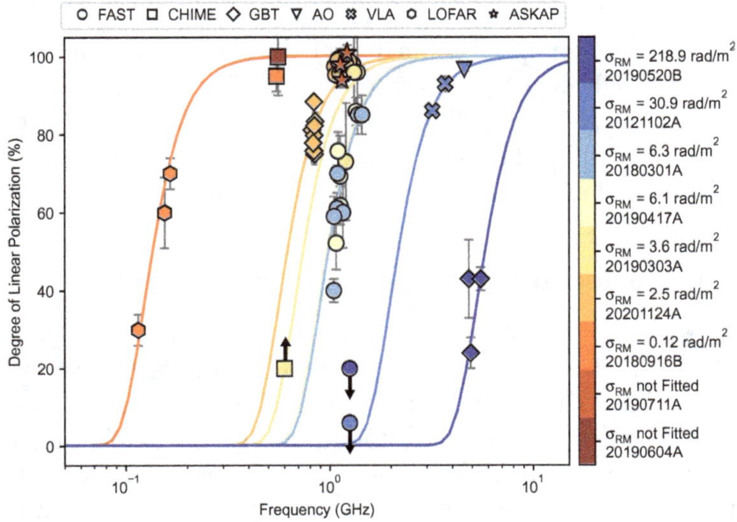

Fig. 5. The degree of linear polarization for FRB sources. Symbol shapes indicate the telescope used for each observation. Different sources are presented by different colors. The figure is quoted from Feng *et al.* (2022).

complexity of magnetized environments associated with repeating FRBs. FRB sources with higher $\sigma_{\rm RM}$ can have higher RM magnitude and scattering timescales. These properties seem to indicate that there is a very complex magnetized environment near the FRB sources, such as ch as young stellar populations.

3.3. *The complex magneto-ionic environment of FRB 20201124A*

FRB 20201124A, was reported to enter a high-level active state in March 2021 (Chime/Frb Collaboration, 2021). Triggered by this alert, the FRB source was monitored and 1863 polarised bursts were detected by the FAST from 2021 April 1 to June 11 with a 96.9 h total observation time (Xu *et al.*, 2022). A large number of radio burst samples covering 54 days of observation showed that the Faraday RM of the source had significant, irregular and short-term variation from -887.2 ± 0.7 to $-362.7^{+2.9}_{-1.4}$ $\rm rad\,m^{-2}$ in the first 36 days, and the RM remained constant in the following 18 days, as shown in Figure 6. Surprisingly, a burst with 75% circular polarization degree broke the previous record of 47% (Kumar *et al.*, 2022). For a limited number of bursts, circular and linear degrees of polarization can oscillate with λ^2 (see Figure 7). These features matches a more complicated, dynamically evolving, magnetised immediate environment around the FRB source. Although such λ^2-dependent oscillation can be brought by the radiative transferring effects during the propagation of the radio waves, an intrinsic radiation mechanism is still required to produce the highly circularly polarized waves (e.g., Wang *et al.*, 2022).

3.4. *A newly active epoch of FRB 20201124A*

An almost daily monitoring of FRB 20201124A from September 25 to October 17 in 2021 found several hundreds bursts, indicating that FRB 20201124A enters a newly active epoch. A variety of burst morphology has been found downward, including downward drifting, upward drifting, no drifting and complex structures (Zhou *et al.*, 2022). The emission spectra of the sub-bursts are typically narrow and scintillation bandwidth is varying. The cumulative burst energy distribution can be well fitted by an exponentially connected broken-power law function with the lower and higher-energy

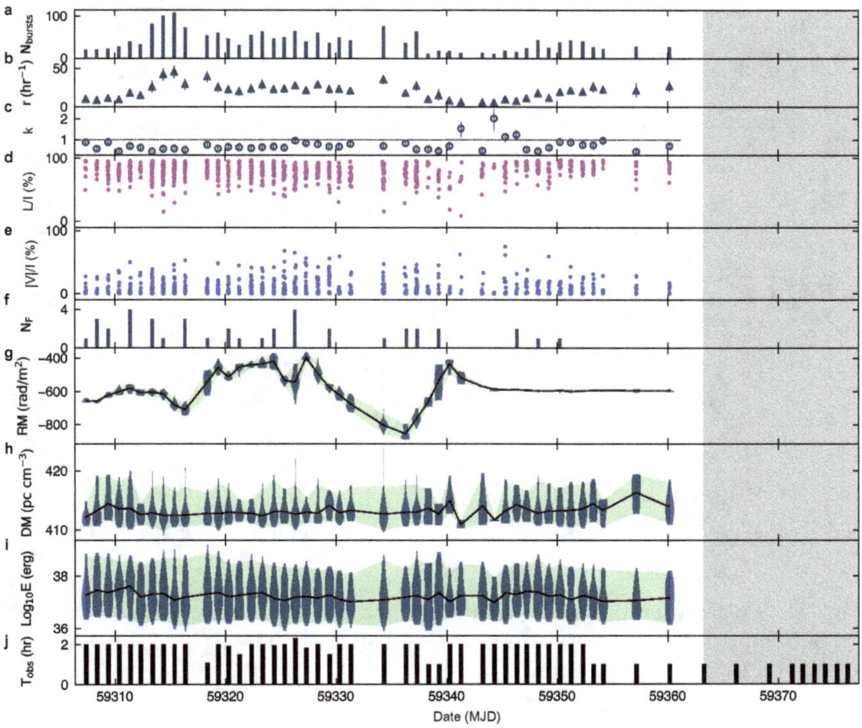

Fig. 6. Overview of our radio observational campaign and temporal variations of the physical parameters for FRB 20201124A. The grey shaded region on the right side of the plot shows the epoch when no burst was detected. The figure is quoted from Xu *et al.* (2022).

indices being -1.22 ± 0.01 and -4.27 ± 0.23 (Zhang *et al.*, 2022). The Faraday rotation measures monotonically dropped from -579 to $-605\,\text{rad}\,\text{m}^{-2}$ in the 4-day window, which is comparable with the previous 4-months observation (Xu *et al.*, 2022), but the RM evolution rate was at least an order of magnitude smaller than the previous reported during the rapid RM-variation phase. Most bursts (more than 90%) are highly polarized with the total degree of polarization greater than 90% (Jiang *et al.*, 2022). For the temporal properties, the waiting time can be modeled by two log-normal functions peaking at 51.22 ms and 10.05 s, and there is no significant periodicity from 1 ms to 100 s (Niu *et al.*, 2022b).

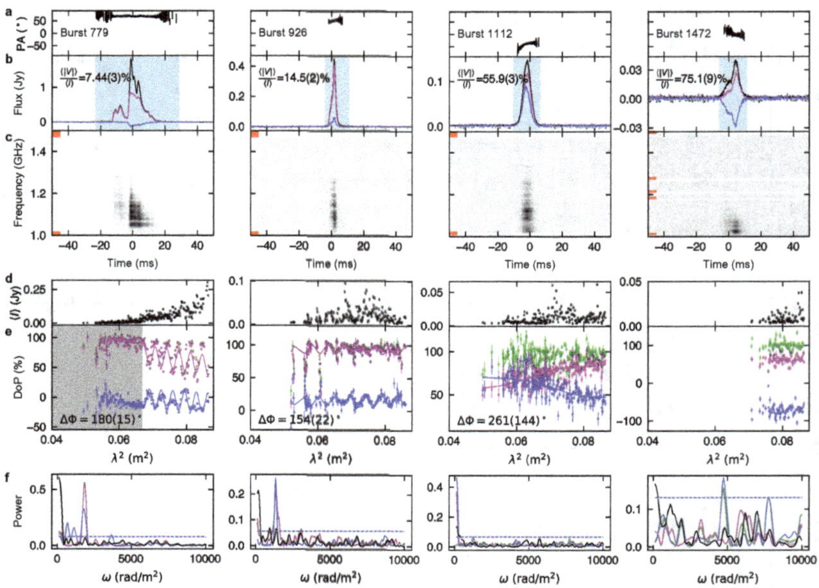

Fig. 7. Polarisation profiles, dynamic spectra, degree of polarisation, and spectra of band-pass of four bursts from FRB 20201124A. The figure is quoted from Xu *et al.* (2022).

4. A Radio-Emitting Magnetar SGR J1935+2154

4.1. *The most urgent limitation on magnetar's radio bursts*

A Galactic magnetar, SRG J1935+2154 was alerted to enter an active episode by emitting intense soft γ-ray bursts. One FRB-like event (the so-called FRB 20200428) was dependently detected by the STARE2 [Bochenek *et al.* (2020)] and the CHIME (CHIME/FRB Collaboration *et al.*, 2020) (with two peaks for the CHIME), in association with a X-ray burst (Li *et al.*, 2021; Mereghetti *et al.*, 2020; Ridnaia *et al.*, 2021; Tavani *et al.*, 2021). The Galactic magnetar was monitored by the FAST for 8 h comprising four sessions, in which 29 soft-γ-ray repeater (SGR) bursts were detected in the third session (Lin *et al.*, 2020). No single dispersed pulsed emission coincident with the arrivals of the SGR bursts and the FRB-like event. A fluence upper limit is placed by the non-detection which is eight orders of magnitude lower than the fluence of FRB 200428. FRB-SGR burst associations are rare possibly, because the physical conditions to generate coherent radio emission is hard to be required, or the FRB emission is highly beamed.

4.2. *A radio pulsar phase of SGR J1935+2154*

The discovery of FRB 20200428 suggests that at least some FRBs originate from magnetars. However, most magnetars are radio-quiet and occasionally emit radio pulsations after X-ray outburst. There are 795 pulses detected in 16.5 hours over 13 days by the FAST, which luminosities are eight decades fainter than FRB 20200428, and the pulses are emitted in a narrow phase window anti-aligned with the X-ray pulsation profile (Zhu *et al.*, 2023). The radio pulses seem to originate from a fixed region within the magnetosphere, while the burst occur in random locations. Combined with X-ray timing observations, a giant glitch seems to happen several days before FRB 20200428 (Ge *et al.*, 2022). The triggered giant glitch likely changed the magnetospheric structure dramatically in favour of FRB generation, and then enters a new epoch of normal radio pulses.

5. Summary and Outlook

The field of FRB is experiencing the era of large sample of source, but the nature of FRBs is still in debate: What is the radiation mechanism and the physical origin of FRBs? As expected, the FAST have made some great achievements on FRB science in these few years. More and more apparently one-off bursts are identified as repeaters, and the deep observation can establish samples with a tremendous amount of burst. Careful polarization measurements can carry more physical information about the radiation mechanism and the source environment. Furthermore, the FAST can also give the most strict upper limit of SGR J1935+2154 during the active phase. The unknown switch mechanism of the emission states (burst state and pulsar state) may provide new opportunities for understanding FRBs and even magnetostars' radiation. All the achievements reflect that the FAST is a superb instrument for monitoring FRB and would be the right tool to reveal the nature of FRB, and one may anticipate more in the future.

Bibliography

Bassa, C. G. and 18 colleagues. (2017). FRB 2020121102A is coincident with a star-forming region in its host galaxy, *ApJL* **843**, 8.

Bochenek, C. D. and 6 colleagues. (2020). A fast radio burst associated with a galactic magnetar, *Nature* **587**, 59–62.

Chatterjee, S. and 24 colleagues. (2017). A direct localization of a fast radio burst and its host, *Nature* **541**, 58–61.

CHIME/FRB Collaboration and 70 colleagues. (2020). A bright millisecond-duration radio burst from a galactic magnetar, *Nature* **587**, 54–58.

Chime/Frb Collaboration. (2021). Recent high activity from a repeating fast radio burst discovered by CHIME/FRB, *The Astronomer's Telegram*, 14497.

Feng, Y. and 25 colleagues. (2022). Frequency-dependent polarization of repeating fast radio bursts — implications for their origin, *Sci.* **375**, 1266–1270.

Ge, M. and 13 colleagues. (2022). A giant glitch from the magnetar SGR J1935+2154 before FRB 200428 *arXiv.2211.03246*.

Gourdji, K. and 6 colleagues. (2019). A sample of low-energy bursts from FRB 121102, *ApJL* **877**, 19.

Hessels, J. W. T. and 25 colleagues. (2019). FRB 121102 bursts show complex time-frequency structure, *ApJ* **876**, 23.

Jiang, J.-C. and 28 colleagues. (2022). FAST observations of an extremely active episode of FRB 20201124A. III. Polarimetry, *RAA* **22**, 124003.

Kumar, P. and 6 colleagues. (2022). Circularly polarized radio emission from the repeating fast radio burst source FRB 20201124A, *MNRAS* **512**, 3400–3413.

Li, D. and 10 colleagues. (2018). FAST in space: Considerations for a multibeam, multipurpose survey using china's 500-m aperture spherical radio telescope (FAST). *IEEE Microwave Magazine* **19**, 112–119.

Li, D. and 31 colleagues. (2021). A bimodal burst energy distribution of a repeating fast radio burst source, *Nature* **598**, 267–271.

Lin, L. and 47 colleagues. (2020). No pulsed radio emission during a bursting phase of a galactic magnetar, *Nature* **587**, 63–65.

Lorimer, D. R., Bailes, M., McLaughlin, M. A., Narkevic, D. J., and Crawford, F. (2007). A bright millisecond radio burst of extragalactic origin, *Sci.* **318**, 777–780.

Luo, R., Lee, K., Lorimer, D. R., and Zhang, B. (2018). On the normalized FRB luminosity function, *MNRAS* **481**, 2320–2337.

Luo, R., Men, Y., Lee, K., Wang, W., Lorimer, D. R., and Zhang, B. (2020). On the FRB luminosity function – II. Event rate density, *MNRAS* **494**, 665–679.

Luo, R. and 36 colleagues. (2020). Diverse polarization angle swings from a repeating fast radio burst source, *Nature* **586**, 693–696.

Macquart, J.-P. and 15 colleagues. (2020). A census of baryons in the universe from localized fast radio bursts, *Nature* **581**, 391–395.

Marcote, B. and 28 colleagues. (2017). The repeating fast radio burst FRB 121102 as seen on milliarcsecond angular scales, *ApJL* **834**, 8.

Mereghetti, S. and 25 colleagues. (2020). Integral discovery of a burst with associated radio emission from the magnetar SGR 1935+2154, *ApJL* **898**, L29.

Metzger, B. D., Margalit, B., and Sironi, L. (2019). Fast radio bursts as synchrotron maser emission from decelerating relativistic blast waves, *MNRAS* **485**, 4091–4106.

Michilli, D. and 33 colleagues. (2018). An extreme magneto-ionic environment associated with the fast radio burst source FRB 121102, *Nature* **553**, 182–185.

Niu, C.-H. and 23 colleagues. (2021). CRAFTS for fast radio bursts: Extending the dispersion-fluence relation with new FRBs detected by FAST, *ApJL* **909**, 8.

Niu, C.-H. and 34 colleagues. (2022a). A repeating fast radio burst associated with a persistent radio source, *Nature* **606**, 873–877.

Niu, J.-R. and 22 colleagues. (2022b). FAST observations of an extremely active episode of FRB 20201124A. IV. Spin period search, *RAA* **22**, 124004.

Price, D. C. and 32 colleagues. (2019). A fast radio burst with frequency-dependent polarization detected during breakthrough listen observations, *MNRAS* **486**, 3636–3646.

Ridnaia, A. and 10 colleagues. (2021). A peculiar hard X-ray counterpart of a galactic fast radio burst, *Nat. Astron.* **5**, 372–377.

Spitler, L. G. and 23 colleagues. (2016). A repeating fast radio burst, *Nature* **531**, 202–205.

Tavani, M. and 62 colleagues. (2021). An X-ray burst from a magnetar enlightening the mechanism of fast radio bursts, *Nat. Astron.* **5**, 401–407.

Tendulkar, S. P. and 23 colleagues. (2017). The host galaxy and redshift of the repeating fast radio burst FRB 121102, *ApJL* **834**, 7.

Thornton, D. and 19 colleagues. (2013). A population of fast radio bursts at cosmological distances, *Science* **341**, 53–56.

Xu, H. and 73 colleagues. (2022). A fast radio burst source at a complex magnetized site in a barred galaxy, *Nature* **609**, 685–688.

Wang, W.-Y. and 7 colleagues. (2022). Repeating fast radio bursts: Coherent circular polarization by bunches, *Sci. China Phys. Mech. Astron.* **65**, article id.289511.

Zhang, Y. G. and 6 colleagues. (2018). Fast radio burst 121102 pulse detection and periodicity: A machine learning approach, *ApJ* **866**, 149.

Zhang, B. (2020). The physical mechanisms of fast radio bursts, *Nature* **587**, 45–53.

Zhang, Y.-K. and 19 colleagues. (2022). FAST Observations of an extremely active episode of FRB 20201124A. II. Energy distribution, *RAA* **22**, 124002.

Zhou, D. J. and 18 colleagues. (2022). FAST Observations of an extremely active episode of FRB 20201124A: I. Burst morphology, *RAA* **22**, 124001.

Zhu, W. and 69 colleagues. (2020). A fast radio burst discovered in FAST drift scan survey, *ApJL* **895**, L6.

Zhu, W. and 69 colleagues. (2020). A radio pulsar phase from SGR J1935+2154 provides clues to the magnetar FRB mechanism, *Sci. Adv.* **9**, ead6198.

https://doi.org/10.1142/9789819800643_0008

Chapter 8

The Hard X-ray Modulation Telescope (HXMT)

Mingyu Ge, Wentao Ye, and Fangjun Lu
Institute of High Energy Physics,
Chinese Academy of Sciences, Beijing 100049, China
gemy@ihep.ac.cn

Though the great progress has been understood from the multi-band observations and theoretical work since the discovery of pulsar more than fifty years ago, some basic problems are not solved, such as radiation process in pulsating ultraluminous X-ray source, the origins of fast radio burst and so on. X-ray observations could supply crucial information to understand the characteristics and behaviors of pulsars. Recently, some progress for these basic problems has been achieved via the observations of *Insight*-HXMT.

Keywords: Pulsars, X-ray pulsar, Magnetar, Type I burst, Magnet field

Contents

1. Introduction

Observational X-ray astronomy in China began in the late 1970s with balloon-borne X-ray telescopes to observe bright X-ray pulsars and X-ray binaries (Dai *et al.*, 1987; Lu *et al.*, 1995). Due to atmospheric absorption at the balloon's altitude, only observations in the hard X-ray range can be carried out. The early experiences of Chinese X-ray astrophysicists in space observations are mostly related to hard X-ray detectors such as scintillators and multi-wire proportional chambers.

It is difficult to perform hard X-ray imaging, as hard X-ray photons can not be focused easily. Consequently, hard X-ray astronomy has been less developed in the early days of X-ray astronomy. An innovative image reconstruction method, the direct demodulation method, was thus proposed by Li and Wu to obtain images using non-imaging instruments (Li & Wu, 1993, 1994; Li *et al.*, 1993). It can reach a high spatial resolution and sensitivity. In September 1993, a collimated hard X-ray telescope named HAPI-4 was carried by a balloon to an altitude of 36–38 km, where it scanned the Cygnus region and obtained the image of this region and the spectrum of Cygnus X-1 in the hard X-ray band (Lu *et al.*, 1995). This balloon observation verified that high-quality hard X-ray imaging can be realized with a non-imaging collimated detector by using the direct demodulation method.

Based on the direct demodulation method and the expertise accumulated from balloon observations, the concept of the Hard X-ray

Modulation Telescope (HXMT) was originally conceived in the early 1990s (Zhang *et al.*, 2020). In 1994, HXMT was officially proposed by Tipei Li and Qibin Li to the Chinese Academy of Sciences as a candidate for a major astronomical facility. The main scientific objectives of HXMT were to complete a hard X-ray all-sky survey with both high sensitivity and angular resolution, and to perform pointed observations for studying the temporal and spectral properties of bright hard X-ray sources. Initially, the payload of HXMT included NaI(Tl)/CsI(Na) phoswich detectors with a total detection area of 5100 cm^2, multi wire proportional counters with a total area of 1120 cm^2, and a gamma ray burst monitor (Li, 1995). This scheme was simplified later to 18 phoswich detectors with a total area 5100 cm^2 covering the 20–250 keV energy range, to reduce the weight and cost of the telescope and further improve the feasibility of the project. officially approved approval of the HXMT project was in 2011, nearly 18 years after it was first proposed. By that time, the European INTEGRAL mission and the US Swift mission had already been launched in 2002 and 2004, respectively (Sazonov *et al.*, 2007; Tueller *et al.*, 2008). These two missions, equipped with coded mask hard X-ray imagers, had already completed deep X-ray imaging surveys of the sky even before the approval of HXMT. To enhance the capability of studying X-ray binary systems (XRBs) in their bright states or outbursts over a broad energy band, two sets of X-ray instruments were added to the HXMT satellite, which are the Low Energy X-ray telescope (LE) covering 1–15 keV and the Medium Energy X-ray telescope (ME) covering 5–30 keV (Zhang *et al.*, 2020).

2. Overview of the *Insight*-HXMT Mission

As illustrated in Figure 1, the HXMT satellite consists of two cabins, the payload cabin and the platform cabin. The payload cabin, located on top of the satellite, houses all the telescopes. The telescopes are aligned in one direction to observe the same source simultaneously. To avoid direct irradiation of the sun, a sunshielding board is installed on one side of the satellite. This provides a relatively low and stable temperature environment for the detectors. The total weight of the satellite is about 2500 kg.

Fig. 1. An artist's illustration of *Insight*-HXMT in space.

The Hard X-ray Modulation Telescope (HXMT), also known as Insight-HXMT post-launch, operates in a low earth orbit with an inclination angle of 43°. The altitude of the orbit descends gradually by 0.7 km per year from its initial value of 550 km. The details of the mission profiles are listed in Table 1. Further information regarding the satellite can be found in Zhang *et al.*'s publication (Zhang *et al.*, 2020).

3. Scientific Instruments

3.1. *The high energy X-ray telescope (HE)*

Figure 2 illustrates the structure of the HE telescope, which contains the main detector (HED), high energy collimator (HEC), auto-gain control detector (HGC), anti-coincidence shield detector (HVT), particle monitor (HPM), data processing and control box (HEB), and power box (HEA). The main characteristics of the HE are listed in Tables 2 and 3 (Liu *et al.*, 2020).

Table 1. Mission profiles of *Insight*-HXMT.

Item	Value
Orbit	550 km, 43°
Total mass	2500 kg
Payload mass	1000 kg
Nominal lifetime	4 years
Observation modes	With collimator: pointed, scanning Without collimator: gamma-ray monitoring
Scanning speed	$0.01° s^{-1}$, $0.03° s^{-1}$, $0.06° s^{-1}$
Pointing accuracy (3σ)	±0.1°
Attitude measurement accuracy (3σ)	±0.01°
Attitude stability	$\pm 0.005° s^{-1}$
Telemetry rate	120 Mbps
ToO response time	~5 h

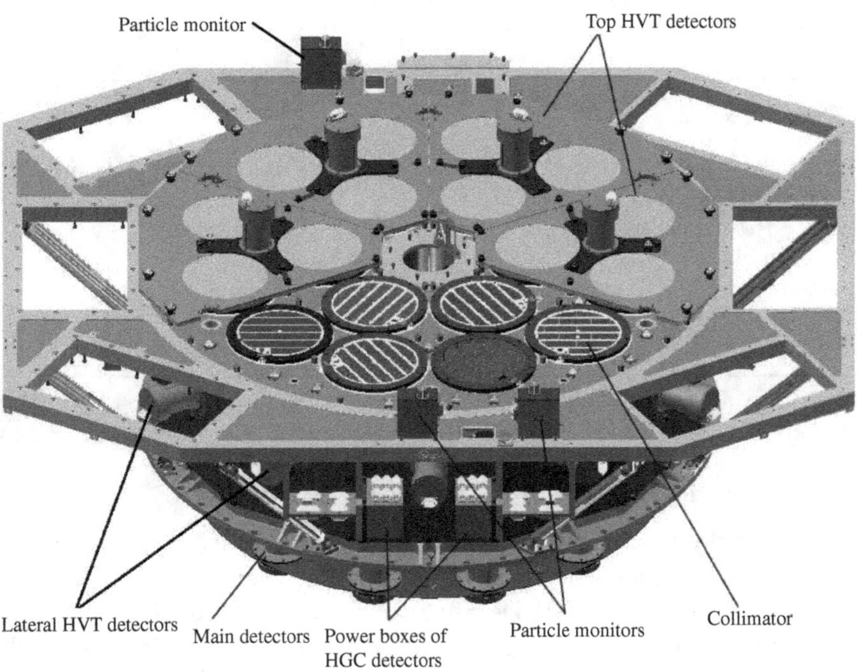

Fig. 2. Illustration of the HE telescope onboard the *Insight*-HXMT astronomy satellite.

Table 2. The main characteristics of the high-energy telescope (HE) onboard the *Insight*-HXMT astronomy satellite.

Parameters	Values
Energy Band	20–250 keV (normal mode)
	0.2–3 MeV (low gain mode for GRB monitoring)
Geometric area	5100 cm^2
Main detector	NaI(Tl)/CsI(Na)
	~3.5 mm/40 mm
Timing accuracy	$<10\ \mu_s$
Dead time	$<10\ \mu_s$
Combined FOV (FWHM)	$5.7^\circ \times 5.7^\circ$
	(15 $1.1^\circ \times 5.7^\circ$, 2 $5.7^\circ \times 5.7^\circ$, blind)
Energy resolution (FWHM)	~17%@60 keV
Maximum count rate	>30,000 cnts/sec

Table 3. The main characteristics of the medium-energy telescope (ME) onboard the *Insight*-HXMT astronomy satellite.

Items	Specifications
Energy Range	5-30 keV
Detector Type	Si-PIN
Pixel Size	12.5 mm × 4.5 mm, 1 mm thick
Total Detection Area	952 cm^2
Energy Resolution (FWHM)	~3 keV at 20 keV
Time Resolution	255 μs
Detector Working Temperature	$-50^\circ C - -5^\circ C$
Maximum count rate allowed	2.2×10^4 cts/s
Power	155 W

3.2. *The medium energy X-ray telescope (ME)*

The ME telescope consists of three medium energy detector boxes (MEDs) and one electric control box (MEB) (Cao *et al.*, 2020). Data collected by the three MEDs are transmitted to the MEB through LVDS bus, while the MEB establishes communication with the satellite platform using both LVDS and 1553B bus. The LVDS bus is mainly for the science data, whereas the 1553B bus for the engineering data and commands. The main technical specifications of ME are summarized in Table 3. MED can be further divided into three main components, which are collimators, detectors, and electronics.

3.3. *The low energy X-ray telescope*

The LE telescope contains three identical detector boxes (LEDs) and one electric control box (LEB) (Chen *et al.*, 2020). The three LEDs are aligned and mounted on the payload's main plate. Their nominal pointing axis is the same as that of HE and ME. The functions of LEB are data management, communication, as well as power supply and distribution. Moreover, it is capable of automatically adjusting the on-board noise threshold of each detector, and processing the count rate and spectrum onboard. It is also the interface between the LEDs and the satellite platform, facilitating telemetry, remote control, as well as instruction and data transmission. A summary of the LE characteristics is listed in Table 4.

4. Performance and Response of the Instruments

Before launch, the three types of telescopes were calibrated with X-ray beams and radioactive sources. The in-orbit performance of these instruments in the early phase is similar to the on-ground ones, but long-term slow evolution has been also detected, which has been taken into account in the released calibration database.

Table 4. The main characteristics of the low-energy telescope (LE) onboard the *Insight*-HXMT astronomy satellite.

Item	Performance
Detector area	384 cm^2
Detector type	Swept Charge Device (SCD)
Energy range	0.7–13 keV
Collimator FOV	$1.6° \times 6°$, $4° \times 6°$, blocked FOV, $50–60° \times 2–6°$ (trapezoidal)
Energy resolution	140 eV@5.9 keV
Time resolution	0.98 ms
Sensor operating temperature	$-80°C \sim -30°C$

4.1. *Response and performance of the high-energy X-ray telescope*

4.1.1. *Pulse shape discrimination*

Because the main detector of HE has a phoswich structure, with the NaI(Tl) in front of the CsI(Na) scintillator, some incident X-ray photons deposit their energies solely in the NaI(Tl) detector of HED, which are called NaI events or "good events"; some deposit their energies only in the CsI(Na), which are called CsI events; and a small fraction of photons deposit their energies in both crystals, which are called Compton events or mixed events. In normal mode, only the NaI events are good science events to be recorded, while the CsI events and mixed events are treated as shielding events. These different types of events are distinguished by the PSD using the different decay time constants of the NaI(Tl) and CsI(Na) scintillators. The PSD adopted in the HEB uses the front-rear-edge discrimination technique (Li and Li, 2003). A point is selected at the front edge and rear edge of a HED signal pulse, for which the amplitudes relative to the peak value of the signal pulse are exactly equal to a preset ratio. The time interval between these two points is defined as the pulse width of this signal. Theoretically, the pulse width remains constant regardless of incident X-ray photon energy, enabling efficient determination of whether an event qualifies as a good event (Li and Li, 2003). Figure 3 shows a typical pulse width spectrum of a ^{133}Ba radioactive source

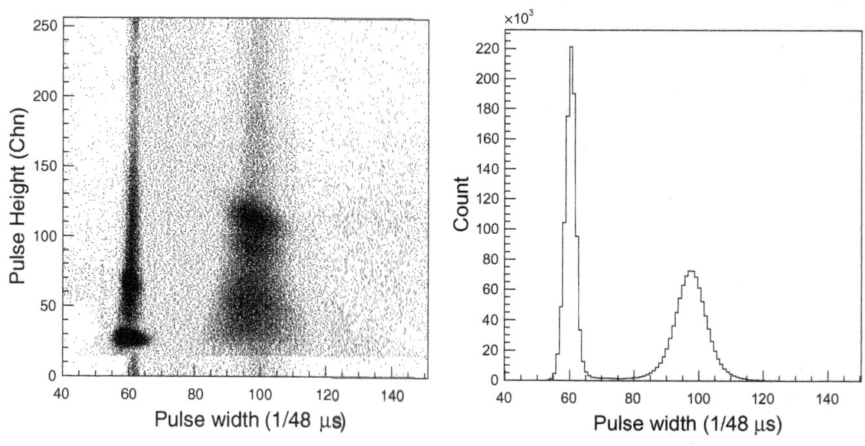

Fig. 3. Pulse width spectrum of ^{133}Ba in normal mode.

obtained from calibration experiments conducted in normal mode. The NaI and CsI events can be easily discriminated.

$$\eta(E) = \frac{\sqrt{a + bE + cE^2}}{E}. \tag{1}$$

4.1.2. *Energy resolution*

Due to the statistical fluctuation of the scintillation light yield and light collection, the monochromatic X-ray photons generate a peaked spectrum with a certain width in HED, which can be accurately described by fitting it with a Gaussian function. For an ideal linear system, the intrinsic energy resolution η of a detector follows a power-law function of energy with an index of -0.5 (Bissaldi *et al.*, 2009) (Chen *et al.*, 2011). However, the energy resolution of a realistic detection system can be impacted by several factors and may deviate from this power-law behavior (Bissaldi *et al.*, 2009; Frontera *et al.*, 1997). Specifically, around the Iodine K-edge, there is a discontinuous change in the energy resolution. Thus, the relationship between the energy resolution and energy was fitted separately below and above the Iodine K-edge. The adopted fitting function is provided in Eq. (1),

Figure 4 illustrates the energy resolution of HED measured at various energies and its corresponding fit to Eq. (1).

Since the geometric area of the scintillator is significantly larger than the photocathode area of the PMT, it is important to measure the non-uniformity of HED's response at different positions. To do this, a total of 192 sampled points on the HED detector plane were sequentially exposed with the X-ray beam at 50 keV. The measurements show that the relative standard deviation of the full energy peak, energy resolution, and detection efficiency are smaller than 2%, 2.5%, and 2.5%, respectively.

4.1.3. *Detection efficiency*

The detection efficiency of the HED in normal mode was calibrated with a double-crystal monochromator, by comparing the measured count rate with that of a well-calibrated HEPGe detector. Figure 5 shows the full-energy-peak detection efficiencies of 18 HEDs in normal mode. The red curve represents the result obtained from

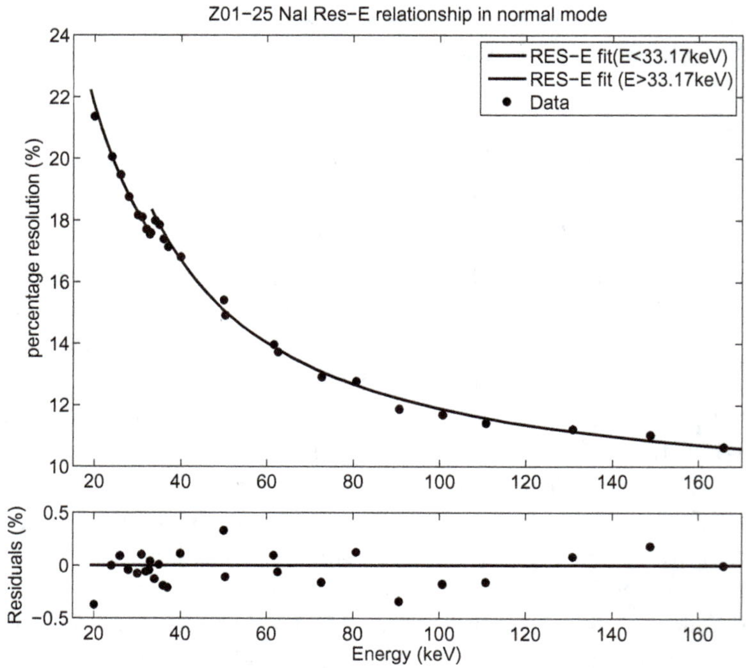

Fig. 4. Energy resolution of HED Z01-25 as a function of energy at normal mode.

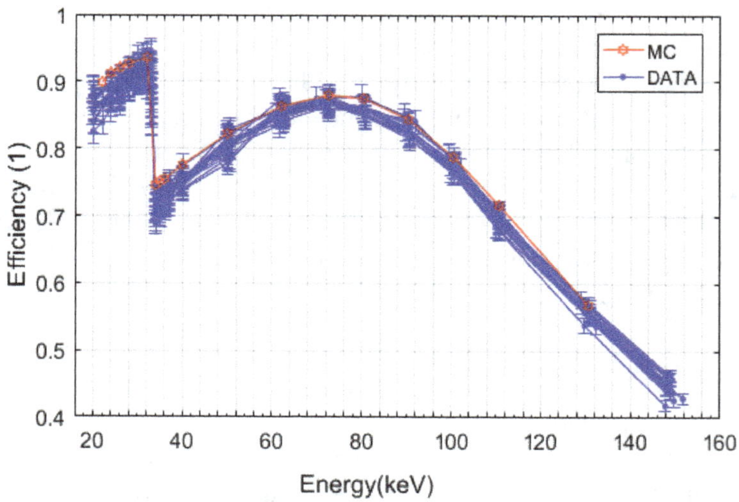

Fig. 5. Detection efficiencies of 18 HEDs derived from calibration experiments; the red curve is from simulation.

Monte-Carlo simulation, while the blue lines depict the experimental result obtained from 18 flight HED modules, which are very consistent with each other and consistent with the trend of the simulation result.

4.2. *Response and performance of the medium energy X-ray telescope*

A series of experiments and tests have been carried out for the ME to calibrate its responses at different temperatures, because the ME detectors only have passive cooling and thus their working temperature varies with the outside thermal conditions. The energy response, temperature response, and dead time of ME are calibrated with radioactive sources and a double-crystal monochromator. Two pixels of the Si-PIN detector with similar readout electronics underwent testing on the PANTER X-ray test facility of the Max Planck Institute for Extraterrestrial Physics in 2013. The calibration results of ME shown here are mostly performed with radioactive sources (^{241}Am and ^{57}Co) and the double-crystal X-ray monochromator at IHEP, but the quantum efficiency (QE) results were obtained on PANTER.

4.2.1. *E-C relation*

The E-C relations of one pixel at temperatures ranging from −30 to 0°C with a step of 5°C are given in Figure 6, which shows that the linearity is quite good and the signal amplitude decreases with temperature as expected.

The values of slope and offset of the 1728 Si-PIN pixels at different temperatures are shown in Figure 7. The slope increases with temperature and the offset remains relatively stable when temperature varies. The signal amplitude decreases by 0.6% per degree.

4.2.2. *Energy resolution*

The energy ranging from resolution is mainly determined by the noise of the detector and readout electronics, with the latter being larger. At lower temperatures, the leakage current of Si-PIN detectors decreases, the energy resolution also improves as expected (Figure 8).

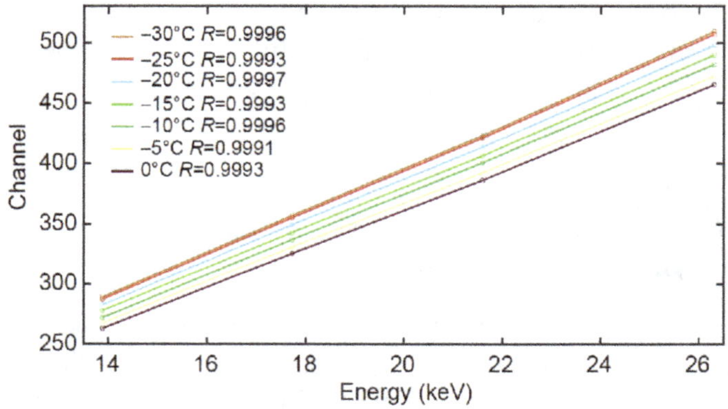

Fig. 6. The E–C relations of a Si-PIN pixel at temperatures ranging from −30°C to 0°C with a step of 5°C (from top to bottom).

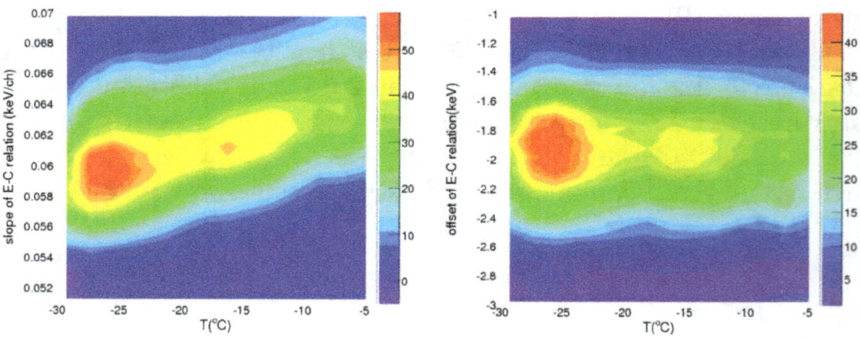

Fig. 7. On the left figure: Distribution of the slope of the E–C relation for the 576 Si-PIN pixels in one MED at different temperatures. The colors from blue to red represent the pixel numbers. On the right figure: Distribution of the offset of the E–C relation for the 576 Si-PIN pixels in one MED at different temperatures. The colors from blue to red represent the pixel numbers.

The typical energy resolution of ME is around 3 keV at the working temperatures.

4.2.3. *Quantum efficiency*

The quantum efficiency was obtained from an experiment carried out at the PANTER facility. The experiment used a CCD type detector as the standard monitor of the beam line, whose quantum efficiency had been calibrated (Stefanie *et al.*, 2010). Using the results of the this experiment, we simulated the quantum efficiency of ME

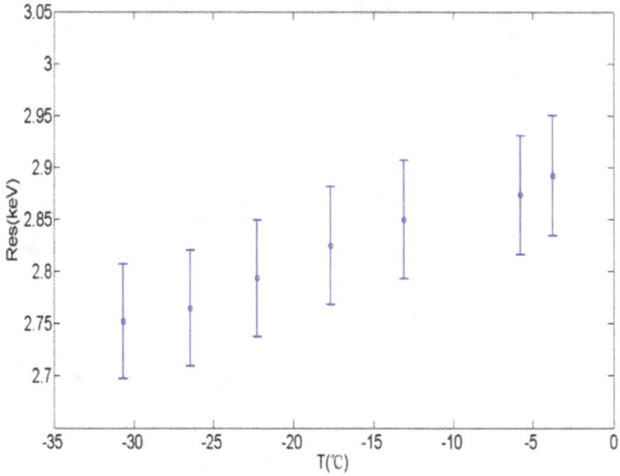

Fig. 8. Change of the full-width-half-maximum (FWHM) at 17.73 keV with temperature for one ME detector.

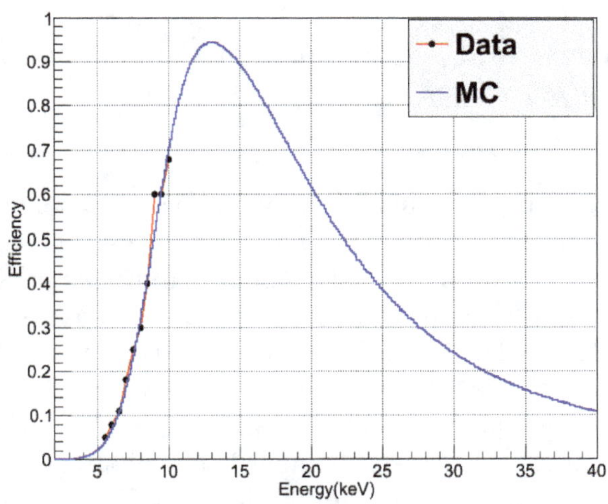

Fig. 9. Simulated quantum efficiency of ME. The data points are obtained from the experiment on PANTER.

in the 5 to 40 keV energy range, as shown in Figure 9, the low efficiency below 10 keV is due to the internal parameters of the ASIC chips, such as VTHR. However, these parameters cannot significantly change the quantum efficiency above 10 keV, the energy range in which the quantum efficiency is mainly related to the thickness of the detector.

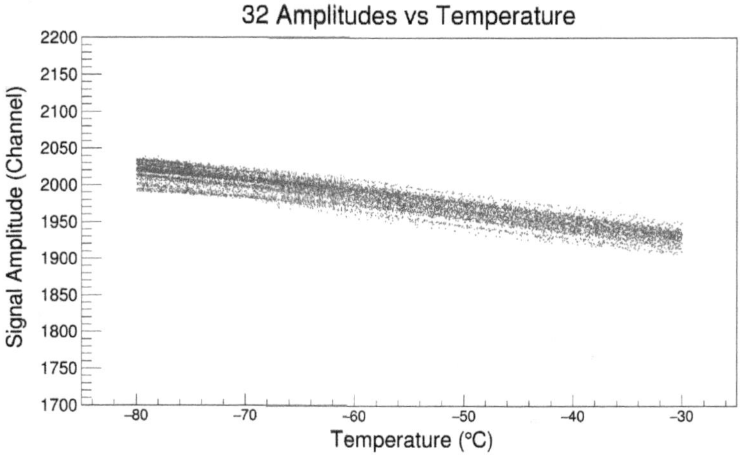

Fig. 10. Temperature-dependent Cu-Kα signal amplitude curves of 32 channels, obtained in the vacuum thermal tests of the LE flight model.

4.3. *Response and performance of the low energy X-ray telescope*

4.3.1. *E–C relationship*

Similar to the ME telescope, LE only has passive cooling, and thus the working temperature of its SCD detectors varies between -80 and -30°C. The E-C relation and the signal amplitudes of X-rays with a given energy also changes with temperature. Figure 10 presents the Cu-Kα (8.048 keV) signal amplitude curves of the 32 readout channels, in which the temperature dependency is obvious. Figure 11 shows an example spectrum obtained in the thermal test of LE flight model, where a copper target is bombarded by photons from an X-ray tube. The differences of the amplitudes between different readout channels are within $\pm 1\%$, and the integral nonlinearity of the E–C relationship is around 0.1%.

4.3.2. *Readout noise and energy resolution*

The merged spectrum of 32 channels is similar to the spectrum of each single channel. As shown in Figure 12, the full-width-half-maximum

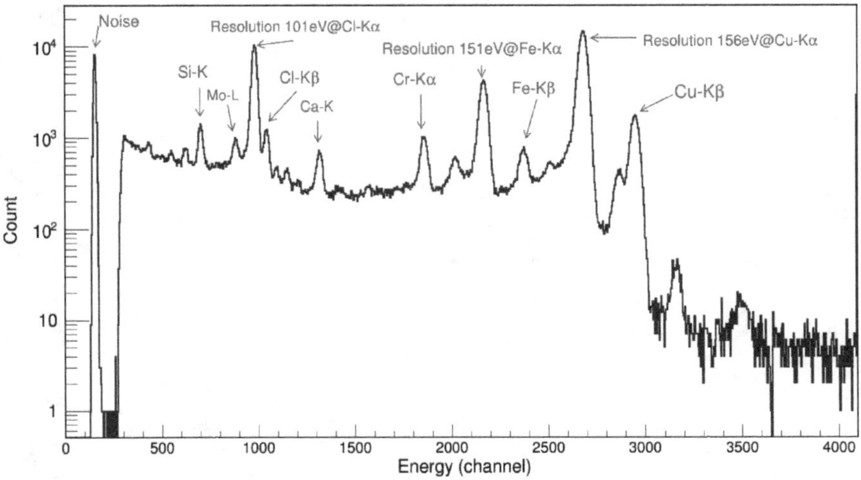

(a)

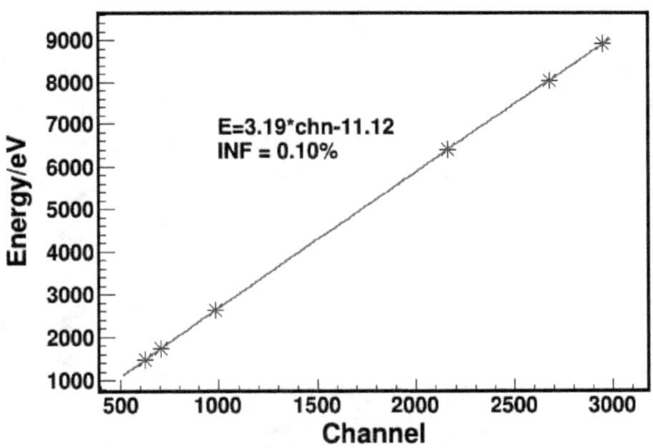

(b)

Fig. 11. (a) An example spectrum obtained in the thermal test of LE flight model. (b) The integral linearity (INL) of LE.

(FWHM) of the merged spectrum is 158 eV@8.0 keV, degrading slightly compared with that from a single channel.

The temperature dependencies of the readout noise and FWHM of LE are shown in Figure 12, where the temperature of the sensor changes from −80 to −30°C. When the temperature is below −50°C,

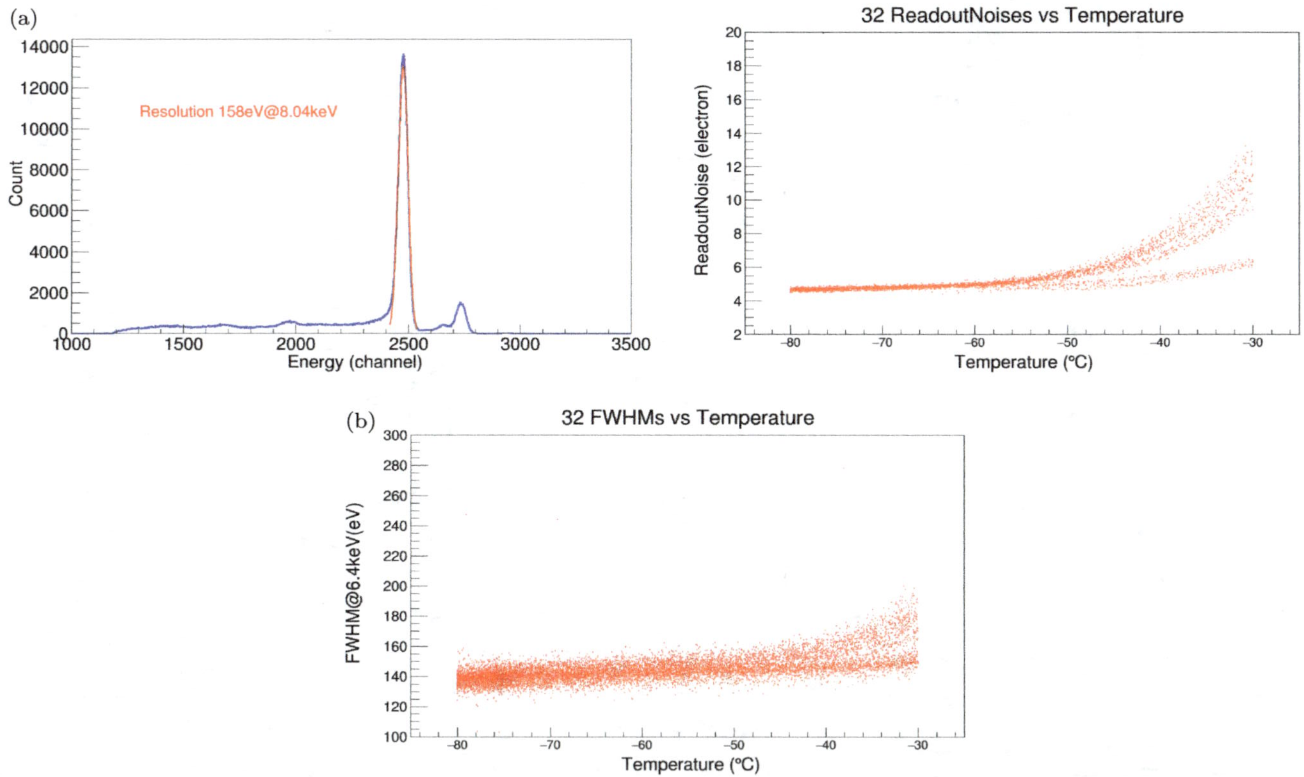

Fig. 12. The merged spectrum of 32 channels, obtained from the flight model. Temperature-dependent readout noise (a) and energy resolutions (FWHM, b) of the 32 detector channels at the Fe Kα energy.

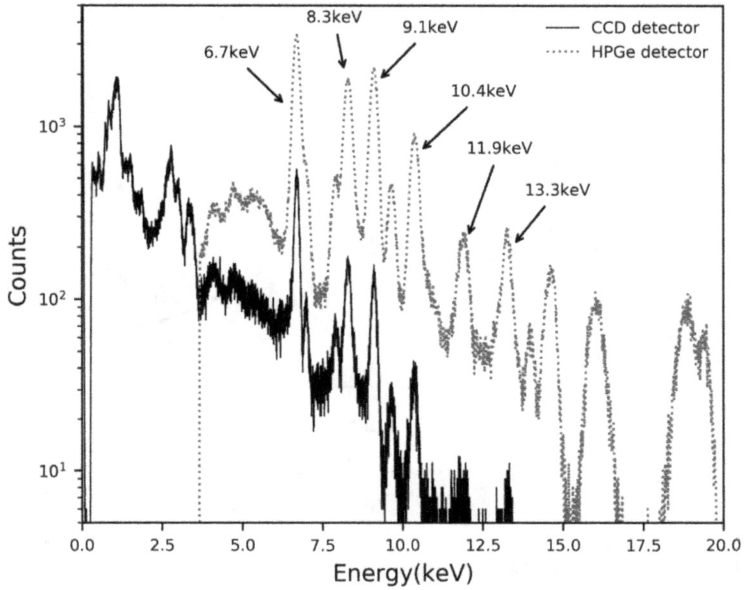

Fig. 13. The energy spectra of the radiation from an EBIT by CCD236 (−100°C) and HPGe (Liu *et al.*, 2016).

the readout noise is almost constant, and the differences between channels are negligible.

4.3.3. *Quantum efficiency*

The relative efficiency of LE has been evaluated with the electron-beam ion trap (EBIT) facility at Fudan University in Shanghai. Figure 13 shows a comparison of the energy spectra measured between the LE sensor operating at −100°C and the standard sensor (High Purity Germanium (HPGe) sensor) is shown. Both spectra contain six peaks whose energies are 6.7, 8.3, 9.1, 10.4, 11.9, and 13.3 keV, respectively. By comparing the total counts of these peaks between the CCD236 and HPGe spectra, the relative quantum efficiency of CCD236 to HPGe, can be obtained. For energies below 3.5 keV, the quantum efficiency of CCD236 is much higher, because the entrance window of the HPGe sensor is much thicker.

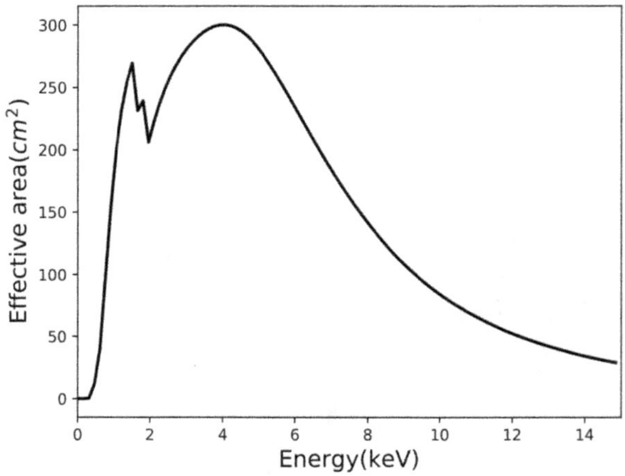

Fig. 14. Calculated total on-axis effective area of LE (including both narrow and wide FOVs SCDs).

Calibrating the SCD efficiency in the low energy range on ground is difficult. One can obtain the efficiency by extrapolating the above data to the low energy range with the help of simulations and calculations, and by correcting the deviations with in-orbit calibrations. The calculated total on-axis effective area of LE (including both the narrow and wide FOV SCDs) is shown in Figure 14, which takes into account detector efficiency and filter transmittance.

4.3.4. *Time response*

Since the charge packets in the pixels are transferred one by one, the time responses of LE depend strongly on the charge transfer process, which can be obtained well under the long exposure mode (Chen *et al.*, 2020), which is shown in Figure 15. In principle, the response is proportional to the area of each L-shaped electrode. For the CCD 236, the area of the electrode increases with its distance to the readout end, and so there are more signal counts as the readout time increases. The recorded readout time follows a triangle-like distribution strictly as shown in Figure 15, and the exact time of each photon cannot be obtained more accurately than 0.98 ms (Tuo *et al.*, 2022).

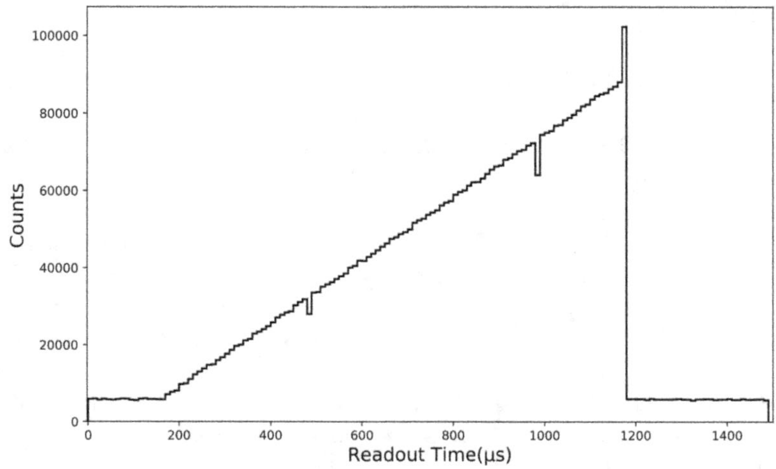

Fig. 15. Time response of LE obtained under long exposure mode.

5. Summary

In this chapter, we introduce the design and performance of the *Insight*-HXMT satellite and the three telescopes onboard. The satellite has been operational for over four years and all the payloads have been functioning normally and steadily. However, the energy resolution of LE, which is sensitive to radiation damage, has been degraded to about 300 eV@6 keV by the charged particles. Using its large effective area, high time resolution, broad energy coverage, and moderate spectral resolution, *Insight*-HXMT opens a new window to study the hard X-ray variabilities and spectra of neutron star and black hole X-ray binaries. It also serves as one of the most effective Gamma-Ray Burst monitors in space. The technologies and experiences obtained in the development of *Insight*-HXMT have been used on its successor, the Enhanced X-ray Timing and Polarimetry observatory, that will be launched in the late 2020's (Zhang *et al.*, 2019).

Bibliography

Bissaldi, E. *et al.* (2009). Ground-based calibration and characterization of the Fermi gamma-ray burst monitor detectors, *Exp. Astron.* **24**, 47.

Cao, X. *et al.* (2020). The medium energy X-ray telescope (ME) onboard the Insight-HXMT astronomy satellite, *Sci. Sin. Phys. Mech. Astron.* **63**, 249504.

Chen., B. X. and Zhang, Z. (2011). *Nuclear Radiation Physics and Detection*, Arbin Engineering University Press, Harbin.

Chen, Y. *et al.* (2020). The low energy X-ray telescope (LE) onboard the insight-HXMT astronomy satellite, *Sci. Sin. Phys. Mech. Astron.* **63**, 249505.

Dai, C. *et al.* (1987). A hard X-ray telescope and its observation of the Crab pulsar, *Chinese Astronomy and Astrophysics* **11**, 179.

Ebermayer, S. *et al.* (2010). High energy, optical, and infrared detectors for astronomy IV, in *SPIE Astronomical Telescopes + Instrumentation 2010*, Vol. 7742, International Society for Optics and Photonics.

Frontera, F. *et al.* (1997). The high energy instrument PDS on-board the BeppoSAX X–ray astronomy satellite, *Astron. Astrophys. S.* **122**, 357.

Li, T. (1995). Direct demodulation method and its application to hard X-ray imaging, *Exp. Astron.* **6**, 63.

Li, T.-P. & Wu, M. (1993). A direct restoration method for spectral and image analysis, *Astrophysics and Space Science* **206**, 91.

Li, T.-P. and Wu, M. (1994). Reconstruction of objects by direct demodulation, *Astrophys. Space Sci.* **215**, 213.

Li, T.-P. *et al.* (1993). Modulation imaging based on a direct deconvolution technique, *Astrophys. Space Sci.* **205**, 381.

Li, YanGuo and Li, TiPei (2003). New pulse shape discriminator, *Nuclear Electronics & Detection Technology* **23**, 5.

Liu, C. *et al.* (2020). The high energy x-ray telescope (HE) onboard the insight-HXMT astronomy satellite, *Sci. Sin. Phys. Mech. Astron.* **63**, 249503.

Liu, X. Y., *et al.* (2016). The relative detection efficiency calibration of CCD detector, *Nucl. Electron. Detect. Technol.* **36**, 144–149.

Lu, Z., *et al.* (1995). Hard X-ray imaging with a slat collimated telescope, *Nucl. Instrum. Methods Phys. Res. A* **362**, 551.

Sazonov, S. *et al.* (2007). Hard X-ray luminosity function and absorption distribution of nearby AGN: INTEGRAL all-sky survey, *Astron. Astrophys.* **462**, 57.

Tueller, J. *et al.* (2008). Swift BAT survey of AGNs, *Astrophys. J.* **681**, 113.

Tuo, Y., *et al.* (2022). In-orbit timing calibration of the insight-hard X-ray modulation telescope, *Astrophys. J. Suppl. Ser.* **259**, 14.

Zhang, S. *et al.* (2019). The enhanced X-ray timing and polarimetry mission — eXTP, *Sci. China Phys. Mechan. Astron.* **62**, 29502.

Zhang, S.-N., *et al.* (2020). Overview to the hard X-ray modulation telescope (Insight-HXMT) Satellite, *Sci. China Phys. Mechan. Astron.* **63**, 249502.

https://doi.org/10.1142/9789819800643_0009

Chapter 9

Advances in Neutron Stars from *Insight*-HXMT

Mingyu Ge

Institute of High Energy Physics, Chinese Academy of Sciences, Beijing 100049, China

gemy@ihep.ac.cn

Though great progress has been achieved from the multi-band observations and theoretical work since the discovery of pulsars more than 50 years ago, some basic problems, such as radiation process in pulsating ultraluminous X-ray source, the origins of fast radio burst and so on. X-ray observations could supply crucial information to understand the characteristics and behaviors of pulsars. Recently, some progress for these basic problems has been achieved via the observations of *Insight*-HXMT.

Keywords: Pulsars, X-ray pulsar, Magnetar, Type-I burst, Magnetic field

Contents

1. Introduction

Insight-HXMT was successfully launched as China's first X-ray astronomy satellite on June 15, 2017 at the Jiuquan Satellite Launch Center, which runs in a low earth orbit with an altitude of 550 km and an inclination angle of 43°. After the launch, HXMT was officially dubbed as Hui-Yan, or Insight-HXMT in honor of the famous Chinese physicist, Prof. He ZeHui, the founder of high-energy astrophysics in China (Zhang *et al.*, 2020).

As shown in Figure 1, *Insight*-HXMT carries out sensitive galactic plane surveys in high cadence and is expected to build a galactic source catalogue with sources showing up their behavior at short time scales at hard X-rays. *Insight*-HXMT will help understand the low/hard state of X-ray Binary (XRB) outbursts, thanks to the large geometrical area in 20–250 keV (Liu *et al.*, 2020; Zhang *et al.*, 2020). At soft X-rays, *Insight*-HXMT can reach energies as low as 1 keV and, with the fast-readout capability of the non-imaging CCDs, Insight-HXMT/LE can ease the problems of photon saturation and pile-up effects in observing bright sources (Chen *et al.*, 2020). Therefore, the source pointed observation of *Insight*-HXMT can provide broadband data with good statistics for sources that are bright persistently or undergoing an outburst. In addition, *Insight*-HXMT/HE serves as an excellent wide-field monitor for high-energy transient sources, i.e. gamma-ray bursts (GRBs), with the thick CsI detectors of HE originally designed only for active shielding (Liu *et al.*, 2020). Therefore, the core functions of the Insight-HXMT are as follows: (1) to scan

Fig. 1. An artist's view of the Insight-HXMT satellite in space. This figure is adopted from Zhang *et al.* (2020).

the galactic plane and find new transient sources and to monitor the known variable sources; (2) to observe X-ray binaries in broad energy band and study the dynamics and emission mechanism in strong gravitational or magnetic fields; (3) to observe GRBs in a relatively rarely explored band from a few hundred keV to a few MeV.

In the following sections, we summarise the process in the neutron star based on the *Insight*-HXMT observations, such as magnetic fields, type-I bursts and high-energy counterpart of fast radio bursts (FRBs).

2. Advances in X-ray Accreting Pulsars

In high mass X-ray binaries (HMXBs), a compact object accretes matter from a binary companion with a mass above $\sim 10 M_{\odot}$, releasing the gravitational potential energy into X-rays (see, e.g., Kretschmar *et al.*, 2019; Reig, 2011). In most cases, the compact object is a highly magnetized neutron star (NS) with a magnetic field $B \gtrsim 10^{12} - 10^{13}$ G. The accretion flow will be truncated by the magnetosphere, within which the matter is channelled by the magnetic field onto the magnetic poles of the neutron star, resulting in coherent pulsations if there is a misalignment between the spin and the magnetic axises. A full and detailed description of the physics of accreting X-ray pulsars (AXPs) can be found in the review paper by Mushtukov and Tsygankov (2022).

In observations, most of AXPs are transient sources with only a few exceptions, such as Her X-1 and Cen X-3. There are two types of outbursts, i.e., type-I outbursts that normally is coincident with the periastron passage and type-II outbursts which are much stronger and long-lasting. In general, the spectrum of AXPs during outbursts can be approximately described as a phenomenological power-law model with a cutoff of tens of keV. There are also other spectral structures, such as cyclotron resonant scattering features (CRSFs; for details, see the following). On the other hand, the temporal properties of AXPs are basically complex, exhibiting energy- and luminosity-dependent pulse profiles and pulse-to-pulse variations as well. Therefore, in order to extensively investigate AXPs, a broad-band energy coverage and a large effective area are required. This perfectly matches *Insight*-HXMT's capability (Zhang *et al.*, 2019a), which makes it the best satellite onboard for AXP-related studies.

2.1. *Measurement of magnetic fields*

CRSFs are the only direct method of measuring magnetic fields in AXPs. They are absorption features in the spectra, caused by the quantization of the energies of electrons' movement (the so-called Landau levels) under a strong magnetic field. The centroid cyclotron line energy is $E_{\rm cyc} = 11.6nB_{12}(1+z)^{-1}$, where z is the gravitational redshift in the line-forming region. $n = 1$ and $n = 2, 3, 4, \ldots$ correspond to the fundamental and harmonic cyclotron lines, respectively. Since their discovery in the 1970s, CRSFs have been discovered in more than 30 sources among ~350 known X-ray binary pulsars (for a review, see Staubert *et al.*, 2019). In the era of *Insight*-HXMT, the CRSF detection is extended to ~100 keV owning to its unprecedented large effective area at high energies. For example, Ge *et al.* (2020) found a CRSF with a centroid energy of around 90 keV, which is the highest fundamental line energy observed from a phase-averaged spectrum. Another record is made in the first galactic pulsing ultraluminous X-ray source Swift J0243.6+6126. Kong *et al.* (2022) discovered a CRSF line at about 146 keV in the phase-resolved spectrum, which is the highest CRSF known to date (Figure 2). Of importance,

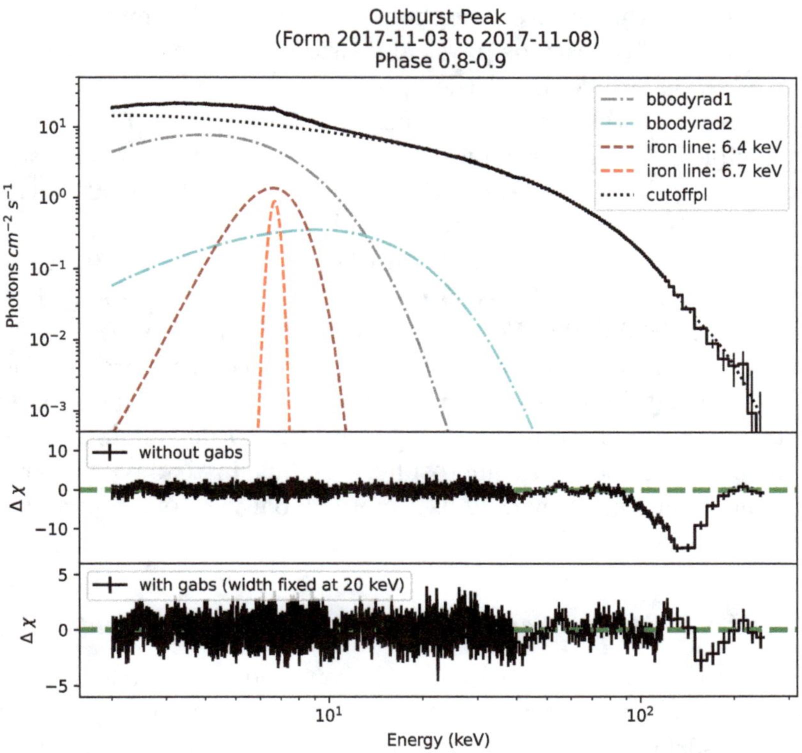

Fig. 2. An example of the spectral fitting in Swift J0243.6+6124. The middle and bottom panels present the residuals without and with the CRSF component, respectively. This figure is adopted from Kong *et al.* (2022).

together with other B measurements, this detection demonstrates the presence of multipole field components close to the surface of the neutron star. The CRSF energy normally does not keep constant. Instead, it varies with the luminosity, the pulse phase, the super-orbital phase and the long-term evolution, reflecting the variation of line-forming regions (Staubert *et al.*, 2014). This has been studied by many authors in different sources with *Insight*-HXMT observations (e.g., Chen *et al.*, 2021; Ding *et al.*, 2021a; Ge *et al.*, 2020; Kong *et al.*, 2021; Liu *et al.*, 2022b; Nabizadeh *et al.*, 2021; Xiao *et al.*, 2019). In particular, Kong *et al.* (2021) discovered both positive and negative correlations with luminosity and the hysteresis effect in 1A 0535+262, which was only reported in V0332+53 previously (Doroshenko *et al.*, 2017).

There are other indirect methods to estimate the magnetic fields of AXPs, based on various physical considerations, e.g., torque models, the propeller effect and the appearance of the accretion column on the neutron star surface. All these methods rely on high cadence and flexible observations, which is in line with *Insight*-HXMT's strategy. *Insight*-HXMT has obtained a large number of observations during outbursts of Be X-ray binaries and achieved the estimations of magnetic fields in Swift J0243.6+6124 (Zhang *et al.*, 2019b), 4U 1901+03 (Ji *et al.*, 2020a; Tuo *et al.*, 2020), 2S 1417-624 (Ji *et al.*, 2020b) and GRO J1008-57 (Wang *et al.*, 2021). Unlike the detection of cyclotron lines, these methods depend on specific theoretical models with poorly known parameters, and therefore have large uncertainties. But we note that some methods, e.g., torque models, depend on the processes at the magnetospheric radius, instead of the surface of the neutron star, providing us a more global view of magnetic field configurations in AXPs.

2.2. *Accretion process under strong magnetic fields*

The accretion matter is channelled from the magnetosphere onto polar caps of the neutron stars. The accreting structure on the neutron star surface is believed to depend on the luminosity (L) (Basko and Sunyaev, 1976; Becker *et al.*, 2012; Mushtukov *et al.*, 2015). If the L is larger than a critical value L_{crit}, where $L_{\text{crit}} \approx 1.5 \times 10^{37}$ $B_{12}^{16/15}$ erg/s and B_{12} is the surface magnetic field strength in units of 10^{12} G, an opacity-thick accretion column appears, which prevents photons passing it vertically. The photons can only escape from the "wall" of the accretion column, perpendicular to B, forming a so-called "fan" beam. When $L < L_{\text{crit}}$, the accreted matter can free fall down to the neutron star surface and the emergent radiation is transported (and scattered) through the infalling matter, forming a "pencil" beam parallel to the magnetic field lines. Therefore, the keys to understanding the accretion under strong magnetic fields are the properties of the accretion flow and the radiative transfer processes in the vicinity of polar caps.

In HMXBs, AXPs are mainly fed by stellar winds from the donor star. Outside the magnetosphere, the accretion might be spherically symmetrical or forming an accretion disk depending on the angular momentum of the accreted matter. Thanks to the large effective area and no pile-up effect, *Insight*-HXMT is able to study the QPO features and the line emissions that originate from the reprocessing, which reveals the accretion geometry of AXPs (Ding *et al.*, 2021b; Ji *et al.*, 2021; Liu *et al.*, 2022c; Ma *et al.*, 2022). In addition, according to the spin-up evolution, the disc versus wind accretion is studied in GX 301-2 by Liu *et al.* (2021).

Insight-HXMT is also the best instrument for studying broadband pulse profiles, which directly indicate the accretion process near the neutron star. For instance, Doroshenko *et al.* (2020) discovered two transitions of pulse profiles with luminosity in Swift J0243.6+6124, of which one is associated with the variation of the power spectra (Figure 3). The transitions can be explained as the appearance of the accretion column (i.e., between sub- and super-critical luminosity states) and the changes of the accretion disk, from a gas pressure dominated (GPD) state to a radiation pressure-dominated (RPD) state.

The complex spectral and timing properties of Swift J0243.6+6124 are studied by Kong *et al.* (2022) and Liu *et al.* (2022a), which support the scenario of the RPD disk as well. Soon after, RPD disks are also reported in 2S 1417–624 and RX J0209.6–7427 (Hou *et al.*, 2022; Ji *et al.*, 2020b), which suggest that they are universal in

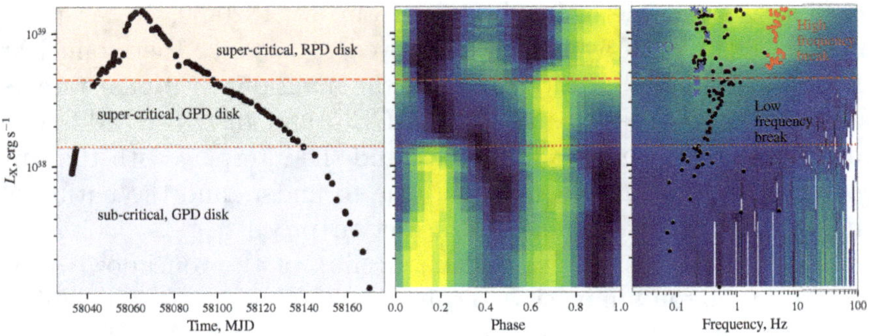

Fig. 3. Left panel: the light curve observed by *Insight*-HXMT (black points) with color-shading showing the different states of the source. The horizontal lines present transitions between different accretion states. Middle panel: the evolution pulse profiles. Right panel: the evolution of power spectra multiplied by frequency. The black and red points represent the break and quasi-periodic oscillation (QPO) frequencies seen in the power spectra. This figure is taken from Doroshenko *et al.* (2020).

the high luminosity state. In addition, the evolution of pulse profiles is investigated in different sources. For example, Wang *et al.* (2020) studied pulse fractions at super-Eddington luminosity of Swift J0243.6+6124 and found the hysteresis effect. Ji *et al.* (2020a) found that the pulse profile during flares can be significantly different from that of the persistent emission with a phase shift in some cases, which clearly reflects changes of beam patterns.

3. Advances on Thermonuclear X-ray Bursts

In the light curves with binsize approximately seconds of low-mass X-ray binary (LMXB) hosting a neutron star, some spikes emerge with peak flux up to the Eddington luminosity and timescale of seconds to minutes. These events are named type-I X-ray bursts, or thermonuclear X-ray bursts because of these behaviors, including spectral softening during decay phase and α parameter (the fluence ratio of the burst and the outburst) within the range of the relative efficiency of accretion and thermonuclear burning. Most bursts are single-peaked except the brightest ones which show photospheric radius expansion (PRE; due to radiation pressure). (For reviews, see Cumming, 2004; Galloway *et al.*, 2008; Lewin *et al.*, 1993; Strohmayer and Bildsten, 2006).

Since its first detection in 1975 from 3A 1820-30, so far there are 118 galactic X-ray binaries observed to produce thermonuclear bursts[1]; the bursts have been one of the main observational targets for X-ray telescopes, e.g., *RXTE*, *NICER*, and *Insight*-HXMT, due to the fact that their characteristics and the interplay with the surrounding circumstance could be a clue to understand the nature of the neutron star (NS) and the accretion process, e.g., taking the burst oscillation as the NS's spin, the flux of the touchdown time as the Eddington luminosity to constrain the NS's mass and radius, the enhancement of the persistent emission at soft X-ray band (Bult *et al.*, 2021; Chen *et al.*, 2019; Ji *et al.*, 2014a; Worpel *et al.*, 2013, 2015), the deficit of the persistent emission at hard X-ray band (Chen *et al.*, 2012, 2022a; Ji *et al.*, 2013; Maccarone and Coppi, 2003) and

[1]See https://personal.sron.nl/~jeanz/bursterlist.html.

a bump peaking at 20–40 keV and/or discrete emission (Ballantyne and Strohmayer, 2004; Degenaar *et al.*, 2018; in't Zand *et al.*, 2013; Keek *et al.*, 2014) to derive the influence of the burst emission on the accretion process.

The accretion circumstance could also have impacts on the burst emission, e.g., the obscuration of the burst emission by the disk (Shaposhnikov and Titarchuk, 2004; Shaposhnikov *et al.*, 2003) and the comptonization of the burst emission by the corona/boundary layer (Chen *et al.*, 2022b).

In the first five years in-orbit of *Insight*-HXMT, we have assembled a sample of ∼200 bursts from observations of several accreting NSs. Thanks to the large detection area and wide bandwidth of the *Insight*-HXMT, we present the spectral evolution of the bursts and explore their interaction with the accretion circumstances.

3.1. *Hard X-ray shortage of the accretion emission during bursts*

The shortage in the hard X-ray of the continuum emission during bursts is reported on several bursters, i.e., IGR J17473-2721 (Chen *et al.*, 2011, 2012), Aql X-1 (Chen *et al.*, 2013; Maccarone and Coppi, 2003), 4U 1636-536 (Güver *et al.*, 2022a; Ji *et al.*, 2013), GS 1826-238 (Ji *et al.*, 2014a), KS 1731-260 (Ji *et al.*, 2014b), 4U 1705-44 (Ji *et al.*, 2014b), 4U 1728-34 (Kajava *et al.*, 2017) and 4U 1724-30 (Kashyap and Chakraborty, 2022), based on *RXTE*, *INTEGRAL*, *Insight*-HXMT, *AstroSat* and *NuSTAR* observations.

It is very hard to detect the deficit up to $5\,\sigma$ in single short burst because of the relatively small detection area of previous missions at hard X-ray, and hence, these firm detection significance are aggrandized by stacked tens to hundreds of bursts. For example, for 4U 1636-536 the deficit detection is based on 36 burst results and on RXTE/PCA observations.

In a burst of 4U 1636-536 detected by *Insight*-HXMT, we find for the first time the hard X-ray deficit/shortage via the single short type-I burst showing up during the outburst in 2018. As plotted in Figure 4, the hard X-ray shortage is detected with a significance of $6.2\,\sigma$ in 40–70 keV, which is consistent with those derived previously from stacking a large amount of bursts detected by RXTE/PCA

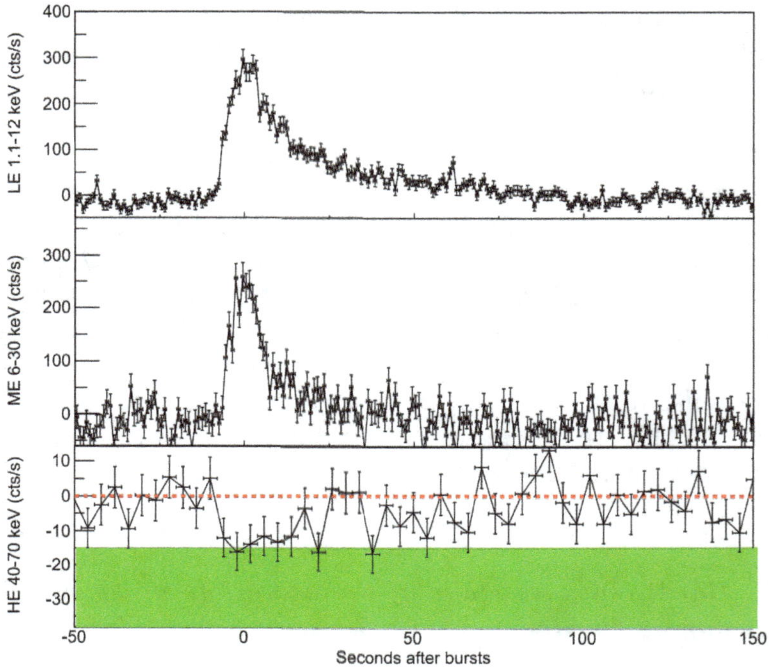

Fig. 4. Left panel shows the LE, ME, and HE light curves of the burst in 1.1–12 keV, 5–30 keV, and 40–70 keV, respectively. The time bin for LE and ME is 1 s and HE is 4 s, the green zone in the bottom panel indicates the background level for HE detectors. This figure is taken from Chen *et al.* (2018).

within a series of flares of 4U 1636-536. Moreover, the broad bandwidth of *Insight*-HXMT also allows us for the first time to infer the burst influence upon the continuum spectrum by performing the spectral fitting of the burst, which ends up with the finding that hard X-ray shortage appears at around 40 keV in the continuum spectrum.

In the scenario that a burst cools the corona to a lower temperature, the fraction of the deficit should have a larger value in a higher energy band. For instance, assuming that the corona temperature is cooled by an amplitude of 10% during the bursts (from 10.8 to 9.7 keV), the count rates in 50–60 keV and 30–40 keV will be dropped by 50% and 30%, respectively. The above estimation is consistent with the observational results from MAXI J1816-195. By stacking the *Insight*-HXMT light curves of 66 bursts from MAXI J1816-195, a hard X-ray shortage is detected with a significance of 15.7 σ in 30–100 keV (Chen *et al.*, 2022a). As expected, the shortage fraction is energy-dependent: larger in a higher energy band. We also note that there is a clue that the fraction of the deficit drops at ~65–85 keV. This maybe related to another hard X-ray generator — the accretion column of the magnetic pole, which is not affected by the burst.

3.2. *Burst photons upscattered by the corona*

For the persistent emission of low/hard or high/soft state of LMXBs, the spectrum is well fitted with an absorbed convolution thermal Comptonization model (with an input seed photon spectrum diskbb), available as thcomp (a more accurate version of nthcomp) (Zdziarski *et al.*, 2020) in XSPEC, which is described by the optical depth τ, electron temperature kT_e, and scattered/covering fraction f_{sc}.

Since the burst photons could also be affected by the corona, we thus check if the model used in the persistent emission could be same with the burst emission. By taking the pre-burst emission as background emission, the burst spectra are fitted by the model tbabs*thcomp*bb, in which the thcomp parameters are fixed at the persistent emission fit results. Thus, convolution thermal Comptonization model (with an input seed photon spectra blackbody) has the same d.o.f with the canonical blackbody model, and a more d.o.f. than the f_a model. bb and thcomp represent the burst emission from the NS photosphere and a corona influence on the burst emission. This model allows us to evaluate the contribution from both the upscattered by the corona/boundary layer and direct photons from the NS surface.

In the PPE phase of 4U 1608-52 (Chen *et al.*, 2022b) and 4U 1730-22, this model provides the best fit and yields physically acceptable spectral parameters. We find that this convolved thermal-Comptonization model provides equally good results with the f_a model but with a more d.o.f. and is statistically preferred to the f_a model in the middle of PRE phase (with the coolest blackbody temperature). However, in the rising and decay parts, such model has a bigger reduced χ_v^2 than the f_a model and even the canonical blackbody model, which may indicate that the burst emission suffers few Comptonization during this phase. Under this scenario, the radius of the photosphere is underestimated with the canonical blackbody model or the f_a model.

3.3. *Disk obscuration in burst emission*

In the observations of the time-resolved spectra of the PRE bursts (Galloway *et al.*, 2008), the vast majority of the peak flux reach a (local) maximum close to the time of peak radius, which does follow the theory's prediction. However, in the bright PRE burst of 4U 1608-52 and 4U 1730-22 detected by *Insight*-HXMT, the bolometric flux of the just lift-up time and the touch-down time is different and the peak flux is not at the time when the photosphere reached its peak radius. The flux shortage during the rise PRE phase is believed to correspond to the obscured lower part of the NS surface by the disk (Chen *et al.*, 2022b; Shaposhnikov and Titarchuk, 2004; Shaposhnikov *et al.*, 2003).

The anisotropy of the burst emission is expected to be accretion-state-dependent since the accretion disk in the low/hard state is believed to be truncated at a larger distance, but there is no observational evidence before 2022 for the accretion-state-dependent anisotropy(Galloway *et al.*, 2020).

Since the bursts in low/hard state and high/soft state have different accretion circumstances, e.g., a different inner disk radii, as expected, the bursts emission of the two groups should have different anisotropic degrees: the flux shortage for the bright PRE bursts detected in the high accretion-rate state with a small inner disk radius is absent in the PRE burst in the low accretion-rate state without the accretion disk detection, which are derived from bursts' results of 4U 1608-52 by *Insight*-HXMT observations, as shown in Figure 5. However, there is only one single PRE burst occurring in the low/hard state, and the universality of the conclusion needs further verification.

4. Progresses of Magnetar and Rotation-Powered Pulsars

4.1. *X-ray counterpart of fast radio bursts*

A fast radio burst (FRB) is an extremely intense radio burst phenomenon with a very short duration, usually only a few milliseconds, and a very high energy release, with typical peak fluxes on the order of Yankee, and the energy released in a millisecond can even exceed the total energy released by the solar system in a month.

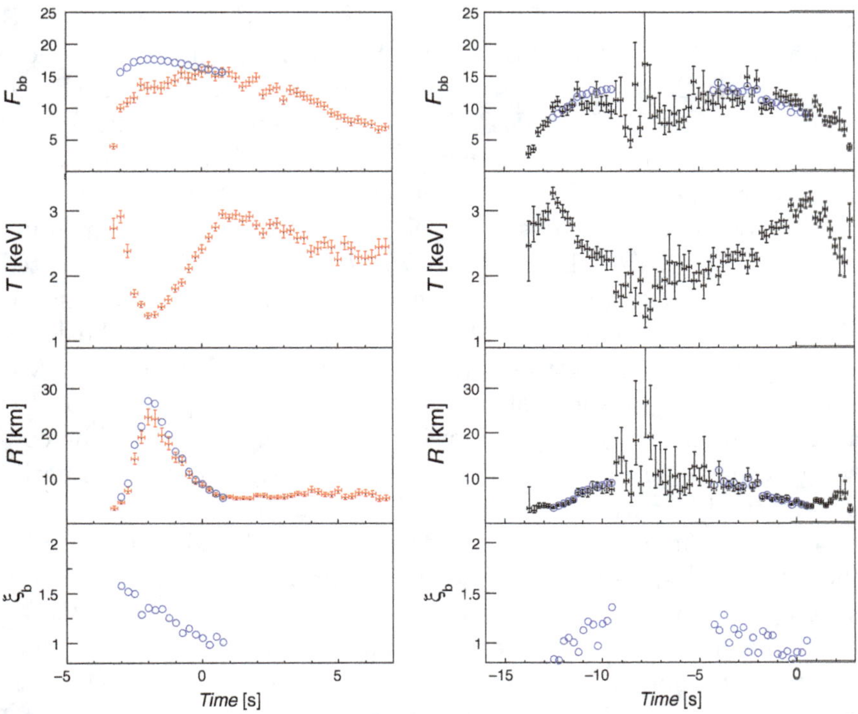

Fig. 5. For burst (left) in the high/soft state and burst (right) in the low/hard state of 4U 1608-52, the model predicted time evolution of the blackbody bolometric flux $F_{\rm bb}$, the radius R of NS photosphere at 4 kpc, anisotropic degree $\xi_{\rm bb}$ are marked in blue; for comparison, the corresponding observed parameters, including the temperature $kT_{\rm bb}$, are also given. Note for burst in the low/hard state, lacking LE detection, we ignore the data points with larger error bars in the middle of the PRE phase with temperature <2 keV, which are beyond the ME and HE's detection.

Although many FRBs have been discovered, their physical origin and production mechanism have not been clearly established for a long time. Astronomers have constructed many theoretical models to explain the observed FRB phenomenon. In order to further study and discriminate these models, besides relying on the continuous improvement of FRB detection technology and data analysis, multi-wavelength detection of FRBs, especially the search for gamma-ray radiation associated with FRBs, has become particularly important.

On April 28, 2020, significant progress was made in FRB observation. The *Insight*-HXMT satellite was under an observation of

Target Of Opportunity (ToO) pointing to the active magnetar (SGR J1935+2154) when the *CHIME* and *STARE2* detected a fast radio burst (FRB 200428) from the direction of the magnetar (Bochenek *et al.*, 2020; CHIME/FRB Collaboration *et al.*, 2020), and *Insight*-HXMT detected a unique X-ray burst from the magnetar 8.6 s before the FRB. The 8.6-second time delay is exactly the time it takes for the radio photos to travel through the interstellar medium more than the X-rays. Interestingly, the *Insight*-HXMT first detected a bimodal structure in the light curve of the X-ray burst, which is almost identical to the bimodal structure of the FRB. The time-integrated spectrum of this burst is derived jointly using HE, ME and LE data. Since the best fit and statistically acceptable model is a cutoff power law, the spectrum is dominated by a power law covering at least the 1–100 keV range, and thus, this burst is primarily non-thermal in nature, as shown in Figure 6.

Based on unique collimator design, *Insight*-HXMT can localize the burst within its field of view and confirm that it came from the magnetar SGR J1935+2154, as shown in Figure 7 (Li *et al.*, 2021), which means that for the first time, a magnetar was found to produce fast radio bursts, and it is the first time in human history that the origin of FRBs has been solved. At the same time, the detection of high-energy radiation corresponding to FRBs is crucial for a deeper understanding of the physical mechanism of FRBs and magnetar outbursts.

4.2. *X-ray properties of the Crab pulsar*

The Crab pulsar is one of the most widely studied celestial objects. It was born in 1054, has a spin period of 33 ms, and is bright over the full electromagnetic spectrum from radio to high-energy γ-rays. At all wavelengths, this pulsar shows a double-pulse structure, with the main pulse (P1) and the interpulse (P2) separated by a phase of $\sim$0.4, and the exact pulse morphology varies as a function of photon energy, which means the spectra evolves with phase (Ge *et al.*, 2012).

The energy dependence of the X-ray pulse profile and the phase-resolved spectra (PRS) of the Crab pulsar have been obtained using observations from NICER and *Insight*-HXMT. The fitted parameters of the PRS from the log-parabola model are displayed in Figure 8. The phase-resolved spectral properties of the two pulses, as can be

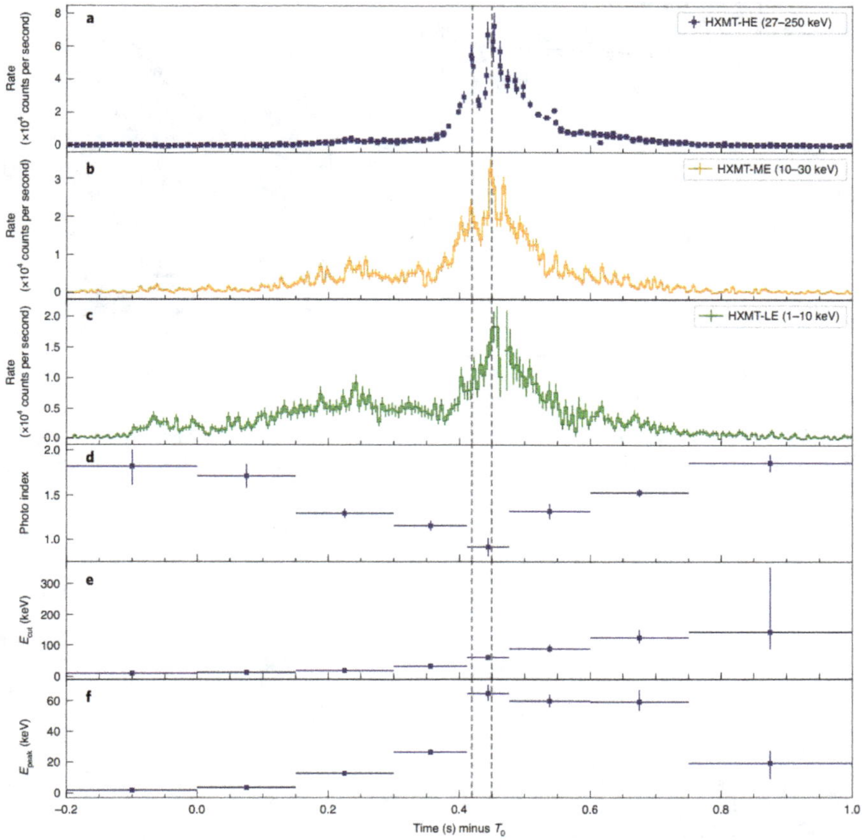

Fig. 6. Light curves and spectra evolution of the burst using *Insight*-HXMT HE, ME and LE data. This figure is taken from Li *et al.* (2021).

seen in Figure 8(a), are quite different: The spectral index α around P1 increases first and then falls off, with the maximum at the peak, and α keeps growing from the leading wing to the trailing wing of P2. In the bridge region, α has a down-and-up trend and changes smoothly with phase. The values of the curvature parameter β for all spectra are positive (Figure 8(b)), implying that the spectral index increases with energy at all phases. β has a W shape in phase and reaches a local minimum around two turning points of the pulse profiles. Both α and β in 2–250 keV show phase evolution trends similar to those in Ge *et al.* (2012), but with higher precision. The value of β in 2–250 keV is smaller than that in 3–60 keV, indicating that

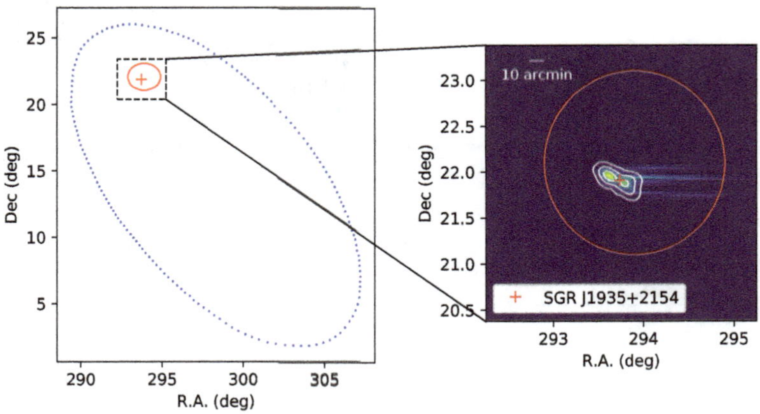

Fig. 7. Localization of the burst using *Insight*-HXMT HE, ME and LE data. The red crosses mark the known position of SGR J1935+2154. This figure is taken from Li *et al.* (2021).

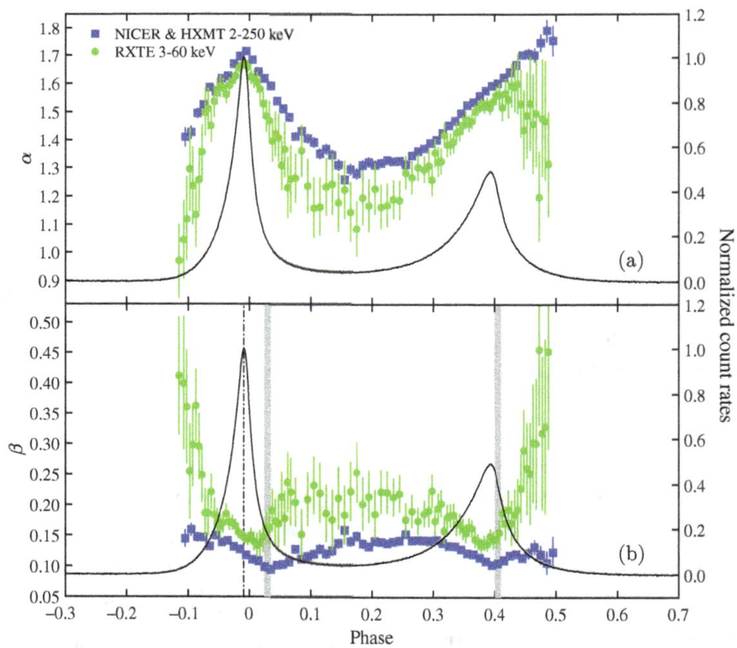

Fig. 8. Fitting results of the PRS of the Crab pulsar using a log-parabola model, where these blue squared data points are the joint fitting results from *NICER* and *Insight*-HXMT's observations which cover 2–250 keV and the green circled data points are the fitting results for *RXTE*'s observations presented in Ge *et al.* (2012). The dash-dotted line marks the peak location of the primary pulse P1, and the two gray bars mark the shape turnover phase intervals. This figure is taken from Yan *et al.* (2022).

the photon index changes more slowly at higher energies. The phase turning points of β in this work (gray phase regions in Figure 8(b)) are not aligned with those in Ge *et al.* (2012), probably due to the spectral complexity of this pulsar, because the energy ranges of the two studies are different.

The X-ray pulse profile and the pulsed flux variations are examined at the two glitches of the Crab pulsar in 2017 and 2019, using data from *Insight*-HXMT, NICER, and Fermi-GBM. For the first time, we have reached precisions comparable to or even higher than the spin-down rate increase at the glitches. These results show that the permanent component of the spin-down rate increase in the 2017 glitch is unlikely due to an enhancement of the overall magnetic field at a significance of $2.8\,\sigma$. The X-ray pulse profiles are also very stable after the 2019 glitch, with an rms as small as 0.35% for the residual distribution of the profile subtraction (Zhang *et al.*, 2022).

5. Future Prospects

Since the launching of *Insight*-HXMT, many outbursts of AXPs, magnetars and rotation-powered pulsars have been monitored. Considering most of AXPs are transient with a recurrence time of years or even tens of years, *Insight*-HXMT will continue observing more new sources or/and new outbursts from known AXPs. We expect potential breakthroughs in the following areas: (1) the detection of CRSFs, especially at high energies, and the CRSF evolution with luminosity; (2) the timing analysis, including the spin evolution, the morphology of pulse profiles and features in the power spectra; (3) the broadband spectroscopy with high cadence observations, which reveals the accretion process when physical models are applied. Magnetars will be concerned intensively and one of the most important target of *Insight*-HXMT to study more properties of FRBs and their couterparts together with other instruments. In addition, we note that recently IXPE-*Imaging X-ray Polarimetry Explorer* opens a new polarization window that perfectly complements *Insight*-HXMT and the spectro-polarimetric analysis will certainly become an important direction. This strongly encourages the development of polarimetric projects in China, such as the eXTP-*Enhanced X-ray Timing and Polarimetry Mission* (Weisskopf *et al.*, 2016; Zhang *et al.*, 2019a).

Acknowledgments

This work is supported by the National Key R&D Program of China (2021YFA0718500) from the Minister of Science and Technology of China (MOST). This work is also supported by the International Partnership Program of Chinese Academy of Sciences (Grant No.113111KYSB20190020).

Bibliography

Ballantyne, D. R. and Strohmayer, T. E. (2004). The evolution of the accretion disk around 4U 1820-30 during a superburst, *Astrophys. J. Lett.* **602**, L105.

Basko, M. M. and Sunyaev, R. A. (1976). The limiting luminosity of accreting neutron stars with magnetic fields, *Mon. Not. R. Astron. Soc.* **175**, 395.

Becker, P. A. *et al.* (2012). Spectral formation in accreting X-ray pulsars: Bimodal variation of the cyclotron energy with luminosity, *Astron. Astrophys.* **544**, A123.

Bochenek, C. D. *et al.* (2020). A fast radio burst associated with a Galactic magnetar, *Nat.* **587**, 59.

Bult, P. *et al.* (2021). On the impact of an intermediate duration X-ray burst on the accretion environment in IGR J17062-6143, *Astrophys. J.* **920**, 59.

Cao, X. *et al.* (2020). The medium energy X-ray telescope (ME) onboard the Insight-HXMT astronomy satellite, *Sci. Sin. Phys. Mech. Astron.* **63**, 249504.

Chen, Y. P. *et al.* (2011). The 2008 outburst of IGR J17473-2721: Evidence for a disk corona, *Astron. Astrophys.* **534**, 101.

Chen, Y. P. *et al.* (2012). Type-I X-ray bursts reveal a fast co-evolving behavior of the corona in an x-ray binary, *Astrophys. J. Lett.* **752**, 34.

Chen, Y. P. *et al.* (2013). The hard x-ray behavior of Aql X-1 during type-I bursts, *Astrophys. J.* **777**, 9.

Chen, Y. P. *et al.* (2018). Insight-HXMT observations of 4U 1636-536: corona cooling revealed with single short type-I X-ray burst, *Astrophys. J. Lett.* **864**, L30.

Chen, Y. P. *et al.* (2019). Insight-HXMT observation on 4U 1608-52: Evolving spectral properties of a bright type-I X-ray burst, *J. High Energy Astrophys.* **24**, 23.

Chen, Y. *et al.* (2020). The low energy X-ray telescope (LE) onboard the Insight-HXMT astronomy satellite, *Sci. Sin. Phys. Mech. Astron.* **63**, 249505.

Chen, X. *et al.* (2021). Relation of cyclotron resonant energy and luminosity in a strongly magnetized neutron star GRO J1008-57 observed by insight-HXMT, *Astrophys. J.* **919**, 33.

Chen, Y. P. *et al.* (2022a). The prolific thermonuclear X-ray bursts from the outburst of the newly discovered millisecond pulsar MAXI J1816-195 observed by Insight-HXMT and NICER, arXiv:2208.12124.

Chen, Y. P. *et al.* (2022b). Insight-HXMT observation of 4U 1608-52: Evidence of interplay between a thermonuclear burst and accretion environment, *Astrophys. J.* **936**, 46.

CHIME/FRB Collaboration, Andersen, B. C. *et al.* (2020). A bright millisecond-duration radio burst from a Galactic magnetar, *Nature* **587**, 54.

Cumming, A. (2004). Thermonuclear X-ray bursts: Theory vs. observations, *Nucl. Phys. B Proc. Suppl.* **132**, 435.

Degenaar, N. *et al.* (2018). Accretion disks and coronae in the X-ray flashlight, *Space Sci. Rev.* **214**, 15.

Ding, Y. Z. *et al.* (2021a). Timing and spectral variability of the high-mass X-ray pulsar GX 301-2 over orbital phases observed by Insight-HXMT, *Mon. Not. R. Astron. Soc.* **506**, 2712.

Ding, Y. Z. *et al.* (2021b). QPOs and orbital elements of X-ray binary 4U 0115+63 during the 2017 outburst observed by Insight-HXMT, *Mon. Not. R. Astron. Soc.* **503**, 6045.

Doroshenko, V. *et al.* (2017). Luminosity dependence of the cyclotron line and evidence for the accretion regime transition in V 0332+53, *Mon. Not. R. Astron. Soc.* **466**, 2143.

Doroshenko, V. *et al.* (2020). Hot disc of the Swift J0243.6+6124 revealed by Insight-HXMT, *Mon. Not. R. Astron. Soc.* **491**, 1857.

Galloway, D. K. *et al.* (2008). Thermonuclear (type I) X-ray bursts observed by the Rossi X-ray timing explorer, *Astrophys. J. Suppl. Ser.* **179**, 360.

Galloway, D. K. *et al.* (2020). The Multi-INstrument Burst ARchive (MINBAR), *Astrophys. J. Suppl. Ser.* **249**, 32.

Ge, M. Y. *et al.* (2012). X-ray phase-resolved spectroscopy of PSRs B0531+21, B1509-58, and B0540-69 with RXTE, *Astrophys. J. Suppl. Ser.* **199**, 32.

Ge, M. Y. *et al.* (2020). Insight-HXMT firm detection of the highest-energy fundamental cyclotron resonance scattering feature in the spectrum of GRO J1008-57, *Astrophys. J. Lett.* **899**, L19.

Güver, T. *et al.* (2022a). Burst-disk interaction in 4u 1636-536 as observed by NICER, *Astrophys. J.* **935**, 154G.

Hewish, A. *et al.* (1968). Observation of a rapidly pulsating radio source, *Nature* **217**, 709.

Hou, X. *et al.* (2022). Fan-beamed X-ray emission from 1 to above 130 keV from the ultraluminous X-ray pulsar RX J0209.6-7427 in the small magellanic cloud, *Astrophys. J.* **938**, 149.

in't Zand, J. J. M. *et al.* (2013). A bright thermonuclear X-ray burst simultaneously observed with Chandra and RXTE, *Astron. Astrophys.* **553**, A83.

Ji, L. *et al.* (2013). X-ray bursts as a probe of the corona: The case of XRB 4U 1636-536, *Mon. Not. R. Astron. Soc.* **432**, 2773.

Ji, L. *et al.* (2014a). A state-dependent influence of type I bursts on the accretion in 4U 1608-52, *Astrophys. J.* **791**, L39.

Ji, L. *et al.* (2014b). Possible hard X-ray shortages in bursts from KS 1731-260 and 4U 1705-44, *Astron. Astrophys.* **564**, A20.

Ji, L. *et al.* (2020a). Switches between accretion structures during flares in 4U 1901+03, *Mon. Not. R. Astron. Soc.* **493**, 5680.

Ji, L. *et al.* (2020b). Timing analysis of 2S 1417-624 observed with NICER and Insight-HXMT, *Mon. Not. R. Astron. Soc.* **491**, 1851.

Ji, L. *et al.* (2021). X-ray reprocessing in accreting pulsar GX 301-2 observed with Insight-HXMT, *Mon. Not. R. Astron. Soc.* **501**, 2522.

Kajava, J. J. E. *et al.* (2017). Variable spreading layer in 4U 1608-52 during thermonuclear X-ray bursts in the soft state, *Mon. Not. R. Astron. Soc.* **472**, 78

Kashyap, U. and Chakraborty, M. (2022). Probing spectral and temporal evolution of the neutron star low-mass X-ray binary 4U 1724-30 with AstroSat, *Mon. Not. R. Astron. Soc.* **512**, 6180.

Keek, L. *et al.* (2014). X-raying an accretion disk in realtime: The evolution of ionized reflection during a superburst from 4U 1636-536, *Astrophys. J. Lett.* **797**, L23.

Kong, L. D. *et al.* (2021). Luminosity dependence of the cyclotron line energy in 1A 0535+262 observed by Insight-HXMT during the 2020 giant outburst, *Astrophys. J.* **917**, L38.

Kong, L. D. *et al.* (2022). Insight-HXMT discovery of the highest-energy CRSF from the first galactic ultraluminous X-ray pulsar swift J0243.6+6124, *Astrophys. J. Lett.* **933**, L3.

Kretschmar, P. *et al.* (2019). Advances in understanding high-mass X-ray binaries with INTEGRALand future directions, *New Astron. Revi.* **86**, 101546.

Lewin, W. H. G., van Paradijs, J., and Taam, R. E. (1993). X-ray bursts, *SSRv* **62**, 223.

Li, C. K. *et al.* (2021). HXMT identification of a non-thermal X-ray burst from SGR J1935+2154 and with FRB 200428, *Nat. Astron.* **5**, 378.

Liu, C. (2020). The high energy X-ray telescope (HE) onboard the Insight-HXMT astronomy satellite, *Sci. Sin. Phys. Mech. Astron.* **63**, 249503.

Liu, J. *et al.* (2021). Disc versus wind accretion in X-ray pulsar GX 301-2, *Mon. Not. R. Astron. Soc.* **504**, 2493.

Liu, J. *et al.* (2022a). Super-Eddington accretion of the first galactic ultra-luminous X-ray pulsar Swift J0243.6+6124, *Mon. Not. R. Astron. Soc.* **512**, 5686.

Liu, Q. *et al.* (2022b). Variations of cyclotron resonant scattering features in Vela X-1 revealed with Insight-HXMT, *Mon. Not. R. Astron. Soc.* **514**, 2805.

Liu, Q. *et al.* (2022c). Detection of a quasi-periodic oscillation at 40 mHz in Cen X-3 with Insight-HXMT, *Mon. Not. R. Astron. Soc.* **516**, 5579.

Maccarone, T. J. and Coppi, P. S. (2003). Spectral fits to the 1999 Aql X-1 outburst data, *Astron. Astrophys.* **399**, 1151.

Ma, R. *et al.* (2022). High energy millihertz quasi-periodic oscillations in 1A 0535 + 262 with Insight-HXMT challenge current models, *Mon. Not. R. Astron. Soc.* **517**, 1988M.

Mushtukov, A. and Tsygankov, S. (2022). Accreting strongly magnetised neutron stars: X-ray pulsars, arXiv:2204.14185.

Mushtukov, A. A. (2015). The critical accretion luminosity for magnetized neutron stars, *Mon. Not. R. Astron. Soc.* **447**, 1847.

Nabizadeh, A. *et al.* (2021). Spectral evolution of X-ray pulsar 4U 1901+03 during the 2019 outburst based on Insight-HXMT and NuSTAR observations, *Astron. Astrophys.* **652**, A89.

Reig, P. (2011). Be/X-ray binaries, *Astrophys. Space Sci.* **332**, 1.

Shaposhnikov, N. and Titarchuk, L. (2004). On the nature of the flux variability during an expansion stage of a type I X-ray burst: Constraints on neutron star parameters for 4U 1820-30, *Astrophys. J.* **606**, L57.

Shaposhnikov, N., Titarchuk, L., and Haberl, F. (2003). The bursting behavior of 4U 1728-34: The parameters of a neutron star and the geometry of a neutron star-disk system, *Astrophys. J. Lett.* **593**, L35.

Staubert, R. *et al.* (2014). Long-term change in the cyclotron line energy in Hercules X-1, *Astron. Astrophys.* **572**, A119.

Staubert, R. *et al.* (2019). Cyclotron lines in highly magnetized neutron stars, *Astron. Astrophys.* **622**, A61.

Strohmayer, T. and Bildsten, L. (2006).New views of thermonuclear bursts, *Compact Stellar X-Ray Sources*, Cambridge, UK: Cambridge University Press, 113–156.

Tuo, Y. L. *et al.* (2020). Insight-HXMT insight into switch of the accretion mode: The case of the X-ray pulsar 4U 1901+03, *J. High Energy Astrophys.* **27**, 38.

Wang, P. J. *et al.* (2020). Insight-HXMT observations of Swift J0243.6+6124: The evolution of RMS pulse fractions at super-Eddington luminosity, *J. High Energy Astrophys.* **497**, 5498.

Wang, W. *et al.* (2021). Insight-HXMT observations of Swift J0243.6+6124: The evolution of RMS pulse fractions at super-Eddington luminosity, *J. High Energy Astrophys.* **30**, 1.

Weisskopf, M. C. *et al.* (2016). The imaging X-ray polarimetry explorer (IXPE), *SPIE* **9905**, 990517.

Worpel, H., Galloway, D. K., and Price, D. J. (2013). Evidence for accretion rate change during type I X-ray bursts, *Astrophys. J.* **772**, 94.

Worpel, H., Galloway, D. K., and Price, D. J. (2015). Evidence for enhanced persistent emission during sub-eddington thermonuclear bursts, *Astrophys. J.* **801**, 60.

Xiao, G. C. *et al.* (2019). Constant cyclotron line energy in Hercules X-1 - Joint Insight-HXMT and NuSTAR observations, *J. High Energy Astrophys.* **23**, 29.

Yan, L.-L. *et al.* (2022). A study on the X-ray pulse profile and spectrum of the crab pulsar using NICER and Insight-HXMT's observations, *Astrophys. J.* **928**, 183.

Zdziarski, A. A., P. (2020). Spectral and temporal properties of Compton scattering by mildly relativistic thermal electrons, *Mon. Not. R. Astron. Soc.* **492**, 5234.

Zhang, S. *et al.* (2019a). The enhanced X-ray timing and polarimetry mission eXTP, *Sci. Sin. Phys. Mech. Astron.* **62**, 29502

Zhang, Y. *et al.* (2019b). Insight-HXMT observations of swift J0243.6+6124 during its 2017-2018 outburst, *Astrophys. J.* **879**, 61.

Zhang, S.-N. *et al.* (2020). Overview to the hard X-ray modulation telescope (Insight-HXMT) satellite, *Sci. Sin. Phys. Mech. Astron.* **63**, 249502.

Zhang, Y. H. *et al.* (2022). Invariable X-ray profile and flux of the crab pulsar during its two glitches, *Astrophys. J.* **932**, 11.

Zhang, B. (2020). The physical mechanisms of fast radio bursts, *Nature* **587**, 45.

https://doi.org/10.1142/9789819800643_0010

Chapter 10

HXMT Views on the Accreting Millisecond X-ray Pulsars

Zhao-Sheng Li
Key Laboratory of Stars and Interstellar Medium,
Xiangtan University, Xiangtan 411105, Hunan, P. R. China
lizhaosheng@xtu.edu.cn

The Hard X-ray Modulation Telescope (HXMT), named "Insight", is China's first X-ray astronomy satellite. There are three main payloads onboard Insight-HXMT, the high energy X-ray telescope (20–250 keV), the medium energy X-ray telescope (5–30 keV), and the low energy X-ray telescope (1–15 keV). The main purpose of this chapter is to introduce scientific achievements on accreting X-ray pulsars including binaries and accretion-powered millisecond pulsars detected by Insight-HXMT. Especially, the most recent observational and interpretational results on accreting X-ray pulsars, i.e., Aql X–1, Swift J1756.9–2508 and MAXI J1816–195, are reviewed to help people understand the physics of these sources.

Keywords: X-ray timing, Neutron star, Pulsar, Accretion disk

Contents

1. Introduction

Accreting millisecond X-ray pulsars (AMXPs) are rapidly spinning, old, recycled neutron stars (NSs) hosted in low-mass X-ray binaries (LMXB). AMXPs are usually confirmed by its coherent pulsation during outburst, with spin periods of a few milliseconds and orbital periods between 40 min and 5 h based on the current sample (see Campana and Di Salvo, 2018; Di Salvo and Sanna, 2022; Papitto *et al.*, 2020, for reviews). As a binary system evolves through phases of accretion onto the NS, it gains angular momentum from the accreted material, which is sufficient to spin-up the NS to a rotation period equilibrium in the millisecond range, that is, the recycling of old radio-pulsars to millisecond periods (Alpar *et al.*, 1982; Falanga *et al.*, 2005b; Papitto *et al.*, 2013; Wijnands and van der Klis, 1998). This picture has been verified by the discoveries of AMXPs starting with SAX J1808.4–3658 by RXTE (Wijnands and van der Klis, 1998), discovery of the pulsar spin-up induced by accretion (Falanga *et al.*, 2005b), and by swinging between accretion- and rotation-powered states in IGR J18245–2452 (Papitto *et al.*, 2013). For reviews of the properties of these objects, we refer to Patruno and Watts (2021), Campana and Di Salvo (2018), and Papitto *et al.* (2020).

Between X-ray outbursts, long-term monitoring shows some AMXPs to exhibit spin-down in quiescence (see e.g., Hartman *et al.*, 2009; Papitto *et al.*, 2010; Patruno *et al.*, 2010). The appearance of X-ray pulsation, the spin-down during the quiescent state, and the spin-up during accretion process allow us to measure the dipole magnetic field of AMXPs in the order of 10^8-10^9 G.

Insight-HXMT (Insight Hard X-ray Modulation Telescope, Zhang *et al.*, 2020) is the first Chinese X-ray telescope, and is equipped with three slat-collimated instruments: the Low Energy X-ray telescope (LE, 1–12 keV; Chen *et al.*, 2020), the Medium Energy X-ray telescope (ME, 5–35 keV; Cao *et al.*, 2020) and the High Energy X-ray telescope (HE, 20–350 keV; Liu *et al.*, 2020), providing capabilities for broadband X-ray timing and spectroscopy.

In this chapter, we introduce the progress of AXMPs that have been made by Insight-HXMT and several other telescopes, including the X-ray pulsations from MAXI J1816–195, Swift J1756.9–2508 and the thermonuclear X-ray bursts properties.

2. The Broadband X-ray Pulsations

2.1. *MAXI J1816–195*

The new X-ray transient MAXI J1816–195 was discovered by MAXI/GSC on June 7 during its 2022 outburst. Using the Swift localization, NICER detected pulsations and thermonuclear type I X-ray bursts from MAXI J1816–195, confirming that the source is an AMXP with a spin frequency of 528 Hz, an orbital period of 4.83 hr and a projected semi-major axis of 0.26 lt-s (Bult *et al.*, 2022a). The mass of the companion star was determined to be in the range 0.10–0.55 $M_\odot$ (Bult *et al.*, 2022a). Meanwhile, Insight-HXMT found X-ray pulsations from MAXI J1816–195 in the hard X-ray/soft γ-ray band. Bult *et al.* (2022a) proposed the flux-bias model, which accounts for the effects of the accretion torque and/or the wandering of the hot spot on the NS surface, to explain the structures left in the timing residuals across the outburst.

Fitting the complete Insight-HXMT HE ToA dataset with a constant frequency model resulted in a very poor fit with an unacceptable χ^2. Ignoring 22 ToA's from the first ∼5 days of the outburst yielded reasonable fits assuming a constant frequency model and a model including a frequency derivative term. The constant frequency model is fully consistent at 1σ confidence with the corresponding model shown in Table 1 of Bult *et al.* (2022a). The model with both ν and $\dot{\nu}$, however, improves upon the constant frequency model at 4.3σ confidence reducing the reduced χ^2 from

Table 1. Positional, orbital and spin parameters derived in this work and used as fixed values from the literature for MAXI J1816–195.

Parameter	Values	Units
α_{2000}	$18^{h}16^{m}52^{s}41168(12)$	
δ_{2000}	$-19°37'57''40138(473)$	
JPL Ephemeris	DE405	
$P_{\rm orb}$	17402.5786(60)	s
$a_{\rm x}\sin i$	0.262 948(18)	lt-s
e	0 (fixed)	
$T_{\rm asc}$	59738.875 632(4)	MJD (TDB)
	Constant frequency model	
Validity range	59741.9 – 59760.6	MJD (TDB)
t_0 (Epoch)	59741.0	MJD (TDB)
ν	528.611 105 832(4)	Hz
χ^2/d.o.f	$86.78/(47-1) = 1.886$	
	Spin-up model	
Validity range	59741.9–59760.6	MJD (TDB)
t_0 (Epoch)	59741.0	MJD (TDB)
ν	528.611 105 774(12)	Hz
$\dot{\nu}$	$(9.0 \pm 2.1) \times 10^{-14}$	Hz s^{-1}
χ^2/d.o.f	$68.18/(47-2) = 1.515$	

1.886 to 1.515 adding one degree of freedom. The measured $\dot{\nu}$ of $(9.0 \pm 2.1) \times 10^{-14}$ Hz s^{-1} can be interpreted as a spin-up across the time interval MJD 59741.9–59760.6.

The pulse-phase residuals of the complete sets of ToA's from all three instruments combined (each determined for a different energy band) after folding upon the (Insight-HXMT/HE 20–35 keV based) spin-up model are shown in Figure 1. Irrespective the energy band each instrument showed a quite similar behaviour: during the first ~5 days of the outburst - the rising flux part -there is a structured residual feature which can be described by a (short-duration) spin-down model. This feature is the cause of the very poor χ^2 when the full ToA datasets are considered (see also Table 1 of Bult *et al.*, 2022a, indicating unacceptable χ^2 fit values) Also, the steep positive excursions for the NICER ToAs at the early beginning and very end of the outburst are striking in Figure 1. It is clear that even a

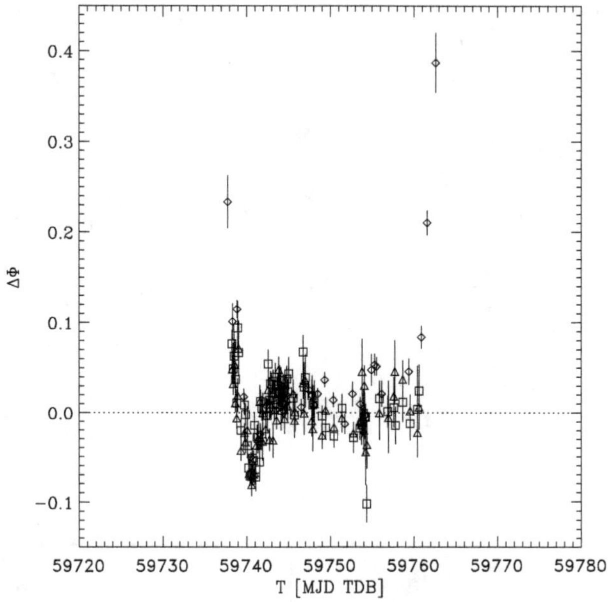

Fig. 1. Pulse phase residuals of Insight-HXMT HE (open triangles; 20–35 keV), Insight-HXMT ME (open squares; 10–20 keV) and NICER (open diamonds; 1.8–10 keV) ToAs with respect to a two-parameter timing model (including a spin-up term) covering MJD 59741.9–59760.6 based on solely Insight-HXMT HE ToAs.

(bolometric) flux bias model cannot describe the full set residuals, consistent with Bult *et al.* (2022a).

Panels (a)–(c) of Figure 2 show the NICER pulse profiles for the energy intervals 0.8–1.8, 1.8–3, and 3–10 keV, respectively, while panels (d)–(f) show the Insight-HXMT ME profiles for the 5–10, 10–20 and 20–35 keV bands. The NuSTAR profiles are displayed in panels (g)–(k) covering the 3–79 keV energy range. Noteworthy is that comparing the NICER and NuSTAR 3–10 keV distributions, shown in panels (c) and (g), respectively, a small phase shift of 0.065 ± 0.020 is detectable which is just compatible with the absolute timing accuracy of NuSTAR (Bachetti *et al.*, 2021). Pulsed emission is detected significantly up to ∼60 keV, with a non-uniformity significance of 5.4σ for the 35–60 keV band (panel j) adopting a Z_2^2-test (Buccheri *et al.*, 1983). This test applied to the 60–79 keV band (panel k) yielded an insignificant result of 1.3σ. Finally, panels (l)–(p) show the Insight-HXMT HE profiles for energies of 20–30, 30–50, 50–65, 65–95 and

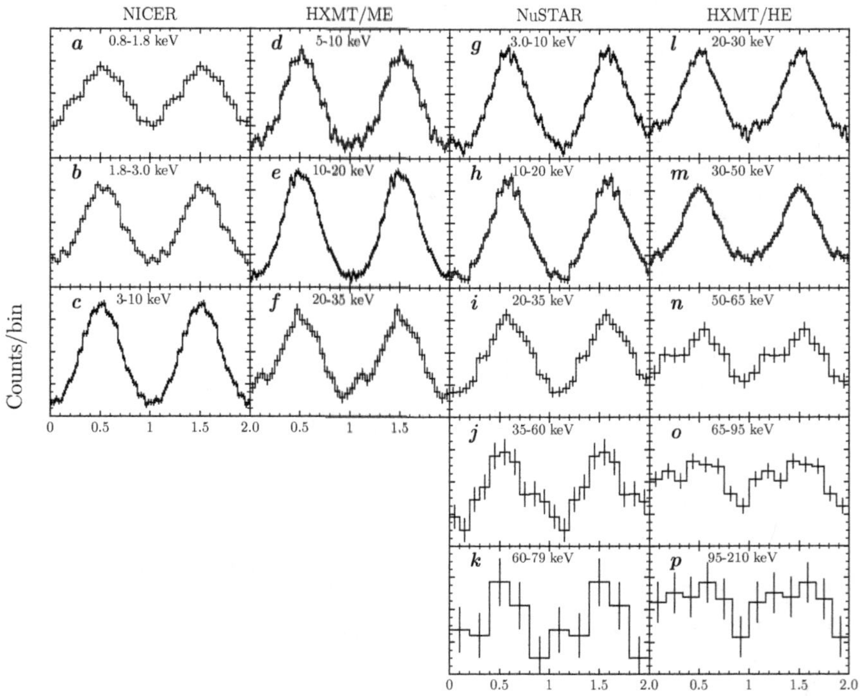

Fig. 2. The 0.8–210 keV broadband pulse profiles of MAXI J1816–195 observed by NICER (panels a–c, 0.8–10 keV), NuSTAR (panels g–k, 3–79 keV), and Insight-HXMT (panels d–f, 5–35 keV for ME; panels l–p, 20–210 keV for HE). In order to improve clarity, two cycles are shown. The error bars represent 1σ errors. The morphology is almost unchanged with energy. All profiles reach their maximum near phase ~0.5.

95–210 keV, respectively. Interestingly, pulsed emission is detected significantly up to ~95 keV. The 65–95 keV band (panel o) shows a non-uniform significance of 4.4σ applying a Z_2^2-test with indications for a second weaker pulse near phase 0.15 that seems to pop-up at energies above ~50 keV (panels n–o). The non-uniformity significance for the distribution shown in panel (p) for the 95–210 keV band is about 1.1σ, and so insignificant.

From Insight-HXMT observations hard X-ray pulsations have been detected significantly up to ~95 keV. The pulse profiles in 0.8–95 keV band are well described by a truncated Fourier series. The phase angles of the fundamental and the first overtone are nearly

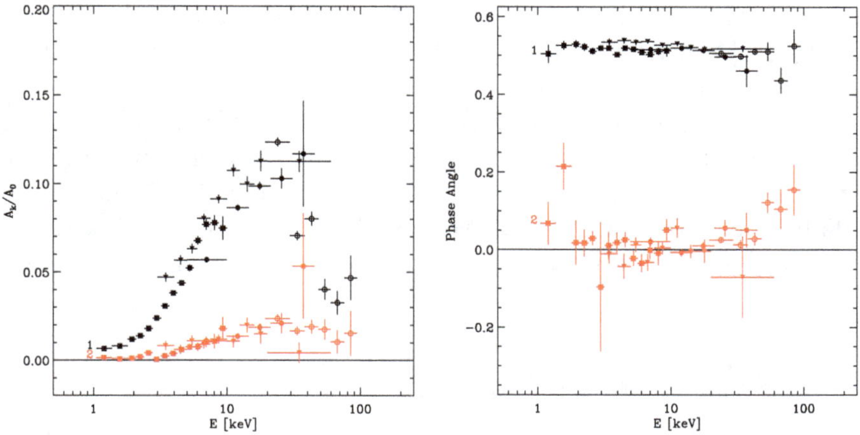

Fig. 3. (a) The fractional amplitude of various harmonics: (1) fundamental (black crosses) and (2) first overtone (red crosses) as a function of energy using background-corrected data from NICER (filled squares), NuSTAR (triangles), Insight-HXMT ME (filled circles) and Insight-HXMT HE (open circles). (b) The phase angle (in units radians/2π) as a function of energy for the two Fourier components, the fundamental (in black) and the first overtone (in red), which are independent of the background subtraction method.

constant for energies below ~60 keV, implying that most photons are likely emitted from the same region at the NS surface with a very similar emission pattern (Figure 3).

The stability of the pulse profiles across a broad energy range allows us to determine the absolute timing accuracy of Insight-HXMT in detail. Cross-correlating Insight-HXMT ME and NICER pulse profiles for the overlapping 5–10 keV energy band yielded a phase shift $\delta\Phi$ of only -0.008 ± 0.005, consistent with zero, suggesting that both instruments are well aligned within 14 μs, with the Insight-HXMT ME events arriving a little bit later than NICER ones. A similar correlation analysis now between Insight-HXMT ME and HE pulse profiles for the 20–35 keV band resulted in a difference of only 0.9 μs with an uncertainty of 13.9 μs.

The ToAs from the different (independent) instruments were well consistent with each other in spite of their different bandpass. The timing residuals showed erratic behavior (or V-shaped) during the rising part of the outburst in MJD 59737–59742, similar to the results shown in Bult *et al.* (2022a). For the decaying part of the outburst between MJD 59742–59761, a significant spin-up $\dot{\nu}$ of $(9.0 \pm 2.1) \times 10^{-14}$ Hz s^{-1} fitting a two parameter timing model has been detected, including ν and $\dot{\nu}$, yielding an acceptable χ^2.

Such spin-up (or spin-down) terms can arise when an erroneous (or inaccurate) source location is used in the barycentering process, however, the sub-arcsecond accurate VLA-location for MAXI J1816–195 has been used, and so believe that the measured value is genuine and due to the gain of angular momentum of NS from the accreted matter.

2.2. *Swift J1756.9–2508*

Swift J1756.9–2508 was discovered by Swift/BAT during its 2007 outburst. Coherent X-ray pulsations at a frequency of ∼182 Hz confirmed the compact object to be an AMXP. In the following observations, carried out by Swift and RXTE, an orbital period of 54.7 min was measured and, based on its mass function, the mass of the companion star — a highly evolved white dwarf — was determined to be in the range of $0.0067-0.03M_\odot$ (Krimm *et al.*, 2007b).

Swift J1756.9–2508 was shown to have undergone another three outbursts: in 2009, 2018 (Bult *et al.*, 2018a; Sanna *et al.*, 2018c), and 2019. For the first three recurrent outbursts between 2007 and 2018, there was no orbital period variation detected (Bult *et al.*, 2018a; Krimm *et al.*, 2007b; Patruno *et al.*, 2010; Sanna *et al.*, 2018c). Moreover, the upper limit on the orbital period derivative is $|\dot{P}_{\rm orb}| < 7.4 \times 10^{-13}$ s s^{-1} from the total 11-year span of the observations, which is consistent with the prediction of a conservative mass transfer in the binary system (Bult *et al.*, 2018a). Swift J1756.9–2508 exhibited long-term spin-down behavior with a spin frequency derivative of $|\dot{\nu}| \lesssim 3 \times 10^{-13}$ Hz s^{-1}, measured by Patruno *et al.* (2010), $|\dot{\nu}| {\sim} 7.3 \times 10^{-16}$ Hz s^{-1} by Bult *et al.* (2018a) and a smaller value $|\dot{\nu}| {\sim} 4.8 \times 10^{-16}$ Hz s^{-1} by Sanna *et al.* (2018c). Assuming that the rotational energy loss due to the magnetic dipole emission dominated the spin evolution of Swift J1756.9–2508, the magnetic field strength at the stellar magnetic poles was constrained to the range of $(3 - 6) \times 10^8$ G (see, e.g., Bult *et al.*, 2018a; Patruno *et al.*, 2010; Sanna *et al.*, 2018c, and references therein).

Pulsed emission has been detected from Swift J1756.9–2508 in the energy range of 1–10 keV using NICER and XMM-Newton data, in 3–60 keV using NuSTAR during both outbursts, and in the 5–100 keV range using Insight-HXMT during the 2018 outburst. No pulsed emission was detected below 1 keV using NICER and XMM-Newton data. Folding INTEGREL-ISGRI barycentered data of the 2018 outburst for the 20–60 keV band did not result in a detection of the pulsed emission in line with the expectations given the moderate total outburst flux, low exposure, and low pulsed fraction of $\lesssim$8%. However, with the Insight-HXMT/HE instrument, significant pulsed emission has been detected up to ~100 keV (see lower right panel of Figure 4) in spite of relatively low exposure. The significance of X-ray pulsations are reached 4.9σ and 5.1σ for ME and HE, respectively. We detected an increase of the pulse fraction from 4% to 7.5% from 1 to 5 keV saturating at higher energies. We found no evidence for a change in the spin frequency from our data set (Table 2). Comparing the observed times of ascending node passage

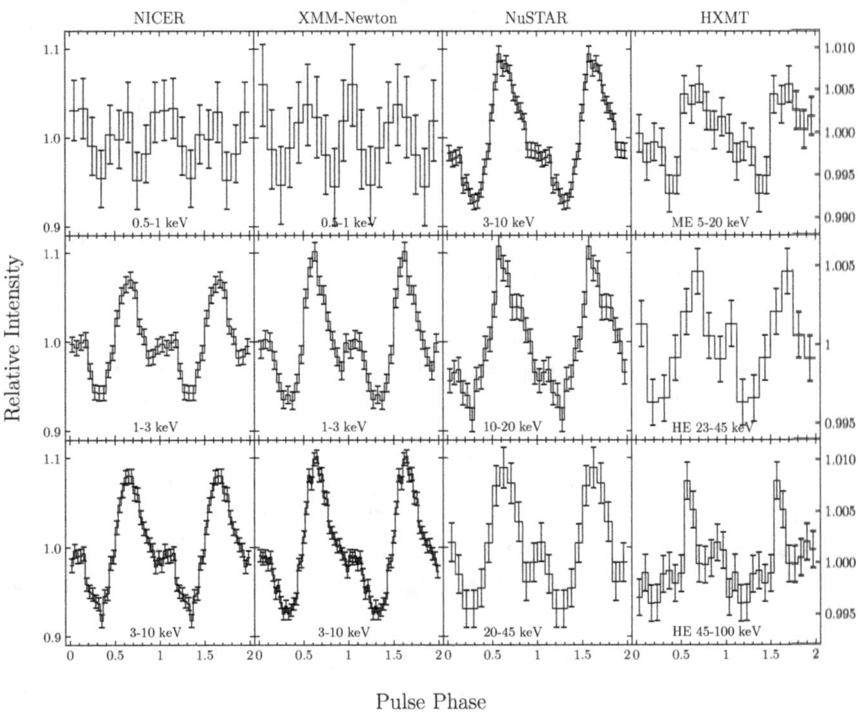

Fig. 4. Pulse profiles of Swift J1756.9–2508 observed by NICER, XMM-Newton, NuSTAR, and *Insight*-HXMT during the 2018 outburst.

Table 2. Positional, rotational, and orbital parameters used from other works and derived in this work for Swift J1756.9–2508.

Parameter	Units	Values
α_{2000}		$17^{\mathrm{h}}56^{\mathrm{m}}57^{\mathrm{s}}350$
δ_{2000}		$-25°06'27''80$
JPL Ephemeris		DE405
P_{orb}	s	3282.3515
$a_{\mathrm{x}} \sin i$	lt-s	0.00597
e		0.00
	Outburst — 2018	
ν	Hz	182.065 803 84(3)
Epoch, t_0	MJD; TDB	58216.0
Validity range	MJD; TDB	58211–58219
$T_{\mathrm{asc,2018}}$	MJD; TDB	58211.017 52(6)
	Outburst — 2019	
ν	Hz	182.065 803 4(2)
Epoch, t_0	MJD; TDB	58654.0
Validity range	MJD; TDB	58653–58657
$T_{\mathrm{asc,2019}}$	MJD; TDB	58655.996 57(12)

with the predicted values, the binary system had a constant orbital period since the first outburst in 2007.

3. Thermonuclear Burning of Accreted Matter on AXMPs

The accreted hydrogen and/or helium can be depleted via stable, unstable, or marginally stable burning processes, which depend on the accretion rate, surface temperature, and abundance (see Galloway *et al.*, 2008, for review). The unstable thermonuclear burning of the hydrogen and helium, also known as type I X-ray burst, usually has the duration of $\sim$10–100 s with a typical energy release of 10^{39} erg, recurs from few hours to days, and ignites at a column depth of $y \approx 10^8$ g cm^{-2} (see Galloway *et al.*, 2008; Galloway *et al.*, 2020; Lewin *et al.*, 1993). There were a few instances of NS LMXBs that exhibited intermediate duration bursts, which can emit $\sim 10^{41}$ erg in $\sim 10^2 - 10^3$ s. In more rare cases, superbursts have been identified from the total energy $\sim 10^{42}$ erg and the duration of $> 10^3$ s, which are believed due to burning carbon at an ignition column depth $\sim 10^3$ times deeper than typical type I X-ray bursts (Cumming and Bildsten, 2001). The recurrence time of superbursts ranges from days to years (e.g., in't Zand, 2017).

Between stable and unstable burning, marginally stable burning on the NS surface manifests as the mHz of quasi-periodic oscillations (QPOs) of the X-ray light curves. Several atoll sources, including 4U 1608–52, 4U 1636–53, and Aql X–1, have been discovered as mHz of QPOs in the frequency range ∼4–15 mHz (Revnivtsev *et al.*, 2001), corresponding to the timescales of 1–4 minutes. Heger *et al.* (2007) explain the mHz of QPOs in multizone numerical models of the NS envelope and proposed that the oscillation frequency depends on the accreted hydrogen fraction and the NS surface gravity, suggesting another way to probe the still unknown NS equation of state.

Superburst provides a unique way to understand the transition of stable, via possible marginally stable, to unstable burning of accreted matter on the NS surface. Keek and Heger (2011) and Keek *et al.* (2012) simulated the generation of superbursts, and discussed the consequences of NS surface cooling. After a superburst, the NS envelope is sufficiently hot, and all accreted hydrogen and helium are burning stably. Afterwards, the envelope cools down, and the marginally stable burning may occur. As the envelope continues to cool, weak type I X-ray bursts with short recurrence times return. Cumming and Macbeth (2004) estimated the quenching time of one to several weeks after a superburst. However, for NS accreting matter with a solar composition, Keek *et al.* (2012) proposed that the quenching time is five times shorter compared with those from Cumming and Macbeth (2004), if both 3α and the CNO breakout reactions contribute to the stability of thermonuclear burning. Due to the lack of contiguous observations and the rare occurrence of superbursts, the transition of burning behavior has not been observed, and only upper limits of the quenching time have been constrained.

3.1. *The burning transition of Aql X–1*

The transient NS LXMB Aql X–1 is composed of a fast spinning pulsar of ∼550.27 Hz and a main-sequence K4 spectral-type companion.

Aql X–1 exhibited frequent outbursts and type I X-ray bursts. Two superbursts from Aql X–1 have been observed by the Monitor of All-sky X-ray Image (MAXI) on the Modified Julian Day (MJD) 56493.3 (2013 July 20) and 59130.7 (2020 October 8), which indicates the recurrence time of ≈7.2 yr. The second superburst has been partially observed by Neutron Star Interior Composition Explorer, which is also its first superburst observation.

In the top left panel of Figure 5, it showed the superburst and type I X-ray bursts light curve of Aql X–1 from MAXI, NICER, Swift/XRT and Insight-HXMT. Almost all NICER, Swift/XRT and Insight-HXMT observations showed multiple data gaps due to South Atlantic Anomaly passages, Earth occultations, and other constraints. Most observations have multiple, nearly contiguous segments, and each of them lasts ∼200–1800 s. The hardness ratio

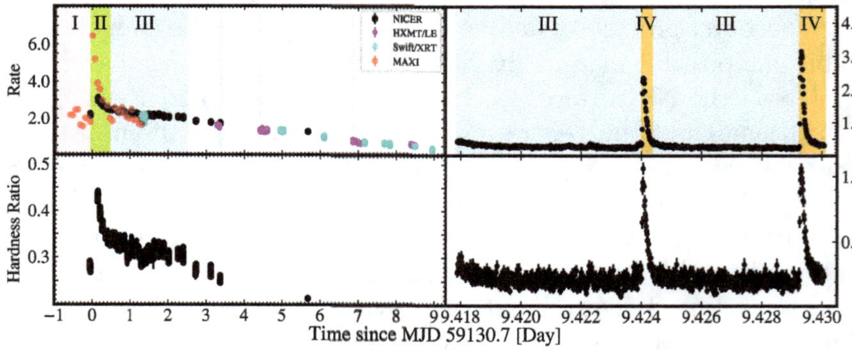

Fig. 5. Top left panel, the light curve of the Aql X-1 superburst from NICER (16 s, 0.5–10 keV, and indicated by black points), Swift/XRT (16 s, 0.3–10 keV, and cyan points), Insight-HXMT/LE (16 s, 1–10 keV, and magenta points) and MAXI (∼250 s, 2–6 keV, and red points). Top right panel, NICER observed (ObsID 305034015) two type I X-ray bursts from Aql X-1 since the superburst, and the 1 s light curve in the energy range 0.5–10 keV is shown. The rate of NICER, Swift/XRT and Insight-HXMT/LE light curves are in units of 10^3 c s^{-1}, 80 c s^{-1}, and 10^2 c s^{-1}, respectively. The MAXI data are in units of c s^{-1} cm^{-2} and are multiplied by 2.5 to make a better visualization. It is not surprising that the time of the available NICER and MAXI observations are well overlapped since both instruments are installed on ISS. The regions I, II, III and IV mark the time intervals of the pre-superburst, the superburst, the detection of mHz QPOs and the return of type I X-ray bursts, respectively. In the bottom left (right) panel, the hardness ratio between the NICER 2.5–10 keV and 0.5–2.5 keV 16 s (and 1 s) light curves are displayed.

between 2–10 and 0.5–2 keV of NICER observations are shown in the bottom left panel of Figure 5. Before the superburst, the hardness ratio is around 0.28. During the decay phase, the hardness ratio decreased from 0.44 to 0.3 due to the cooling of the superburst emission.The post-superburst emission was harder than the pre-superburst. Assuming that the source accretion rate was not changing during the superburst period in a half-day, the hardening just after the superburst is a consequence of the accretion interacting with the still very hot plasma around the NS; therefore, the source changed to a harder spectrum. The light curves are divided into four intervals: the pre-superburst (I), the superburst (II), the detection of mHz QPOs (III), and the return of type I X-ray bursts (IV), respectively.

After the superburst, two type I X-ray bursts 9.44 days after the peak of the superburst from NICER were observed. On MJD 59140.518, 0.389 days after the second NICER burst, the third type I X-ray bursts from Swift/XRT has been found. In the right panels of Figure 5, we plot the 0.5–10 keV light curve of two type I X-ray bursts and show the hardness ratio between 2.5–10 and 0.5–2.5 keV at 1 s resolution. During these two type I X-ray bursts, the hardness ratios evolve as the same trend of their light curves that is a fast rise and exponential decay, which are typical behaviors of a thermonuclear burst. We determine the start time for each burst is when the intensity reached to 10% of the peak above the persistent intensity level. Then, we define the rise time as the time between the start of the burst and the time at which the intensity reached 90% of the peak burst intensity. For these two bursts, the rise time is around 2 s. The recurrence time of these two type I X-ray bursts is 7.6 minutes.

The Lomb–Scargle periodogram has been applied to all segments to search for mHz QPO from NICER, Swift/XRT and Insight-HXMT/LE, in the 0.5–10 keV, 0.3–10 keV, and 1–10 keV light curves, respectively. The frequencies of mHz QPO range from 2.7 to 11.3 mHz, corresponding to the timescales of 1.47–6.17 minutes from these three instruments, which indicated the occurrence of the marginal stable burning of accreted matter on the NS surface in Aql X–1, see region III in Figure 5.

After the superburst from Aql X–1, the transition from superburst, via marginally stable burning, to unstable burning, has been observed for the first time. We then compare the observed quantities, including the quenching time, the recurrence of type I X-ray bursts after the superburst, and the frequencies of mHz QPO, with the simulations provided in Keek *et al.* (2012). Keek *et al.* (2012) simulated the production and subsequent impacts of a superburst for a NS of $1.4M_\odot$ with a 10 km radius accreting composition by a solar, helium-rich and a mixture of carbon and iron (models H, He and C, respectively) at a constant rate of $\dot{M} = 5.25 \times 10^{-9} M_\odot \ \mathrm{yr}^{-1}$, corresponding to 30% of the Eddington-limit rate. They found that the superburst occurs 1.28 yr after the start of accretion. For the case of Aql X–1, the NS accretes the composition from the companion similar to the model H in Keek *et al.* (2012). However, there are few differences between the simulations and the observations in Aql X–1. Aql X–1 is a transient source that has been observed X-ray outbursts recurring around one year, each of them lasting ∼1–6 months (Degenaar *et al.*, 2019). Considering that the superburst recurrence time is $t_{\mathrm{rec}} \propto \dot{M}^{-1}$, we obtain $t_{\mathrm{rec}} \approx 1.28(0.0854\dot{M}_{\mathrm{Edd}}/0.3\dot{M}_{\mathrm{Edd}})^{-1}(79/365)^{-1}$ yr ≈ 20.8 yr, if we take a typical value of the averaged accretion rate of $0.0854\dot{M}_{\mathrm{Edd}}$ in total 79 days from the 2013 outburst in Aql X–1 (Ootes *et al.*, 2018). Hence, the recurrence time of the superburst in Aql X–1 should be larger than the value we observed. However, the energy release from the inner crust may deduce the recurrence time due to frequent outbursts in Aql X–1, similar to the case of GX 17+2 (see also Keek *et al.*, 2006).

When the NS cooling continues, with the quenching time of 1.1 days (model H) and 11.3 days (model He) after the superburst, type I X-ray bursts are returned. We find that two weak type I X-ray bursts with a short recurrence time of 7.6 minutes appear 9.44 days after the superburst of Aql X–1. We suggest that these two bursts are likely the first appearance of unstable burning from Aql X–1 since the superburst. There are data gaps, i.e., 0.625 days of total exposure with the time span of 8.924 days, between the end of superburst and the detection of type I X-ray bursts, which may miss bursts. However, we have 47 continuous segments lasting at least 500 s (NICER, Insight-HXMT/LE and Swift/XRT data), i.e., longer than the recurrence time of 7.6 min, so more bursts should be detected. If the recurrence time is longer than 7.6 min, then we can estimate the probability of the number of bursts for a 0.625 days exposure in the total duration of 8.924 days. By applying a binomial distribution, the probability P of detecting n bursts out of N is $P = C(N,n)\eta^n(1-\eta)^{(N-n)}$, where $\eta \approx 0.07$ is the observation efficiency (see Keek *et al.*, 2012, for more details), and n is zero in our case. So, we have possibilities higher than 90% with the total number of bursts $N \leq 1$. There is a little chance that we missed potential bursts. Moreover, the properties of the two type I X-ray bursts are consistent with the simulated burst resumption. We propose that this is the first accurate measurement of the quenching time, even though data gaps exist. This is also the shortest quenching time in all superbursts so far, except for the upper limit 2.2 days from GX 17+2 that apparently has a higher accretion rate. If the simulated quenching time is corrected with the relation $t_{\rm quen} \propto \dot{M}^{-3/4}$ proposed by Cumming and Macbeth (2004) (and note that the pre-superburst emits at $\sim$10.5%$\dot{M}_{\rm Edd}$ and drops to $\sim$1.4%$\dot{M}_{\rm Edd}$ till the return of type I X-ray bursts), we will obtain a quenching time, i.e., $\sim$3.7 days assuming an averaged accretion rate of 6%$\dot{M}_{\rm Edd}$. Moreover, the rapid rotation of Aql X–1 can delay the appearance of type I X-ray bursts, leading to a longer quenching time (Keek *et al.*, 2009; Keek *et al.*, 2012). The simulated and observed quenching time can be matched.

When the type I X-ray bursts return, the recurrence time of 7.6 minutes has been observed. The bolometric flux between the two type I X-ray bursts is $F_{\rm pers} = 1.45 \pm 0.01 \times 10^{-10}$ erg cm^{-2} s^{-1}. The luminosity is $L_{\rm pers} = 4\pi D^2 F_{\rm pers} \approx 4.3 \times 10^{36}$ erg, assuming the distance of 5 kpc to Aql X–1. The local accretion rate per unit area onto the NS is $\dot{m} = L_{\rm pers}(1+z)(4\pi R^2(GM/R))^{-1} \approx 2.4 \times 10^3$ g cm^{-2} s^{-1}, where G is the gravitational constant, the gravitational redshift is $1+z = 1.31$ for a NS with a mass of $1.4M_\odot$ and there is a radius of $R = 10$ km. We then estimate the ignition column depth $y_{\rm ign} = 4\pi f_{\rm b} D^2(1+z)/(4\pi R^2 Q_{\rm nuc}) \approx 1.0 \times 10^7$ g cm^{-2}, where $Q_{\rm nuc} \approx 1.31 + 6.95\bar{X} - 1.92\bar{X}^2$ MeV nucleon^{-1} ≈ 5.23 MeV nucleon^{-1} is the nuclear energy produced for solar composition, i.e., $\bar{X} = 0.7$ (Goodwin *et al.*, 2019). From the ignition column depth and local accretion rate, the recurrence time between the two bursts is calculated, $t_{\rm rec} = (y_{\rm ign}/\dot{m})(1+z) \approx 90.3$ min, significantly larger than the observed value. For a helium burst, the calculated recurrence time is four times longer. Therefore, the observed recurrence time is too short to accrete enough matter to trigger the second X-ray burst. Type I X-ray bursts with short recurrence times have also been observed in several sources, e.g., EXO 0748–676 and 4U 1636–536. Several possibilities that can reduce the recurrence time. First, the resumption of type I X-ray bursts may ignite at a shallow column depth if the NS envelope is sufficiently hot (see e.g., Galloway *et al.*, 2008). Second, the accreted matter on the NS surface is inhomogeneously distributed, causing a larger local accretion rate. Third, the first type I X-ray burst only depletes a considerably small fraction of accreting matter. Fourth, the rapid rotation of Aql X–1 can induce mixing of freshly accreted fuel to explain the recurrence time on a timescale of 10 minutes (Fujimoto, 1993; Galloway *et al.*, 2018; Keek *et al.*, 2009).

3.2. The non-detection type I X-ray burst from Swift J1756.9−2508

More than seven thousand bursts have been observed to date (see, e.g., Galloway *et al.*, 2008; Galloway *et al.*, 2020). However, seven sources (including Swift J1756.9−2508) out of 22 known AMXPs, including 20 AMXPs mentioned in Table 1 from Di Salvo and Sanna (2022) as well as the newly confirmed AMXP IGR J17494−3030 (Ng *et al.*, 2021), have not shown type-I X-ray bursts during their outbursts.

The non-detection of bursts in Swift J1756.9−2508 is likely a consequence of a low burst rate coupled with a low observational duty cycle in the X-ray observations, such that any bursts that might occur would fall within data gaps. The thermonuclear burst rate depends on the composition of the accreted fuel and the local mass accretion rate, $\dot{m}$, which, in turn, determine how much fuel must be accreted prior to ignition and the ignition depth. The known AMXPs can be divided into two distinct groups based on their orbital periods; either around 40 min or within the range of 2–11 h (see e.g., Campana and Di Salvo, 2018). Stellar evolution models predict that AMXPs with an orbital period longer than one hour should host a highly evolved, hydrogen-rich brown dwarf companion star, and the heating from steady burning of the accreted H fuel prior to ignition will lead to burst recurrence times of hours to days (see e.g., De Falco *et al.*, 2017a,b; Falanga *et al.*, 2011; Ferrigno *et al.*, 2011; Galloway and Cumming, 2006; Heger *et al.*, 2007; Kuiper *et al.*, 2020; Li *et al.*, 2018). However, "ultracompact" X-ray binaries (UCXBs) with orbital periods <80 min are expected to have low-temperature

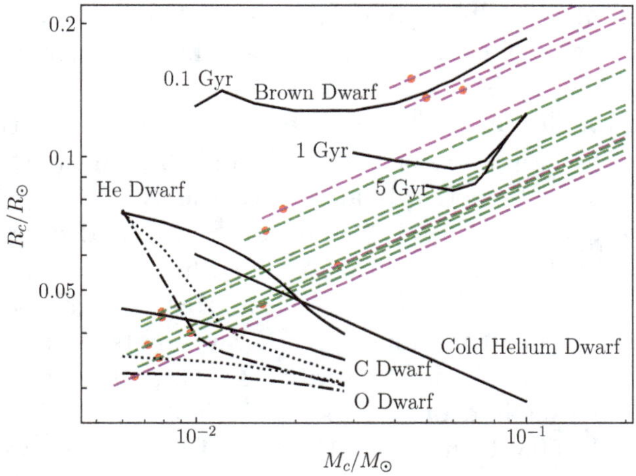

Fig. 6. Assumption of a Roche lobe-filling companion implies a mass-radius relation $R_c = 0.082\,(M_c/0.1M_\odot)^{1/3}(P_{orb}/40\ \text{min})^{2/3}\,R_\odot$, shown in logarithmic scale for different AMXPs with orbital periods around 40 min. From bottom up, the sources are IGR J17062−6143 (Strohmayer *et al.*, 2018), XTE J1807−294 (Campana *et al.*, 2003), XTE J1751−305 (Papitto *et al.*, 2010), XTE J0929−314 (Galloway *et al.*, 2002), MAXI J0911−655 (Sanna *et al.*, 2017), IGR J16597−3704 (Sanna *et al.*, 2018a), Swift J1756.9–2508 (Krimm *et al.*, 2007b), NGC 6440 X-2 (Altamirano *et al.*, 2010), IGR J17494−3030 (Ng *et al.*, 2021), HETE J1900.1−2455 (Kaaret *et al.*, 2006), IGR J17379−3747 (Sanna *et al.*, 2018b), SAX J1808.4−3658 (Wijnands and van der Klis, 1998), and IGR J00291+5934 (Galloway *et al.*, 2005).

C, O or He white dwarf companions, that is, hydrogen-poor, highly evolved dwarfs. The mass-radius relation of AMXPs are shown in Figure 6, by the assumption of a Roche lobe-filling companion. The corresponding M_c versus R_c relations for different stellar evolution models are also provided in Figure 6. In the figure, the magenta dashed lines show the AMXPs with detected type-I X-ray bursts, while the green dashed lines show the AMXPs without type-I X-ray bursts. The red points mark the inclination angle at 60°. We note that from all AMXPs with the orbital period shorter than 1 hr, except MAXI J0911−655 and IGR J17494−3030, no type-I X-ray bursts were detected. We also show the mass-radius relations for low mass regime H-poor dwarf equation of states for O (dot-dash lines), C (dotted lines), and He (solid lines) as well as low (10^4 K, lower lines) or high (3×10^4 K, upper lines) central temperatures. The expected burst recurrence time for such systems, even if reaching the same accretion rates during outbursts as the H-rich systems, are substantially lower, significantly reducing the chance of detecting bursts (e.g., Cumming, 2003).

We can roughly estimate the likely burst recurrence time based on the persistent flux of the source during the outburst. The local mass accretion rate per unit area onto the NS is $\dot{m} = 4\pi d^2 F_{\rm bol}(1+z)(4\pi R_{\rm NS}^2(GM_{\rm NS}/R_{\rm NS}))^{-1}$, i.e., $\dot{m} \approx 3.7 \times 10^3 d_8^2$ g cm^{-2} s^{-1} and $\dot{m} \approx 1.9 \times 10^3 d_8^2$ g cm^{-2} s^{-1} for the 2018 and 2019 outbursts, respectively. If the accreted matter is pure helium, the expected recurrence time for type-I X-ray bursts is $t_{\rm rec} \approx y/\dot{m} \approx 116$ d $(y/10^{10}$ g cm$^{-2})(\dot{m}/10^3$ g cm^{-2} s$^{-1})^{-1}$, where y is the ignition depth of a helium burst (see e.g., in't Zand *et al.*, 2007). For Swift J1756.9–2508 during its 2018 and 2019 outbursts, we estimate a burst recurrence time at the peak of the outburst of 30 d, longer than the total duration of the outbursts. If the accreted matter is composed of C and O, the X-ray bursts would be expected to occur at even larger ignition depths. The recurrence time should be much longer. Given recurrence times in excess of the outburst duration, it is questionable whether sufficient material would accrete to produce even a single burst.

While no type-I burst events were detected during outbursts of transient ultracompact AMXPs, including Swift J1756.9–2508, we note that some UCXBs can sustain accretion persistently and exhibit type I bursts (e.g., Cumming, 2003; Falanga *et al.*, 2008; in't Zand *et al.*, 2007). This property may explain the recent detection of a presumed type I X-ray burst in the AMXP MAXI J0911−655, and two intermediate duration X-ray bursts in IGR J17062−6143 (Degenaar *et al.*, 2013; Keek *et al.*, 2017). The transient AMXPs spend most of the time in a quiescent phase, with X-ray luminosities of $10^{31}-10^{33}$ erg s^{-1} corresponding to accretion rates that are likely far too low to produce thermonuclear bursts.

4. Future Prospects

Since the operation of Insight-HXMT, it has been observed several AXMPs during their X-ray outbursts. In next few years, Insight-HXMT will continue to observe known AXMPs or find new AXMPs. From the well performed broadband spectral and timing capabilities of Insight-HXMT, we will study the location of X-ray pulsations for different energy bands. Moreover, we will carry out the pulse profile modelling to measure the mass and radius of NS and to constrain the equation of state of condense matter.

Bibliography

Alpar, M. A. *et al.* (1982). *Nature* **300**, 728.

Altamirano, D., Patruno, A., Heinke, C. O. *et al.* (2010). *ApJ Letter* **712**, L58.

Bachetti, M., Markwardt, C. B., Grefenstette, B. W. *et al.* (2021). *ApJ* **908**, 184.

Boirin, L., Keek, L., Méndez, M. *et al.* (2007). *A&A*, **465**, 559.

Buccheri, R., Bennett, K., Bignami, G. F. *et al.* (1983). *A&A* **128**, 245.

Bult, P., Altamirano, D., Arzoumanian, Z. *et al.* (2018a). *ApJ* **864**, 14.

Bult, P., Altamirano, D., Arzoumanian, Z. *et al.* (2022a). *ApJ Letter* **935**, L32.

Campana, S. and Di Salvo, T. (2018). in ASSL, Vol. 457, The Physics and Astrophysics of Neutron Stars, (Berlin, Heidelberg: Springer), pp. 149–184.

Campana, S. *et al.* (2003). *ApJ Letter* **594**, L39.

Cao, X., Jiang, W., Meng, B. *et al.* (2020). *SCPMA* **63**, 249504.

Chen, Y., Cui, W., Li, W. *et al.* (2020). *SCPMA* **63**, 249505.

Chen, Y.-P., Zhang, S., Ji, L. *et al.* (2022). *ApJ Letter* **936**, L21.

Cumming, A. (2003). *ApJ* **595**, 1077.

Cumming, A. and Bildsten, L. (2001). *ApJ Letter* **559**, L127.

Cumming, A. and Macbeth, J. (2004). *ApJ Letter* **603**, L37.

Cumming, A., Macbeth, J., in't Zand, J. J. M., and Page, D. (2006). *ApJ* **646**, 429.

De Falco, V., Kuiper, L., Bozzo, E. *et al.* (2017a). *A&A* **603**, A16.

De Falco, V., Kuiper, L., Bozzo, E. *et al.* (2017b). *A&A* **599**, A88.

Degenaar, N., Miller, J. M., Wijnands, A. C. *et al.* (2013). *ApJ Letter* **767**, L37.

Degenaar, N., Ootes, L. S., Page, D. *et al.* (2019). *MNRAS* **488**, 4477.

Di Salvo, T. and Sanna, A. (2022). in *ASSL* **465**, Millisecond Pulsars, 87–124.

Falanga, M., Bonnet-Bidaud, J. M., Poutanen, J. *et al.* (2005a). *A&A* **436**, 647.

Falanga, M., Chenevez, J., Cumming, A. *et al.* (2008). *A&A* **484**, 43.

Falanga, M., Kuiper, L., Poutanen, J. *et al.* (2011). *A&A* **529**, A68.

Falanga, M., Bonnet-Bidaud, J. M., Poutanen, J. *et al.* (2005). *A&A* **436**, 647.
Ferrigno, C., Bozzo, E., Falanga, M. *et al.* (2011). *A&A* **525**, A48.
Fujimoto, M. Y. (1993). *ApJ* **419**, 768.
Galloway, D. K. *et al.*, (2002). *ApJ Lett.* **576**, L137.
Galloway, D. K. and Cumming, A. (2006). *ApJ* **652**, 559.
Galloway, D. K. *et al.* (2005). *ApJ Letter* **622**, L45.
Galloway, D. K., Muno, M. P., Hartman, J. M., *et al.* (2008). *ApJS* **179**, 360.
Galloway, D. K., in't Zand, J. J. M. *et al.* (2018). *ApJ Letter* **857**, L24.
Galloway, D. K., in't Zand, J., Chenevez, J. *et al.* (2020). *ApJS* **249**, 32.
Goodwin, A. J., Heger, A., and Galloway, D. K. (2019). *ApJ* **870**, 64.
Hartman, J. M., Galloway, D. K., and Chakrabarty, D. (2011). *ApJ* **726**, 26.
Hartman, J. M., Patruno, A., Chakrabarty, D. *et al.* (2008). *ApJ* **675**, 1468.
Hartman, J. M., Patruno, A., Chakrabarty, D. *et al.* (2009). *ApJ* **702**, 1673.
Heger, A., Cumming, A. *et al.* (2007). *ApJ Letter* **671**, L141.
in't Zand, J. J. M., Jonker, P. G., and Markwardt, C. B. (2007). *A&A* **465**, 953.
in't Zand, J. (2017). In 7 years of MAXI: monitoring X-ray Transients.
Kaaret, P., Morgan, E. H., Vanderspek, R., and Tomsick, J. A. (2006). *ApJ* **638**, 963.
Keek, L., in't Zand, J. J. M., and Cumming, A. (2006). *A&A* **455**, 1031.
Keek, L., Langer, N., and in't Zand, J. J. M. (2009). *A&A* **502**, 871.
Keek, L. and Heger, A. (2011). *ApJ* **743**, 189.
Keek, L., Heger, A., and in't Zand, J. J. M. (2012). *ApJ* **752**, 150.
Keek, L., Iwakiri, W., Serino, M., *et al.* (2017). *ApJ* **836**, 111.
Keek, L., Arzoumanian, Z., Bult, P. *et al.* (2018). *ApJ Letter* **855**, L4.
Krimm, H. A., Markwardt, C. B. *et al.* (2007). *ApJ Letter* **668**, L147.
Kuiper, L., Tsygankov, S. S., Falanga, M. *et al.* (2020). *A&A* **641**, A37.
Lewin, W. H. G., van Paradijs, J., and Taam, R. E. (1993). *Space Science Reviews* **62**, 223.
Li, Z., De Falco, V., Falanga, M. *et al.* (2018). *A&A* **620**, A114.
Liu, C., Zhang, Y., Li, X. *et al.* (2020). *SCPMA* **63**, 249503.
Ng, M., Ray, P. S., Bult, P. *et al.* (2021). *ApJ Letter* **908**, L15.
Ootes, L. S., Wijnands, R., Page, D., and Degenaar, N. (2018). *MNRAS* **477**, 2900.
Papitto, A., Riggio, A., di Salvo, T. *et al.* (2010). *MNRAS* **407**, 2575.
Papitto, A., Riggio, A., Burderi, L. *et al.* (2011). *A&A* **528**, A55.
Papitto, A., Ferrigno, C., Bozzo, E. *et al.* (2013). *Nature* **501**, 517.
Papitto, A., Falanga, M. *et al.* (2020). *New Astronomy Reviews* **91**, 101544.
Patruno, A., Rea, N., Altamirano, D. *et al.* (2009b). *MNRAS* **396**, L51.

Patruno, A., Altamirano, D., Hessels, J. W. T. *et al.* (2009). *ApJ* **690**, 1856.

Patruno, A., Altamirano, D., and Messenger, C. (2010). *MNRAS* **403**, 1426.

Patruno, A. (2010). *ApJ* **722**, 909.

Patruno, A. and Watts, A. L. (2021). Timing Neutron Stars: Pulsations, Oscillations and Explosions, *ASSL* **461**, 143–208.

Pan, Y. Y., Li, Z. S., Zhang, C. M., and Zhong, J. X. (2022). *MNRAS* **513**, 6219.

Poutanen, J. and Gierliński, M. (2003). *MNRAS* **343**, 1301.

Psaltis, D. and Chakrabarty, D. (1999). *ApJ* **521**, 332.

Revnivtsev, M., Churazov, E., Gilfanov, M., and Sunyaev, R. (2001). *A&A* **372**, 138.

Shapiro, S. L. and Teukolsky, S. A. (1983). Black holes, white dwarfs, and neutron stars: The physics of compact objects, New York: Wiley.

Sanna, A., Bahramian, A., Bozzo, E. *et al.* (2018a). *A&A* **610**, L2.

Sanna, A., Ferrigno, C., Ray, P. S. *et al.* (2018b). *A&A* **617**, L8.

Sanna, A., Papitto, A., Burderi, L. *et al.* (2017). *A&A* **598**, A34.

Sanna, A., Pintore, F., Riggio, A. *et al.* (2018c). *MNRAS* **481**, 1658.

Strohmayer, T. E., Arzoumanian, Z. *et al.* (2018). *ApJ Letter* **858**, L13.

Wijnands, R. and van der Klis, M. (1998). *Nature* **394**, 344.

Zhang, S.-N., Li, T., Lu, F. *et al.* (2020). *SCPMA* **63**, 249502.

https://doi.org/10.1142/9789819800643_0011

Chapter 11

Pulsars at Shanghai Tian Ma 65 m Radio Telescope

Zhen Yan*, Zhiqiang Shen†, Yajun Wu, Rongbing Zhao, and Qinghui Liu

Shanghai Astronomical Observatory, CAS, Shanghai 200030, China

**yanzhen@shao.ac.cn*

†zshen@shao.ac.cn

Shanghai Tianma Radio telescope (TMRT) is a newly-built full-steerable 65-m antenna which can work from 1.3 to 50 GHz with observing efficiencies better than 50% using eight sets of low-noise receiver and active surface systems. Pulsars are not only important space laboratory of testing physical laws for extreme conditions (such as ultra-strong gravitational, magnetic and electronic fields) and detecting gravitational waves, but also can be used for deep-space autonomous navigation, time keeping, etc. So, pulsars were chosen as important scientific research objects at the TMRT. Besides baseband backends for Very Long Baseline Interferometry (VLBI) observations, a digital backend system (DIBAS) supporting commonly-used observation modes for pulsars and a real-time Fast Radio Burst monitoring backend were equipped at the TMRT to meet demands of pulsar-related observations. Taking advantages of its high sensitivity and multi-frequency working ability, it has supported a series of pulsar studies, such as magnetar monitoring, high frequency pulsar observations, pulsar timing, pulsar VLBI, etc. More and more interesting research results will be obtained as we will go on devoting more time and efforts at the TMRT.

Keywords: Pulsar, Radio telescope, VLBI, Timing

Contents

1. Introduction

Pulsars are highly magnetized fast-rotating compact stars that emit lighthouse-like beams of electromagnetic radiation from their magnetic poles. Their radiation can be detected only when the emission beams sweep across the Earth. Because of their extremely strong gravitational, electric and magnetic fields, pulsars are very good space laboratories that can be used to test physical laws under these extreme conditions (Cordes *et al.*, 2004; Xu, 2006). Compared with other celestial objects, pulsars show special radiation properties, because their signals are a train of narrow pulses with relatively high degree of polarization and stable inertial flux density on average. Radio signals from pulsars are significantly affected by propagation effects in the tenuous plasma of the interstellar medium (ISM). These effects include interstellar dispersion, Faraday rotation and scintillation, which exhibit as frequency-dependent propagation time delay, linear polarization plane rotation and flux density fluctuation, respectively. Based on observations of phenomena caused by these effects, pulsars can be used as important probes to study the distribution, turbulence and magnetic field of ISM (Han *et al.*, 2006; Rickett, 1990). In addition, a pulsar timing array (PTA) can be constructed with regular high-precision timing observations of a widely distributed array of millisecond pulsars (MSPs). The PTA uses sets of extremely well-timed pulsars as a Galaxy-scale detector with arms extending between Earth and each pulsar, making it very sensitive to detect gravitational waves at nanohertz frequencies (Agazie *et al.*, 2023; EPTA Collaboration *et al.*, 2023; Reardon *et al.*, 2023; Xu *et al.*, 2023). Besides the important meanings to exploring basic physical laws, pulsar studies also have many practical application values. The extreme stability of MSPs allows them to be used in establishing ephemeris time. Furthermore, MSPs are the potential lighthouses for deep-space spacecraft navigation.

Since the first pulsar discovered in 1967, more than 3359 pulsars have been discovered by now. According to pulsar catalog (PSRCAT) (Manchester *et al.*, 2005), more than 90% of known pulsars can only be detected in the radio band, so radio telescopes are very important tools to do pulsar research. Pulsars are normally weak radio sources. For the current sample of pulsars listed in the PSRCAT with their flux densities measured at 1400 MHz (S_{1400}), the median value of S_{1400} is 0.35 mJy with a range between 3 μJy and 1.05 Jy (Manchester *et al.*, 2005).

2. Pulsar Observation Systems

The Shanghai Tianma Radio Telescope (TMRT) is a full-steerable antenna with a diameter of 65-m built as a major cooperative project between the Chinese Academy of Sciences (CAS) and Shanghai Municipality, which was jointly funded by the CAS, Shanghai Municipality, and Chinese Lunar exploration project. The TMRT received its current name as it is located around the Tianma mountain in the suburban Songjiang district of Shanghai. After two phases of on-site construction and testing, the TMRT reached high operating levels in the frequency-coverage from 1.3 to 50 GHz using eight sets of low-noise receiver systems. In the first period of construction (2010–2013), most of the operational works were focused on improving the performance of TMRT in four low-frequency bands. In the next stage of construction (2013–2017), we devoted most time to

Fig. 1. An aerial photograph of the TMRT taken in 2018.

building and testing its high-frequency observation systems, such as its active surface, high-frequency receivers, etc. After a series of testing and adjusting, the TMRT can keep efficiency butter than 50% at all its working bands ranging from 1.3 to 50 GHz (Dong *et al.*, 2018; Wang *et al.*, 2017). Figure 1 shows an aerial photograph of the TMRT taken in 2018.

2.1. *Receiver systems*

As what mentioned previously that pulsars are comparatively weak radio sources, a telescope with high-sensitivity is indispensable to do high-quality pulsar observations. Besides the effective collection area of the antenna, the bandwidth and system noise of its receiver systems are also important parameters that affect the observation sensitivity of the corresponding telescope. In order to do distinctive and feasible studies on pulsars with the TMRT, we collected parameters of full-steerable large radio telescopes in the world, including Green Bank 100 m (GBT), Effelsberg 100 m (EF), Parkes 64 m (PKS), Jodrell Bank 76 m (JB) and Sardinia 64 m (SRT). For the convenience of judging their sensitivities at different

Table 1. The parameters of low frequency receivers of the TMRT and other full-steerable large telescopes.

B_{para}\ANT		GBT	EF	PKS	JB	SRT	TMRT
L	*FR* (GHz)	1.15-1.73	1.27-1.73	1.2-1.8	1.25-1.75	1.3-1.8	1.35-1.75
	SEFD (Jy)	10	13	31	35-40	67	39
S	*FR* (GHz)	1.73-2.6	2.20-2.30	2.2-2.5	—	—	2.2-2.3
	SEFD (Jy)	12	306	25	—	—	46
C	*FR* (GHz)	3.95-5.85	5.75-6.75	4.5-5.1	6.0-7.0	5.7-7.7	4.0-8.0
	SEFD (Jy)	10	25	61	80	50	30
X	*FR* (GHz)	8.00-10.1	7.9-9.0	8.1-8.7	—	—	8.2-9.0
	SEFD (Jy)	15	18	170	—	—	48

bands, we further calculated their system equivalent flux densities ($SEFD = 2k_{\rm B}T_{\rm sys}/A_{\rm e}$) considering both the system temperature ($T_{\rm sys}$) and the effective aperture ($A_{\rm e}$). Table 1 lists some useful parameters at four low-frequency bands ($B_{\rm para}$) that are more commonly used in pulsar observations, including the *SEFD* and frequency range (*FR*) of these antennas (ANTs) (Yan *et al.*, 2018, 2017). It can be seen that the TMRT has comparative sensitivity as the other large telescopes in the world. Especially at the C band, the TMRT has its advantages because of its wider bandwidth and lower system temperature. In addition, the S/X dual-band feed system of the TMRT makes it have a ability of receiving signals at S and X bands simultaneously.

2.2. *Backends*

Besides the large diameter, active surface and low-noise receiver of a telescope, a high-performance backend is necessary for good quality pulsar observations, as it accomplishes tasks of analog-to-digital converting, digitizing, sampling, processing and recording. The digital backend system (DIBAS) of the TMRT is a FPGA-based spectrometer built upon the design of Versatile GBT Astronomical Spectrometer (VEGAS) with pulsar modes that provide much the same capabilities as Green Bank Ultimate Pulsar Processing Instrument (GUPPI) (Ransom *et al.*, 2009). For pulsar observations, DIBAS can well support both online-folding and searching observation modes. Due to the fact that radio waves of higher frequency travel faster through the ISM, radio pulses from pulsars also arrive earlier at higher frequencies (usually called interstellar dispersion). Using the DIBAS, the dispersion effect of ISM suffered by pulsar signals can also be removed with either the coherent or incoherent dedispersion method according to the demands of observers. Although coherent dedispersion can remove dispersion effects more thoroughly, it needs more computing resources than incoherent dedispersion. Eight pairs of graphics processing unit (GPU) on eight parallel-working high-performance computer (HPC) are used to meet the demands of massive calculation in coherent dedispersion. Each GPU & HPC unit can maximally support the computation demand of coherent dedisperion on 125 MHz observation data. So, the maximum bandwidth is 1 GHz for the coherent dedispersion observation mode of the DIBAS. For the incoherent dedispersion observation mode, the bandwidth of each of digitizer channel is up to 2 GHz. Since three pairs of digitizer (one digitizer for each polarization) are currently available, a maximum bandwidth of 6 GHz can be supported using first three pairs of HPC. The DIBAS supports full Stokes-parameter pulsar observations and writes out the data in 8-bit PSRFITS format (Hotan *et al.*, 2004). In addition to the DIBAS, a real-time Fast Radio Burst monitoring backend and baseband backends for Very Long Baseline Interferometry (VLBI) observations were also built and installed at the TMRT (Yan *et al.*, 2018).

3. Some Pulsar Observation Projects and Related Results

As mentioned previously, the TMRT is located in a suburban district of a metropolis, so its low-frequency observations are inevitably affected by radio frequency interferences (RFIs). On the other hand, the TMRT can keep its high sensitivity from low to high frequencies because of its low-noise receivers and active surface, making it have comparative advantages at high frequencies. As it is equipped with some dual-frequency receiver systems (such as the S/X-band receiver) and the DIBAS that has the ability of parallel working, it has special capabilities of performing simultaneous multifrequency observations. Considering of these, a series of research projects related to pulsars have been carried out at the TMRT, such as magnetar monitoring, high-frequency observations, long-term timing, pulsar VLBI, etc. Most of these projects make full use of its advantages mentioned above and try to avoid its disadvantages at low frequencies.

3.1. *Magnetar monitoring*

Unlike most pulsars that are normally thought to be rotation-powered neutron stars, magnetars are special neutron stars that sometimes exhibit extremely strong X-ray or γ-ray bursts which can not be afforded by losing their rotational energies (Duncan and Thompson, 1992). Their radiations are mainly powered by the decay of extremely strong magnetic fields, which are hundreds times stronger than usual pulsars. Magnetars also have longer rotation periods and show more irregular rotational properties than usual pulsars (Camilo *et al.*, 2006; Gao *et al.*, 2016; Tong *et al.*, 2013). By now, there are only about 30 magnetars, which are discovered as soft γ-ray repeater (SGRs), anomalous X-ray pulsars (AXPs) and their candidates. But, only in 6 of known magnetars have been detected radio pulsations (Olausen and Kaspi, 2014).

The magnetar PSR J1745−2900 is famous as the only pulsar known within a few arcmins from Sgr A*. It was first accidentally detected as an X-ray flare thought to be from Sgr A* by the Swift, and was identified as a magnetar by follow-up observations (Mori *et al.*, 2013). Although the TMRT was still at the stage of site commissioning, we did six epochs of observation on PSR J1745−2900 at a frequency of 8.6 GHz between June and October, 2014. We found that both the flux density and integrated profile of PSR J1745−2900 changed dramatically from epoch to epoch, showing that it was in its 'erratic' phase. The integrated profile and phase-time plots of the region around the pulse for each observation are shown in Figure 2. Clearly, the pulse profile varied greatly over the span of the observations, although on the two adjacent days (MJDs 56911 and 56912) the profiles were the same within the noise. We found the flux density of this magnetar was about 8.7 mJy on MJD 56836, which was ten times brighter than that reported at the time of its discovery. Because of its high flux density on this day, we did single-pulse analyses on it and found its emission to be dominated by narrow 'spiky'

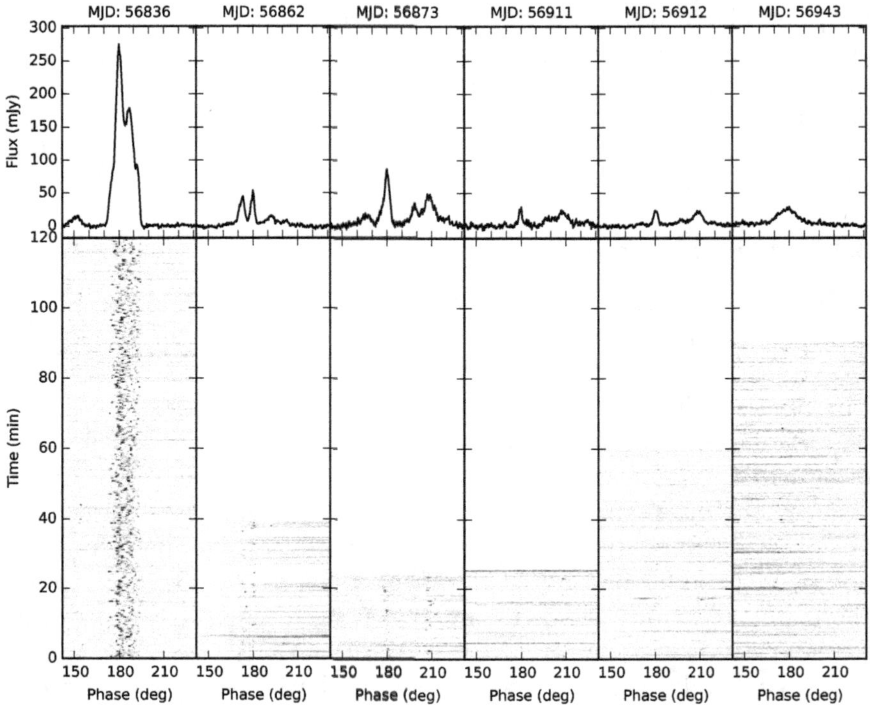

Fig. 2. Integrated profiles and the phase-time plots of the magnetar PSR J1745−2900 during six days of TMRT observations. The abscissa of each subplot is the rotational phase of the pulsar, with just the region around the pulse being shown. The MJD for the day of observation is given at the top of each subplot. All the plots use the same grayscale which is set according to the observations of MJD 56911 where the flux density is relatively low. The horizontal striping visible in the phase-time plots results from low-level radio-frequency interference (RFI).

pulses following a log-normal distribution in their peak flux density. From 1913 pulses, we detected 53 bright pulses with their peak-flux densities ten times greater than that of the integrated profile. It was found that these bright pulses were concentrated in phase position at the peaks of the corresponding integrated profile. Besides no evidence of subpulse drifting, we found no clear correlation between the width and peak flux density of these pulses. Relatively strong spiky pulses showing same properties were also detected in the other five epochs of observation. Based on the above analyses, we drew a conclusion as that these strong spiky pulses cannot be classified as 'giant' pulses but were more closely related to normal pulse emission (Yan *et al.*, 2015).

Shortly after the discovery of Swift J1818.0−1607, which was not only the fifth magnetar known with periodic radio pulsations but also the fastest rotating one, we also did a long-term (from MJD 58936 to 59092) simultaneous 2.25/8.60 GHz monitoring with the TMRT. We obtained its spin-frequency (ν) and first-order derivative ($\dot{\nu}$) with piece-wise fitting method as it was affected by unstable timing. We found that the amplitude of $\dot{\nu}$ fluctuations decreased with time, and the long-term declining trend of ν discovered previously continued in our observations. We obtained its more accurate long-term $\dot{\nu} = -2.25 \times 10^{-11} s^{-2}$ based on our data spanning as long as 156 days, and derived its characteristic age (τ) to be about 522 yr which supported the viewpoint that this magnetar may be older than initially thought at the time of discovery. We also found that its flux density increased at both 2.25 and 8.60 GHz accompanying the spectra getting flatter. In addition, we also showed how Swift J1818.0−1607 switched between the bright and quiet (B-Q) emission modes at 2.25 and 8.60 GHz. We found that Swift J1818.0–1607 switched between the B and Q modes synchronously at 2.25 and 8.60 GHz in 3 days of our observations. Though the mode switching detected by us took place only 3 days later than that reported by the Parkes, they were very different as it previously varied between two longitudinally distinct modes (P and M mode) instead of obvious flux density variations (Huang *et al.*, 2021).

3.2. *High-frequency observations*

The pulsar radiation process is still an open field of investigation. Multi-band observation of pulsar radiation is an effective way to test and constrain theoretical models. By analyzing multi-frequency pulse profiles, the location, geometry and spectrum of pulsar emission can be determined, and thus to constrain models. In order to improve the signal-to-noise ratio (S/N) of pulsar observations, a number of individual pulses are added synchronously with the rotational period and allows to obtain the corresponding integrated profiles. For a given pulsar, even though the individual pulses show different shapes, its integrated pulse profile at a given frequency is usually very stable. For different pulsars, the integrated pulse profiles are morphologically different (Lyne and Manchester, 1988). So, the integrated profile is thought of as 'fingerprints' of the corresponding pulsar. At different frequencies, the shape of the integrated pulse profiles often changes substantially. These variations might be related to the different ways in which the line of sight cuts across the radiation beam and intrinsic asymmetry within the individual pulsar beams. Therefore, studying integrated pulse profiles at multiple frequencies helps to investigate both the geometry and the mechanism of pulsar radiation. However, the well-organized radio pulse profiles obtained by integrating some hundreds of individual pulses conceal a rich diversity of behavior among these individual pulses. When sufficient S/N is available, high time resolution observations of individual pulses provide a wealth of information about the emission process.

Limited sample of pulsars have been observed at relatively high frequencies because of their steep spectra (mostly following $S_\nu \propto \nu^\alpha$ with a typical spectral index $\alpha = -1.8$) (Sieber, 1973). Using the TMRT, we made a series of pulsar observations at frequencies above 5.0 GHz. We presented integrated pulse profiles at 5 GHz for 71 pulsars, including 8 MSPs. Their mean flux densities and pulse widths were also measured. For 19 normal pulsars and one MSP, these were the first detections at 5 GHz and for a further 19, including 5 MPSs, the profiles have a better S/N than previous observations (Zhao *et al.*, 2019). There were much fewer pulsars observed in a frequency above 8 GHz. The total number of pulsars detected and published above 8 GHz was less than 50 by then (e.g., Johnston *et al.*, 2006; Kramer *et al.*, 1997; Morris *et al.*, 1981). Using the TMRT, we presented integrated pulse profiles, mean flux densities and pulse widths at 8.6 GHz for a sample of 26 pulsars. For 11 pulsars, these were the first high-frequency observations and for a further four, our observations had better S/N than previous observations. Based on our and previous low-frequency observations, we discussed correlations of the power-law spectral index with the characteristic age, radio pseudo-luminosity and spin-down luminosity. The frequency dependence of component separations was investigated for pulsars in our samples and three groups were found: in ∼50% pulsars the separation between the outmost leading and trailing components decreases with frequencies, roughly in agreement with radius-to-frequency mapping; in the remaining ~50% pulsars the separation was nearly constant; in some of cases the separation between the outmost components increases with frequencies (Zhao *et al.*, 2017, 2019). In addition, pulsar radio radiation models (particularly the inverse Compton scattering model) were studied based on pulse profiles' evolution over a wider frequency range (Lu *et al.*, 2016; Shang *et al.*, 2017).

However, integrated profiles for some pulsars show sporadic changes between two (or more) quasi-stable states, which are called mode changing or switching. This phenomenon has been seen in a few dozens of pulsars in subsequent years (e.g., Lyne, 1971; Wang *et al.*, 2007). Some of the pulsars that show mode changes also exhibit drifting subpulses (van Leeuwen *et al.*, 2002). But the connection between these two phenomena is not well understood yet. Simultaneous multi-frequency single-pulse observations are the most direct way to study whether the mode changing and subpulse drifting are wide band phenomena and their intrinsic relationships. We investigated the mode-changing properties of PSR B0329+54 using 31 epochs of simultaneous 2.25/8.60 GHz single-pulse observations obtained with the TMRT. Compared with previous studies, our observations had much wider frequency-span and higher time-resolution. PSR B0329+54 was found in the abnormal emission mode 17 times, accounting for ~13% of the 41.6 hours observation time in total. Figure 3 shows the typical integrated profiles obtained with normal and abnormal mode observation data at 2.25 and 8.60 GHz, respectively. Single pulse analyses indicated that mode changes took place simultaneously at 2.25/8.60 GHz within a few rotational periods. We divided the pulsar radiation window into three components (C1, C2 and C3) corresponding to main peaks of the integrated profile (see Figure 3). The bright pulses preferentially occurred at pulse phases corresponding to peaks of C2 and C3. Fluctuation spectra showed that C2 had excess red noise in the normal mode, but broad quasi-periodic features with central frequencies around 0.12 cycles/period in the abnormal mode. At 8.60 GHz, C3 had a stronger quasi-periodic modulation centered around 0.06 cycles/period in the abnormal mode. Although there were some asymmetries in the two-dimensional fluctuation spectra, we found no clear evidence for systematic subpulse drifting (Yan *et al.*, 2018).

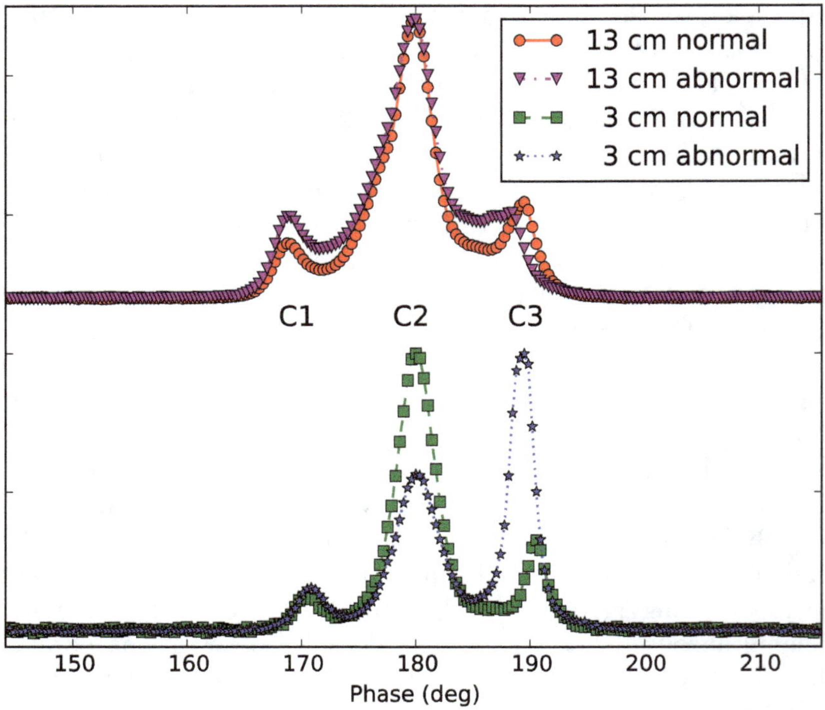

Fig. 3. Typical normal and abnormal mode integrated profiles of PSR B0329+54 at 2.25 & 8.60 GHz.

3.3. *Pulsar timing*

The pulsar timing is regular monitoring of the rotation of neutron stars by tracking the arrival times of their radio pulses. Accurate pulsar timing permits exquisitely sensitive measurements of quantities such as the power of gravitational radiation emitted by a pulsar in a binary system, neutron-star masses, general relativistic effects in strong gravitational fields, orbital perturbations from binary companions as light as planets, and potentially the distortions of interstellar space produced by long-wavelength gravitational radiation from the mergers of super-massive black holes throughout the universe. So, long-term pulsar timing is also chosen to be one of important regular observation projects at the TMRT as most other large radio telescopes did. Considering the RFI condition at the TMRT and the power-law spectra of pulsars, the 2.25 GHz is chosen as

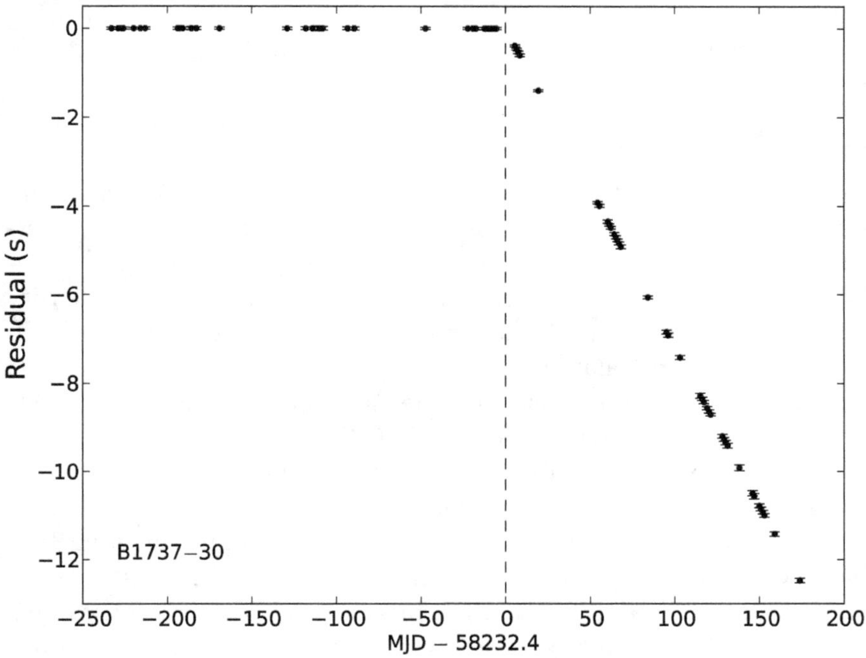

Fig. 4. Timing residual plot of PSR B1737−30: the dashed vertical line implies the epoch of the glitch.

the normally-used frequency to do pulsar timing observations at the TMRT. It is indicated by testing observation that the TMRT has an ability to monitor the time properties of more than 350 pulsars (including 10 MSPs) at 2.25 GHz.

Using timing observation data from MJD 57999 to 58406 obtained with the TMRT, one large glitch was detected in PSR B1737−30, which was a young pulsar showing frequent glitches. As shown in Figure 4, the glitch took place around MJD 58232.4 when the pulsar underwent an increase in the rotation frequency of $(\Delta\nu)$ about 1.38×10^{-6} Hz, corresponding to a fractional step change of $\Delta\nu/\nu \sim 8.39 \times 10^{-7}$. This was the fourth largest glitch of this pulsar among 36 glitches reported since July 1987. Post-glitch ν gradually decreased to the pre-glitch value. It was indicated that the glitch size distribution could be well described by a power-law with the index of 1.13. Furthermore, the distribution of the interval between two adjacent glitches were found to follow a Poissonian probability density function. While no correlation was discovered between the glitch size and the interval since previous glitch, the interval was found being prone to be long after a large glitch for this pulsar (Liu *et al.*, 2019).

3.4. *Pulsar VLBI*

Accurate distance and proper motion measurements of pulsars are important for studying a series of fundamental astrophysical questions, such as confining kick effects in supernova explosions, tracing back to their birth places, optimizing Galactic electron density distribution models, etc (Chatterjee *et al.*, 2005). Distances of pulsars are usually estimated with dispersion measures (DMs) based on the Galactic electron density distribution models, such as NE2001 (Cordes and Lazio, 2002) and YMW16 (Yao *et al.*, 2017), but they sometimes have unavoidably large errors. Though it is possible to fit pulsar distance and proper motion with high-sensitivity pulsar timing observations, it's difficult for pulsars with timing irregularities, because there are many other parameters coupled with these astrometry parameters to be fitted. High-precision VLBI astrometry offers a powerful way to directly measure the distances and proper motions of pulsars. In return, VLBI astrometry results can improve the pulsar timing accuracy, especially for pulsars showing timing irregularities. Figure 5 presents examples of optimizing effects on timing fitting of two pulsars (B0136+57 and B2351+61), which show large timing noises. It can be clearly seen that timing noises are suppressed by using model-independent astrometry results obtained with the VLBI. In addition, we also found that the glitch epoch of PSR B0154+61 measured with a previous ephemeris was about 13 days later than the result obtained with updated VLBI astrometric solutions (Liu *et al.*, 2020).

Doing pulsar astrometry with the VLBI is a challenging work because of weak pulse signals, ionospheric effects, etc. Some pulsar astrometry projects with the VLBI have been successfully accomplished as steady progresses in hardware, observation and correlation techniques (Brisken *et al.*, 2002). A larger pulsar astrometry program called $PSR\pi$ is being carried out with the Very Long Baseline Array (VLBA). It has finished astrometry observations of more than 60 pulsars (Deller *et al.*, 2019). As an important station of domestic and international VLBI networks, such as Chinese VLBI Network (CVN), European VLBI Network (EVN) and East Asian VLBI

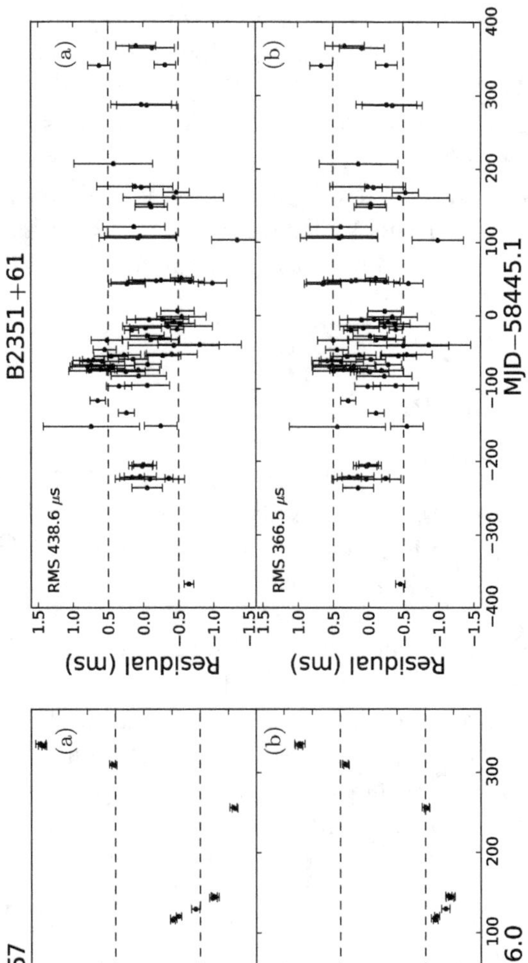

Fig. 5. Timing residuals after fitting ν and $\dot{\nu}$ based on timing (a) and VLBI-updated (b) ephemeris. Axis scales of two panels are same for each pulsar.

Network (EAVN), the TMRT also joins a serious of pulsar observations. For pulsars located at high declination (DEC $> 45°$), the resolution of the VLBA in right ascension (RA) will be affected because of the limited length of projected baseline along the East-West direction. The participation of TMRT can lengthen the baseline twice in East-West direction and effectively improve the UV-coverage. Joint pulsar observations with the VLBA and TMRT were also successfully carried out at the L-band with data rates up to 2 Gbps (Yan *et al.*, 2018).

4. Conclusions and Prospects

Overall, pulsars are chosen as one of the most important scientific observation targets at the newly-built TMRT. A series of pulsar related research projects have been arranged at the TMRT, such as magnetar monitoring, high-frequency observations, long-term timing, pulsar VLBI, etc. Based on these projects, very valuable high-frequency and multi-frequency pulsar observation results have been published. Besides pulsar research results mentioned above, there are also some other research projects carried out with the TMRT. Taking advantages of the simultaneous dual-frequency observation ability, the researchers at the TMRT are developing a technology that can do mutually phase calibration between observation data obtained at these frequencies, and they have detected much more pulsars at high frequencies (especially at 8.60 GHz). In addition, a research project that propose to do deeply pulsar search at relatively higher frequencies (above 8.60 GHz) around the supermassive black hole in the center of the Milky Way. Furthermore, some interesting results about interstellar scintillation of pulsars are also preliminarily obtained at the TMRT. If we can keep on putting more efforts on this area, it is foreseeable that more and more interesting achievements on pulsars can be made with the TMRT.

Acknowledgments

This work is supported in part by the National SKA Program of China (Grant No. 2020SKA0120104), the National Natural

Science Foundation of China (Grant Nos. U2031119, 12041301, U1631122), the National Key R&D Program of China (Grant No. 2022YFA1603104), and the Natural Science Foundation of Shanghai (Grant No. 20ZR1467600). Besides all the members of the TMRT team, we also want to express our thanks to many external cooperators, such as R. N. Manchester, Xinji Wu, Renxin Xu, Kejia Lee, Na Wang, Jinlin Han, Jianping Yuan, Hao Tong, etc.

Bibliography

Agazie, G., Anumarlapudi, A., Archibald, A. M. *et al.* (2023). The NANOGrav 15 yr data set: Evidence for a gravitational-wave background, *Astrophys. J. Lett.* **951**, 8.

Brisken, W. F., Benson, J. M., Goss, W. M. *et al.* (2002). Very long baseline array measurement of nine pulsar parallaxes, *Astrophys. J.* **571**, 906–917.

Camilo, F., Ransom, S. M., Halpern, J. P. *et al.* (2006). Transient pulsed radio emission from a magnetar, *Nature* **442** (7105), 892–895.

Chatterjee, S., Vlemmings, W. H. T., Brisken, W. F. *et al.* (2005). Getting its kicks: A VLBA parallax for the hyperfast pulsar B1508+55, *Astrophys. J. Lett.* **630**, L61–L64.

Cordes, J. M., Kramer, M., Lazio, T. J. W. *et al.* (2004). Pulsars as tools for fundamental physics and astrophysics, *Nature* **48**(11–12), 1413–1438.

Cordes, J. M. and Lazio, T. J. W. (2002). NE2001.I. A new model for the galactic distribution of free electrons and its fluctuations, http://arxiv.org/abs/arXiv:astro-ph/0207156.

Deller, A. T., Goss, W. M., Brisken, W. F. *et al.* (2019). Microarcsecond VLBI pulsar astrometry with PSRπ II. Parallax distances for 57 pulsars, *Astrophys. J.* **875**(2), 100.

Dong, J., Fu, L., Liu, Q. *et al.* (2018). Measuring and analyzing thermal deformations of the primary reflector of the Tianma radio telescope, *Exp. Astron.* **45**(3), 397–410.

Duncan, R. C. and Thompson, C. (1992). Formation of very strongly magnetized neutron stars: Implications for gamma-ray bursts, *Astrophys. J. Lett.* **392**, 9.

EPTA Collaboration, InPTA Collaboration, Antoniadis, J. *et al.* (2023). The second data release from the European Pulsar Timing Array. III. Search for gravitational wave signals, *Astron. Astrophys.* **678**, 50.

Gao, Z. F., Li, X. D., Wang, N. *et al.* (2016). Constraining the braking indices of magnetars, *MNRAS* **456**(1), 55–65.

Han, J. L., Manchester, R. N., Lyne, A. G. *et al.* (2006). Pulsar rotation measures and the large-scale structure of the galactic magnetic field, *Astrophys. J.* **642**, 868–881.

Hotan, A. W., van Straten, W., and Manchester, R. N. (2004). PSRCHIVE and PSRFITS: An open approach to radio pulsar data storage and analysis, *Publ. AStron. Soc. Aust.* **21**, 302–309.

Huang, Z.-P., Yan, Z., Shen, Z.-Q. *et al.* (2021). Simultaneous 2.25/8.60 GHz observations of the newly discovered magnetar — Swift J1818.0-1607, *MNRAS* **505**(1), 1311–1315.

Johnston, S., Karastergiou, A., and Willett, K. (2006). High-frequency observations of southern pulsars, *MNRAS* **369**, 1916–1928.

Kramer, M., Xilouris, K. M., Jessner, A. *et al.* (1997). Origin of pulsar radio emission I. High frequency data, *Astron. Astrophys.* **322**, 846–856.

Liu, J., Yan, Z., Shen, Z.-Q. *et al.* (2020). New timing measurement results of 16 pulsars, *Publ. Astron. Soc. JPN.* **72**(4), 70.

Liu, J., Yan, Z., Yuan, J.-P. *et al.* (2019). One large glitch in PSR B1737-30 detected with the TMRT, *Res. Astron. Astrophys.* **19**(5), 073.

Lu, J. G., Du, Y. J., Hao, L. F. *et al.* (2016). Multi-frequency radio profiles of PSR B1133+16: Radiation location and particle energy, *Astrophys. J.* **816**(2), 76.

Lyne, A. G. (1971). The comparative properties of the pulsars, In R. D. Davies and F. G. Smith (eds.), *IAU Symp 46* (Reidel), pp. 182–194.

Lyne, A. G. and Manchester, R. N. (1988). The shape of pulsar radio beams, *MNRAS* **234**, 477–508.

Manchester, R. N., Hobbs, G. B., Teoh, A., and Hobbs, M. (2005). The Australia telescope national facility pulsar catalogue, *Astron. J.* **129**(4), 1993–2006.

Mori, K., Gotthelf, E. V., Zhang, S. *et al.* (2013). NuSTAR discovery of a 3.76 s transient magnetar near Sagittarius A*, *Astrophys. J. Lett* **770**(2), 23.

Morris, D., Graham, D. A., Seiber, W., Bartel, N. *et al.* (1981b). Observations of the polarization of average pulsar profiles at high frequency, *Astron. Astrophys. Sup.* **46**, 421–472.

Olausen, S. A. and Kaspi, V. M. (2014). The McGill magnetar catalog, *Astrophys. J. Suppl. S.* **212**(1), 6.

Ransom, S. M., Demorest, P., Ford, J., *et al.* (2009). GUPPI: Green Bank Ultimate pulsar processing instrument, In *American Astronomical Society Meeting Abstracts #214, American Astronomical Society Meeting Abstracts*, Vol. 214, p. 605.08.

Reardon, D. J., Zic, A., Shannon, R. M. *et al.* (2023). Search for an isotropic gravitational-wave background with the Parkes pulsar timing array, *Astrophys. J. Lett.* **951**, 6.

Rickett, B. J. (1990). Radio propagation through the turbulent interstellar plasma, *Annu. Rev. Astron. Astr.* **28**, 561–605.

Shang, L.-H., Lu, J.-G., Du, Y.-J. *et al.* (2017). Investigating the multifrequency pulse profiles of PSRs B0329+54 and B1642-03 in an inverse Compton scattering model, *MNRAS* **468**(4), 4389–4398.

Sieber, W. (1973). Pulsar spectra — a summary, *Astron. Astrophys.* **28**, 237–252.

Tong, H., Xu, R. X., Song, L. M. *et al.* (2013). Wind braking of magnetars, *Astrophys. J.* **768**(2), 144.

van Leeuwen, A. G. J., Kouwenhoven, M. L. A., Ramachandran, R. *et al.* (2002). Null-induced mode changes in PSR B0809+74, *Astron. Astrophys.* **387**, 169–178.

Wang, J., Zuo, X., Michael, K. *et al.* (2017). TM65 m radio telescope microwave holography, *Sci. Sin-Phys. Mech. As.* **47**(9), 099502.

Wang, N., Manchester, R. N., and Johnston, S. (2007). Pulsar nulling and mode changing, *MNRAS* **377**, 1383–1392.

Xu, R.-X. (2006). Pulsars and quark stars, *Chinese. J. Astron. AST. Sup.* **6**(2), 279–286.

Xu, H., Chen, S., Guo, Y. *et al.* (2023). Searching for the nano-hertz stochastic gravitational wave background with the Chinese pulsar timing array data release I, *Res. Astron. Astrophys.* **23**, 075024.

Yan, Z., Shen, Z.-Q., Manchester, R. N. *et al.* (2018). Simultaneous 13 cm/3 cm single-pulse observations of PSR B0329+54, *Astrophys. J.* **856**(1), 55.

Yan, Z., Shen, Z.-Q., Wu, X.-J. *et al.* (2015). Single-pulse radio observations of the galactic center magnetar PSR J1745-2900, *Astrophys. J.* **814**(1), 5.

Yan, Z., Shen, Z.-Q., Wu, Y.-J. *et al.* (2018). Pulsar research with the newly built shanghai tian ma radio telescope, *URSI Radio Science Bulletin* **366**, 10–18.

Yan, Z., Shen, Z. -Q., Wu, Y. -J. *et al.* (2017). Pulsar studies with Shanghai tian ma radio telescope, In *XXXIInd General Assembly and Scientific Symposium of the International Union of Radio Science (URSI GASS)*, Montreal, QC, Canada, **2017**, pp. 1–4.

Yao, J. M., Manchester, R. N., and Wang, N. (2017). A new electron-density model for estimation of pulsar and FRB distances, *Astrophys. J.* **835**(1), 29.

Zhao, R.-S., Wu, X.-J., Yan, Z. *et al.* (2017). TMRT observations of 26 pulsars at 8.6 GHz, *Astrophys. J.* **845**(2), 156.

Zhao, R.-S., Yan, Z., Wu, X.-J. *et al.* (2019). 5.0 GHz TMRT observations of 71 pulsars, *Astrophys. J.* **874**(1), 64.

https://doi.org/10.1142/9789819800643_0012

Chapter 12

The Large High-Altitude Air Shower Observatory (LHAASO)

Xin-Hua Ma[*,†,§] **and Zhen Cao**[*,†,‡,¶]

[*]*Key Laboratory of Particle Astrophysics, Institute of High Energy Physics, Chinese Academy of Sciences, Beijing 100049, China*
[†]*TIANFU Cosmic Ray Research Center, Chengdu, Sichuan 610000, China*
[‡]*University of Chinese Academy of Sciences, Beijing 100049, China*
[§]*maxh@ihep.ac.cn*
[¶]*caozh@ihep.ac.cn*

The Large High-Altitude Air Shower Observatory (LHAASO) is a unique and new generation cosmic ray experiment at high altitude. It takes the function of hybrid technology to detect extensive air showers (EAS) and to upgrade greatly the resolving power between gamma ray and cosmic ray. LHAASO can study gamma ray sources over almost 4 orders of magnitude of energy to study various types of Galactic and extra-Galactic gamma ray emissions. On the other hand, the detection range of cosmic ray by LHAASO spans 6 orders of magnitude, fully covering the knee region and extending close to the ankle area. Moreover, LHAASO can search the new physical phenomena, and perform as an interdisciplinary research platform to provide clues and evidence for related topics in solar physics, atmospheric physics, and geophysics. Here experimental setup and physics targets of LHAASO are introduced overall.

Keyword: LHAASO

Contents

1. Introduction

Since the century discovery of cosmic ray, the origin of cosmic ray has still been a mystery. The study on the origin of high-energy cosmic ray is in an inter-discipline between gamma ray astronomy and the cosmic ray physics. The Large High-Altitude Air Shower Observatory (LHAASO) (Cao, 2010; He, 2018; Ma *et al.*, 2022) is a unique and new generation cosmic ray experiment with the advantages of high altitude, all-weather, and large-scale. It takes the function of hybrid technology to detect extensive air showers (EAS) and to upgrade greatly the resolving power between gamma rays and cosmic rays. The scientific core objective of LHAASO is to explore the origin of high-energy cosmic ray and carry out the relevant studies on the basic sciences, such as the high-energy radiation, celestial body evolution, dark matter distribution, and so on. The particular scientific objectives of LHAASO are given as follows:

(1) To make the full sky survey of gamma ray emission, to accumulate the statistical samples of various sources, to explore their high energy radiation mechanisms, including the mechanism to produce the violent time-variation phenomena, to study the evolutionary rules of active galactic nuclei (AGN) with the central super-massive black holes, to catch out the high-energy Gamma Ray Burst (GRB) events in the universe, and to explore their eruption mechanisms.

(2) To measure accurately the broad range energy spectra of gamma ray sources, especially by searching the cosmic PeVatrons in the energy range above 100 TeV, to study the particle features of high-energy radiation sources, to explore the evidence for the existence of hadron accelerator in the Galaxy, and to make a breakthrough in the aspect of cosmic ray source detection.
(3) To measure accurately the energy spectrum and composition of cosmic ray, and to study the cosmic ray origin, acceleration and propagation mechanisms.
(4) To search the new physical phenomena, such as the dark matter, quantum gravity, Lorentz invariant violation, etc., and to find new physical rules.

2. Experimental Setup

LHAASO (Ma *et al.*, 2022) (Figure 1) is located at Mt. Haizi (4410 m a.s.l., 600 g/cm^2, 29° 21' 27.56" N, 100° 08' 19.66" E) in Daocheng, Sichuan province, P.R. China. LHAASO consists of 1.3 km^2 array (KM2A) of electromagnetic particle detectors (ED) and muon detectors (MD), a water Cherenkov detector array (WCDA) with a total active area of 78,000 m^2, 18 wide field-of-view air Cherenkov telescopes (WFCTA) and a newly proposed electron-neutron detector array (ENDA) covering 10,000 m^2. Each detector is synchronized with all the other through a clock synchronization network based on the White Rabbit protocol. The observatory includes an IT center which comprises the data acquisition system and trigger system, the data analysis facility, and the high altitude infrastructure.

2.1. *KM2A*

KM2A (Figure 2) focuses on discovering galactic cosmic ray sources by searching for galactic gamma ray sources above 30 TeV in the northern sky and measuring cosmic rays in the energy range from 10 TeV to 100 PeV. The observation of a gamma-ray source with spectrum extending above 100 TeV represents an unambiguous identification of the 'PeVatrons', i.e., the sources of cosmic rays for energies up to the knee region of the cosmic ray spectrum. KM2A can detect about 1000 gamma rays above 100 TeV from a Crab-like source

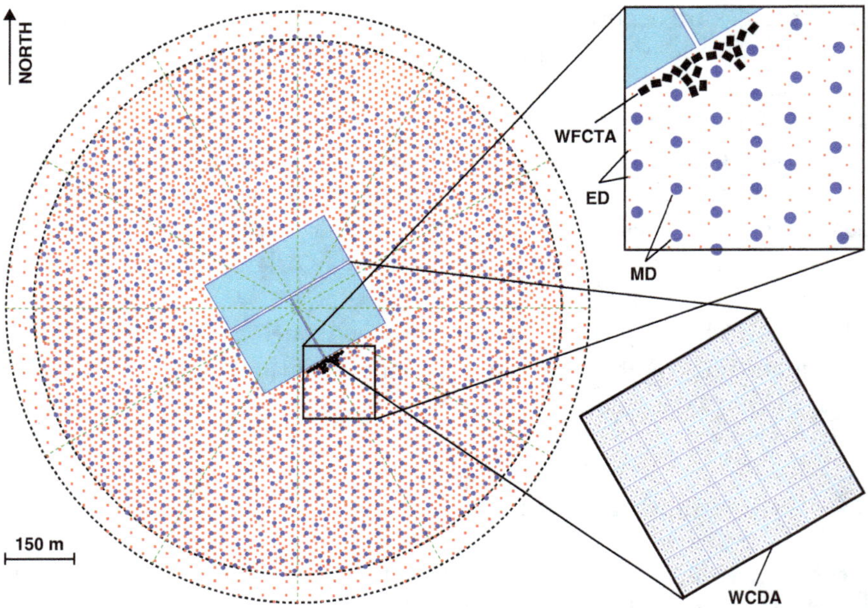

Fig. 1. Layout of LHAASO (Ma *et al.*, 2022).

in one year. For gamma ray detection, KM2A has following performance: effective area can reach 0.8 km^2 at 30 TeV, angular resolution is 0.4° at 30 TeV and 0.2° at 1 PeV, core position resolution is 7 m at 30 TeV and 2 m at 1 PeV, and energy resolution is 28% at 30 TeV. KM2A will reach a sensitivity of about 1% of a Crab-like source for energy of 100 TeV in one year of observation (Cui *et al.*, 2014). From observation of the Crab Nebula as a standard candle, It is obtained that the angular resolution is 0.26° and 0.18° at the 30 TeV and 300 TeV respectively, and pointing error of KM2A is less than 0.1° (Aharonian *et al.*, 2021a).

The ED array is distributed over 1.3 km^2. It is divided into two parts: the central array consisting of 4911 EDs in a triangle grid with 15 m spacing in a circle area with radius of 575 m, and the guarding ring array consisting of 305 EDs with 30 m spacing surrounding the central one with outer radius of 635 m. A detection-sensitive unit of ED consists of a plastic scintillator tile, 12 wavelength shifting fibers, a layer of tyvek wrapper and black cloth. The tile has high detection efficiency for charged particles, quick response time and good stability. The tile size is 100 cm × 25 cm × 1 cm. Twelve fibers with length

Fig. 2. Photo of LHAASO in August 2021. WCDA is the building at the center. The soil bumps are MDs. The white or green small boxes are EDs. The blue boxes beside WCDA are WFCTs (Ma *et al.*, 2022).

of 2.7 m and diameter of 1.5 mm are routed in 24 grooves on the tile. Each tile is wrapped with a layer of Tyvek to improve photon collection efficiency. The tile assembly is covered with black cloth to avoid external and inter-tile optical cross-talk. One ED is composed of four detection sensitive units, a 1.5 inch photo-multiplier tube (PMT), an electronics system, a power supply system and a shell. Four detection sensitive units are installed into the steel shell symmetrically to cover the 1 m^2 detection area. All ends of 96 fibers are bunched together to be polished and coupled directly to the PMT photocathode. In order to increase detection efficiency for secondary γ and absorb charged particles with low energy to improve the angular resolution, a 5 mm thick lead layer is placed on the surface of each ED to convert γ into electron-positron pairs.

The MD array is composed of 1188 water Cherenkov tanks deployed on the same area as the ED array in a triangle grid with a spacing of 30 m. MD area is optimized to 36 m^2 by balancing the total area and spacing of MDs. Soil is the cheapest material to shield EAS electromagnetic components and reduce the so-called punch-through effect, i.e., EAS electromagnetic components generate sub-showers in the shielding soil, and the products (mainly gamma rays, e^+ and e^-) have a certain probability to penetrate the soil and to be misidentified as muon. The water Cherenkov detector acts as a calorimeter by measuring the total number of Cherenkov photons yielded by a particle in water, which is approximately proportional to the total deposited energy of particle in water. If the total punch-through energy of EAS electromagnetic components is smaller than the total deposited energy of muons in water, shielding soil can provide further rejection power to EAS electromagnetic components. the overburden soil has thickness of 2.5 m so that most of muons have energies ~1 GeV. A MD consists of a concrete tank, a thermal layer, a water liner, pure water and an eight-inch PMT.

2.2. *WCDA*

WCDA is a survey instrument sensitive to gamma rays with energies between 100 GeV and 30 TeV. In the energy regime of TeV it can attain the world's best survey sensitivity of < 0.1 flux intensity of Crab nebula. From observation of the Crab Nebula as a standard candle, It is obtained that angular resolution of WCDA for the gamma ray arrival direction is better than $0.4°$ above 3 TeV, the pointing accuracy was better than $0.05°$, and the energy resolution is found to be 33% for gamma rays around 6 TeV (Aharonian *et al.*, 2021b). Besides observation of Galactic gamma ray sources, WCDA has the discover potential and is sensitive in monitoring extra-galactic variant sources. WCDA will give an observation with high significance for studying the origin and acceleration of cosmic rays, multi-waveband radiation mechanisms of GRBs, AGNs, EBL, intergalactic magnetic field, galactic evolution, etc. Meanwhile, the WCDA has an advantage for observing extended sources.

WCDA (Figure 3) covering an area of 78,000 m^2, is composed of 3,120 water Cherenkov detectors (WCDs), divided into three separate arrays. Each array is a single water pool with water depth of 4.4 m. Two pools with effective area of 150×150 m^2 contain 900 WCDs individually. The third pool with area of 300×110 m^2 contains 1,320 WCDs. Each WCD has a 5×5 m^2 area separated by black plastic curtains with area of 3.3 m $\times$ 4.7 m vertically hung in water to absorb scattered light. Side walls and bottom of the pools are lined with 2 mm thickness HDPE film to keep water from leaking outside. To guarantee an attenuation length of longer than 15 m for the near-ultra-violet light, a water purification and re-circulation system is operated. The first pool has 900 WCDs. Each WCD is equipped by an 8-inch PMT for timing and pulse height in low range and a 1.5-inch PMT for pulse height in high range at the center of WCD and 4 m beneath water surface. Dynamic range of WCD is enlarged very much by using the small PMT. This configuration enables measurement of detailed particle density distribution in EAS cores without significant saturation even for energetic EAS up to 10 PeV, and reaches core location resolution better than 3 m over a wide energy range. This is designed for identification of cosmic ray composition and measurement of energy spectrum. LHAASO's upgrading design is to replace the 8-inch PMTs by 20-inch PMTs in the other two pools. The customized design of the 20-inch PMTs using multi-channel-plate instead of the traditional dynodes enables good uniformity between PMTs as well as Transit Time Spreads less than 7 ns and Cathode Transit Time Distribution less than 2 ns. In order to compensate loss of energy dynamic range, a 3-inch PMT is installed beside this large PMT in WCD, read out only pulse height by a simplified version of FEE covering three orders of amplitudes in number of photoelectrons.

2.3. *WFCTA*

With the advantage of high altitude, LHAASO possesses multiple instruments capable of measuring the complete and 3D observations on the longitudinal and transverse developments of EAS, and enhances greatly accuracy in energy determination and power in identifying the cosmic ray composition. The shape of the EAS Cherenkov image is observed by WFCTA, including the ratio between the major and minor image axes and the angular distance from the image centroid to the arrival direction of the shower, are closely related to the position when the EAS is developed longitudinally to the maximum. The total photo-electrons measured by WFCTA can be used to measure the cosmic-ray energy. Energy at the core region recorded by WCDA contains the hadron information in the early development of EAS. Muon component recorded by MD reflects the hadron information in EAS. WFCTA is an array of Cherenkov telescopes, a movable design idea is adopted, different

Fig. 3. Photo inside one hall of WCDA without water and curtains (Ma *et al.*, 2022).

array configurations are used for the different observation stages to realize the accurate measurement of energy spectrum of different composition from 10^{13} to 10^{18} eV, and the energy scale transfer from the direct space measurement to the extremely high energies. In total, three stages are employed by the WFCTA to perform the composition and energy spectrum measurement of cosmic rays. In the first observation stage, WFCTA aims to observe the primary energy range from 10 to 300 TeV, one Cherenkov telescope will point at the zenith angle of 0°. In the second observation stage, WFCTA aims to observe the primary energy range from 50 TeV to 10 PeV, six Cherenkov telescopes will point at the zenith angle of 30° and each telescope points at different azimuth angles, in order to cover the sky with zenith angle from 22° to 38° and azimuth angle from 0° to 156°. In the third observation stage, WFCTA aims to observe the primary energy range from 1 to 100 PeV. To make sure that the position of shower maximum is far from the observation surface, the main axis of 18 Cherenkov telescopes will point at 45° in zenith angle. Cherenkov telescopes will point at different azimuth angles in order to cover 0° to 360° in azimuth.

Fig. 4. Bird view of WFCTA (Ma *et al.*, 2022).

WFCTA (Figure 4) is composed of 18 wide-field Cherenkov telescopes (WFCTs). Each WFCT consists of an array of 32×32 Silicon photomultipliers (SiPMs), a 5 m^2 spherical aluminized mirror, a portable and adjustable container, power supply system and slow control system. It has a field of view (FOV) of $16° \times 16°$ with a pixel size of approximately $0.5° \times 0.5°$. Mirrors are mounted at the back of the container. The SiPM camera is located at the focal plane which is 2870 mm away from the center of the mirror. The container is mounted on a truck frame with a pitching rotation system that allows the container to be lifted up-and-down manually, from 0° to 90° with a step of 0.1°. The SiPM camera consists of 64 sub-clusters. Each sub-cluster is composed by 4×4 SiPMs, each SiPM coupled with a light funnel (Winston cone), a pre-amplifier board, a voltage and temperature compensation loop board, two analogue boards, a digital board and a power regulator board. To monitor the gain of the SiPM camera, six UV-LEDs are mounted at the center of the mirror.

2.4. *Other facilities*

When arriving at Earth, high energy cosmic rays interact with the air nuclei thus inducing EAS. They consist of a core of high energy hadrons that continuously feed the electromagnetic part of the shower through fast decays of neutral pions, kaons and etas. Nucleons and other high energy hadrons contribute to the hadronic cascade. High energy hadrons, which constitute the EAS skeleton as

G.T. Zatsepin called it, carry important information for multi-parameter correlation studies. For example, hadronic observables, primarily the hadron number/electron number correlation, depend on the nature of the particle inducing the shower. Thus, the detection of high energy hadrons, addressed to improve the discrimination power in these analysis, is highly advisable. A way to deal with this problem avoiding the usage of huge and expensive HCALs was brought out is detection of thermal neutrons generated by EAS hadrons interacting with ambient matter (air, building, ground, etc.). This idea led to the development of the electron-neutron detector (EN-detector), based on a mixture of the well-known inorganic scintillator ZnS(Ag), capable of recording both thermal neutrons and relativistic charged particles. A novel type of ZnS(Ag) scintillator alloyed with B_2O_3 with the ^{10}B isotope about 19% is developed. An EN-Detector Array (ENDA) composed of 16 EN-detectors has been built at LHAASO since 2017 for test, and it will be extended to 64 detectors (ENDA-64) soon. After achieving good results, ENDA will be extended up to 400 detectors (ENDA-400) with array area of 10,000 m^2 to study energy spectrum and mass composition at the energy region from 100 TeV up to 100 PeV in conjunction with LHAASO.

To guarantee the angular resolution of reconstruction, a 500 ps RMS overall synchronization precision must be achieved among the thousands of wildly spread detectors. Due to the large quantity of detector nodes, a combined timing and data link could be more cost effective than the conventional solution of using dedicated clock link and data link, which also greatly simplifies the cable deployments. Since the most part of the timing network is installed and worked in the unmanned field at very high altitude with simple mechanical housing, strong requirements are also raised for covering wide temperature range, reliability, easy of replacement and prevention from dust. WR (White Rabbit) is a technology originally proposed by CERN and GSI and developed by tens of institutes and universities from all over the world. It aims to provide sub-nano-second synchronization accuracy and pico-seconds precision over few thousands nodes in the distance of up to 20 km distance. To avoid the bi-directional delay asymmetry caused by physical link length difference, WR uses single fiber with wave multiplex for upstream and downstream. For certain types of fiber, the speed difference of two wavelengths can be calibrated. WR has the advantage of using one single fiber for both Ethernet data transmission up to 1 Gbps and sub-ns synchronization, that makes it an ideal solution for LHAASO timing system. The scalability, long distance features are also of importance.

The data acquisition system (DAQ) resides between electronics system and off-line analysis system. It is mainly a software system running on computing nodes, responsible for data collection from electronics, software trigger, and event storage. KM2A and WCDA do not have event trigger (triggerless) at the on-line electronics level, and by Front-End Electronics signals are digitalized and packed as "hit packet" sent to DAQ. Each telescope of WFCTA has a hardware trigger built in each camera so that data are packed as event sent to DAQ. All Electronics systems communicate with DAQ through a fiber network. The DAQ system is required to collect all these data from readout electronics, perform software trigger and store events to disks. Readout nodes are used to collect and buffer data from electronics. CPUs on both readout and compute nodes can be assigned to perform online processing tasks including software trigger. Aim of the trigger system is to remove amount of background and select real events which meet requirements of scientific motivations of LHAASO. LHAASO events are recorded in three modes: trigger mode, low multiplicity mode and single particle mode. The different trigger modes include different contributions from the arrays, and collect different measurement variables from the arrays. It leads that the different trigger modes have very different event rates and data rates. One physical motivation may be correlated with several event types, or one event type may contribute to several physical motivations. Moreover, the trigger modes on the basis of software control in DAQ are so flexible that, if any new physical motivation is proposed, the present trigger modes can be adjusted conveniently, and even a new trigger mode can be supplemented into the online software.

LHAASO **O**ffline **D**ata Processing **S**of**t**ware Fr**a**mewo**r**k (LodeStar) is developed for the LHAASO offline data processing, including Monte Carlo (MC) data production, experimental data processing as well as physics analysis. The typical data processing is to loop over all events with some calculations applied on the event data. The usual procedure for one event during event looping can be divided into three steps: read in an event, process the event, and write out the results. LodeStar consists of an underlying framework and experiment specific applications such as event generator, detector simulation, electronics simulation, calibration, reconstruction and physics analysis tools. The framework plays very important roles for the performance and functionalities of LodeStar. It provides most of the common functionalities for the offline data processing, such as data management, data processing controlling, data input/output, common services and tools, as well as friendly user interfaces. The framework also builds the unified software environments and computing platform for the developers, and defines the standard interfaces between different applications. Therefore many developers can synchronously develop their data processing applications. All applications developed with standard interfaces can be smoothly plugged into the framework.

The data analysis facility for LHAASO is designed to build a integrated platform including a large-scale data storage and computing system to support the running of analysis software and finally provides physicists with a uniform experiment data analysis environment. The data analysis facility provides fundamental computing and network services, such as high performance computing (HPC), massive data storage, high-speed transmission network, distributed data sharing and analysis system to support detector monitoring, data transmission, storage, decoding, calibration, reconstruction and analysis. The facility is composed of a small on-site data center at Daocheng, a large data center at Beijing, high-speed fiber network from Daocheng to Beijing, remote operation center and Information Technology Service platform and so forth.

LHAASO has two sites in Daocheng. One is in Mt. Haizi and the other is in the downtown of Daocheng. The distance between these two sites is about 50 km. The site in the Daocheng downtown is the living base including dormitory and the other two buildings at 3,750 m a.s.l. The total area of the living base is about 13,000 m^2. The other site in Mt. Haizi at 4,410 m a.s.l. is the observatory where all detectors are installed in arrays. The observatory is about 10 km away from the Daocheng airport. The total length of roads in the observatory is 21.9 km including 5.9 km paved. The observatory also includes assembling hall, pumping station and 35 KV transformer station. In the LHAASO project, the WCDA needs about 400 thousand tons of pure water to fill all three pools up and the MD needs about 50 thousand tons of ultra-pure water to fill up all the 1171 tanks. The original water is pumped from a small lake which is connected to the nature river about 300 m away. The water is simply filtered by a filter bag to reduce dust in the pumping station before being pumped to the water purification factories. Four control rooms are arranged on the second floor of those buildings. Central timing, clock distributing WR switches, data collecting network, high voltage power suppliers and water transparency monitoring system are hosted in the control room. The solo entrance of each pool with a door way equipped with air blowers is also located in the control room.

3. Physics Topics Studied by LHAASO

LHAASO can study gamma ray sources over almost 4 orders of magnitude of energy. Figure 5 shows the integrated sensitivity in one year measurement, obtained by simulating the response of the detector to gamma ray flux from a source like the Crab nebula one. In the same figure the Crab nebula spectrum is also shown as a reference flux. The LHAASO sensitivity curve is the combination of two components: the first relative to WCDA, operating in the energy range 0.3–10 TeV, and the second relative to the KM2A array, sensitive to energies above 10 TeV. The sensitivity curves of other experiments (some in operation, some in project) are also reported. It should be noted that for a general convention the sensitivity of air shower detectors is reported for one year of the whole sky and full-time observation, while that of Cherenkov telescopes is relative to 50 hours of "on source" measurement.

There are various types of gamma-ray emissions in the Galaxy, including: (1) Supernova remnants (SNRs). A supernova is the end point of a part of stellar evolution. The star can throw most or even almost all of its matter outward at a high speed through explosion to collide with surrounding interstellar matter to generate shock waves, forming a shell-like structures known as SNRs. SNR has long been considered to be the main source of galactic cosmic rays. (2) Pulsar Wind Nebula (PWN). With super-strong magnetic field, the pulsar's high-speed rotation can accelerate the charged particles to relativistic speeds to rush into the interstellar medium, and generate a strong shock wave, in which the magnetized particle stream

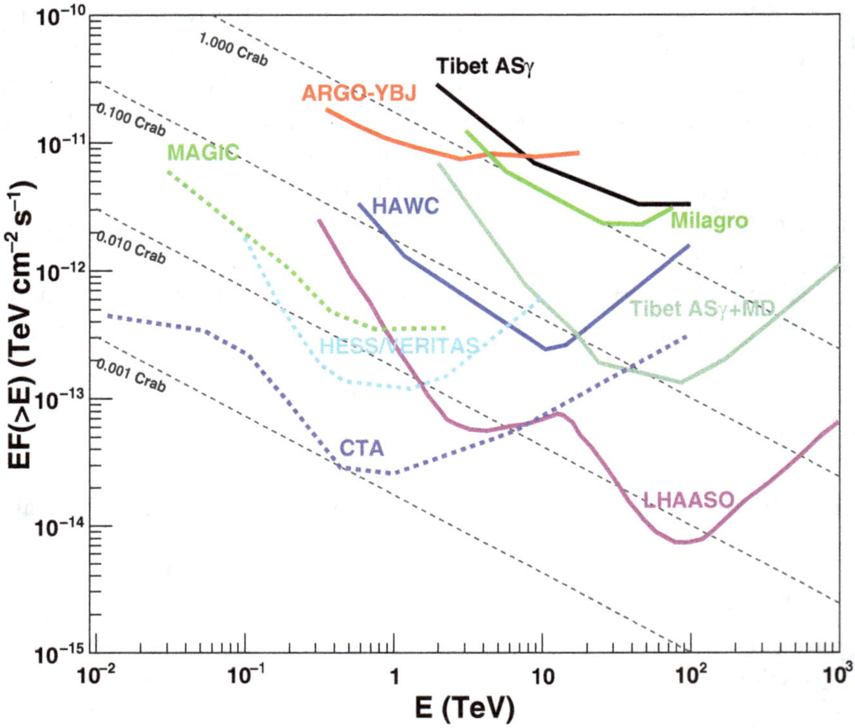

Fig. 5. The integrated sensitivity of LHAASO (in pink) in one year measurement from a source like the Crab nebula one, compared with other experiments. (Wang *et al.*, 2022).

is accelerated and ejected. PWNs are the most abundant type in the galactic sources in the TeV energy region. Currently, there are 34 PWNs observed in the TeV energy region, of which 13 are in the LHAASO field of view (Liu *et al.*, 2016) (Chen *et al.*, 2022). (3) The galactic dispersive gamma radiation can be used to study the propagation of galactic cosmic rays, and to study the distribution and interaction of cosmic rays and interstellar medium gases in different regions. (4) Gamma radiation is generated by the interaction of high-energy particles with interstellar matter or radiation fields. These high-energy particles escape from the acceleration region, so many very high-energy gamma-ray sources behave as extended sources. In addition, most of the diffuse gamma radiation on the galactic plane comes from the radiation generated by the interaction of cosmic rays with interstellar matter and radiation fields.

Gamma-ray emissions outside the Galaxy include: (1) AGNs are usually accreting super massive black holes, characterized by significant flux changes. Due to the very long distance of the source, through joint observations with other electromagnetic wave bands, radiation changes can be used to explore many fundamental astrophysical problems involving large-scale matter distribution, extragalactic background light distribution, and quantum gravitational effects. (2) There is a class of spatially extended sources, such as a fairly bright gamma-ray source located in the direction of the constellation Cygnus, whose current intensity far exceeds the minimum sensitive current intensity that can be detected by the Cherenkov telescope, but They were not detected by these narrow field-of-view telescopes. This fully reflects the advantages of wide-field scanning detectors in discovering sources with extended characteristics, and also leaves a broad space for LHAASO to discover stronger sources. (3) GRBs are the most violent stellar-level high-energy explosions in the universe. In observation, GRBs appear as sudden increases in the flux of gamma rays from a certain direction in the universe in a short period of time. We still don't know how high-energy the photons generated by GRBs can reach, and the high-energy radiation observations of GRBs provided by LHAASO (Wang *et al.*, 2022) are crucial for a comprehensive understanding of this violent explosion phenomenon, which is important for black hole and compact star-related physics, high-energy cosmic rays It can also be used to limit key parameters such as the Lorentz factor of GRBs and the magnetization of the source region.

At present, it is believed that the knee region is due to acceleration limitation on energy of the cosmic ray source in the Galaxy. There are many differences in the characteristics of the knee region predicted by various theoretical models of the origin of cosmic rays. Only the precise component energy spectrum can give a clear judgment on different theoretical models. The detection range of LHAASO is from 300 GeV to 1 EeV, spanning 6 orders of magnitude, fully covering the knee area and extending close to the ankle area. LHAASO is to achieve high-precision detection on the energy spectrum of cosmic rays of 6 orders of magnitude, deeply study composition at the knee region, and chain the direct detection results in the lower energy region with the results in the highest energy region (such as the Pierre Auger project with an array of 3,000 square kilometers) by unifying energy scaling (Piazzoli *et al.*, 2022).

In new physics, LHAASO's measurements of gamma rays and cosmic rays can be used to constrain the parameter space of various dark matter and axion theoretical models (Bi *et al.*, 2022). Some theories predict that dark matter can produce high-energy gamma-ray pairs through annihilation, superimposing an excess on the diffuse gamma-ray energy spectrum, and LHAASO has unprecedented sensitivity in this energy region. In addition, in relativity and quantum field theory, Lorentz symmetry is a basic symmetry, and some theories expect that under the high-energy conditions of Planck scale and quantum gravity conditions, Lorentz symmetry may have certain destruction. An important prediction of Lorentz's destruction is that the propagation velocity of high-energy photons is reduced relative to low-energy photons. The present data of LHAASO show that the current gamma ray spectrum continues to high energies above PeV, and no "mysterious" disappearance of high-energy gamma incidents has been found, indicating that the Lorentz symmetry is approaching the general trend. The Rank energy subscript is still correct. The research results were published in Physical Review Letters on February 3, 2022. (Cao *et al.*, 2022).

Cosmic rays can produce a variety of particles during the generation, acceleration, and propagation, including cosmic rays themselves, gamma rays, and neutrinos. These particles are the detectable "messengers" emitted by cosmic ray sources. The reason why high-energy neutrinos are important is because high-energy neutrinos must be produced in the process of interaction between cosmic rays and space medium, and the source of neutrinos must be the source of cosmic rays. Like gamma rays, neutrinos may also have internal and extragalactic sources, point sources and extended sources, but neutrinos are very weak. To capture neutrinos, it is necessary to build large volumes of transparent water or large-scale experiments targeting ice, such as the IceCube in Antarctica. So far, no neutrino source has been definitively discovered by IceCube. By continuously observing nearby blazar and starburst galaxies, LHAASO can provide more detailed information on candidate neutrino sources (Li *et al.*, 2022). In addition, LHAASO can do a job on its own by detecting horizontal neutrino events by detecting showers flying laterally from a large angle, especially the rare tau neutrino, which is significantly different from other showers. The unique characteristics of LHAASO, facilitate the exclusion of background.

As an interdisciplinary research platform, LHAASO can also contribute in other fields of physics (Jia *et al.*, 2022). The sun is the closest star to us, constantly emitting particles of different energies to form the solar wind. Strong coronal mass ejections may cause huge disturbances to the earth's magnetic field and trigger geomagnetic storms, causing sudden changes or even catastrophes in "space weather", affecting our lives. These solar events cause changes in low-energy radiation on the ground, including changes in the electromagnetic composition and neutron flux, that cause changes in the count rate in the detectors. On the other hand, changes in the electric field in the Earth's atmosphere, such as thunderstorms, can also induce changes in the flux of low-energy cosmic rays. Interestingly, crustal movements (such as earthquakes) can also affect the release of radioisotopes in the crust (including soil, rock, concrete, etc.), thereby changing the flux of neutrons. LHAASO has a variety of detection methods and uninterrupted massive cosmic ray data, which can be used for in-depth tracking, monitoring and research on these changing phenomena, providing important clues and evidence for related topics in solar physics, atmospheric physics, and geophysics.

4. Summary

LHAASO Started complete operation in July 2021. LHAASO adopts a variety of detection technologies, which can measure secondaries of gamma ray and cosmic ray in the atmosphere in a large field of view, all-weather, wide energy band and omni-direction, and reconstruct their basic information. LHAASO's unique hybrid detection method, the ability to cross-check each other, and the strong ability to distinguish gamma-ray and cosmic ray will ensure the accuracy and reliability of the measurement results. Combining the advantages of high altitude and long-term stable operation, LHAASO can discover a large number of gamma ray sources, and achieve world-leading sensitivity in the research of ultra-high-energy gamma ray sources and cosmic ray composition and energy spectra. LHAASO also provides an important experimental platform for conducting cutting-edge, cross-scientific research on the atmosphere, environment, and space weather, and becomes a scientific platform for multilateral and international cooperation to jointly conduct high level research.

Acknowledgements

This work was supported in China by the Science and Technology Department of Sichuan Province (grant number 2021YFSY0031) and the National Natural Science Foundation of China (NSFC) (grant number U2031103). The authors would like to thank all staff members who work at the LHAASO site above 4400 meters above sea level year-round to maintain the detector and keep the water recycling system, electricity power supply and other components of the experiment operating smoothly. We are grateful to Chengdu Management Committee of Tianfu New Area for the constant financial support for research with LHAASO data.

Bibliography

Aharonian, F. *et al.* (2021). Observation of the crab Nebula with LHAASO-KM2A — A performance, *CPC* **45**, 025002.

Aharonian, F. *et al.* (2021). Performance of LHAASO-WCDA and observation of crab Nebula as a standard candle, *CPC* **45**, 8, 085002.

Bi, X.-J. *et al.* (2022). Chapter 5: Dark matter and new physics beyond the standard model with LHAASO, *CPC* **46**, 3, 030005.

Cao, Z. (2010). A future project at Tibet: The large high altitude air shower observatory (LHAASO), *CPC* **34**, 249–252.

Cao, Z. *et al.* (2022). Exploring Lorentz invariance violation from ultrahigh-energy γ rays observed by LHAASO, *Phys. Rev. Lett.* **128**, 051102.

Chen, Y. *et al.* (2002). Chapter 2: Galactic gamma-ray sources, *CPC* **46**, 3.

Cui, S.-W. *et al.* (2014). Simulation on gamma ray astronomy research with LHAASO-KM2A, *Astropar. Phys.* **54**, 86–92.

He, H.-H. (2018). Design of the LHAASO detectors, *Radia. Detect. Technol. Methods* **2**, 7.

Jia, H.-Y. *et al.* (2022). Chapter 7: Solar and heliospheric physics and interdisciplinary research with LHAASO, *CPC* **46**(3), 030007.

Li, Z. *et al.* (2022). Chapter 6: Multimessenger physics, *CPC* **46**(3), 030006.

Liu, Y. *et al.* (2016). Expectation on observation of supernova remnants with the LHAASO project, *Astrophys. J.* **826**(1), 63.

Ma, X.-H. *et al.* (2022). Chapter 1: LHAASO instruments and detector technology, *CPC* **46**(3), 030001.

Piazzoli, B. D. *et al.* (2022). Chapter 4: Cosmic-ray physics, *CPC* **46**(3), 030004.

Wang, X.-Y. *et al.* (2022). Chapter 3: Extra-galactic gamma-ray sources, *CPC* **46**(3), 030003.

https://doi.org/10.1142/9789819800643_0013

Chapter 13

The Highest Energy Gamma Ray from Pulsars Detected by LHAASO

Xin-Hua Ma[*,†,§] **and Zhen Cao**[*,†,‡,¶]

[*] *Key Laboratory of Particle Astrophysics, Institute of High Energy Physics, Chinese Academy of Sciences, Beijing 100049, China*

[†] *TIANFU Cosmic Ray Research Center, Chengdu, Sichuan 610000, China*

[‡] *University of Chinese Academy of Sciences, Beijing 100049, China*

[§] *maxh@ihep.ac.cn*

[¶] *caozh@ihep.ac.cn*

One of main physical targets of LHAASO is to search for gamma ray sources and then discover origin of cosmic ray. LHAASO starts certainly to explore the high-energy TeV emission from pulsars, PWNs and TeV halos, which are expected to be one kind of Galactic cosmic ray sources, helping to find a full sample of them and set constraints for theoretical modeling. LHAASO adopts the working mode of construction, operation, and analysis simultaneously. Before the completion of the full array, some important physics results had been obtained: 12 PeVatrons in the Galaxy were discovered, and the energy spectra of some of the PeVatrons were detected up to PeV range. In the vicinity of the extended ultra-high energy (UHE) sources, one can find potential counterparts for both the gamma ray production regions and the nearby particle accelerators. In this chapter, some of the PeVatrons possibly correlated to pulsars, PWNs and TeV halos are described.

Keywords: LHAASO, Cosmic Ray, Gamma Ray Source, Extensive Air Shower

Contents

1. Introduction

Pulsars are fast-rotating neutron stars, and most of the rotational energy of a pulsar is released in a form of the pulsar wind. The high-energy, relativistic particles in the pulsar wind interact with the ambient medium around a pulsar forming a terminal wind shock. Particles at the shock emanate synchrotron radiation, making the pulsar wind nebula (PWN) bright from radio to X-ray energies. It is speculated that the Inverse Compton (IC) scattering process gives rise to emission up to GeV/TeV range with electrons of Lorentz factor up-scattering background infrared photons. The modeling of a broad-band spectrum of a PWNe allows us to study its particle population, magnetic field, and dynamical evolution. Especially, when a PWN enters the last evolutionary stage in which the pulsar escaped from its parent supernova remnant (SNR) or its parent SNR almost disappeared, the original PWN will first dim in the low energy band, leaving an extended TeV emission region called "TeV halo".

PWN is expected to be a kind of Galactic cosmic ray source. The particles with energy below the knee region in cosmic ray energy spectrum can be effectively trapped by the magnetic fields of PWNe. Thus, PWNe could store a large amount of energy in relativistic protons, if PWNe could continuously produce energetic protons up to PeV. Hence, such PWNe may also be so-called PeVatrons which are the accelerators can accelerate cosmic ray to PeV energies (Ohira *et al.*, 2018). If a PWN is located in a dense environment and contains relativistic protons, the hadronic emission from the energetic protons may have a significant contribution to GeV–TeV gamma ray. Identification of PeVatrons is a prime objective toward understanding the origin of cosmic ray in the Galaxy. TeV halos could be indicators for invisible pulsars (Linden *et al.*, 2017). Extended TeV halos could be implications for those misaligned pulsars. This is especially meaningful for relatively old pulsars whose host SNRs already faded. Followup observations in multiwavelengths would then provide further evidence about the origin of the halos. If Energy spectra of more halos are measured, the relationship between the acceleration limit of PWNe and other parameters, such as the ages of the pulsars, will be revealed.

So far, more than 200 pulsars have been detected by Fermi (Acero *et al.*, 2015). Among them, more than 30 PWNe or candidates have been detected at TeV energies, and Fermi has been able to detect a few of them (Acero *et al.*, 2013). Particularly, the first astrophysical source detected above 100 TeV is the prototype PWN Crab, which was achieved by the Tibet AS+MD (Amenomori *et al.*, 2019) and HAWC (Abeysekara *et al.*, 2019) experiments. More recently, HAWC released a >56 TeV catalog (Abeysekara *et al.*, 2020) containing nine sources which all are likely related to pulsars and may be TeV halos. Especially, HAWC observations of extended very-high-energy (VHE) gamma-ray emission from two middle-aged, isolated pulsars, Geminga and Monogem (Abeysekara *et al.*, 2017). The spatial morphologies of the gamma-ray emission indicate that the diffusion of particles in the regions around those pulsars is much slower than the average value to give enough secondary particle yields (Yuan *et al.*, 2017) (Aguilar *et al.*, 2016).

LHAASO starts certainly to explore the high-energy TeV emission from pulsars, PWNs and TeV halos, helping by finding a full sample of them and setting constraints for theoretical modeling (Chen *et al.*, 2022).

2. The Current Results of LHAASO

LHAASO adopts the working mode of construction, operation, and analysis simultaneously. Before the completion of the full array, some important physics results had been obtained:

2.1. *12 PeVatrons in the galaxy*

The extension of the cosmic ray spectrum beyond 1 PeV indicates the existence of the PeVatrons. It is needed to locate and identify such objects to find the origin of Galactic cosmic ray. The principal signature of both electron and proton PeVatrons is Ultrahigh-energy (UHE) gamma ray radiation. Although gamma ray with energies slightly higher than 100 TeV have been reported from a few objects in the Galactic plane, unbiased identification and in-depth exploration of PeVatrons requires detection of gamma ray with energies

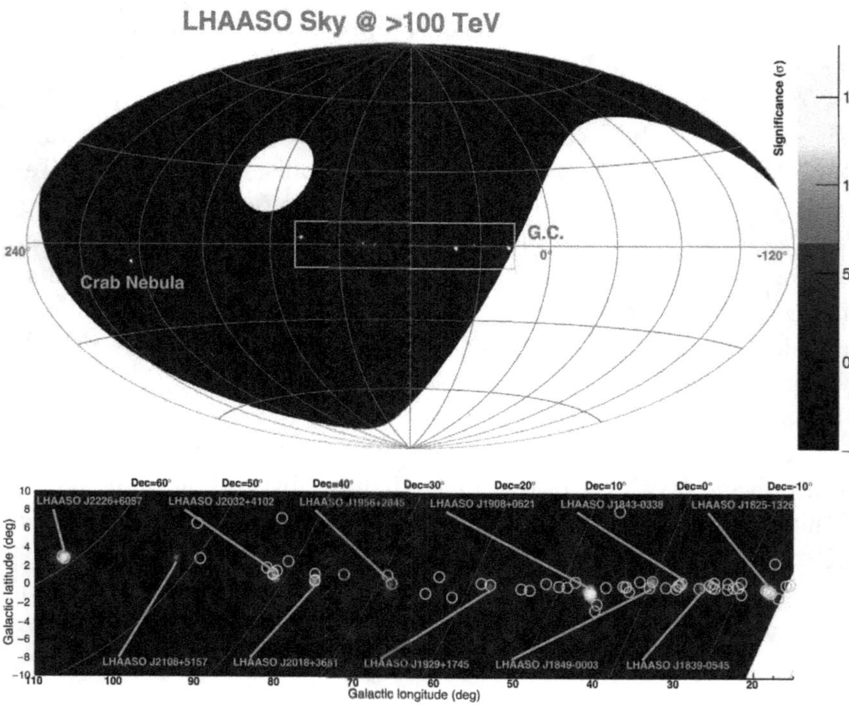

Fig. 1. LHAASO sky map at energies above 100 TeV. The circles indicate the positions of known VHE gamma ray sources (Cao *et al.*, 2021b).

well above 100 TeV. Based on the 1/2-scale KM2A that has been built, LHAASO discovered twelve sources of gamma rays with signals above 7σ at energies above 100 TeV in the Galaxy (Cao *et al.*, 2021b) (Figure 1). Recorded gamma photons with an energy of 1.4 PeV is the highest energy photon observed historically.

Of LHAASO J2226+6057, LHAASO J1908+0621 and LHAASO J1825-1326, three of these sources, the two-dimensional images extend to at least 1°, implying that gamma ray emitters occupy huge regions in the Galactic plane. Although gamma ray emission itself indicates the presence of active or recent particle accelerators inside or in the proximity of gamma-ray-emitting regions, the localization and identification of particle accelerators is not a trivial task and requires deep theoretical and phenomenological studies based on comprehensive multi-wavelength data. Not surprisingly, in the vicinity of the extended UHE sources, one can find potential counterparts for both the gamma ray production regions and the nearby particle accelerators. The candidates are potentially responsible for the electron and proton PeVatrons in the Galaxy: pulsars and PWNe, SNRs and young massive star clusters. Detailed studies of the spectral and morphological features of UHE sources and their astrophysical implications by using the data from both WCDA and KM2A, acquired over three decades from 1 TeV to 1 PeV, will be made source by source in the following steps.

2.2. *The Crab nebula*

In the 12 PeVatrons in the Galaxy found by LHAASO, the only firmly identified source is the Crab Nebula, a representative of PWNe. The Crab Nebula, known as the "standard candle" for high-energy astronomy, is a bright source of gamma ray powered by the Crab Pulsar's rotational energy through formation and termination of a relativistic electron-positron wind. The Crab Nebula differs from other members of PWNe by the Crab pulsar's huge spin-down luminosity, compact size and large nebular magnetic field. The latter makes conversion of energy of relativistic electrons to IC gamma ray inefficient, as low as 0.01%. Thanks to the large spin-down luminosity, however, the Crab Nebula is a luminous gamma ray source. Compared with it, rotational powers of other pulsars are smaller by orders of magnitude. On the other hand, the magnetic field strength in the surrounding nebulae is typically smaller than 10 μG. Large nebular magnetic field of the Crab Nebula dramatically enhances the gamma ray production efficiency, which scales as B^{-2}, and thus compensates for the relatively modest rotational power of pulsars (Figure 2).

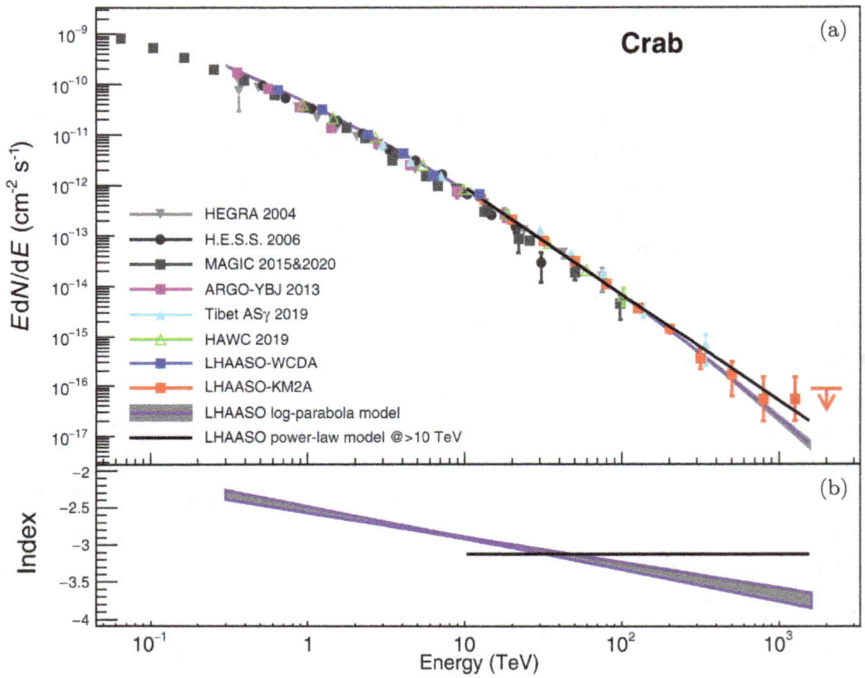

Fig. 2. Gamma ray flux of the Crab measured by LHAASO and spectral fitting. (a) TeV to PeV gamma ray fluxes of the Crab plotted as EdN/dE. The red squares and blue squares are the spectral points measured using KM2A and WCDA, respectively. The spectral points above 100 TeV were obtained in the signal-dominated regime, with 89 detected gamma ray and two events expected from cosmic ray-induced (hadronic) air showers after the muon cuts. No events were detected in the 1.6 PeV to 2.5 PeV bin, where an arrow indicates the flux upper limit at 90% confidence level. The purple line shows the fitting using a log-parabola model in the 0.3 TeV to 1.6 PeV interval. For comparison, the black line shows the fitting using a simpler power law model in the 10 TeV to 1.6 PeV interval. Also plotted are previous observations of the Crab by other facilities. (b) Energy-dependent local power law index G derived by the log-parabola model fitting, as indicated by the purple band. For comparison, the black line shows the photon index 3.12 ± 0.03 derived from the simpler power law model fitting. Error bars represent one standard deviation (Cao *et al.*, 2021a).

LHAASO reported the accurate detection of gamma ray from this source with energies from 5×10^{-4} PeV to 1.1 PeV with a spectrum showing gradual steepening over three energy decades (Cao *et al.*, 2021a). The UHE photons imply the presence of a PeVatron in the nebula, with an acceleration rate exceeding 15% of the theoretical limit. It was constrained that the PeVatron's size is between 0.025 and 0.1 parsecs and the magnetic field is about 110 μG. The production rate of PeV electrons, 2.5×10^{36} ergs per second, constitutes 0.5% of the pulsar spin-down luminosity, although a contribution of PeV protons to the production of the highest-energy gamma ray cannot be excluded.

It is considered whether the detection of PeV photons agrees with predictions from the standard Magnetohydrodynamic (MHD) paradigm of the Crab Nebula, which assumes that non-thermal emission from X-rays to multi–TeV gamma rays is produced by electrons accelerated at the termination of the pulsar wind. The Crab's multi-wavelength radiation is modeled within the idealized synchrotron-IC one-zone model, assuming a homogeneous spatial distribution of the magnetic field and electrons. Within the one-zone model, the IC gamma ray spectrum can be precisely calculated. Whereas the KM2A spectral points from 10 TeV to 1 PeV agree with the one-zone model within the statistical uncertainties, there are possible deviations from its predictions. Between 60 and 500 TeV, the two differ with a significance of 4σ, with the observational data having a steeper spectrum than the one-zone model predictions. The possible excess around 1 PeV indicates an opposite tendency — a hardening of the spectrum. A hardening of the electron spectrum is difficult to accommodate theoretically with plausible assumptions. Any detection of gamma rays well beyond 1 PeV would require a non-leptonic origin for the extra component of radiation, e.g., multi–PeV protons and atomic nuclei in the nebula. Because of the limited energy budget available for acceleration of protons, hadronic interactions cannot be responsible for the overall broad-band gamma ray luminosity. However, the contribution of hadronic interactions at PeV energies could be non-negligible.

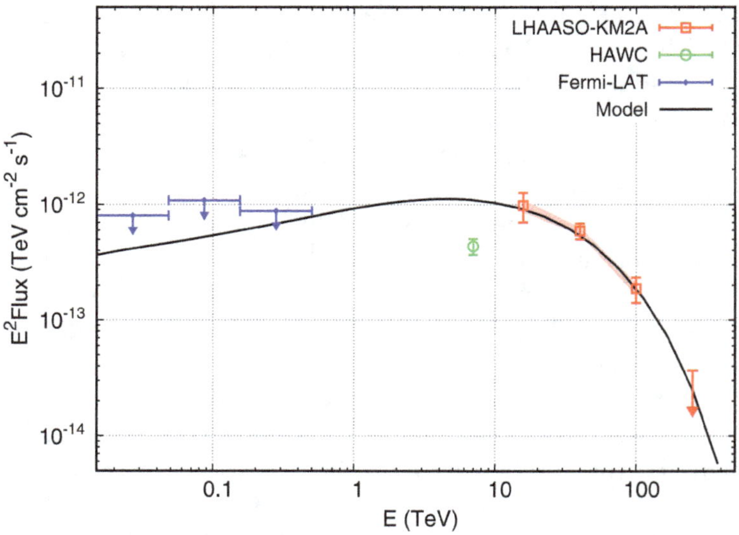

Fig. 3. The spectrum of LHAASO J0621+3755. The error bars represent statistical uncertainties, and the shaded band shows the systematic uncertainties. The HAWC measurement of 3HWC J0621+382 [23] and the Fermi-LAT 95% upper limits are also shown. The line shows the prediction based on the pulsar halo model (Aharonian *et al.*, 2021).

2.3. *LHAASO J0621+3755*

By using about ten months of data recorded with the half array of the LHAASO-KM2A, a gamma-ray extension source LHAASO J0621+3755 was discovered around the location of the middle-aged (207.8 kyr) pulsar PSR J0622+3749, confirming that high-energy particles can diffuse out of the medium of the pulsar halo (Aharonian *et al.*, 2021). No clear extended multiwavelength counterpart of the LHAASO source has been found from the radio to sub-TeV bands (Figure 3).

PSR J0622+3749 is a middle-aged pulsar, similar to those of Geminga and Monogem: No clear PWN emission is visible in the radio, infrared, optical, and X-ray bands. The lack of PWN emission associated with PSR J0622+3749 can be understood in terms of its larger distance. The multi-wavelength observational results are consistent with the scenario that the VHE gamma ray emission comes from the electrons and positrons diffusing out from the pulsar that then scatter the interstellar radiation field. Given the current data statistics, the energy dependence of the diffusion coefficient cannot be robustly determined yet. If LHAASO J0621+3755 is interpreted as the pulsar halo scenario, inferred for electrons with median energies of about 160 TeV, the diffusion coefficient is consistent with those obtained from the extended halos around Geminga and Monogem and much smaller than that derived from cosmic ray secondaries.

The wide band gamma ray SED of LHAASO J0621+3755 is modeled with an IC scattering model of e$\pm$ diffusing out from the pulsar or PWN. A two-zone diffusion model is assumed with a slow-diffusion region size of 50 pc, and a diffusion coefficient 300 times larger outside. The model flux is slightly higher than the flux measured by HAWC of 3HWC J0621+382 at median energy of about 7 TeV. The HAWC flux was derived assuming a disk extension of 0.5°, which is smaller than the extension measured by LHAASO (0.70°). It is expected that more dedicated analysis of HAWC and LHAASO-WCDA data will be very helpful in clarifying the spectral behavior below 10 TeV energies and in better constraining the model parameters. The discovery of LHAASO J0621+3755 enriches the class of pulsar halos and confirms that high-energy particles generally diffuse very slowly in the disturbed medium around pulsars, and may help to establish that the diffusion of cosmic rays in the Galaxy is in-homogeneous.

2.4. *LHAASO J2108+5157*

LHAASO J2108+5157, Another UHE gamma ray source was discovered by analyzing the first 11 months of data from the KM2A half-array (Cao *et al.*, 2021c) (Figure 4). This source is not significantly favored as an extended source with an angular extension smaller than the point-spread function of KM2A. A harder spectrum is demanded at lower energies considering the flux upper limit set by Fermi-LAT observations. The position of the gamma-ray emission is correlated with a giant molecular cloud, which favors a hadronic origin. In the absence of VHE gamma ray counterparts, this is the first gamma-ray source directly discovered in the UHE band. The source is correlated with the molecular cloud [MML2017]4607. Colliding between cosmic ray protons and the ambient gas may produce the observed radiation. Other possible scenarios, such as a PWN, can also be invoked to explain the gamma-ray emission. So far, no conclusion about the

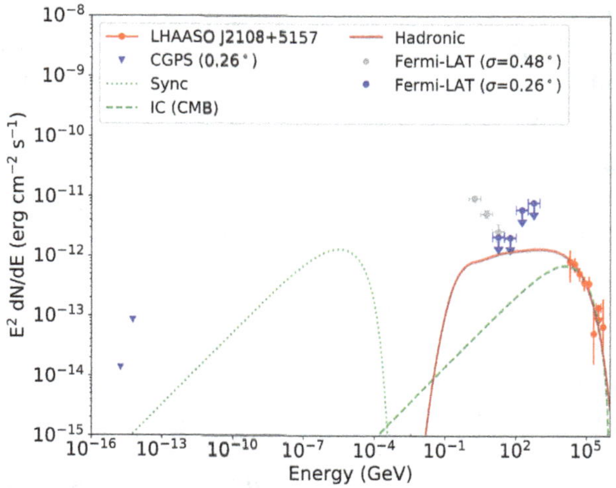

Fig. 4. Multi-wavelength SED of LHAASO J2108+5157 with hadronic and leptonic models. The red points and arrows are the LHAASO-KM2A observations. The blue triangles are the radio fluxes. The gray points and blue arrows are the Fermi-LAT spectral points and upper limits (Cao *et al.*, 2021c).

origin of its UHE emission can be achieved. Deeper multi-wavelength observations will help to cast new light on this intriguing UHE source. The forthcoming observations by LHAASO will reduce the uncertainties of the spectral points and allow us to extend the measured energy range.

2.5. *LHAASO J0341+5258*

LHAASO J0341+5258 was discovered after the 12 PeVatrons shown above (Cao *et al.*, 2021d) (Figure 5). The source is positionally coincident with the GeV gamma ray source 4FGL J0340.4+5302 observed by Fermi-LAT. The upper limit of the latter at 10 GeV demands a hard spectrum of LHAASO J0341+5258 (photon index <2) when extrapolating toward low energies. Together with the fast drop of the gamma ray fluxes above 50 TeV, in the phenomenological fitting in the leptonic and hadronic scenarios, the power-law requires the indices of the parent electrons and protons to be 2.5 and 1.5, with cutoffs of about 200 TeV and 190 TeV, respectively. The most likely realization of the leptonic scenario is the extended emission of a PWN and/or a pulsar halo. The lack of an energetic pulsar

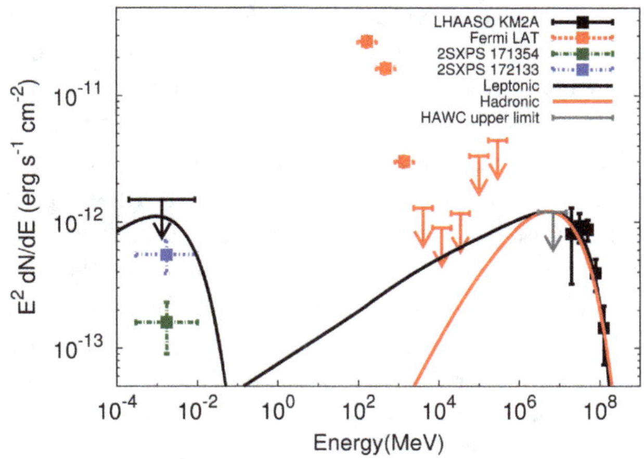

Fig. 5. The multiwavelength SEDs of LHAASO J0341+5258. The black squares are the LHAASO-KM2A observations, the red points and arrows are the Fermi-LAT spectral points and upper limits, and the green and blue points are the X-ray flux for 2SXPS 171354 and 2SXPS 172133 from the 2RXPS catalog, respectively. The black arrow labels the upper limit derived from Chandra observations. The gray arrow is the HAWC upper limit from the online interactive tools (see the text for details). Also shown is the phenomenological fitting in both leptonic (black curves) and hadronic scenarios (red curves) (Cao *et al.*, 2021d).

and a young SNR inside or in the vicinity of LHAASO J0341+5258 challenge, but do not exclude, both the leptonic and hadronic scenarios of gamma-ray production. Interestingly, such a pulsar could be the gamma-ray source 4FGL J0340.4+5302 with a characteristic spectrum below 1 GeV. The detection of pulsed radio emission from this source would support the IC origin of the UHE gamma-ray emission. The hadronic origin of the UHE emission can be interpreted as an "echo" from a molecular cloud(s) — the result of interactions of protons with dense gas regions in the proximity of an old (currently invisible) SNR.

3. Expectation

Regardless of the nature of objects associated with the sources detected by LHAASO, the photons far beyond 100 TeV prove the existence of Galactic PeVatrons. Moreover, it is likely that the Galaxy is full of these perfectly designed particle accelerators. The acceleration of protons to PeV energies requires extreme physical conditions, representing a challenge for any Galactic source population including PWNe, as suspected major contributors to Galactic cosmic rays. LHAASO Started complete operation in July 2021. In the following years, observations with LHAASO will reduce the flux detection threshold by at least an order of magnitude. It is expected that LHAASO will discover more pulsars and the other types of gamma ray sources, and fulfill a deep observation of them by Extending the spectra without an indication of a cutoff beyond several PeVs to reveal the sites of super-PeVatrons, the cosmic ray factories in the Galaxy responsible for the locally observed flux of cosmic rays well above the 'knee'.

Acknowledgements

This work was supported in China by the Science and Technology Department of Sichuan Province (grant number 2021YFSY0031) and the National Natural Science Foundation of China (NSFC) (grant number U2031103). The authors would like to thank all staff members who work at the LHAASO site above 4400 meters above sea level year-round to maintain the detector and keep the water recycling system, electricity power supply and other components of the experiment operating smoothly. We are grateful to Chengdu Management Committee of Tianfu New Area for the constant financial support for research with LHAASO data.

Bibliography

Abeysekara, A. U. *et al.* (2019). Measurement of the crab Nebula spectrum past 100 TeV with HAWC, *ApJ* **881**(2), 134.

Abeysekara, A. U. *et al.* (2020). Multiple galactic sources with emission above 56 TeV detected by HAWC, *Phys. Rev. Lett.* **124**(2), 021102.

Abeysekara, A. U. *et al.* (2017). Extended gamma-ray sources around pulsars constrain the origin of the positron flux at Earth, *Sci.* **358**, 911.

Acero, F. *et al.* (2015). Fermi large area telescope third source catalog, *ApJS* **218**, 23.

Acero, F. *et al.* (2013). Constraints on the galactic population of TeV pulsar wind nebulae using fermi large area telescope observations, *ApJ* **773**(1), 77.

Aguilar, M. *et al.* (2016). Precision measurement of the boron to carbon flux ratio in cosmic rays from 1.9 GV to 2.6 TV with the alpha magnetic spectrometer on the international space station, *Phys. Rev. Lett.* **117**, 231102.

Aharonian, F. *et al.* (2021). Extended very-high-energy gamma-ray emission surrounding PSR J0622 + 3749 observed by LHAASO-KM2A, *Phys. Rev. Lett.* **126**, 241103.

Amenomori, M. *et al.* (2019). First detection of photons with energy beyond 100 TeV from an astrophysical source, *Phys. Rev. Lett.* **123**(5), 051101.

Cao, Z. *et al.* (2021). Peta–electron volt gamma-ray emission from the crab Nebula, *Sci.* **373**, 425–430.

Cao, Z. *et al.* (2021). Ultrahigh-energy photons up to 1.4 petaelectronvolts from 12 gamma-ray galactic sources, *Nature* **594**, 33–36.

Cao, Z. *et al.* (2021). Discovery of the ultrahigh-energy gamma-ray source LHAASO J2108+5157, *ApJ* **919**, L22.

Cao, Z. *et al.* (2021). Discovery of a new gamma-ray source, LHAASO J0341+5258, with emission up to 200 TeV, *ApJL* **917**, L4.

Chen, Y. *et al.* (2022). Chapter 2: Galactic gamma-ray sources, *CPC* **46**(3), 030002.

Linden, T. *et al.* (2017). Using HAWC to discover invisible pulsars, *Phys. Rev. D* **96**(10), 103016.

Ohira, Y., Kisaka, S., and Yamazaki, R. (2018). Pulsar wind Nebulae inside supernova remnants as cosmic-ray PeVatrons, *MNRAS* **478**(1), 926–931.

Yuan, Q., Lin, S.-J., Fang, K., and Bi, X.-J. (2017). Propagation of cosmic rays in the AMS-02 era, *Phys. Rev. D* **95**, 083007.

Part 3

QTT: Qitai Radio Telescope; JRT: Jingdong 120 m Radio Telescope; eXTP: X-Ray Timing and Polarimetry Mission; GECAM: Electromagnetic Counterpart All-sky Monitor Mission; Hubbs: Hot Universe Baryon Surveyor Space-based X-ray Telescope

https://doi.org/10.1142/9789819800643_0014

Chapter 14

The Qitai Radio Telescope (QTT)

Na Wang
Xinjiang Astronomical Observatory,
Chinese Academy of Sciences Urumqi 830011, China
Key Laboratory of Radio Astronomy,
Chinese Academy of Sciences Nanjing, 210008, China
Xinjiang Key Laboratory of Radio Astrophysics
Urumqi 830011, China
na.wang@xao.ac.cn

The QiTai radio Telescope (QTT) is a fully steerable Gregorian-type telescope with the main reflector aperture 110 m in diameter. QTT adopts an umbrella support, homology-symmetric lightweight design, and the main reflector is active adjustable through actuators. QTT will operate from 150 MHz to 115 GHz frequency and ultra-wideband receivers and large field-of-view multi-beam receivers will be equipped. A multi-function signal processing system based on RFSoC and GPU processor will be developed. QTT will allow high sensitivity observations on pulsars, spectral line, continuum and Very Long Baseline Interferometer (VLBI) observing modes, and form a world-class observational platform in these areas. This chapter briefly introduces the engineering design and the science goals of QTT.

Keywords: Radio telescope, Astronomical observations, Receiver, RFI, Site testing

Contents

1. Introduction

It is always crucial to detect weak signals from the remote objects and to study their detailed structures. Hence, enhancing the sensitivity and resolution of radio telescopes are the key factors (Cheng, 2009). Building larger telescope is the major and effective approach. Large aperture single-dish antennas are demanding in structural design, but are relatively flexible in terms of upgrading the feeds and receivers. To further improve the sensitivity, broad-band low-noise cryogenic receiver and fast data sampling are crucial. In the last decade, the leading technology of the ultra-wideband (UWB) receiver has significantly increased the bandwidth to about several to ten frequency octaves, and the Phased Array Feed (PAF) is becoming the main option for multi-beam receiver through which large field of view can be obtained. In this chapter, we describe the design of the QiTai radio Telescope (QTT), a 110-meter-aperture steerable radio telescope, which is proposed to be built in Qitai County of Xinjiang Uygur Autonomous Region of China (Wang *et al.*, 2023). The planned observing frequencies for the QTT cover a wide range from 150 MHz to 115 GHz. To ensure high performance, especially for the higher frequency bands, the panels of the main reflector are adjustable to overcome the gravity deformation. As the world's largest fully steerable radio telescope, QTT will be able to cover more than 75% of the sky down to a declination of about −40 deg, including a large fraction of the plane of our Milky Way (Naddaf, 2023).

2. Technical Scheme

2.1. *The QTT structure*

The QTT antenna is in Gregorian parabolic style with active main surface to overcome the deformation caused by gravity. The antenna has an azimuth-elevation structure on a wheel-and-track mount. Figure 1 shows the QTT structure, including the backup structure (BUS) of the 110-m main reflector, the 12-m sub-reflector and its quadruped supporting system. The connection between the reflector

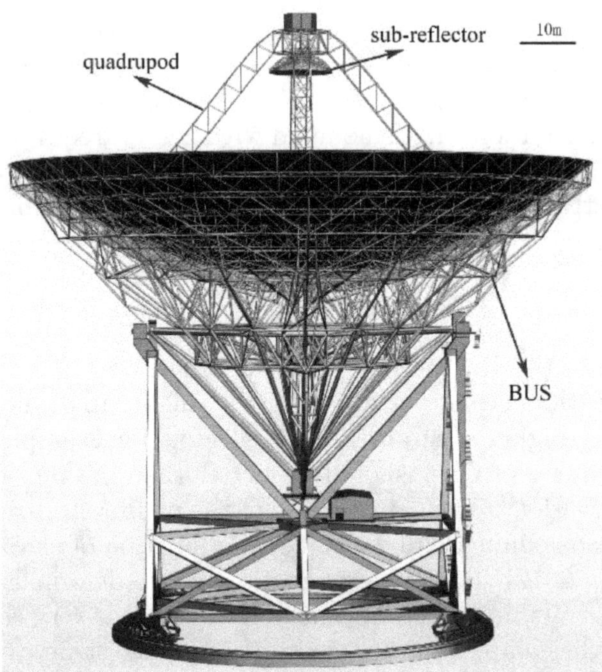

Fig. 1. QTT antenna design. The main reflector is actively adjustable through actuators.

back frame and antenna supporting frame is similar to the umbrella-like structure of the Effelsberg telescope in Germany (Wielebinski *et al.*, 2011), in which the BUS adopts a rotational symmetric design which contains concentric radial trussed girders supported by the umbrella rods. The sub-reflector is connected to the truss system via the head of quadruped supporting legs with length about 60 m. In order to reduce dead load and the thermal deformation, the sub-reflector will be fabricated using the carbon fiber-reinforced plastic.

The optical layout is a compromise between the main reflector focal ratio and the diameter of the sub-reflector, which are taken to be 0.33 and 12 m, respectively. In order to realize multi-beam observations at both primary and secondary foci, standard paraboloids are designed for both primary and secondary reflectors. And for single-beam low frequency observations, the antenna efficiency is expected to be improved through active shaping of the main reflector.

For the primary focus mode, a prime focus platform (PFP) is designed to carry the receivers. In operation mode, the PFP is pulled up to the horizontal position and the observing receiver is moved to the prime focus, while in stow mode, PFP is retracted to align with one of the quadruped legs (see Figure 2). For the Gregorian focus mode, a 5-m rotation turret mounted with higher frequency receivers is installed in the receiver cabin (see Figure 3).

The aperture efficiency of the telescope can be described by Ruth formula:

$$\eta = e^{-(4\pi\sigma/\lambda)^2}, \tag{1}$$

in which σ is the RMS error of the main surface, λ is the observing wavelength. The shorter wavelength or higher frequency observations require higher accuracy of the reflector, for example the main surface accuracy of QTT should be better than 0.2 mm (RMS) for 3 mm observations. The deformation of the main reflector caused by gravity can be eliminated by a correction function of elevation angle (Süss *et al.*, 2012). Real time correction of the deformation caused by environmental influences such as sun exposure is very challenging, because of the complex service environment, huge structure and high precision requirement. A high precision and rapid photogrammetry system (Konyakhin *et al.*, 2008; Li *et al.*, 2020a) based on image sensor is designed for QTT.

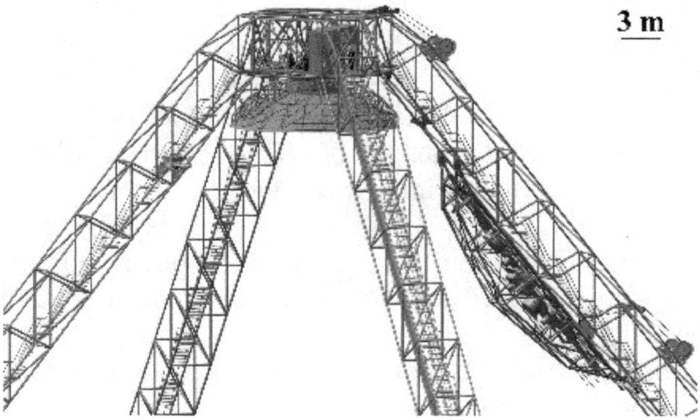

(a)

(b)

Fig. 2. PFP in stow mode (a) and in operation mode (b).

Fig. 3. The feed switching mechanism at the Gregorian focus, positions for multiple receivers are reserved.

The position of the sub-reflector is adjustable through a heteroid Stewart-platform system. This is a modified 6-degree-of-freedom (6-DOF) parallel hexapod mechanism that provides static support to the sub-reflector as well as correcting its position and orientation through changing the length of each support arm. This allows fine adjustment to further improve the optical performance of the antenna (Kazezkhan *et al.*, 2020) through better alignment of the main reflector and sub-reflector. The position of the sub-reflector is expected to be measured through an optical system with multiple laser beams.

The azimuth track is connected to the foundation by anchor bolts forming a high precision and reliable platform to support the weight of the 6000-ton antenna and in the meantime resisting wind and earthquake loads. The concrete foundation consists of 59 rock-socketed piles with diameter of 2 m and average depth of 23.3 m, and the unevenness settlement is required to be less than 2 mm within 5 years. The wheel track of QTT is designed to be a continuous-welded circular rail with a 76 m in diameter. The base material of the track is high strength alloy steel and the track flatness is within 0.3 mm.

2.2. *Servo control and accurate pointing*

Astronomical observations require high-precision pointing of the radio telescope. The pointing accuracy is proportional to $1.22\lambda/D$, where λ is the observing wavelength, and D is the aperture of the antenna. For QTT operating at 3 mm, ideally it should be better than 1.5 arcsecond. The meet such restrictive requirements the antenna servo system should have repeatable higher tracking accuracy, which usually is 2/3 of the pointing accuracy. External disturbances, such as wind load and structural flexibility are the main factors that contribute to the tracking error of the servo system (Karcher, 2021). In addition, the high inertia and low resonant frequency of the antenna increase the complexity of servo control (Gawronski, 2007; Ranka *et al.*, 2016). Hence, the servo control must be robust on anti-disturbance and vibration-suppression performances. Apart from the above issues, the antenna is also required to have good dynamic performance with slight overshoot and shorter settling time.[1]

Traditionally, the antenna servo control uses model-free proportional integral-derivative (PID) algorithm as the position-loop and velocity-loop controllers. Further compensation can be obtained through Disturbance Observer (DO) or Extended State Observer (ESO) to reduce the positioning error. For example, the PID controller combined with ESO is also known as Active Disturbance Rejection Control (ADRC) (Han, 2009), which is probably an approach aiming to enhance the QTT tracking performance operated in disturbance condition. Another important type of advanced control algorithm is model-based, for example H_∞ (Huang *et al.*, 2022) and Quantitative Feedback Theory (QFT) (Garcaia-Sanz, 2017), and their application on QTT are under assessment.

2.3. *Receiver arrangement*

QTT will be equipped with broadband receivers at 40 cm and 1.3 cm, ultra wide-band receivers (UWB) at 15 cm and 5 cm, phased array feeds (PAF) at 20 cm and a multi-beam receiver at 3 mm. All receivers are low-noise cryogenic and linearly polarized, and their specifications are shown in Table 1. The receivers mounted at the primary focus will be directly digitized at radio frequency (RF) without down-conversion, whereas receivers mounted at the Gregorian focus will be down-converted to multiple intermediate frequency (IF) bands to achieve full-band acquisition.

For the 15 cm UWB for instance, to achieve the bandwidth spanning six frequency octaves within a limited size, the feed adopts a quad-ridged horn (Dunning *et al.*, 2015). As shown in Figure 4, the equalization characteristics of the entire working band are achieved by adding ripple grooves and medium materials to the top and middle of the horn. Probes are located at four ports at the bottom of the

[1]https://safe.nrao.edu/wiki/bin/view/GB/PTCS/ServoImprovementsHome, "GBT High Frequency Observing and the Precision Telescope Control System Project".

Table 1. Specifications of the receivers.

Type	Band (cm)	RF frequency (GHz)	Focus	T_{sys} (K)
Single Beam	40	0.27–1.8	Primary	20
Single Beam	15	0.7–4	Primary	16
Single Beam	5	4–16	Gregory	18
Single Beam	1.3	16–30	Gregory	20
PAF	20	0.7–1.8	Primary	20
Multi-beam	0.3	67–115	Gregory	60

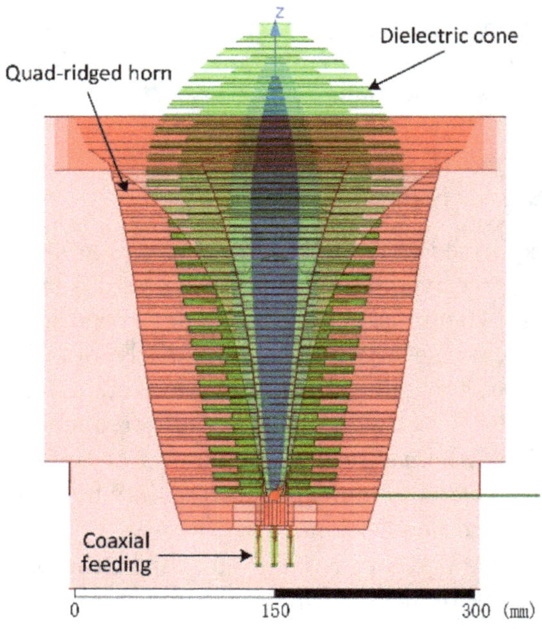

Fig. 4. Ultra-wideband quad-ridged horn feed. The part in red is the quad-ridged horn, and in green is the dielectric cone.

horn, and the dual-linear polarization is divided into three segments to output at the same time.

The cryogenic front-end of the PAF receiver consists of 96 Vivaldi antenna units which are connected to low noise amplifiers. The 192 signals of the dual linear polarization outputs are sampled at the receiver cabin by a signal acquisition and pre-processing unit with a 12-bit sampling rate of 2.048 giga-samples per second (GSPS). The digitized signals are then transmitted through the 100 Gigabit Ethernet (GbE) optical fiber for beam-forming and post-processing. Around 36 beams will be formed for a maximum instantaneous bandwidth of 500 MHz selected arbitrarily from the RF passband. The 3 mm band multi-beam receiver contains 36 beams, covering a frequency range of 67–115 GHz, the system noise temperature is less than 60 K, and the signal processing bandwidth reaches 4 GHz. The feed array has a 6×6 layout and adopts an array network design to improve machining and assembly accuracy. The receiver reduces the system noise temperature through the overall cooling of the feed array and the low-noise amplifiers. The signal processing system realizes multi-beam ultra-high-speed real-time processing through FPGA+GPU heterogeneous distributed cluster.

2.4. *The backend and data archive*

To obtain high-fidelity observational signals and meet different scientific goals, a multi-function digital backend is designed with cutting-edge technologies, from which RF signals are sampled directly at the receiver front-end and processed in flexible modes. This design will effectively reduce the fluctuations of signal gain and phase caused by environmental changes during transmission. The system is designed based on RFSoC (RF System on Chip), CPU and GPU heterogeneous architecture. The signal acquisition and pre-processing unit adopts high-performance, low-power consumption RFSoC technology with integrated high-speed ADCs (Kocz *et al.*, 2014; Pei *et al.*, 2022). The maximum sampling rate is 4 GSPS with 12-bit precision. The scale of CPU and GPU computing cores is adjustable through multiple high performance computer (HPC) nodes according to the bandwidth and calculation complexity, making the system highly flexible and expandable. The design of QTT data management system performs online archiving from the observational data acquisition system, remote backup of original data and the metadata extraction for data release. The database will be handled according to the latest International Virtual Observatory Alliance (IVOA) standards.

Users will be allowed to access the online data retrieval platform through web browsers, standard virtual observatory (VO) tools, various scripts, etc.

2.5. *Software control center*

The software control center of QTT is designed to adopt a modular distributed microservice framework (Liu *et al.*, 2019c) Each module provides an independent service interface through Remote Procedure Calls (RPC) and message queues in microservice mode. The user interface will be developed through web technology. Users can observe, control or view the telescope status through an arbitrary web browser remotely. The main functions of the software control center include antenna driving, main reflector adjustment, feed switching, receiver and backend configuration and system health monitoring. The environmental conditions are collected via meteorological and electromagnetic monitoring systems, which provide criteria for safe operation and dynamic scheduling. The system will rapidly measure the telescope efficiency and gain of each band and conduct the telescope pointing correction. The system monitoring module mainly provide updated information of the vulnerable hardware or fatigue-prone devices, to prompt early warning and reduce the impact of device failures and potential safety hazards.

2.6. *RFI mitigation*

RFI is usually caused by mobile communications, radio and digital television, satellite communications, and radio navigation systems, and mainly appears in the low frequency band, i.e., below 3 GHz. Large telescopes are very sensitive to the signals radiated by various electronic components. The radio telescope itself is an integration of a large number of electronic components, and their radiation coupled with outside RFI makes the electromagnetic environment of the telescope very complicated. Hence the electromagnetic compatibility of the telescope has important impact to the performance of the system. In order to limit the influence of RFI to a minimum, electromagnetic compatibility (EMC) design according to the protection criteria for radio astronomical measurement ITU-R RA.769 recommendation is formulated by setting thresholds of allowable radiation levels for equipment located on QTT sites (Liu *et al.*, 2019b).

Multi-layer shielding is a further approach to suppress RFI generated by various electronics at the site. In particular, most of the electronics will be sheltered in a high performance shielding chamber in the basement under the telescope. Otherwise, the electronic components distributed in the laboratory are arranged in shielded cabinets or boxes. The buildings at the site are shielded to reduce the RFI generated by lamps, switches etc. Earlier experiments show the shielding efficiency of this scheme is about 30 dB at 0.1–1 GHz (Liu *et al.*, 2019a). In addition, a dedicated reference antenna will be installed for monitoring and further mitigating the RFI during the data sampling process in future observations. The digital backend system also provides the possibility of adaptive RFI filtering.

3. Science Goals

The science goals of QTT include nanoHertz gravitational wave detection through Pulsar Timing Array (PTA) (Hobbs *et al.*, 2009), monitoring of pulsars. The UWB receiver with bandwidth from 0.7 to 4 GHz and digital signal processor with high time and frequency resolution will be a powerful tool. Identifying and monitoring Fast Radio Bursts (FRBs) and learn their origin through polarization observations is of high scientific value. A pulsar clock can be established based on pulsars that make up a PTA. This independent time system can be further used to detect variations in atomic time and to study the rotation of the Earth and small changes in the Solar System ephemeris. Pulsars with both radio and X-ray radiation are suitable candidates for deep space autonomous navigation. Regular timing observations by QTT will contribute to monitoring their rotation status and to updating the Pulsar Catalogue.

Searching for known and unknown molecular line emission in star-forming regions, asymptotic giant branch (AGB) star envelopes, and planetary nebulae from the 1 cm band to the 3 mm band, is of great significance for understanding the origin of interstellar molecules and the chemical composition. The CO (1–0) line in galaxies at redshifts higher than 1.4, CS (2–1), HCN (1–0), and HCO+ (1–0) lines in galaxies at redshifts higher than 0.9 are detectable at 50 GHz (7 mm) or lower bands. Establishment of an association between a supernova shock wave and a molecular cloud is helpful for the study of cosmic ray acceleration (Vink, 2012). It is worth investigating the association between gamma rays and molecular clouds associated with supernova remnants, through the spectral imaging of OH masers at 1720 MHz and other spectral lines such as SiO and CS at 43.42 and 48.99 GHz.

The VLBI observation with sub-milliarcsecond resolution makes it possible to obtain detailed images of the relativistic jet near black holes. In particular, through the VLBI polarization measurements, we are able to study the physical mechanisms of the jet formation, acceleration and collimation processes, and reveal the density of electrons around the source and the magnetic field configuration. Higher resolution and more sensitive observations are still the key approach to reveal the interaction and coevolution of the central black hole and its host galaxy. The gravitational lensing effect of the foreground dark matter halo acts on distant galaxies, providing a unique and powerful way to detect the distribution of dark matter. High resolution and high sensitivity VLBI observations of gravitational lensing systems can provide useful clues to the evolution of galaxies, the galaxy-dark matter relation and the distribution of dark matter in the universe (Koopmans, 2005). Establishing accurate connection between the radio and optical reference frames is important for astrometry and space positioning. Especially accurate link between International Celestial Reference Frame (ICRF) and Gaia Celestial Reference Frame (GCRF) is essential for fundamental astrometry in Gaia era. VLBI astrometry of optical stars with radio emissions will significantly contribute to this. Large telescopes are much needed for accurate VLBI measurements of very weak radio stars.

4. The Site

The QTT site is located in Eastern Tianshan Mountains, 46 km south of Qitai County, Hui Autonomous Prefecture of Changji, Xinjiang, with an elevation of about 1800 m. The statistical analysis of the annual data reveals that the monthly wind speed distribution of the site in autumn and winter is weaker than that in spring and summer

(Li *et al.*, 2020b). The wind speed proportion of lower than 2 m/s (wind scale 2) is around 68%, lower than 4 m/s (wind scale 3) is 91%, lower than 6 m/s (wind scale 4) is 97%. For the extreme case, the proportion of wind greater than 17 m/s (wind scale 8) is less than 0.1%. As for operation experience of the GBT, the QTT site is suitable for ensuring the safe operation and pointing accuracy of the antenna in terms of wind load influence.

The atmosphere is not completely transparent to electromagnetic waves, especially at millimeter and sub-millimeter wavelengths. One important limiting factor for observation is the water vapor absorption caused by the atmosphere and the extra noise induced by atmospheric thermal emission (Kendrew *et al.*, 2010). Our analysis show that the PWV content of QTT site is 6.64 mm in average, with a mid-value of 5.11 mm. The largest value appears in July, about 14.44 mm, and the minimum is in January, about 1.95 mm. About 80% of the time the PWV level is below 10 mm. Taking 90% as the threshold of atmosphere transmittance for 3 mm electromagnetic waves, the PWV levels are required to be lower than 5 mm in winter and 7.2 mm in summer, resulting in about 180 observable days at QTT site (Li *et al.*, 2020b).

The QTT site has been especially selected to be as free of external terrestrially generated RFI as possible over a large radio frequency range. The QTT site is in a basin surrounded by mountains, and the terrain provides natural shielding which blocks the external radio interference to a great extent. The site will be maintained as such and will be established as a radio quiet zone (RQZ). The RQZ is an area with a radius of 30 km, which consists of the core zone, restricted zone and coordination zone from inside to outside with reducing control levels.

Figure 5 shows the geographical distribution of the major radio telescopes in the European VLBI Network (ENV), East-Asia VLBI Network (EAVN) and Chinese VLBI Network (CVN). Located in the blank area of the VLBI network in the center of Eurasia, the QTT will contribute significantly to the EVN, improving the uv-coverage while increasing the system sensitivity by 30% to 70% in multiple frequency bands. The QTT will also significantly improve the overall sensitivity of the CVN and EAVN. With the QTT joining the VLBI network, breakthroughs are expected in black hole and galaxy studies.

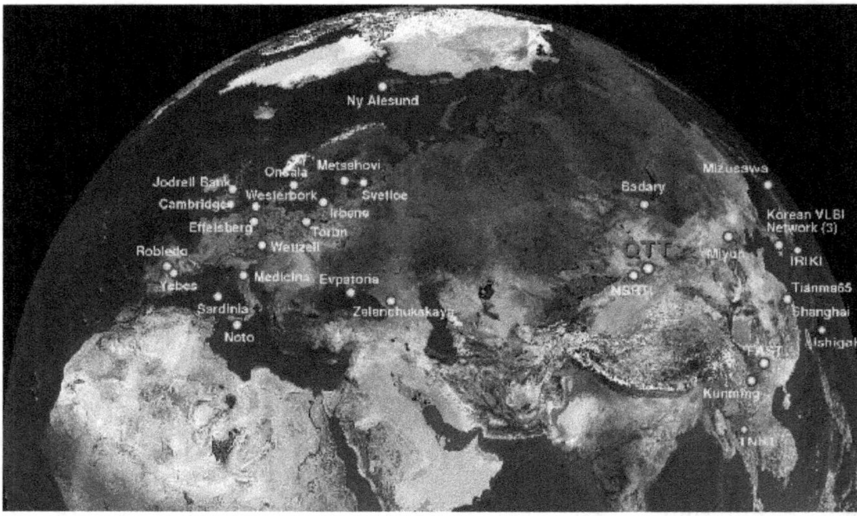

Fig. 5. The geographical distribution of the major radio telescopes in the world.

5. Conclusion

Currently, there are only two 100-meter fully steerable radio telescopes operating in the world, the Effelsberg telescope (Wielebinski *et al.*, 2011) and GBT (Srikanth *et al.*, 1999). As introduced in Section 4, the QTT adopts a symmetric structure via umbrella-like supporting frame. The advantage of such design is the weight of the telescope can be significantly reduced, while the disadvantage is the sub-reflector inevitably blocks the main reflector, results in a reduction in telescope efficiency. The GBT gave priority to avoiding aperture blockage. Hence, an asymmetric structure was designed in which the sub-reflector is off-axis arranged relative to the paraboloid main reflector. The trade-off of such design is the increasing complexity of the structure of the supporting truss and frame, and the weight of the telescope is hard to control. Other than that, QTT and GBT can both operate at frequency bands up to 3 mm, and the high precision of the active main reflector are adjustable through actuators.

At 3 mm wavelength the ALMA Wootten and Thompson (2009) is the most powerful interferometer with an unprecedented resolution of 0.04 arcsecond at the longest baseline, and roughly equivalent sensitivity to a hundred-meter single dish. The larger beam of the

QTT will enable us to conduct extensive and fast pilot surveys with limited resolution, which can guide follow-up high-resolution observations with ALMA. Furthermore, while the interferometer is good at zooming into the compact emission regions, it also loses information about the large-scale diffuse emission. The QTT will be able to fill in this gap. Combining observations with the interferometer with those of the QTT will allow more complete determination of the appearance of the sky from small to large scales.

The entire radio frequency band covers from meter to submillimeter wavelengths. Operating at 70 MHz 3GHz, FAST has extremely high sensitivity through forming a 300-m aperture instantaneous paraboloid in the observation direction by active control (Nan *et al.*, 2011). The Commensal Radio Astronomy FAST Survey (CRAFTS) (Cruces *et al.*, 2021) and the Galactic Plane Pulsar Snapshot (GPPS) (Han *et al.*, 2021) survey have now found more than 600 pulsars, and its powerful detection capabilities could lead to the discovery of rare objects such as black holes and more unknown phenomena in the future (Cui *et al.*, 2021). The follow-up observation and verification of the new objects very much needs a joint effort with QTT, the largest steerable telescope with tracking ability (Hobbs *et al.*, 2019). As the QTT has much wider frequency coverage, coordinated observations with FAST can be made to investigate the frequency-related physics properties.

Real-time processing and storage of massive data quantities could be challenging in the future. Taking pulsar observations as an example, in order to correct for interstellar medium propagation effects, pulsar observation data require extremely high time and frequency resolution and sampling intervals of tens of microseconds. Some special high-precision baseband observation systems require Nyquist rate sampling in the full band of several GHz, which requires extremely high rates of online data processing. The electromagnetic wave of celestial radiation is extremely weak, so it is crucial to improve the sensitivity and resolution of radio telescopes to discover fainter celestial objects and study finer structures. Usually two methods are applied for increasing the antenna receiving area to improve the detection sensitivity. The first is to build a single aperture large radio telescope, the second is to break up the whole, to build an array antenna with a diameter 10 to 20 m. Both approaches have advantages and also limitations. The former has high requirements for antenna structure, but it is relatively simple and flexible in feeds and receiver mounting, which is conducive to system update and timely adoption of the latest receiving technology. For the latter, the challenge on the structural of large aperture antennas are broken down into small antennas, the effective receiving area is not limited in principle, the signal accumulation is transformed into an electronic technology issue, and the complexity of data reception and processing is greatly increased. The field of view (FoV) of the array telescope is usually much smaller compared to the single telescope, hence can accurately measure the position of celestial objects, while the single telescope can provide relatively large FoV required for sky survey program.

Acknowledgments

This work was supported by the National Key Research and Development Program of China (Grant Nos. 2021YFC2203501, 2021YFC2203502, 2021YFC2203503 and 2021YFC2203600), the National Natural Science Foundation of China (Grant Nos. 12173077, 11873082, 11803080 and 12003062), the Scientific Instrument Developing Project of the Chinese Academy of Sciences (Grant No. PTYQ2022YZZD01).

Bibliography

Cheng, J. Q. (2009). *The Principles of Astronomical Telescope Design*, New York: Springer. I ASSL. **360**, 1–634.

Cruces, M., Champion, D. J., Li, D. *et al.* (2021). FAST early pulsar discoveries: Effelsberg follow-up, *MNRAS* **508**(1), 300–314.

Cui, X. H., Zhang, C. M., Wang, S. Q. *et al.* (2021). Fast radio bursts: Do repeaters and non-repeaters originate in statistically similar ensembles? *MNRAS* **500**, 3275–3280.

Dunning, A., Bowen, M., Bourne, M. *et al.* (2015). An ultra-wideband dielectrically loaded quad-ridged feed horn for radio astronomy, *2015 IEEE-APS Topical Conference on Antennas and Propagation in Wireless Communications (APWC)*, 787–790.

Garcaia-Sanz, M. (2017). *Robust Control Engineering: Practical QFT Solutions*, CRC Press, London.

Gawronski, W. (2007). Control and pointing challenges of large antennas and telescopes, *Control Syst. Technol. IEEE Trans.* **15**(2), 276–289.

Han, J. Q. (2009). From PID to active disturbance rejection control, *IEEE Trans. Ind. Electron.* **56**(3), 900–906.

Han, J. L., Wang, C., Wang, P. F. *et al.* (2021). The FAST galactic plane pulsar snapshot survey: I. Project design and pulsar discoveries, *Res. Astron. Astrophys.* **21**(5), 107.

Hobbs, G. B., Bailes, M., Bhat, N. D. R. *et al.* (2009). Gravitational-wave detection using pulsars: Status of the Parkes pulsar timing array project, *Publ. Astron. Soc. Aust.* **26**(2), 103–109.

Hobbs, G. B., Dai, S., Manchester, R. N. *et al.* (2019). The role of FAST in pulsar timing arrays, *Res. Astron. Astrophys.* **19**(2), 20.

Huang, G. Q., Wen, C., Lu, J. Y. *et al.* (2022). Application of fixed-order H_∞ controllers in antenna servo system, *Radio Eng.* **52**(5), 864–871.

Karcher, H. J. (2021). From main axes control to flexible body control of large telescope structures, *Adv. Control Appl.* **3**(4), e88.

Kazezkhan G., Xiang, B. B., Wang, N., and Yusup, A. (2020). Dynamic modeling of the Stewart platform for the Nanshan Radio Telescope, *Adv. Mech. Eng.* **12**(7). 1–10.

Kendrew, S., Jolissaint, L., Brandl, B. *et al.* (2010). Mid-infrared astronomy with the E-ELT: Performance of METIS, *Proc. SPIE, Ground-based and Airborne Instrumentation for Astronomy III* **7735**, 77355F.

Kocz, J., Greenhill, L. J., Barsdell, B. R. *et al.* (2014). A scalable hybrid FPGA/GPU FX correlator, *J. Astron. Telesc. Instrum. Syst.* **3**(1), 1450002-330.

Konyakhin, I., Artemenko, Y., and Timofeev, A. (2008). Control of the deformation for the millimeter wave range radiotelescope mirrors, **7133**, 71333R.

Koopmans, L. V. E. (2005). Gravitational imaging of cold dark matter substructures, *MNRAS* **363**(4), 1136–1144.

Li, R. P., Qin, X. J., Konyakhin, I. *et al.* (2020a). CMOS multi-matrix optoelectronic system for high speed measurement of main mirror deformation large diameter radio telescope, *IEEE Access* **8**, 51821–51829.

Li, M. S., Li, R., Wang, N., and Zheng, X.W. (2020b). Research and application of reanalysis data for radio astronomical site testing, *Res. Astron. Astrophys.* **20**(12), 200.

Liu, Q., Liu, Y., Chen, M. Z. *et al.* (2019a). Shielding engineering progress for the QTT buildings, in *2019 RFI Workshop — Coexisting with Radio Frequency Interference (RFI)*, 1–4.

Liu, Q., Wang, N., Liu, Y., and Chen, M. Z. (2019b). An EMC control method for large-diameter radio telescope, *Sci. Sin.-Phys. Mech.*

Astron. **49**(9), 099511.

Liu, Z. Y., Li, J., Wang, N. *et al.* (2019c). The architecture design of astronomical observation and system monitoring and control software for large radio telescope, *Sci. Sin.-Phys. Mech. Astron.* **49**(9), 099509.

Naddaf, M. (2023). The science events to watch for in 2023, *Nature* **613**(7942), 11–12.

Nan, R., Li, D., Jin, C. *et al.* (2011). The five-hundred aperture spherical radio telescope (fast) project, *Int. J. Mod. Phys. D J.* **20**(6), 989–1024.

Pei, X., Wang, N., Werthimer, D. *et al.* (2022). Design of RFSoC-based digital phased array feed (PAF) and hybrid architecture beamforming system, *Astron. Astrophys. Rev.* **22**(4), 045016.

Ranka, T. M., Garcia-Sanz, M., Symmes, A. H. *et al.* (2016). Dynamic analysis of the Green Bank Telescope structure and servo system, *J. Astron. Telesc. Instrum. Syst.* **2**, 014001.

Srikanth, S., Norrod, R., King, L., and Parker, D. (1999). An overview of the Green Bank Telescope, In *IEEE Antennas and Propagation Society International Symposium. 1999 Digest. Held in Conjunction with: USNC/URSI National Radio Science Meeting (Cat. No. 99CH37010), Orlando, FL, USA, 1999*, **3**, 1548–1551.

Süss, M., Koch, D., and Paluszek, H. (2012). The Sardinia Radio Telescope (SRT) optical alignment, In *Proc. SPIE 8444, Ground-based and Airborne Telescopes IV*, 84442G.

Vink, J. (2012). Supernova remnants: The X-ray perspective, *Astron. Astrophys. Rev.* **20**, 49–120.

Wang, N., Xu, Q., Ma, J. *et al.* (2023). The Qitai radio telescope, *Sci. China Phys. Mech. Astron.* **66**(8), 289512.

Wielebinski, R., Junkes, N., and Grahl, B. H. (2011). The Effelsberg 100-m Radio telescope: Construction and forty years of radio astronomy, *J. Astron. Hist. Herit.* **14**(1), 3–21.

Wootten, A. and Thompson, A. R. (2009). The Atacama large millimeter/submillimeter array, *IEEE Proc.* **97**(8), 1463–1471.

https://doi.org/10.1142/9789819800643_0015

Chapter 15

Hot Universe Baryon Surveyor (HUBS)

Zhao-Sheng Li
Key Laboratory of Stars and Interstellar Medium,
Xiangtan University, Xiangtan 411105,
Hunan, P.R. China
lizhaosheng@xtu.edu.cn

Hot Universe Baryon Surveyor (HUBS) is proposed in China as a major X-ray mission for the next decade. The primary scientific objective is to conduct a census of baryons in the warm hot circum galactic and intergalactic medium and thus to directly address the issue of "missing baryons" in the local universe. The results are expected to impact our understanding of galaxy formation. Secondary objectives are many, including hot interstellar medium, diffuse X-ray background, supernova remnants, as well as charge exchange processes in the solar system. The aspects of the project including scientific motivation, methodology, preliminary design, timetable, waveband of interest of HUBS will be introduced in a comprehensive way, in particular, to demonstrate the impact of this project on pulsar astronomy, comprehensively.

Keywords: Neutron star, Pulsar, Accretion disk

Contents

1. Introduction

The *Hot Universe Baryon Surveyor* (HUBS) mission (Cui *et al.*, 2020) will focus on the major scientific problem of "missing baryons" in the universe. HUBS will develop and launch a large-field-of-view, high-efficiency, high-resolution X-ray imaging satellite to detect the large-scale filament of the universe by spectral observations, and to discover the spatial distribution and physical and chemical properties of the "missing baryons" in the universe. It will fill the gaps in the domestic field of advanced technologies such as superconducting transition-edge sensor microcalorimeters, very low temperature refrigeration, and large field of view X-ray focusing. A preliminary design for the HUBS satellite has been formed with the following main indicators/parameters (Bregman *et al.*, 2023): (1) the band covers 0.1–2 keV, in which hot gas radiation spectra are expected to converge in intergalactic medium and circumgalactic medium; (2) the field of view is 1 square degree; (3) the angular resolution is better than 1 arcminute (half power angular diameter); (4) the effective area is 500 cm^s @ 1keV (field of view averaged); (5) the spectral resolution is 2 eV for main array and 0.6 eV for center sub-array.

2. Neutron Stars

As the fundamental units of galaxies, stars play a key role in the recycling of matter. X-ray observations associated with stars not only deepen understanding of a wealth of astronomical phenomena, but also contribute to the understanding of extreme physical processes.

NS formed by supernova explosions are the most compact objects in the universe. The equation of state for cold and dense matter is still inconclusive with respect to the understanding of the non-perturbative nature of the fundamental interactions between quarks (Lattimer and Prakash, 2007). The equation of state of NS and strangeon star predict different mass-radius relations (Lai and Xu, 2017; Steiner *et al.*, 2013). The accurate measurements of NS mass and radius could put stringent constraints on the equation of state (Özel and Freire, 2016). Mass measurements of massive NS, $M > 2M_{\odot}$, have been already excluded a number of equations of state that predict the maximum mass smaller than $2M_{\odot}$ (Demorest *et al.*, 2010). Although a number of masses of NSs in compact binaries have been measured from radio observations with high precision, radius measurements are much difficult to achieve with a comparable precision.

Usually, NS mass and radius can be measured from type I X-ray bursts occurred in NS low-mass X-ray binaries (LMXBs), pulse profile modelling of X-ray pulsars and so on. NS LMXBs are composed of NS and a main sequence donor orbiting each other. The masses of the NS and its companion can be determined by kinetic methods, with the orbital motion of the star in the NS LMXBs causing its spectral lines to undergo periodic redshifts and blue shifts due to the Doppler effect. The optical and/or near-infrared (NIR) spectroscopic observations can determine the mass function (stellar radial velocity profile) of the star. Over the past decades, optical/NIR observations have shown that this method has the potential to constrain the compact object mass of LMXBs. However, there are also some shortcomings, mainly in that (1) this method requires a relatively bright optical/infrared flux of stars in LMXBs with strong absorption or emission lines, which is difficult with current optical/NIR telescopes for optically faint LMXBs; (2) this method can only measure the stellar mass function, but not the dense stars. If we can use an X-ray telescope with high energy resolution and large effective area, we will be able to measure the velocity profile of dense stars and obtain the mass function of compact stars, which can be combined with optical/NIR observations to measure the binary mass ratio. The masses of compact stars can be constrained more precisely if the stellar masses can be determined from optical observations. This has important implications for the mass spectrum of black holes and neutron stars, and for the solution of the "mass gap" problem. Even if the stellar masses cannot be determined, the mass ratio of the two objects, combined with other measurements, can be used to constrain the binary masses as well.

We carry out a series of simulation by using the ancillary response file, hubs_cc_s_v20181102.arf, and the redistribution matrix file, hubs_cc_s_v20181109.rmf. The data are produced by using the command fakeit in Xspec 12.13.0c.

2.1. *The eclipsing low-mass X-ray binaries*

Eclipsing low-mass X-ray binaries are high inclination systems with the inclination angle larger than 75°. Zhang *et al.* (2012) suggested that absorption lines from accretion disk winds are redshifted or blueshifted due to the Doppler effect of orbital motion. These spectral features are produced in the vicinity of compact objects and trace their motion, which can constrain the mass of compact objects in LMXBs. This approach was subsequently applied to the eclipsing NS LMXB MXB 1659–298, but the uncertainties of the measured radial velocities are large because the energy resolution of Chandra and NuSTAR is not high enough (Ponti *et al.*, 2018). In general, the maximum radial velocity of compact objects in X-ray binaries is in the order of 100 km/s, which causes a spectral shift of order 10^{-4}. Therefore a high energy resolution of the detector is required. The energy resolution of HUBS has the possibility to measure the Doppler shift of the spectral lines with high precision.

We perform the HUBS simulations for MXB 1659–29. This source is an eclipsing NS LMXB, which has an orbit period of 7.1 hr. Based on the results from Ponti *et al.* (2018), we adopt the continuum spectrum which has an absorbed flux of 5×10^{-10} erg cm^{-2} s^{-1} in the energy range of 0.1–2 keV and a blackbody temperature of 3 keV. The Mg XIII Lyα absorption line centering at 1.472 keV, which is covered by HUBS, is also added. The exposures of 1 ks, 10 ks and 100 ks are simulated for each observation. For the whole orbit period, we carry out 12 observations with uniform separations. In order to measure the radial velocities at different orbital phases, we focus on the Mg XIII Lyα line and determine its shift. The left panel in Figure 1 shows the spectral lines at the orbital phases of 0 and 0.75, in blue and orange, respectively. The vertical dotted line is the central energy of the spectral line at that phase. The right panel in Figure 1 presents the simulation data with errors and the solid lines for the

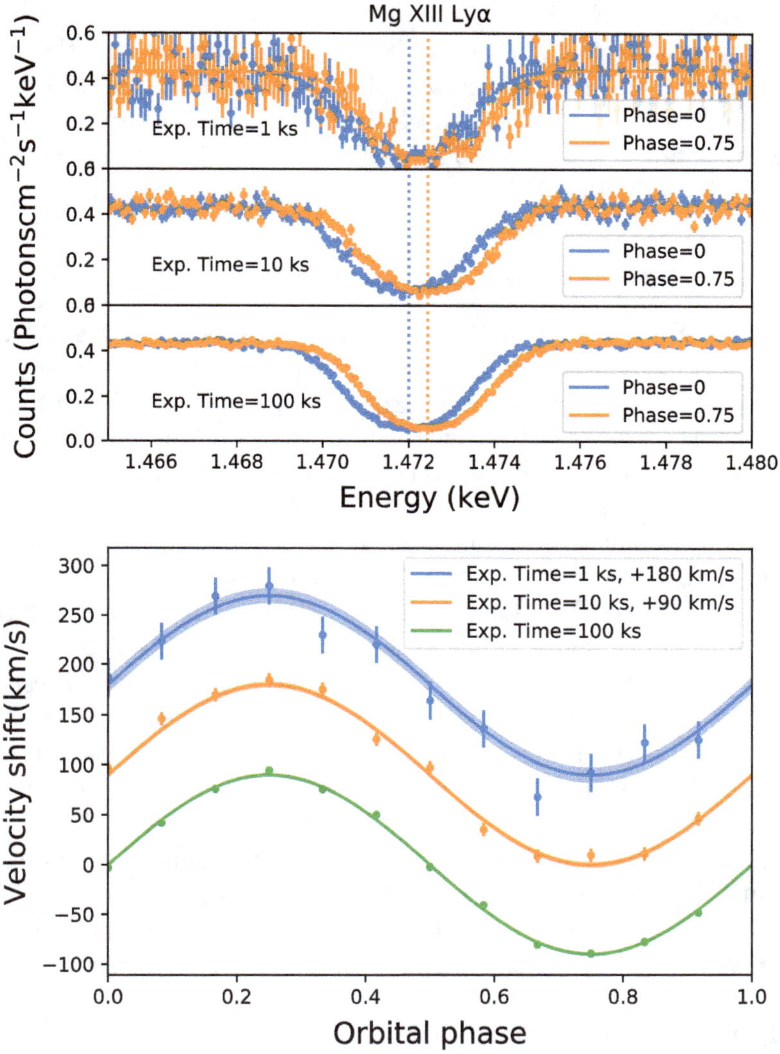

Fig. 1. Top panel, from top to bottom, the spectral simulation of MXB 1659–298 center around the Mg XIII Lyα line with exposure times of 1 ks, 10 ks, and 100 ks, respectively. The blue vertical dotted line is the spectral line Mg XIII Lyα in the rest frame, and the orange vertical dotted line is blueshifted as a factor of 3×10^{-4}. Bottom panel, the radial velocities and uncertainties observed at different orbit phases.

best fitting model. For the case of 10 ks exposure, the maximum radial velocity is constrained at 89.84 ± 2.73 km/s, and the mass function of NS can be achieved with $f = 0.022 \pm 0.002 M_{\odot}$. Besides MXB 1659–298, there are many eclipsing binaries can perform such studies, such as, XB 1832–330, MXB 1728–34, GX 13+1, 4U 1626–67, 4U 1538–52, 4U 1916–05 and 2S 0921–630.

2.2. *Symbiotic X-ray binaries*

Symbiotic X-ray binaries (SXBs) are a special subclass of NS LMXBs, which host a neutron star accreting matter from a M-type giant companion, rather than a dwarf. Usually, the orbital periods of SXBs and the spin periods of the neutron star are both long. Among them, 4U 1954+31 is the longest accreting neutron star with a spin period of about 18300 seconds (Masetti *et al.*, 2006). Why do neutron stars in SXBs have such long spin periods? The standard disk accretion model does not explain it yet. At present, the rotation periods of nearly half of the SXBs have yet to be determined, and it remains to be confirmed whether they are all longer. In addition, we also need to study the spin period evolution of SXB due to accretion, to test the recycling scenario of the traditional millisecond pulsar (Alpar *et al.*, 1982). The possible presence of ultra low-mass neutron stars in such systems, if confirmed, would pose a great challenge to the equation of state of neutron stars (Xu, 2014). On the other hand, if we can measure the masses of neutron stars in these systems, we will be able to refine the neutron star mass spectrum. The vast majority of SXBs do not have measured orbital periods, and if we are able to measure them well, we will also be able to further understand the evolution of LMXBs.

The SXB 4U 1700+24 has a red giant donor. Its X-ray emission is dominated by wind accretion. In the X-ray spectrum of 4U 1700+24, the O viii (hydrogen-like Ly-α) emission line is observed with a central energy of 0.65 keV (Nucita *et al.*, 2014). The emission line is corresponding to a gravitational redshift of 0.009, suggesting that 4U 1700+24 is an ultra low-mass neutron star candidate, which has to be verified by HUBS observations (Xu, 2014).

The continuum spectrum of 4U 1700+24 consists of a blackbody, a Gaussian and a Compton scattering component (Tiengo *et al.*, 2005). The blackbody has a temperature of 1.4 keV and the Comptonization

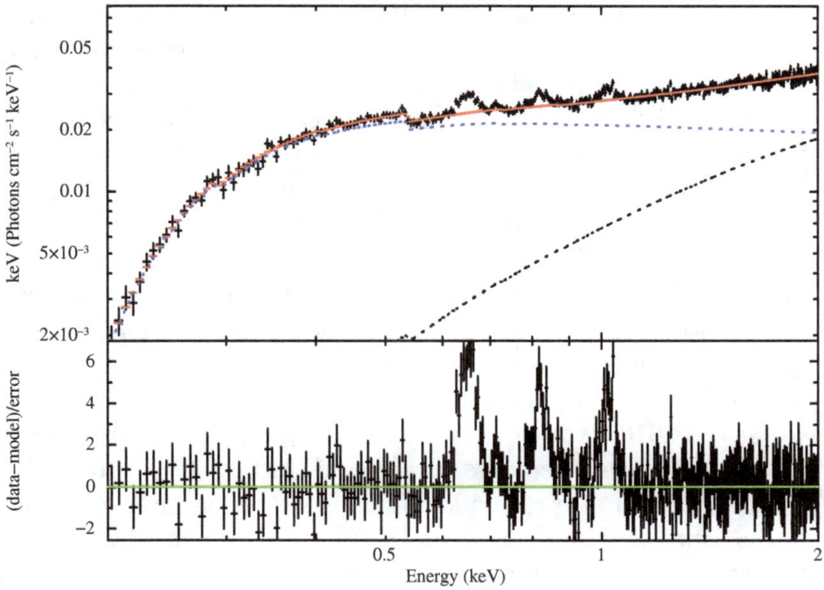

Fig. 2. The simulated X-ray burst spectrum of 4U 1700+24. The continuum model is `phabs*(bbodyrad+compST+gaussian+gaussian+gaussian)`. The exposure time is 10 ks. The central energies of three emission lines are 0.65 keV, 0.82 keV and 1.02 keV, respectively. All the line width are 0.02 keV. The normalization of the Gaussian components are 5×10^{-4} photons cm^{-2} s^{-1} for the 0.65 keV line and 2.5×10^{-4} photons cm^{-2} s^{-1} for the other two lines. The red line fits the spectrum by a blackbody plus compST model.

of the photons leads to an increase in the flux of high-energy photons, which results in a strong flux in the 0.1–2 keV band. To simulate the series of Ly-α lines, we added three Gaussian emission lines whose energies correspond to the energies of each of the Ly-α lines, each with a width of 20 eV. Since the 652.8 eV line has been observed in XMM-Newton, we assume that the intensity of the remaining lines is half of the 652.8 eV line intensity. The exposure time was 10 ks, and the spectrum simulation is shown in Figure 2.

2.3. *Type I X-ray bursts*

The unstable thermonuclear burning of the helium and mixed of hydrogen and helium, also known as type I X-ray burst, usually has the duration of ∼10−100 s with a typical energy release of 10^{39} erg, recurs from few hours to days, and ignites at a column depth of $\sim 10^8$ g cm^{-2} (Galloway and Keek, 2021). In a more rare case, superbursts, which are believed due to unstable burning of carbon, have been identified from the total energy release of $\sim 10^{42}$ erg and the duration of $>10^3$ s (Cumming and Bildsten, 2001). Cottam *et al.* (2002) reported the identification of absorption lines by stacking of XMM-Newton spectra over dozens of type I X-ray bursts from NS LMXB EXO 0748–676, which they claimed were gravitational redshifted Fe and O lines from the stellar surface. However, the spectral lines have not been confirmed by the following observations (Cottam *et al.*, 2008). This particular source is now believed to be rotating rapidly with frequency of 552 Hz from its burst oscillation (Galloway *et al.*, 2010), which makes it challenging to explain the relatively narrow spectral features reported in Cottam *et al.* (2002). in 't Zand *et al.* (2017) also reported the none detection of spectral lines in Rapid Burster from Chandra/HETG observations. Rauch *et al.*, Rauch *et al.* (2008) calculated the possible spectral lines in the soft X-ray band, i.e., Fe and O, that can be generated during type I X-ray bursts. Why the spectral lines during type I X-ray bursts have not been detected by XMM-Newton or Chandra? The spectra could evolve quickly in a normal type I X-ray burst, due to the photosphere expansion and contraction on NS surface. By stacking the spectra in the raise and decay stages separately from dozens of type I X-ray bursts would not increase the signal-to-noise ratio of spectral lines. Another possibility is the effective areas of the high energy resolution detectors on board of XMM-Newton and Chandra are not too large.

The characteristic spectral lines of Fe XXVI, Fe XXV, and O VIII are in the 0.1–2 keV energy band and can be observed by HUBS. In order to confirm the possible appearance of spectral features in the EXO 0748–676 during type I X-ray bursts, we assume that the continuum spectrum is a blackbody emission of $kT \sim 1.8$ keV, which is close to the Eddington luminosity, and the absorption line (FeXXVI) has a central energy of 0.954 keV and a width of 20 eV. We chose an exposure time of 100 s. The simulated X-ray burst energy spectrum is shown in Figure 3. It is worth noting that the 100 s exposure is usually too long for a single burst, since its spectra usually evolve rapidly as mentioned above. The energy of spectral line may change during a normal type I X-ray burst. The spectra can be combined at the same stage from burst samples, such as the peak and

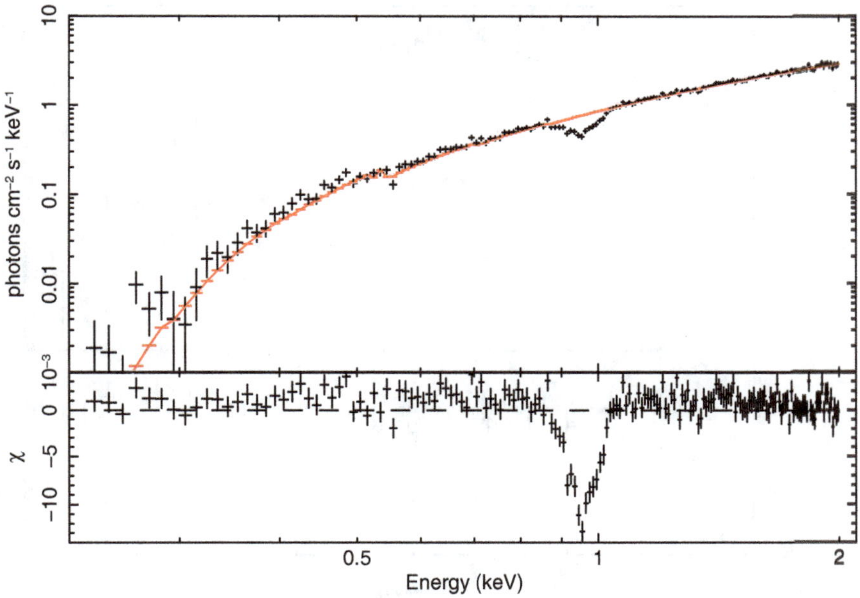

Fig. 3. The simulated X-ray burst spectrum of EXO 0748–676. The continuum spectrum is a blackbody with temperature of 1.8 keV. The exposure time is 100 s. We assume the Eddington limit is reached. We add an absorption line at 0.954 keV and width of 0.04 keV. The red line fits the spectrum by a blackbody model. The absorption line is clearly shown in the residual.

the touchdown moment. Moreover, from a single superburst, we can obtain long exposure spectra with sufficient X-ray photons to produce significant spectral lines.

2.3.1. *The X-ray dimmed isolated neutron stars*

The X-ray dimmed isolated neutron stars (XDINSs) mainly emit blackbody spectra in X-ray bands, and show optical/ultraviolet (UV) excesses (Haberl, 2007). All the seven known XDINSs were found by soft X-ray instruments. RX J1856–3754 is the brightest and closest neutron star. Chandra observations showed that RX J1856–3754 had an almost blackbody spectrum in the soft X-ray band, with no emission or absorption lines (Burwitz *et al.*, 2001). Absorption lines may be present in other XDINSs, which need to be confirmed. It is generally believed that the soft X-ray thermal spectrum of XDINS comes

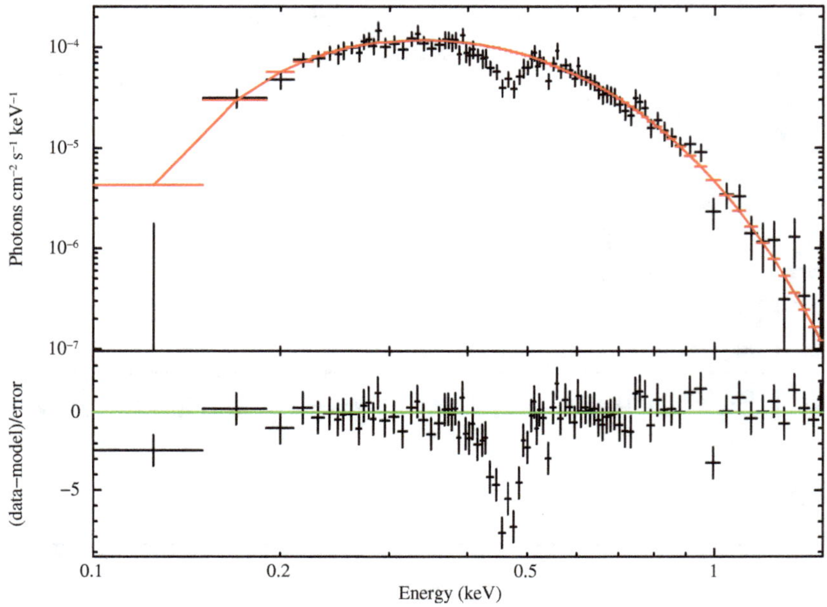

Fig. 4. The simulated thermal spectrum of the XDINS RX J0806.4–4123. The continuum spectrum is a blackbody with temperature of 0.1 keV. The blackbody normalization is 100. The exposure time is 100 ks. We add an absorption line at 0.46 keV and width of 0.04 keV. The solid line fits the spectrum by a blackbody model.

from the surface of the star. The absorption lines in the XDINS spectrum are produced in the magnetic environment of the neutron star. Moreover, the structure of the absorption lines is related to the stellar surface properties. Wang *et al.* (2017) proposed that the accretion of the interstellar medium to form the atmosphere of a strangeon star, and that the accreted atmosphere may produce absorption features. The temperature and radius of the continuum spectrum depend on the equation of state of the dense star (see Ho *et al.* 2005 for neutron stars; Wang *et al.* 2017 for strangeon stars). The mechanisms of absorption line generation are related to the equation of state of compact object, such as the equally spaced cyclotron absorption lines of electrons or protons in a magnetic field, collective oscillations of electrons on the strangeon star surface, etc. The mechanism of absorption line generation can be inferred from the energies of multiple spectral lines. Therefore, XDINS is an excellent target to study the surface properties and the equation of state of neutron stars.

We choose the input model **phabs*bbodyrad*gabs** for the XDINS RX J0806.4–4123. The exposure time is 70 ks. The blackbody spectrum of RX J0806.4–4123 has a temperature of 0.1 keV, a normalization of 100, and a hydrogen column density of 1×10^{20} cm^{-2} (see Kaplan *et al.*, 2011). To describe the absorption line structure, we added a Gaussian absorption component with an energy of 460 eV and width of 40 eV (see Haberl *et al.*, 2004). The simulated data and the blackbody fit are shown in Figure 4.

Besides measuring the NS mass and radius, HUBS can also study the magnetic field of anomalous X-ray pulsars (AXPs) and soft-γ-ray repeaters (SGRs). AXPs and SGRs are slowly rotating, isolated and extremely powerful magnetic field neutron stars (Kaspi, 2007; Mereghetti, 2013). Their X-ray activities, short bursts and outbursts, are powered by magnetic energy. During outbursts, the X-ray spectra of AXPs and SGRs may appear absorption line, which is interpreted as a proton cyclotron feature. HUBS could resolve the absorption features in 0.1–2 keV from AXPs and SGRS, and measure the magnetic field (see e.g., Tiengo *et al.* 2013).

In summary, HUBS will detect spectral features on the neutron star (NS) surface or the surrounding accretion disk to constrain the NS equation of state with its large area and high spectral resolution.

Bibliography

Bregman, J., Cen, R., Chen, Y. *et al.* (2023). arXiv:2307.05672. doi:10.48550/arXiv.2307.05672.

Burwitz, V., Zavlin, V. E., Neuhäuser, R., *et al.* (2001). *A&A* **379**, L35.

Cottam, J., Paerels, F., and Mendez, M. (2002). *Nature* **420**, 51.

Cui, W., Chen, L.-B., Gao, B., *et al.* (2020). *J. Low Temp. Phys.* **199**, 502. doi:10.1007/s10909-019-02279-3.

Cumming, A. and Bildsten, L. (2001). *ApJL* **559**, L127.

Demorest, P. B., Pennucci, T., Ransom, S. M., Roberts, M. S. E., and Hessels, J. W. T. (2010). *Nature* **467**, 1081.

Galloway, D. K. and Keek, L. (2021). in *Astrophysics and Space Science Library*, T. M. Belloni, M. Méndez, and C. Zhang (eds.), **461**, 209–262.

Galloway, D. K., Lin, J., Chakrabarty, D., and Hartman, J. M. (2010). *ApJL* **711**, L148.

Güdel, M. (2004). *A&A Review* **12**, 71.

Güdel, M., Audard, M., Reale, F., Skinner, S. L., and Linsky, J. L. (2004). *A&A* **416**, 713.

Güdel, M. and Nazé, Y. (2009). *A&Ar* **17**, 309.
Haberl, F. (2007). *ApSS* **308**, 181.
Hinkel, N. R., Mamajek, E. E., Turnbull, M. C., *et al.* (2017). *ApJ* **848**, 34.
Ho, W. C. G., Kaplan, D. L., Chang, P., *et al.* (2005). in American Astronomical Society Meeting Abstracts, Vol. 207, *American Astronomical Society Meeting Abstracts* **198**, 03.
in 't Zand, J. J. M., Bagnoli, T., D'Angelo, C., *et al.* (2017). arXiv e-prints, arXiv:1703.07221.
Johnstone, C. P. and Güdel, M. (2015). *A&A* **578**, A129.
Kaspi, V. M. (2007). *ApSS* **308**, 1.
Xiaoyu, L. and Renxin, X. (2017). *J. Phys. Conf. Ser.* **861**, 012027.
Lattimer, J. M. and Prakash, M. (2007). *Phy. Rep.* **442**, 109.
Liefke, C., Fuhrmeister, B., and Schmitt, J. H. M. M. (2010). *A&A* **514**, A94.
Masui, K., Mitsuda, K., Yamasaki, N. Y., *et al.* (2009). *PASJ* **61**, S115.
Mereghetti, S. (2013). *Braz. J. Phys.* **43**, 356.
Mittag, M., Schmitt, J. H. M. M., and Schröder, K. P. (2018). *A&A* **618**, A48.
Mukai, K., Kinkhabwala, A., Peterson, J. R., Kahn, S. M. and Paerels, F. (2003). *ApJ* **586**, L77.
Nordon, R. and Behar, E. (2008). *A&A* **482**, 639.
Nucita, A. A., Stefanelli, S., De Paolis, F., *et al.* (2014). *A&A* **562**, A55.
Özel, F. and Freire, P. (2016). *ARAA* **54**, 401.
Pandel, D., Cordova, F. A., Mason, K. O. and Priedhorsky, W. C. (2005). *ApJ* **626**, 396.
Pizzolato, N., Maggio, A., Micela, G., Sciortino, S. and Ventura, P. (2003). *A&A* **397**, 147.
Ponti, G., Bianchi, S., Muñoz-Darias, T. and Nandra, K. (2018). *MNRAS* **481**, L94.
Rauch, T., Suleimanov, V. and Werner, K. (2008). *A&A* **490**, 1127.
Reiners, A., Schüssler, M. and Passegger, V. M. (2014). *ApJ* **794**, 144.
Sazonov, S., Revnivtsev, M., Gilfanov, M., Churazov, E. and Sunyaev, R. (2006). *A&A* **450**, 117.
Steiner, A. W., Lattimer, J. M. and Brown, E. F. (2013). *ApJ Letter* **765**, L5.
Testa, P., Drake, J. J. and Peres, G. (2004). *ApJ* **617**, 508.
Tiengo, A., Esposito, P., Mereghetti, S., *et al.* (2013). *Nature* **500**, 312.
Wang, W., Lu, J., Tong, H., *et al.* (2017). *ApJ* **837**, 81.
Wright, N. J., Drake, J. J., Mamajek, E. E., and Henry, G. W. (2011). *ApJ* **743**, 48.
Xu, R.-X. (2014). *RAA* **14**, 617.
Zhang, S.-N., Liao, J., and Yao, Y. (2012). *MNRAS* **421**, 3550.

https://doi.org/10.1142/9789819800643_0016

Chapter 16

The 120 m Jingdong Radio Telescope (JRT)

Longfei Hao, Min Wang* and Yonghua Xu
Yunnan Observatories, Chinese Academy of Sciences, Kunming, 650011, China
**wm@ynao.ac.cn*

In Jingdong County (101°E, 24.5°N), Pu'er City, Yunnan Province, the Yunnan Observatories Chinese Academy of Science will construct the 120-meter Jingdong Radio Telescope (JRT) dedicated to pulsar-related science. This chapter introduces the scientific objectives and key technologies of the JRT. The telescope covers the frequency range from 100 MHz to 10 GHz (i.e., from the wavelength 3 m to 3 cm). Once finished, it will be the world's largest fully steerable single-dish decimetre-wavelength radio telescope. The JRT will be driven by three major scientific goals, which include (1) long-term high-precision pulsar timing for time-frequency metrology and nanohertz gravitational wave detection; (2) pulsar astrophysics, fast radio bursts, gravity theory tests, and black hole physics; (3) deep space exploration, geodesy, and celestial reference system. We will also cover the engineering design aspects in this chapter, and present the current plan and design for the telescope mechanical structure, front-end, digital backend, and control system.

Keywords: Pulsars, JRT, Radio telescope

Contents

1. Introduction

Pulsars are believed to be compact stars formed after the supernova explosion. Most of the pulsars are detected in radio bands, and their spin period falls in a wide range of 1 ms to 10 s. The typical mass and radius of a pulsar are 1.4 Solar mass and 10 km. The compactness of pulsars makes their rotation extremely stable, which can be monitored by measuring the time of arrival of pulsars' radio pulses, a technique called pulsar timing. Over 3300 radio pulsars have been found so far (psrcat, 2022).

The radio telescope is the most crucial equipment to observe and characterise the physical properties of pulsars. By timing the binary pulsars and comparing the observation with the model predictions, one can test gravity theories in the strong field regime (Taylor *et al.*, 1991) and probe the equation of states of matter at nuclear density. Observing a group of stable millisecond pulsars and extracting the commonly correlated signals from those widely spaced pulsars further forms the *pulsar timing array* (PTA). Data from a PTA can be used to quantify the local atomic clock stability (Taylor *et al.*, 1991) and study the Solar System dynamics (Guo *et al.*, 2018, 2019). PTA is also the only known method to detect nanohertz gravitational waves (Jenet *et al.*, 2005).

At present, international efforts to detect nanohertz gravitational waves include European Pulsar Timing Array (EPTA) with five 100-m European radio telescopes, the North American Nanohertz

Observatory for Gravitational Waves (NANOGrav) (McLaughlin, 2013) with the Green Bank 100-m Telescope (GBT) and Arecibo 300 m Radio Telescope (Arecibo), and Parkes Pulsar Timing Array (PPTA) with the Parkes 64-m radio telescope (PKS). Recently, India also established the India PTA (Joshi *et al.*, 2018). Chinese Pulsar Timing Array (CPTA) was formed in 2016 (Hobbs *et al.*, 2019; Lee, 2016). All the regional PTAs form the International Pulsar Timing Array (IPTA) at the inter-continental scale (Perera *et al.*, 2019).

FAST telescope (Jiang *et al.*, 2019) passed the national evaluation and acceptance at the beginning of 2020. It is currently the largest decimetre wavelength radio telescope in the world. The ultimate sensitivity makes it among the best telescopes to observe pulsars. Shortly afterward, it delivered scientific outputs in the field of fast radio bursts (Li *et al.*, 2021; Lin *et al.*, 2020; Lu *et al.*, 2020; Luo *et al.*, 2020) and pulsar search (Han *et al.*, 2021).

It is also, currently, the only telescope in China supporting CPTA timing observation to achieve 100-ns level precision. Besides FAST radio telescopes, there are a group of 25-m to 65-m level telescopes in China. The Shanghai Sheshan and Xinjiang Nanshan 25-m radio telescopes, respectively, were constructed in the 1980s and 1990s. The Miyun 50-m, Kunming 40-m, and Shanghai 65-m radio telescopes were built in 2006 and 2012, respectively, with deep space exploration as their major duties, to carry out the observations in lunar exploration project. The 110-m radio telescope which is being constructed in Qitai, Xinjiang (QTT) will cover 150 MHz to 115 GHz frequency range, a much wider band than the current coverage of Chinese radio telescopes. It will be China's largest multi-purpose radio telescope supporting research of gravitational waves, black holes, star formation, molecular clouds, and galaxy genesis (Wang, 2014).

Due to the limited aperture of currently available telescopes, CPTA critically depends on the FAST telescope at present. To reduce the risk and increase the telescope resource of CPTA, new large Chinese radio telescopes dedicated to pulsar science are thus highly desired. We introduce the Jingdong 120-m Pulsar Radio Telescope (JRT) with this paper. The telescope had been financed jointly by the Chinese Academy of Sciences and Yunnan Province and will be built at the radio-quiet site in Jingdong County, Pu'er City, Yunnan Province, as proposed by the Yunnan Observatories, Chinese Academy of Science. JRT covers the frequency range from 100 MHz to 10 GHz. JRT is dedicated to pulsar observations, particularly the long-term, high cadence, and high precision pulsar timing observations, which will provide observational data for the development of China's independent pulsar time-scale system and nanohertz gravitational wave detection. In addition, JRT will support national strategic requirements in fields including pulsar navigation, and deep space exploration. We will discuss in the following, the JRT site environment and major science cases, the technical aspects of JRT, the expectation of pulsar observations and conclusions are also addressed.

2. JRT's Site and Scientific Goals

JRT locates at approximately 101°E 24.5°N. The telescope latitude is lower (towards the south direction) than most of the current 100-m telescopes. As a comparison, we listed the major 100-m level astronomical radio telescopes in the world in Table 1 and the corresponding sky coverage is presented in Figure 1. In this figure, we show the currently known pulsar population along the Galactic plane.

Table 1. Specifications of international large radio telescopes.

Name	Position	Diameter (m)	Efficiency	T_{sys} (K)	Gain (K/Jy)	Frequency (GHz)
FAST[a]	106°51′24.0″E 25°39′10.6″N	500	0.63	20	16.1	1.05–1.45
QTT[b]	89°40′56.99″E 43°36′4.03″N	110	>0.6	25	2.06	0.6–4.0 primary focus
GBT[c]	79°50′23.406″E 38°25′59.236″N	100 × 110	0.7	20	2.0	0.7–0.9 1.1–1.9
PKS[d]	148°15′46.5″E 32°59′52.01″N	64	0.7	21	0.8	0.7–4.2 primary focus
EF[e]	6°52′01.0″E 50°31′29.4″N	100	0.44	56	2.0/1.25	1.0–3.0 primary focus
JRT	101°E 24.5°N	120	0.6	25	2.46	1.0–7.0 primary focus

[a](Jiang *et al.*, 2019);[b](Wang, 2014); [c](Alam *et al.*, 2021); [d](Hobbs *et al.*, 2020a); [e](EF, 2022).

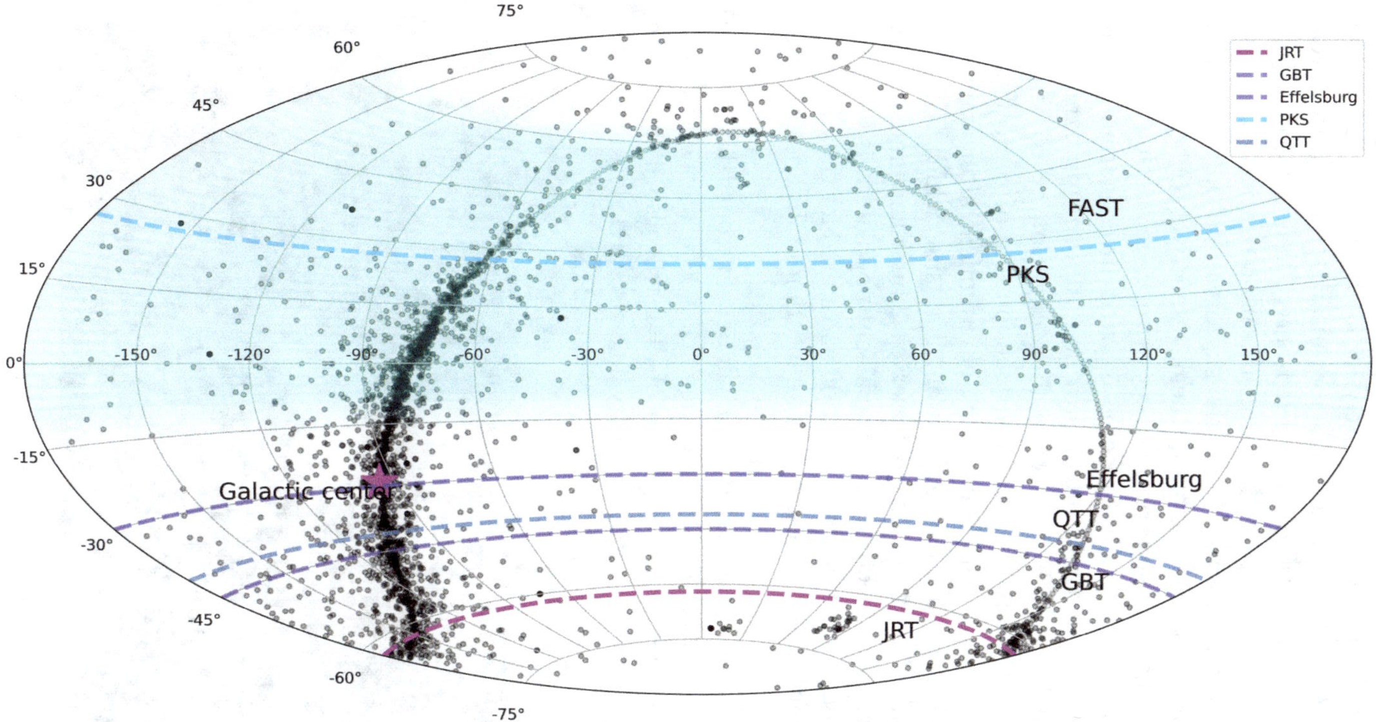

Fig. 1. Pulsar sky map covered by International Large Radio Telescopes (Black dots represent the discovered pulsars, and the blue shadow area represent the sky map of FAST, the blue line segment represents the observation limit of Parkes 64-m radio telescope in the northern sky, The blue, steelblue, navy and purple dashed lines represent the observation limits of EF, GBT, QTT, and JRT in the southern sky, respectively. The five-pointed star represents the position of the center of the Milky Way).

As the galactic center is approximately at 30° S, radio telescopes in the southern hemisphere are desirable for pulsar astronomy. However, large telescopes such as GBT and Effelsberg 100-m telescope (EF) are all located in the northern hemisphere with a relatively high latitude, they cover 75% and 82% of the sky area and only 68% and 76% of known pulsars are visible to them, respectively. Therefore, it is critical to build large-aperture radio telescopes in south.

From 2009, Yunnan Observatories started investigating candidate sites throughout the Yunnan province. Among approximately 20 sites, Xujiaba (101°E, 24.5°N) in Jingdong County of Yunnan Province has been chosen as the construction site of the JRT station with an extraordinary radio-quiet environment after extensive monitoring, comparison, and evaluation. The site, which is in the Ailao Mountain National Nature Reserve's hinterland and has an elevation of roughly 2,500 m and is sparsely populated, is ideal for long-term radio environmental protection and suited for the construction of substantial radio telescopes. The location of JRT is shown in Figure 2.

The Chinese VLBI network (CVN) will benefit from JRT's unique geographic advantages of sky coverage, i.e., 90% of the sky and 91% of known pulsars. The addition of JRT will also increase the international VLBI network's capacity because JRT locates in the

Fig. 2. The site of JRT (The scenery is real, the antenna is fictional).

intersection region of several international VLBI networks (EVN, KVN, LBA, etc).

The current IPTA timing precision (~100 ns) is on the cusp of nanohertz gravitational wave detection (Alam *et al.*, 2021; Perera *et al.*, 2019). The number of pulsars (M), observation frequency (c), timing accuracy (σ), gravitational wave background amplitude (A), data length (T), and other factors (see Eq. 1) (Antoniadis *et al.*, 2022; Bassa *et al.*, 2016; Lam *et al.*, 2019; Pennucci, 2019; Siemens *et al.*, 2013) all have a direct impact on the sensitivity of PTA in identifying nanohertz gravitational waves, i.e.,

$$S/N \propto Mc\frac{A^2}{\sigma^2}T^{13/3}. \tag{1}$$

The pulsar timing precision depends on the telescope's effective reception area ($A_{\rm eff}$), system temperature ($T_{\rm sys}$), observation bandwidth (Δf), and integration time ($t_{\rm obs}$) (Verbiest *et al.*, 2018) with timing accuracy

$$\sigma \propto \frac{T_{\rm sys}}{A_{\rm eff}\sqrt{\Delta f t_{\rm obs}}}. \tag{2}$$

The number of pulsar observations, the frequency of observations, the timing precision, and other factors must all be improved to increase the sensitivity of PTA nanohertz gravitational wave detection. Table 2 shows the performance comparison between the JRT and various major radio telescopes from across the world. Dedicating to pulsar observation, offering more than 6,000 hour of observation time per year, the JRT radio telescope is expected to deliver high-precision and high-cadence timing observations on more than 100 millisecond pulsars ($P < 10\,{\rm ms}$). By utilizing the shortening of the orbital period of the pulsar binary PSR B1913+16 in 1993, Taylor and Hulse proved that gravitational waves exist and that their radiation is consistent with the general relativity prediction (Hulse *et al.*, 1975).

One hundred years after Einstein made his prediction about gravitational waves, LIGO-Virgo discovered a gravitational wave ($10-10^3$Hz) caused by the merging of two-star black holes in the nearby cosmos, ushering in the age of gravitational wave astronomy (Abbott *et al.*, 2016). Only PTA is capable of detecting the nanohertz gravitational waves ($1-100\,{\rm nHz}$) produced by the merger of supermassive binary black holes (10^8–10^{12} $M_\odot$). The gravitational wave puzzle's nanohertz gravitational component is made up of these waves, and they are a crucial tool for understanding the birth and evolution of galaxies and black holes as well as the early universe's expansion history (Mingarelli *et al.*, 2012).

From the international point of view, NANOGrave, EPTA, IPTA, and PPTA have all conducted the discovery of nanohertz gravitational wave background using their own sets of pulsar timing data (see for instance Refs. (Arzoumanian *et al.*, 2020; Chen *et al.*, 2021; Goncharov *et al.*, 2021)). The temporal signal properties of several pulsars are generally comparable with the projected gravitational wave background characteristics, although the confidence level still has to be raised (Antoniadis *et al.*, 2022). In the frequency range between 70 MHz and 3 GHz, FAST is the most sensitive single-dish radio telescope. FAST's time available for timing observation is quite limited, and its primary application is to find fainter pulsars (Yan *et al.*, 2020). JRT spans the FAST observation frequency range, enabling follow-up monitoring of pulsars found by FAST, completing the FAST invisible observation sky area, and boosting the quantity and frequency of pulsar observations. Timing precision will be further enhanced by JRT's high-frequency observations, which will aid in identifying and correcting the interstellar medium's influence. FAST will outperform IPTA in terms of overall performance and is anticipated to achieve a breakthrough in the detection of nanohertz gravitational waves by joining forces with QTT, JRT, and other large-aperture radio telescopes in China to form CPTA. Regarding the quantity, sensitivity, and cadence of pulsar observations, JRT will play a significant role in the nanohertz gravitational wave detection.

Besides gravitational wave detection, the autonomous pulsar time standard is another key scientific objective of JRT. One can create the time standard by monitoring a collection of millisecond pulsars and comparing the timing with local atomic clocks (Hobbs *et al.*, 2020b; Taylor *et al.*, 1991). The pulsar time standard is distinct from the current atomic time system as we expected different systematics. As a result, by combining the pulsar time standard and atomic time standard, better stability is expected. At present, paper time from pulsar time standard using IPTA data achieved annual stability of 5×10^{-15} (Hobbs *et al.*, 2020b; Manchester *et al.*, 2013). The next step is to construct a real-time pulsar time standard. For instance, to undertake precise timing measurements on 18-millisecond pulsars to confirm the viability and usefulness of pulsar time scales servicing the current atomic clock system and Galileo navigation time system, EPTA started the "PulChron" research project (Píriz, 2019) in 2018.

China has also conducted research on the creation and used the pulsar timing system, including studies on the pulsar timing theory by the National Time Service Center (NTSC) (Yang *et al.*, 2016; Yin *et al.*, 2017; Zhao *et al.*, 2018) and the application study on pulsar timing by Yunnan Observatories using the timing data of PSR J1713+0747 and J0437–4715 (Li *et al.*, 2020). We expect that JRT pulsar timing will be key input data for the Chinese pulsar timescale experiment.

When it is commissioned, JRT will be the biggest single-dish, fully steerable intermediate and low-frequency radio telescope ever built. The following fields are those in which their distinct geographic location can be crucial. In the beginning, the JRT will conduct pulsar search activities in the sky regions that cannot be viewed and searched by huge antennas like GBT and EF telescopes. Secondly, the JRT will observe the Milky Way's center for around 3300 hours, which is particularly helpful for examining the characteristics of the black hole at the Milky Way's center and testing relevant theories (Kramer, 2016). Thirdly, JRT will successfully enhance the overall functionality of both domestic and foreign VLBI networks. Last but not least, JRT will also contribute to other sectors such as pulsar navigation, deep space detection, space target monitoring, and radar detection of near-Earth objects.

3. JRT Technical Solutions

JRT will adopt main focus design with a horizontal mounting wheel rail structure. The designed frequency range is from 100 MHz to 10 GHz. The JRT will be built in two phases. The first phase includes a 1−7 GHz ultra-wideband (UWB) cryogenic receiver and a multi-functional digital back-end for data recording. Phased Array Feed (PAF) and multi-beam receivers are planned in the second phase. The system design of JRT is illustrated in Figure 3.

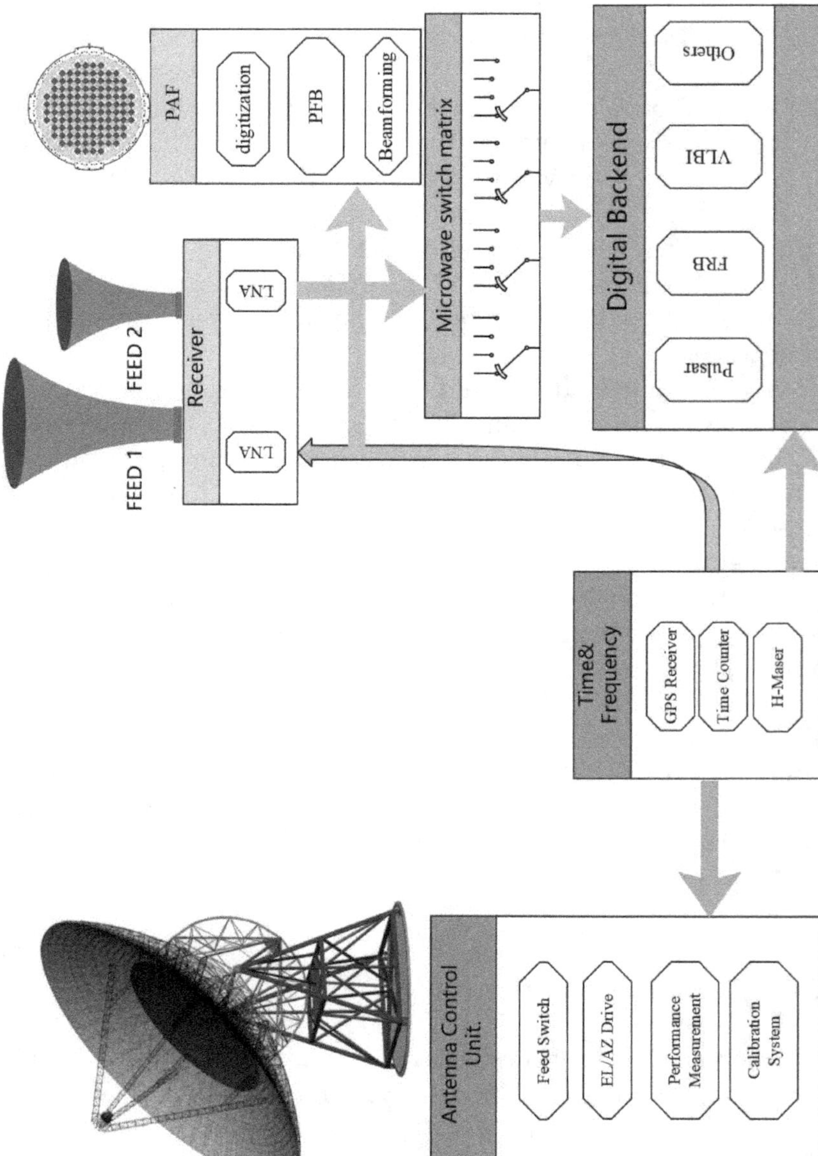

Fig. 3. The system frame of the Jingdong 120 m pulsar radio telescope.

Table 2. The design specifications of JRT.

Item	Details
Main reflector[a]	120 m
Mount	AZ/EL, Wheel-rail type
Feed	Primary focus, f/D = 0.42
Polarization	Dual-linear
Frequency	100 MHz–10 GHz
Panel accuracy	<0.3(r.m.s) for all
System temperature	$25K$
Pointing accuracy	$\leq 10''$
Aperture efficiency	$\geq 60\%$
Moving speed	$0.8^\circ/s$ @AZ, $0.3^\circ/s$ @EL
Moving range	$\pm 270^\circ$ @AZ (Due south is 0°), 5–89° @EL
Weight	$\sim 3400\,t$

Notes: [a]The outer 60 m is a perforated metal plate, and the inner 60 m is a solid panel.

3.1. *Antenna design scheme*

Pulsar research is the first priority target of the JRT. Table 2 lists the design specifications of JRT. The 120 m in-diameter reflectors are composed of inner 60 m solid panel and outer 60 m perforated plates. Such design guarantees the ultra-broadband coverage of 1–7 GHz with an efficiency of 60%.

The JRT adopts the main focus design with a conformal surface. The antenna reflector will consist of main reflector, center body, rear frame, and main focus cabin supporting structure. An umbrella-style support structure is used for the antenna back frame. Sixteen frames are used in the center body, which lessens the impact of pitch bearing, pitches large tooth arc, and counterweight on the uneven deformation of the reflector rear frame and enables the realization of an antenna conformal design.

An LQG controller will be used to implement JRT's servo control since it offers superior dynamic performance and stability over the traditional proportional integral derivative (PID) method. Of course, we're also thinking about whether we could use the $H\infty$ controller (Gawronski, 2001).

The (1–7) GHz frequency range will be covered by a multi-layer dielectric loaded quad-ridged dual polarisation horn. The flange, the loaded quad-ridged structure, and the dielectric cone are mainly a response to waves below 1.8 GHz, 1.8–2.8 GHz, and above 2.8 GHz, respectively. Such a property provides opportunities to further fine-tune each sub-band.

3.2. *Design of cryogenic receiver*

Cryogenic ultra-wideband receivers had been widely used in pulsar astronomy e.g., 0.7–4 GHz (Hobbs *et al.*, 2020c) and 0.6–3 GHz UWB receivers are now installed at the PKS and the EF telescopes for daily pulsar timing measurements. The (0.7–4) GHz UWB cryogenic refrigeration receiver is used by NANOGrav (Ransom *et al.*, 2019).

At present, JRT's design adopts the 1–7 GHz frequency coverage. Figure 4 displays the block diagram of UWB cryogenic receiver. It is comprised of a cryogenic Front-End and a Multi-Channels Down-converter. The wideband low noise design together with high polarization fidelity requirements are expected to be technologically challenging.

To decrease the receiver's front-end noise, the whole feed of JRT will be cooled. The essential component of the JRT receiver is the low temperature microwave unit. The UWB feed horn, cryogenic low-noise amplifier, and other components operate on the 15K platform and are used to receive RF signals between 1 and 7 GHz, separate them into horizontal and vertical polarization signals, and then perform very low noise amplification. The optical transceivers are used to transmit the IF signal and frequency standard signal between the antenna's primary focus and the control room. Single-mode fiber, photoelectric conversion, and electro-optical conversion are the components of the link.

3.3. *Pulsar digital backend*

An essential tool for timing pulsars and tracking search activity is the pulsar digital backend. The demands placed on pulsar digital backends' performance are likewise steadily rising as a result of pulsar research's advancement and the quick development of digital technology. As an illustration, PKS employs the DFB, CASPSR, BPSR, and UWL pulsar digital terminal systems. GUPPI and VEGAS are two

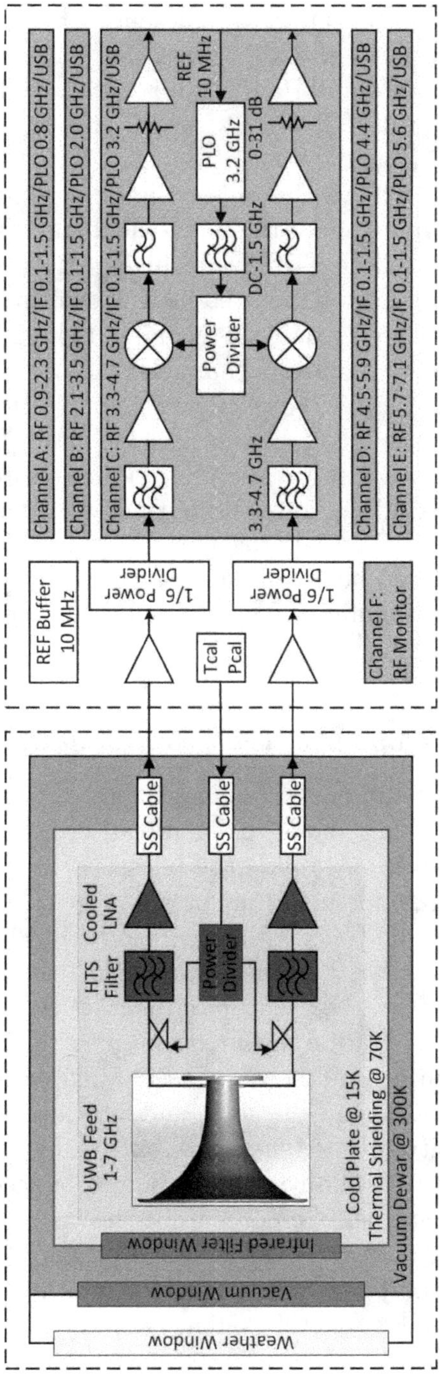

Fig. 4. The block diagram of JRT UWB cryogenic receiver.

components of GBT's pulsar digital backend. EF implements EBPP, PSRIX, etc. (Lazarus *et al.*, 2016). In 2019, pulsar digital backend UWL started to operate at PKS (Hobbs *et al.*, 2020c). The observation bandwidth of JRT is 6 GHz, which encounters the technical obstacle of ultra-wideband data capture, large data transfer, and real-time operation. To achieve the pulsar digital terminal mode design and fulfill the requirements of the ultra-wideband cryogenic receiver system, JRT will combine FPGA and high-performance GPU clusters. The block diagram of the system is shown in Figure 5. High-speed ADC technology is used to achieve high-speed sampling of ultra-wideband radio signals, 5-channel FPGA units are used to gather signals with a 6 GHz observation bandwidth, DPDK (dpdk, 2021) technology is used to realize massive amounts of data, and GPU clusters are used to achieve a digital terminal system with coherent dispersion, RFI mitigation, pulsar timing, and real-time search for transient sources. The heterogeneous terminal system based on FPGA and high-performance computer GPU cluster will fully exploit the benefits of diverse computing processing units and will be simple to cluster expansion.

3.4. *Radio telescope operation control system*

The radio telescope's operation management control system serves as the main control component and completes all control and observational responsibilities, enabling the telescope to operate automatically in an effective and steady manner.

Since computer and network technology have advanced, distributed architecture has become common in telescope control systems (see for instance the Refs. (Kodilkar *et al.*, 2019; Li *et al.*, 2021)). The JRT operation management control system will use a modular distributed middleware-based architecture. The design objective is to support cross-platform and multilingual development, with high scalability and extensibility, and to be simple to create, upgrade, and maintain. The system is designed to accommodate multiple concurrent local and remote user sessions and support local application-, web-, and app-based mobile access functions as well. The flowchart of the operation control system is in Figure 6. The operation management control system is mainly separated into the

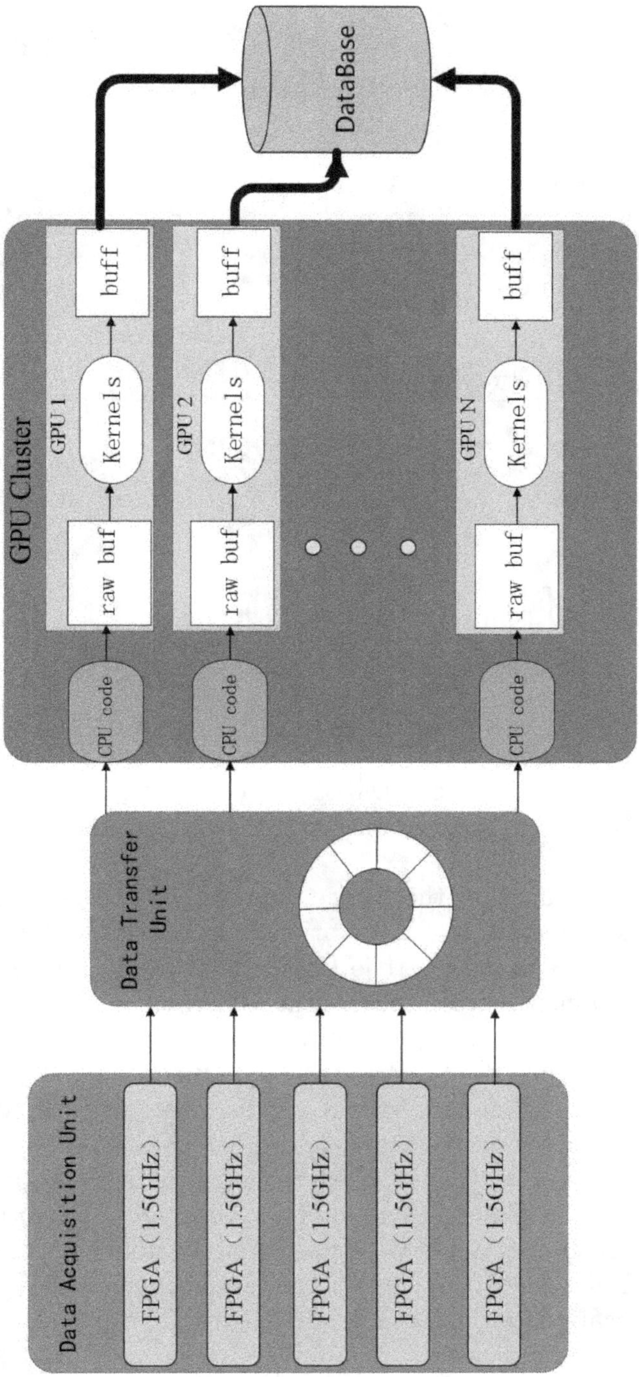

Fig. 5. The multifunctional digital back-end block diagram of JRT.

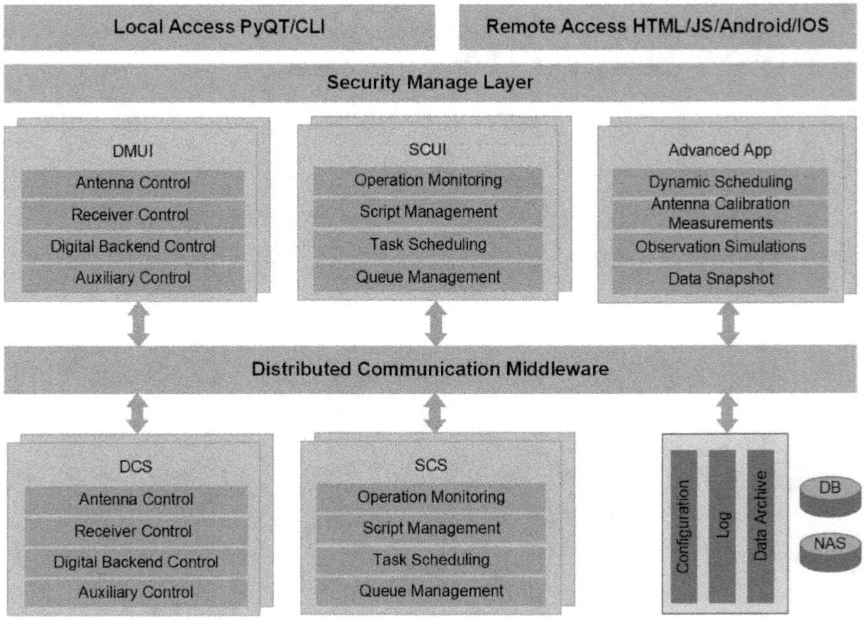

Fig. 6. The block diagram of JRT software system.

Device Control System (DCS), Service Control System (SCS), Device Management User Interface (DMUI), Service Control User Interface SCUI, advanced applications, and distributed communication middleware.

The SCS operates on the device control system and primarily completes the preparation of observation tasks, scheduling execution, and equipment coordination. The DCS primarily realizes the agent control of equipment, directly drives the underlying hardware equipment or software services, issues instructions, and obtains the status through the parameterized operation environment. Users can create and specify observation task scripts, set the observation mode, terminal settings, observation length, and other parameters, and then schedule those tasks to run regularly to conduct observations. The device and service management interface offer cross-platform user interface interaction and remote access based on mobile terminal and browser, and it makes use of the security management layer to do user authentication, authorization, and logging. Through distributed communication middleware, each subsystem communicates with the others. This middleware adopts message-based communication, supports a variety of synchronous and asynchronous communication modes, including request/response, publish/subscribe, routing/processing, etc., and offers high-throughput, low-latency communication interfaces.

4. Performance Evaluation of JRT Pulsar Observation

The pulsar signal is affected by the variation of dispersion, scattering, and scintillation effects due to the interstellar medium fluctuation. Ultra-wideband receivers and multi-band simultaneous observations are important to correct those interstellar medium effects (Lamph *et al.*, 2018; Lee *et al.*, 2014, 2017). The dispersion measure (DM) of pulsars fluctuating by $10^{-4} \sim 10^{-3}\,\mathrm{cm}^{-3}\,\mathrm{pc}$ over several years is one of the most significant source of systematics for high-precision millisecond pulsar timing (Jones *et al.*, 2017). The broadband timing (Liu *et al.*, 2014) is expected to solve the problem of profile evolution and interstellar medium fluctuation at the same time. We expect that the 1−7 GHz ultra-wideband observation of JRT will be critical for correcting above mentioned systematics.

The gain of the JRT is about 2.5 (K/Jy) and the ratio of the effective area of the telescope to the system noise temperature is approximately 452.39 m^2/K, assuming that the antenna efficiency of the JRT is 60% and the system temperature is $25K$. The key element affecting the time of arrival (TOA) measurement accuracy of the pulsar is the observed pulsar signal-to-noise ratio. The value of a signal-to-noise that provides information about the quality of the signal is given by:

$$SNR = \frac{SG\sqrt{BN_p t}}{T_{\mathrm{rec}} + T_{\mathrm{sky}}}\sqrt{\frac{P-W}{W}}, \tag{3}$$

where G is the antenna gain (K/Jy), B is the observation bandwidth (Hz), N_p is the polarization number, t is the integration time, P is the pulse period, W is the pulse width, and T_{rec} and T_{sky}, respectively, are the receiver noise and sky background noise. The signal-to-noise ratio and pulsar profile are closely related to the TOA accuracy, which is frequently calculated as:

$$\sigma \sim \frac{W}{2 * \mathrm{snr}}. \tag{4}$$

For instance, JRT will perform polarimetric observation of millisecond pulsar J1909-3744 with a flux density of around 2 mJy, a pulse width of approximately 40 μs for 15 min, and an effective bandwidth of 1.4 GHz. It has a signal-to-noise ratio of roughly 3600, and the expected equivalent TOA radiometer error is only 5 ns.

The JRT is a dedicated pulsar radio telescope with the capacity to devote around 6,000 hours annually to timing observations of more than 100 millisecond pulsars, which will exceed the observation cascade and the amount of target sources of the current IPTA. Around 2,000 hours of observation time are also available for use in other types of scientific research, like sky surveys and fast radio bursts.

5. Conclusion

We have presented the JRT project's site, system performance, and main scientific objectives. The JRT is anticipated to be the biggest medium and low frequency fully steerable radio telescope devoted to pulsar science in the world. It has unique geographic advantages that allow for observing 90% of the sky. To achieve high-sensitivity pulsar observations, the JRT will adopt the conformal dish together with the main focus design supporting the ultra-broadband observations.

The JRT will be independently developed by the necessary domestic scientific research forces, fostering the growth of China's engineering technology sectors such as large-scale machinery design and production, automatic control, low-temperature electronics, and system integration. To further improve the telescope's observational capabilities, JRT will eventually be equipped with a low-frequency receiver, phased array feed, and multi-beam receiver.

Acknowledgments

This work was funded by the National SKA Program of China (2020SKA0120100), the Special Project of Foreign Science and Technology Cooperation of Yunnan Provincial Science and Technology Department (202003AD150010), National Key R&D Program of China No. 2022YFC2205203, and the National Natural Science Foundation of China (NSFC, Grant Nos. 12073076, 12173087, 12041303 and 12063003), the CAS "Western Light Youth Project", the CAS-MPG LEGACY Project and the Max-Planck Partner Group.

Bibliography

(2021). https://www.dpdk.org/.
(2022). https://www.atnf.csiro.au/research/pulsar/psrcat/.
(2022). https://eff100mwiki.mpifr-bonn.mpg.de/doku.php?id=information_for_astronomers:rx_list.
Abbott, B. P. *et al.* (2016). *Phys. Rev. Lett.* **116**(6), 061102.
Alam, M. F. *et al.* (2021). *Astrophys. J. Ser.* **252**(1), 5.
Antoniadis, J. *et al.* (2022). *MNRAS* **510**(4), 4873–4887.
Arzoumanian, Z. *et al.* (2020). *Astrophys. J. Lett.* **905**(2), L34.
Bassa, C. G. *et al.* (2016). *MNRAS* **456**(2), 2196–2209.
Chen, S. *et al.* (2021). *MNRAS* **508**(4), 4970–4993.
Gawronski, W. (2001). *IEEE Antennas and Propagation Magazine* **43**(4), 52–60.
Goncharov, B. *et al.* (2021). *Astrophys. J. Lett.* **917**(2), L19.
Guo, Y. J. *et al.* (2018). *MNRAS* **475**(3), 3644.
Guo, Y. J. *et al.* (2019). *MNRAS* **489**(4), 5573.
Han, J. L. *et al.* (2021). *Res. Astron. Astrophys.* **21**(4), 107.
Hobbs, G. *et al.* (2019). *Res. Astron. Astrophys.* **19**(2), 020.
Hobbs, G. *et al.* (2020a). *MNRAS* **491**(4), 5951–5965.
Hobbs, G. *et al.* (2020b). *Pub. Astron. Soc. Aus.* **37**, e012.
Hobbs, G. *et al.* (2020c). *Pub. Astron. Soc. Aus.* **37**, e012.
Hulse, R. A. *et al.* (1975). *Astrophys. J. Lett.* **195**, L51–L53.
Jenet, F. A. *et al.* (2005). *Astrophys. J. Lett.* **625**(2), L123–L126.
Jiang, P. *et al.* (2019). *Sci. Chi. Phys. Mech. Astron.* **62**(5), 959502.
Jones, M. L. *et al.* (2017). *Astrophys. J.* **841**(2), 125.
Joshi, B. C. *et al.* (2018). *Jou. Astrop. and Astron.* **39**(4), 51.
Kodilkar, J. *et al.* (2019). *2019 URSI Asia-Pacific Radio Science Conference (AP-RASC)* (IEEE), 1.
Kramer, M. *et al.* (2016). *Inter. Jour. Mod. Phys. D* **25**(14), 1630029–61.
Lam, M. T. *et al.* (2019). *Amer. Astron. Soc. Meet. Abs.* **233**(107), 01.
Lam, M. T. *et al.* (2018). *Astrophys. J.* **861**(2), 132.
Lazarus, P. *et al.* (2016). *MNRAS* **458**(1), 868–880.
Lee, K. J. (2016). *Astron. Soc. Paci. Conf. Ser.* **502**, 19.
Lee, K. J. *et al.* (2014). *MNRAS* **441**(4), 2831–2844.
Lee, K. J. *et al.* (2017). *Sci. Sin. Phys. Mech. Astron.* **47**(5), 059507.
Li, D. *et al.* (2021). *Nature* **598**(7880), 267–271.
Li, J. *et al.* (2021). *Adv. Astron.* **2021**. Article ID: 2655250.

Li, Z. X. *et al.* (2020). *Sci. Sin. Phys. Mech. Astron.* **63**(1), 219512.
Lin, L. *et al.* (2020). *Nature* **587**(7832), 63–65.
Liu, K. *et al.* (2014). *MNRAS* **443**(4), 3752–3760.
Lu, J. *et al.* (2020). *Sci. Sin. Phys. Mech. Astron.* **63**(2), 229531.
Luo, R. *et al.* (2020). *Nature* **586**(7831), 693–696.
Manchester, R. N. *et al.* (2013). *Pub. Astron. Soc. Aus.* **30**, id.e017.
McLaughlin, M. A. (2013). *Cla. Quant. Grav.* **30**(22), 224008.
Mingarelli, C. M. F. *et al.* (2012). *Phys. Rev. Lett.* **109**(8), 081104.
Pennucci, T. T. (2019). *Astrophys. J.* **871**(1), 34.
Perera, B. B. P. *et al.* (2019). *MNRAS* **490**(4), 4666–4687.
Píriz, R. *et al.* (2019). In *Proceedings of the 50th Annual Precise Time and Time Interval Systems and Applications Meeting*.
Ransom, S. *et al.* (2019). *Bulletin of the American Astronomical Society* **51**, 195.
Siemens, X. *et al.* (2013). *Cla. Quant. Grav.* **30**(22), 224015.
Taylor, J. *et al.* (1991). *IEEE Proc.* **79**, 1054–1062.
Verbiest, J. P. W. *et al.* (2018). *Cla. Quant. Grav.* **35**(13), 133001.
Wang, N. (2014). *Sci. Sin. Phys. Mech. Astron.* **44**(008), 783–794.
Yan, J. *et al.* (2020). *Jour. De. Spa. Exp.* **7**(2), 128.
Yang, T. G. *et al.* (2016). *Chin. Astron. Astrop.* **40**(2), 256–265.
Yin, D. S. *et al.* (2017). *Chin. Astron. Astrop.* **41**(3), 430–441.
Zhao, C. S. *et al.* (2018). *Chin. Astron. Astrop.* **42**(2), 291–302.

https://doi.org/10.1142/9789819800643_0017

Chapter 17

The Enhanced X-ray Timing and Polarimetry Mission (eXTP)

Yu-Peng Xu*, Y.P. Xu*,†, A. Santangelo*,†,‡, and Shuang-Nan Zhang*

**Key Laboratory for Particle Astrophysics, Institute of High Energy Physics, CAS, Beijing 100049, China*
†*Institut für Astronomie und Astrophysik, Eberhard Karls Universität, Tübingen 72076, Germany*
‡*andrea.santangelo@uni-tuebingen.de*

eXTP is a flagship space mission for X-ray timing, spectroscopy and polarimetry developed by a large International Consortium led by China. The core objectives of eXTP aim at answering key open questions of fundamental physics under extreme density, gravity and magnetism conditions. In addition, eXTP will be a powerful X-ray observatory, providing unprecedented performance to observe varies objectives including pulsars. The mission has completed phase B study and is expected to be adopted by the end of 2023 and to be launched around 2029.

Keywords: X-ray timing, X-ray polarimetry, Multi-messenger astronomy, Neutron stars, Black holes

Contents

1. Introduction

The enhanced X-ray Timing and Polarimetry mission (eXTP) is China's next-generation flagship X-ray astronomy satellite after the Hard X-ray Modulation Telescope (*Insight*-HXMT). The three core objectives of eXTP aim at answering key open questions of fundamental physics: (1) The physical nature of cold ultra-dense matter; (2) the behaviour of matter and light in the space-time shaped by strong-field gravity; (3) the astrophysics and physics of the strongest magnetic fields in nature. The matter inside neutron stars (NSs), the space-time in the vicinity of the Black Hole (BH) horizon, and the extremely magnetized vacuum close to magnetars and accreting pulsars are uncharted territories of fundamental physics. NSs and BHs provide a unique arena for their exploration. The eXTP mission will revolutionize these fundamental areas of today's research by high precision X-ray observations of NSs across the magnetic field scale and BHs across the mass scale.

In addition to investigating fundamental physics, eXTP will be a very powerful observatory for astrophysics, which will provide observations of unprecedented quality on a variety of galactic and extragalactic objects. In particular, its wide field monitoring capabilities will be highly instrumental in identifying the electro-magnetic counterparts of gravitational wave sources.

What makes eXTP unique in comparison to other existing X-ray missions, including those currently in development, is its unprecedented combination of broad-band large collecting area, polarimetric capability, and spectral resolution. Owing to this combination, eXTP is expected to open an entire new window in X-ray observations: simultaneous spectral-timing-polarimetry. eXTP will thus be complementary to ATHENA in terms of science and payload. The eXTP satellite is planed to be launched to a low earth orbit with an inclination less than 2.5° and an altitude about 550 km around 2029. The mission will continuously monitor the X-ray sky, characterizing the active X-ray Universe on a large range of time scales, and will enable multi-wavelength and multi-messenger studies for gravitational waves and neutrinos sources.

Payload consortium: The SFA and PFA are the responsibility of the IHEP of CAS in China. The LAD is under the main responsibility of Italy (led by INAF). The Wide Field Monitor (WFM) is the responsibility of Spain (led by ICE-CSIC). Other major contributors include Germany (led by the University of Tübingen), Switzerland (led by the University of Geneva), France (CNES, CEA), Poland (CBK), Czech Republic (ASU), Denmark (DTU), The Netherlands (NOVA and SRON), Austria (University of Wien), and Turkey (Sabanci University), as well as several other CAS institutes and universities in China. China will be in charge of the platform and satellite Assembling, Integration, and Testing (AIT), as well as of the launch. The ground segment, operations, and data center infrastructure of the mission will be jointly organized by CAS and several European institutions and agencies.

As shown in Figure 1, the scientific proposal consists of four instruments: the Spectroscopy Focusing Array (SFA), the Large Area Detector (LAD), the Polarimetry Focusing Array (PFA), and the

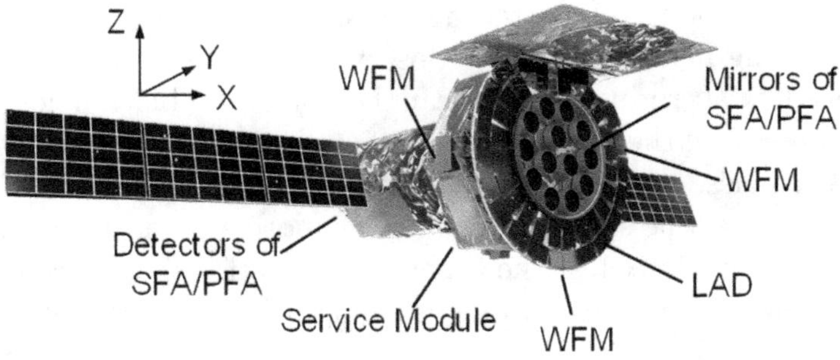

Fig. 1. Artistic view of the eXTP satellite. The science payload consists of four instruments: the focused SFA and PFA telescopes arrays, the large area instrument LAD, and the WFM to monitor a large fraction of the sky.

WFM. The proposed instruments have very large effective area, good spectral resolution, high time resolution and unprecedented polarimetry capabilities in the energy range 0.5–30 keV (up to 80 keV in the expanded mode), so it will be also a powerful X-ray observatory. The instrument configuration and key specifications are summarized in Table 1.

A brief overview of the scientific objectives, instrument configuration, mission profile, and operating principles will be given in this chapter. For more details please refer to the five white papers of the mission (De Rosa, 2019; in't Zand, 2019; Santangelo, 2019; Watts, 2019; Zhang, 2019) published in a special issue of *Science China Physics, Mechanics & Astronomy* and the updated status (Zhang, 2022) published in *Proc. SPIE.*

2. Scientific Goals of the Mission

The three core objectives of the eXTP mission aim at answering key open questions of fundamental physics:

(1) The physical nature of cold ultra-dense matter: To investigate the equation of state (EOS, i.e., pressure-density-temperature relationship) of ultra-dense matter by measuring the mass-radius relationship of neutron stars, to solve the unsolved mystery of whether neutron matter or quark matter dominates the interior of neutron stars, and to test the theory of quantum chromodynamics (QED) at extreme densities.
(2) The behaviour of matter and light in the space-time shaped by strong-field gravity: By observing the X-ray radiation produced by accreting matter near the horizon of black holes, eXTP will measure the mass and spin of stellar-mass and supermassive black holes, test the theory of black hole formation and evolution, and at the same time study the motion of matter near black holes to precisely test the general relativistic effects such as gravitational redshift and frame-dragging under strong gravitational field conditions.
(3) The astrophysics and physics of the strongest magnetic fields in nature: To study the radiation mechanism of magnetars through high-precision X-ray spectral, timing and polarization

Table 1. Full overview of the instrument requirements and goals.

Instrument configuration	Instruments requirements (goals)
Spectroscopy Focusing Array (SFA) 9 telescopes: FL = 5.25 m, Focal plane: 19-cell SDD array	Effective area: $\geq$6,000 (6,600) cm^2 @ 1–2 keV, $\geq$4,400 (4,800) cm^2 @ 6 keV Energy range: 0.5–10 (0.3–10) keV Energy resolution: $\leq$180 (150) eV @ 6 keV Field of view: $\geq\phi 12'$ Angular resolution: HPD $\leq 1'$, W90 $\leq 3'$ Time resolution: $\leq$10 μs Absolute time accuracy: $\leq$2 μs Dead time: $\leq$6% @ 1 Crab Sensitivity: $\leq 3 \times 10^{-12}$erg/s/cm^2 (0.5–10 keV, 5σ, 10s) Background rate: bkg (0.5–10 keV) $\leq$ 0.2 cts/s, bkg (2–10 keV) $\leq$ 0.1 cts/s Throughput: $\geq$1 (2) Crab sustained, $\geq$15 (20) Crab 300 minutes every two days
Large Area Detector (LAD) 40 modules: Capillary plate, collimator, Large area SDD	Effective area: $\geq$3.0 (3.3) m^2 @ 8 keV, $\geq$1.5 (2.0) m^2 @ 2.5 keV, $\geq$0.37 (0.4) m^2 @ 30 keV Energy range: 2–30 (1–40) keV (80 keV extended) Field of view (FWHM): $\leq 65'$ Energy resolution: $\leq$260 (180) eV @ 6 keV (all events), $\leq$200 (160) eV @ 6 keV (single events) Time resolution: $\leq$10(5) μs Background: $\leq$10(5) mCrab Throughput: $\geq$1 Crab sustained, $\geq$15 Crab 300 minutes every two days
Polarimetry Focusing Array (PFA) 4 telescopes: FL = 5.25 m Focal plane: Gas Pixel Detector (GPD)	Effective area: $\geq$300 (500) cm^2 @ 3 keV Energy range: 2–8 (1.5–12) keV Energy resolution: $\leq$25% (20%) @ 6 keV Field of view: $\geq 8' \times 8'$ Angular resolution: HPD $\leq 30''(15'')$ Dead time: $\leq$30% (10%) @ 1 Crab Time resolution: $\leq$10(8) μs Absolute time accuracy: $\leq$4 μs Response stablity: $\Delta P/P \leq$ 5% (2%) Polarimetric sensitivity (MDP): $\leq$3%(1 m Crab, 10^6s) Background rate (2–8 keV): $\leq$0.1(0.05)cts/s/cm^2/det Throughput: $\geq$1(2) Crab sustained
Wide Field Monitor (WFM) 3 pair of cameras: 1.5D coded mask, Position sensitive SDD	Point source localization: $\leq 1'(0.5')$(90%C.L.) Peak Sensitivity (5σ source detection): 5(4) mCrab (50 ks), 1.0(0.8) Crab (1 s) Field of view: $\geq$3.1(4) Sr Angular resolution: $\leq 5'$ (FWHM) Energy range: 2–50(1.5–50) keV Energy resolution: $\leq$500(300) eV @ 6 keV Time resolution: $\leq$10 μs Data downlink delay: $\leq$3 (1.5) hours

observations of neutron stars, and test the theory of quantum electrodynamics (QED) under extreme magnetic field conditions by measuring the vacuum birefringence effect of magnetars and strong magnetic field accretion pulsars with high precision.

In addition to investigating fundamental physics, eXTP will be a very powerful observatory for astrophysics, which will provide observations of unprecedented quality on a variety of galactic and extragalactic objects. In particular, its wide field monitoring capabilities will also be crucial in the context of multi-messenger astronomy by detecting and monitoring the electro-magnetic counterparts of gravitational waves and neutrino cosmic sources. eXTP will operate at the time of operation of major facilities in multi-messenger astronomy, such as the second generation of the GW interferometer network (e.g., aLIGO, aVIRGO, KAGRA, LIGO-India), the neutrino observatories IceCube-Gen2 and KM3NeT, and other multiwavelength facilities including SKA, LOFAR and FAST (radio), HERD (GeV to TeV), CTA and LHAASO (from TeV to PeV), and ALMA (mm).

3. The Science Payload

The current baseline of the scientific payload includes four science instruments: the Spectroscopy Focusing Array (SFA), the Large Area Detector (LAD), the Polarimetry Focusing Array (PFA), and the Wide Field Monitor (WFM). The instrument configuration and specifications are summarized in Table 1.

Spectroscopy Focusing Array (SFA)

The SFA consists of an array of 9 identical Wolter-I grazing-incidence X-ray telescopes (see Figure 2), and is mainly used for spectral and timing observations in the energy range 0.5–10 keV. Each telescope consists of one mirror module and one Silicon Drift Detector (SDD) camera at its focal plane (FP). The angular resolution of the Wolter-I mirror is required to be better than 1 arcmin (Half power diameter, HPD), which can reduce the background well when coupled with the electron deflector. Each SFA FP camera includes a 19-cell SDD array, whose energy resolution is better than 180 eV at 6 keV. The time resolution is 10 s, and the dead time is expected to be less than 5% at

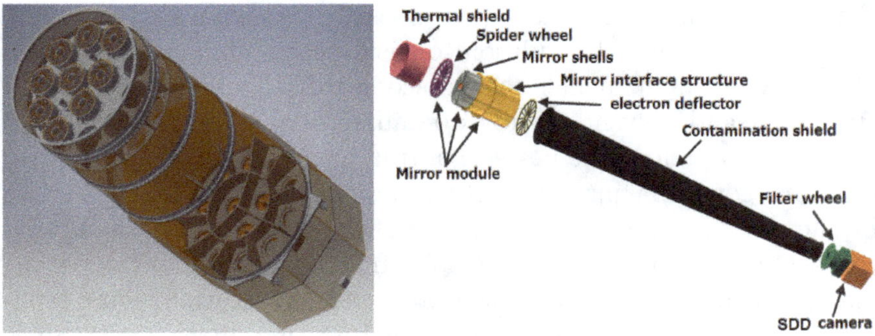

Fig. 2. (a) Layout of the nine SFA telescopes (the four empty holes are reserved for PFA telescopes); (b) Schematic structure of one SFA telescope.

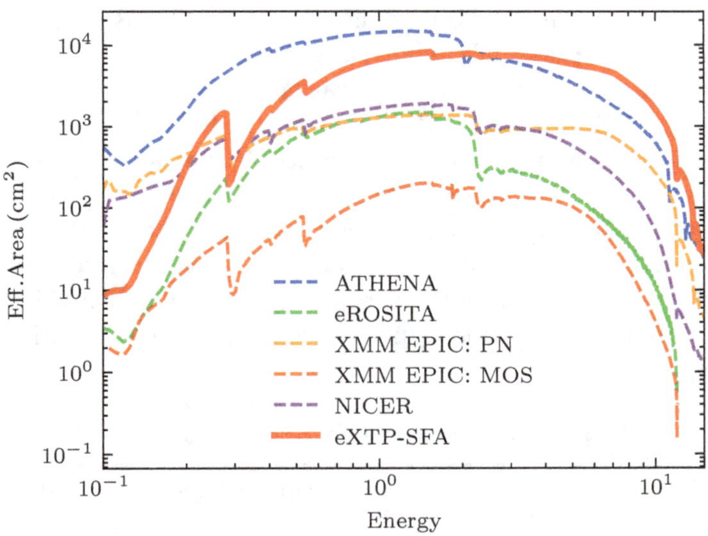

Fig. 3. Comparison of the total effective area between SFA and other relevant X-ray focusing telescopes.

1 Crab. The total effective area of SFA is required to be larger than 6,000 cm^2 at 1–2 keV, and the expected performance in comparison with other X-ray focusing telescopes can be seen in Figure 3.

The mirror assembly (including the electronic deflector) of each telescope is installed on the optical bench of the satellite. The focal plane camera is installed in the payload compartment, located on the other end of the satellite. The two compartments are connected by the telescope tube. The telescope tube is also the main support structure of the platform. In addition to ensuring the alignment accuracy of the two compartments, the telescope tube is also used to avoid stray light and prevent possible contamination.

The optics adopt the nested multi-layer Wolter-I type concept to achieve a good imaging capability and a high effective area (45 shells in one mirror module). Each mirror shell consists of a coaxial and confocal paraboloid and hyperboloid. In consideration of the performance, weight, space environment and cost, the electroforming replication process is employed to manufacture the mirror shells. Since the requirements of the optics in SFA and PFA are nearly the same, the same optical design and interfaces are adopted to simply the engineering implementation and reduce the risk. The main difference is the reflective coating. The baseline design of SFA uses gold coating electroformed nickel shells, while PFA uses electroformed nickel shells directly. The main technical specifications can be seen in Table 2 and the design of the mirror module is shown in Figure 4.

Each SFA focal plane camera consists of 4 parts (see Figure 4): an SDD array, an electronics box, a filter/calibration wheel, and a thermal electric cooler (TEC). The SDD array consists of 19 hexagonal cells with side length of 3.2 mm and deplete layer thickness of 450 m. The total sensitive area of the array is 5 cm^2. The detector box hosts the electronics boards for signal processing, data acquisition, data

Table 2. Main specifications of SFA/PFA mirror module.

Parameters	Specification for one mirror module
Focal length	5250 mm
On-axis collecting area	$\geq$820 cm^2 at 1.5 keV (SFA) $\geq$800 cm^2 at 3 keV(PFA) $\geq$550 cm^2 at 6 keV
Angular resolution (HPD)	$\leq 1'$(SFA), $\leq 30''$ (PFA, goal $15''$)
Energy range	0.3–10.0 keV
Refection surface	gold (SFA), nickel (PFA)
Envelope dimension	Diameter $\leq$ 600 nm
Working temperature	20 ± 2°C
Weight	$\sim$125 kg

Fig. 4. (a) One mirror module; (b) One SFA focal plane camera.

reduction, power supplies and telemetry & tel-command interfaces. The wheel mechanism has 6 positions, which are optical blocking filters with 3 different thicknesses, in-flight calibration source (^{55}Fe), a thin polyimide protection layer for weak source observations, and a metal shutter for internal background measurement and protection from damage during launch and extreme solar events.

Large Area Detector (LAD)

The LAD is designed to perform, on a large collecting area, photon-by-photon observations of X-ray sources in the energy range of

2–30 keV (up to 80 keV in the expanded mode, for out-of-field-of-view burst events). The LAD large area is achieved by a modular and intrinsically highly redundant design, which consists of 40 identical detection modules (called LAD module, see Figure 5). Each module consists of a set of 4×4.

Large area SDDs to measure the energy and time of arrival of the incident photons with time resolution better than $10\,\mu$s, and 4×4 capillary plate collimators to reduce the backgrounds caused by the diffused sources and X-ray backgrounds. The effective area of LAD is required to be larger than 3.0 m^2 at 8 keV, and the expected performance in comparison with other collimated X-ray telescopes can be seen in Figure 6. The details of the LAD design can refer to Feroci (2022) and the references therein.

Polarimetry Focusing Array (PFA)

The PFA consists of 4 identical telescopes optimized for X-ray imaging polarimetry, sensitive in the energy range of 2–8 keV. It offers energy, time, and spatially resolved X-ray polarimetry at high sensitivity, and is the only instrument on eXTP capable of imaging with a resolution of $\leq 30''$. The PFA features a sensitivity in polarimetry down to 3% given a 1 ks observation of the Crab. The total effective area of PFA is required to be larger than 300 cm^2 at 3 keV. The 4 PFA telescopes share the same optical design with SFA. The configuration of the PFA mirror module is also the same as SFA, except the reflection surface is nickel instead of gold, and the required angular resolution is much better. The focal plane camera comprises detector and electronics boxes (see Figure 7). A detector box contains a Gas Pixel Detector (GPD) and filter wheel mechanism. There are two polarized calibration sources, three unpolarized calibration sources, an opened aperture for week source observations, a gray filter for bright source observations, and a metal shutter on the filter wheel. The functions of the electronics box are data acquisition, scientific data packaging and transmission, command receiving and execution, housekeeping, GPD high voltage bias supply, GPD active thermal control, and secondary power supply. The expected performance in comparison with IXPE can be seen in Figure 8.

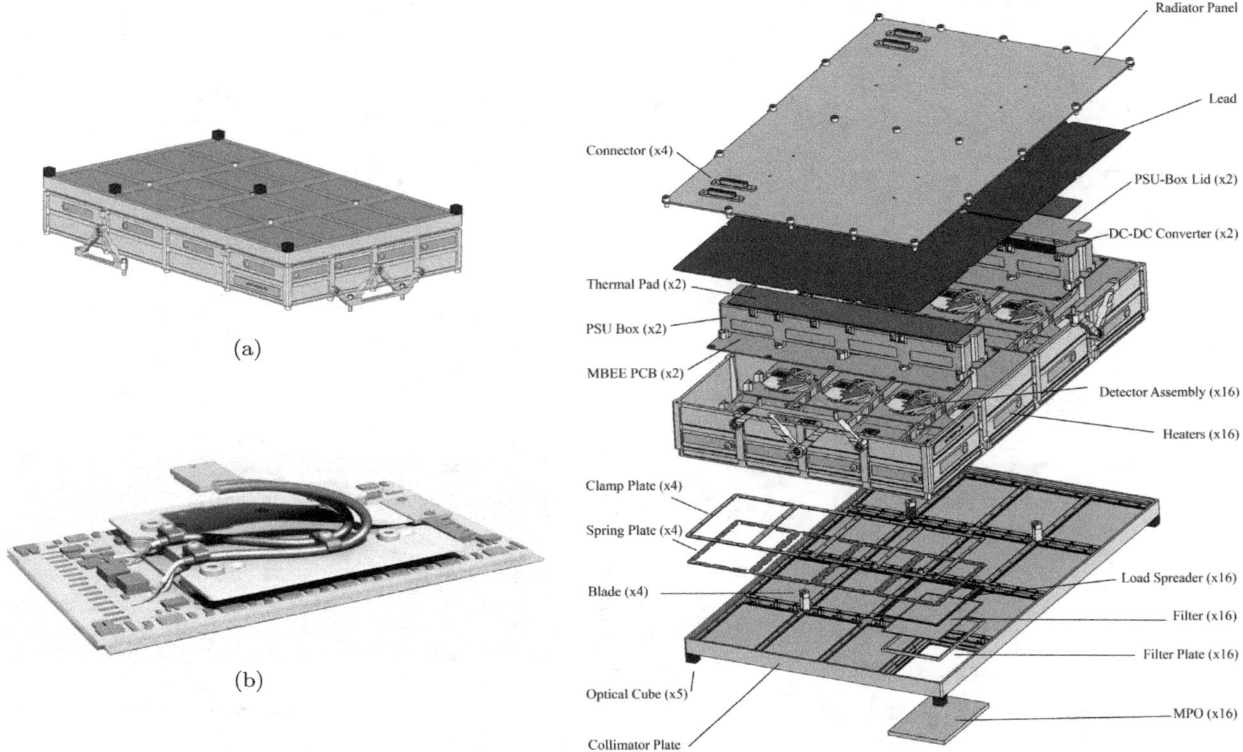

Fig. 5. The current design of the LAD Module. Left panel: LAD module assembled (a), and a detail of the LAD Detector Assembly (b). Right panel: a exploded view of the module.

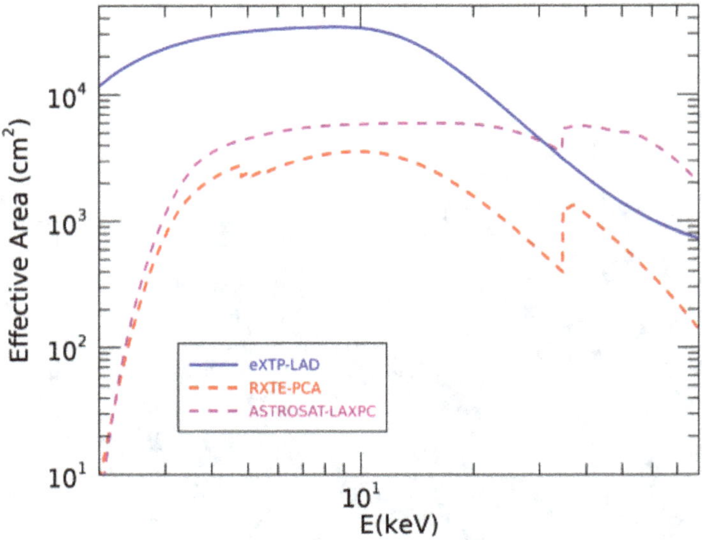

Fig. 6. Comparison of the total effective area between LAD and other collimated X-ray telescopes.

Wide Field Monitor (WFM)

The WFM consists of 6 wide field of view coded mask cameras, equipped with position sensitive detectors and arranged in three camera pairs (see Figure 9). Each camera has an effective field of view of 30° (fully coded) and 90° at zero response. The camera pairs axis are separated by ±60° around the line of sight of the Narrow Field of View Instruments (SFA, LAD and PFA), thus covering 3.1 sr (goal 4.1 sr, ∼33% of the sky) and operating in the energy range 2−50 keV.

The main task of the WFM is to conduct long-term monitoring of targets of interest, and trigger and localize transient events (such as Gamma-Ray Bursts). WFM can provide important information of transient sources for the main telescope arrays, and decide whether to maneuver the platform to perform follow-up observations to transients according to pre-determined strategies. The comparison of the field of view between WFM and other wide field monitors can be seen in Figure 10. The details of the WFM design can refer to Hernanz (2022) and the references therein.

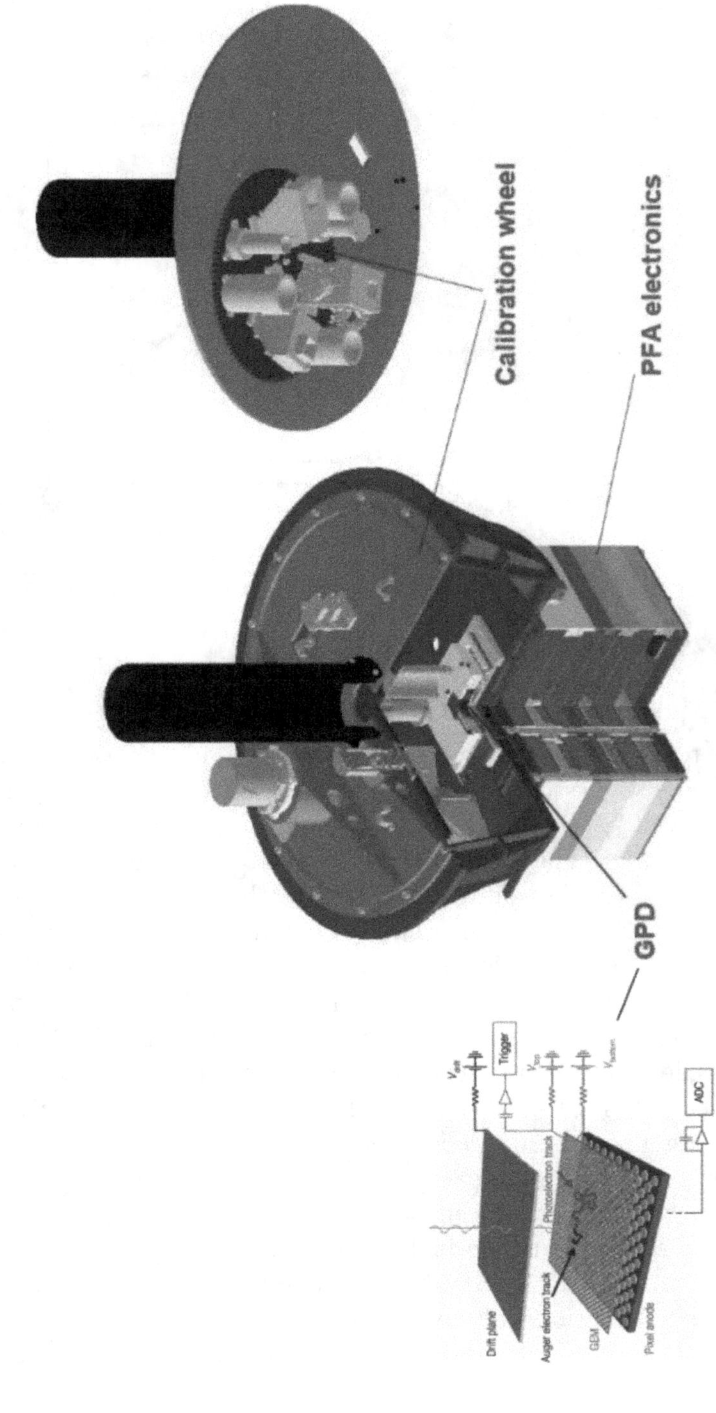

Fig. 7. Focal plane camera of PFA.

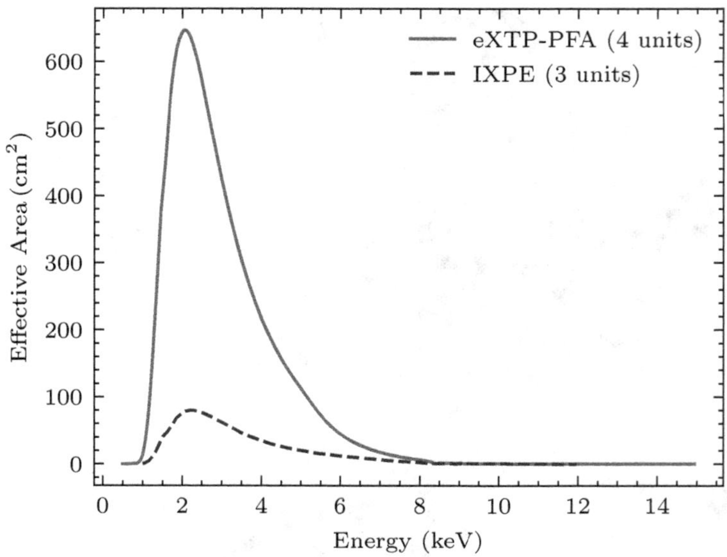

Fig. 8. Comparison of the total effective area between PFA and IXPE.

4. Mission Overview

According to the Space Science Programs of China, the eXTP mission consists of six systems: satellite system, launcher system, launch site system, TT&C system, ground support system, and science application system. The overall eXTP mission will be designed and implemented under China's responsibility with major contributions from the European Consortium. The system architecture is shown in Figure 11. The main technical specifications can be seen in Table 3.

The eXTP satellite will consist of a Chinese platform supporting the payloads. The Chinese side will provide SFA and PFA. The European side will provide LAD and WFM. The three Narrow Field Instruments (NFIs: SFA, LAD and PFA) are installed coaxially to observe a target source, and WFM monitors sources in its very large FoV. In order to ensure the energy resolution of LAD and minimize the particle background of the payload, the orbit is selected as a 550 km circular orbit with an inclination angle of $<2.5°$.

In order to meet the observational requirements, there are three types of communication links between the satellite and the ground station, i.e., S-band data link, X-band data link, Beidou short

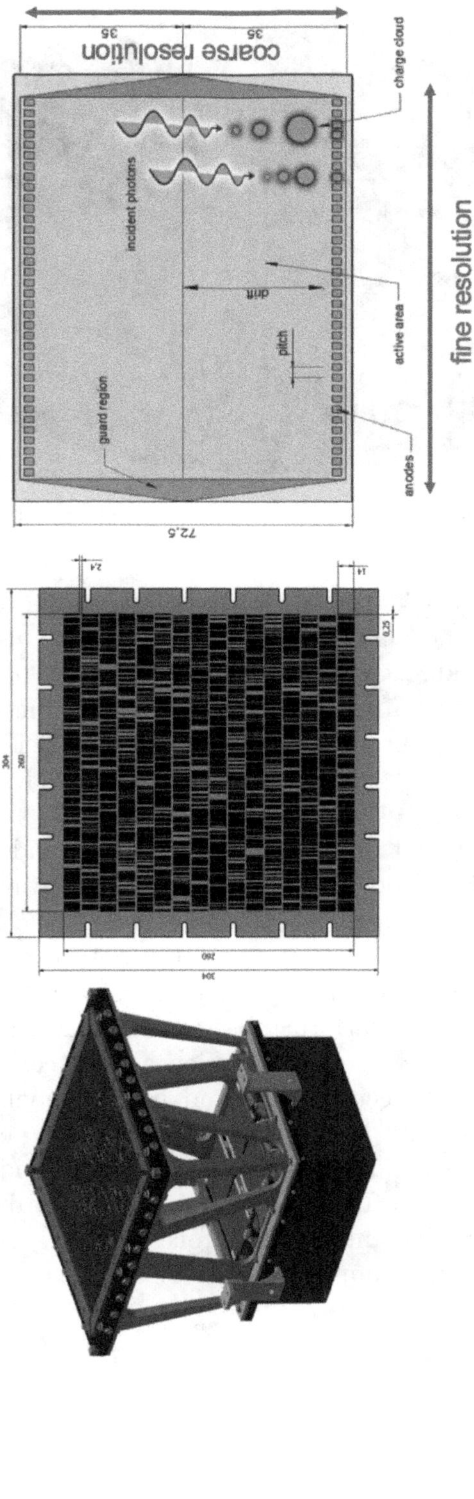

Fig. 9. WFM camera concept.

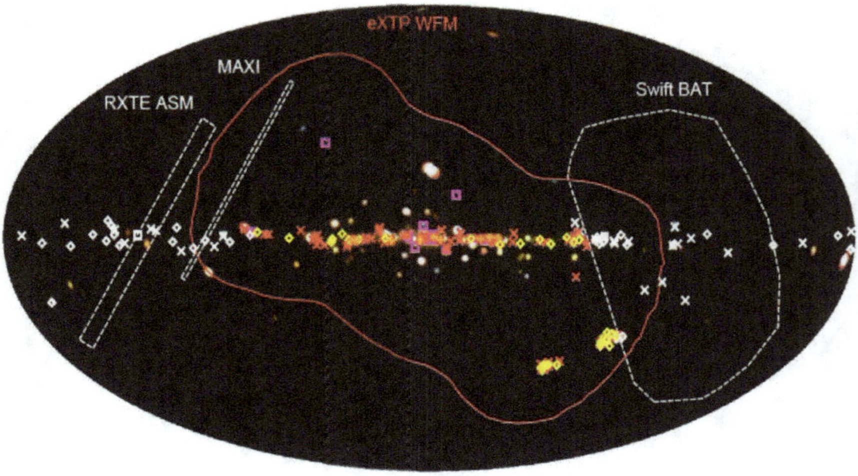

Fig. 10. Comparison of the field of view between WFM and other wide field monitors.

message link and VHF stations (a heritage from the China-France SVOM mission to be launched in 2024). S-band is used for satellite telemetry, tracking, and command (TT&C), X-band for scientific data down transmission, and Beidou short message link for the rapid uplink of ToO observation tasks. Both Beidou and VHF networks will be used for real-time downlink of WFM burst alert information.

The ground segment of eXTP consists of ground support system (including mission operation center, ground stations, and burst alert subsystem) and science application system (including science operation center and China/Europe science data center), as shown in Figure 11. The ground control center is responsible for satellite platform maintenance, platform observation planning and s-band telemetry data reception, etc. The science data center is responsible for observation execution, X-band data receiving, and long-term data archiving and storage. Considering the near-equatorial orbit, ground stations close to the equator are provided by China and European partners. The science data center is responsible for observation planning, scientific data processing, trigger information processing and distribution, and target of opportunity observation planning, etc.

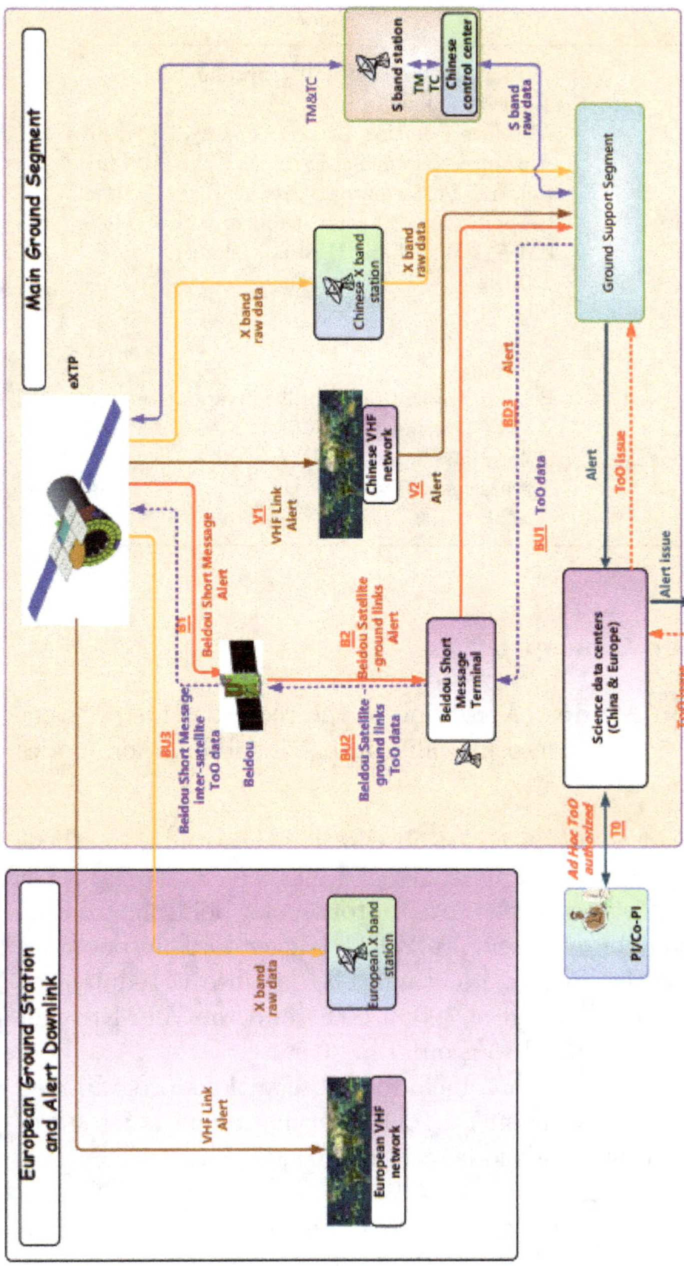

Fig. 11. eXTP system architecture.

Table 3. Mission profiles of eXTP.

Item	Value
Orbit	550 km, inclination $\sim 0°$, circular orbit
Pointing	3-axis stabilized Absolute Pointing Error (APE): $\leq 3'$ (3σ) Absolute Knowledge Error (AKE): $\leq 1'$ (3σ) Relative Performance Error (RPE): $\leq 1'$/s
Maneuverability	Re-pointing $\geq 30°$ and stabilize with in 600 s
Launch	LM-5, @Wenchang (Hainan Island)
Launch mass	5300 kg
Onboard mass storage	5 Tbits
Data telemetry	X-band, 210 Mbps
TT&C	S-Band
Burst alert	Beidou Navigation Satellite System; VHF transmitter (SVOM)
Ground Stations	Colombo (China), Malindi (Italy)
Mission duration	5 years (goal 8 years)
Launch data	$\sim$2029

5. Observation Concept

(1) **Observation Modes:** According to the requirements of scientific observation and on-orbit calibration, the observation modes mainly include:

(a) *Routine observation mode*: observations according to scheduled targets.

(b) *Follow-up observation mode*: autonomous pointing towards a detected burst after the WFM trigger and autonomous return to the current pointing. The satellite is required to re-point with a speed of 30° in 600 s and the WFM trigger should be download to ground in 30 s.

(c) *ToO observation mode*: including observations programmed by the ground segment by ToO observing requests for astronomical events, additional calibrations, with time delay constraints.

(d) *Small sky scan mode*: for PSF or vignetting function calibration of SFA, LAD and PFA, by scanning a $3° \times 3°$ area centered to a known X-ray source.

(2) **Pointing Strategy:** According to the payload constraints on safety and thermal control, the pointing strategy of the satellite is as follows (see the coordinate system defined in Figure 1):

(a) The angle between the Sun and +X direction should be >60°, to avoid stray light and thermal disturbance of all instruments.

(b) The angle between the Sun and +X direction should be <120°, to achieve stable thermal environment for LAD (180°, for observations without strict energy resolution requirement).

(c) The Sun should be kept in the X-Z plane towards +Z. The angle between the Sun and the X-Z plane should be within ±2°.

(3) **Science Operations and Policy:** The observational program of eXTP includes a core program and a guest observer (GO) program. It is estimated that 50% of the observing time is needed for the core program. During the nominal mission phase 50% observing time will be devoted to the GO program and to the Director Discretionary Time (DDT), used for ToOs. Both GO and DDT observations will be open to the whole international community. After the completion of the core program, the full (or most) observing time will be devoted to GO and DDT. Data relative to transient information like alert information regarding transient/outburst sources generated from WFM data, will be released immediately. SFA/LAD/PFA observational data will be released when the data proprietary period expires. All data from DDT observations will be released immediately.

eXTP operations are planned for 5 years, but the mission could be extended to an additional period of 3 years or any additional period that may be agreed by the participating parties. A controlled re-entry at the of lifetime will be carried out to comply with casualty-risk regulations.

6. Summary

In this chapter, we introduced the instrument configuration, expected performance and operation scheme of the eXTP mission. The mission will continuously monitor the X-ray sky, characterizing the active

X-ray Universe on a large range of time scales, and will enable multi-wavelength and multi-messenger studies for gravitational waves and neutrinos sources.

The eXTP phase A and B studies were successfully concluded, and the mission is under evaluation for adoption. The launch of eXTP is planned for 2029.

Acknowledgments

This work was supported by Strategic Priority Research Program of Chinese Academy of Sciences (XDA15020100), DLR of Germany (FKZ 50 OO 1902), the Italian Space Agency (under agreement n. 2020-3-HH.0 and ASI-INAF n.2017-14-H.O) and INFN (project XRO – X-ray Observatories), and MICIN/AEI of Spain (PID2019-108709GB-I00).

Bibliography

De Rosa, A. *et al.* (2019). Accretion in strong field gravity with eXTP, *Sci. China Phys. Mech. & Astron.* **62**(2), 29504.

Feroci, M. *et al.* (2022). The large area detector onboard the eXTP mission, In *Proc. SPIE* **12181**, 121811X.

Hernanz, M. *et al.* (2022). The wide field monitor onboard the Chinese-European X-ray mission eXTP, In *Proc. SPIE* **12181**, 12181Y.

in't Zand, J. J. M. *et al.* (2019). Observatory science with eXTP, *Sci. China Phys. Mech. & Astron.* **62**(2), 29506.

Santangelo, A. *et al.* (2019). Physics and astrophy of strong magnetic field systems with eXTP, *Sci. China Phys. Mech. & Astron.* **62**(2), 29505.

Watts, A. L. *et al.* (2019). Dense matter with eXTP, *Sci. China Phys. Mech. & Astron.* **62**(2), 29503.

Zhang, S.-N. *et al.* (2019). The enhanced X-ray timing and polarimetry mission — eXTP, *Sci. China Phys. Mech. & Astron.* **62**(2), 29502.

Zhang, S.-N. *et al.* (2022). The enhanced X-ray timing and polarimetry mission: eXTP: An update on its scientific cases, mission profile and development status, In *Proc. SPIE* **12181**, 12181W.

https://doi.org/10.1142/9789819800643_0018

Chapter 18

GECAM: A Gamma-Ray Monitor Constellation

Shaolin Xiong[*,‡], **Yanqiu Zhang**[*,†], **Chenwei Wang**[*,†]

[*]*Key Laboratory of Particle Astrophysics, Institute of High Energy Physics, Chinese Academy of Sciences, Beijing 100049, China*

[†]*University of Chinese Academy of Sciences, Beijing 100049, China*

[‡]*xiongsl@ihep.ac.cn*

Gravitational wave high-energy Electromagnetic Counterpart All-sky Monitor (GECAM) is a dedicated constellation of spacecrafts equipped with gamma-ray detectors (about 6 keV to 5 MeV) to explore high energy transients, including Gamma-ray Burst (GRB), Soft Gamma-ray Repeater (SGR), etc. Thanks to the high timing accuracy, all-sky field of view and wide energy band, GECAM can also be used to monitor pulsars all over the sky. GECAM constellation is currently composed of four members on orbit: GECAM-A and GECAM-B micro-satellites launched in December 2020, GECAM-C onboard SATech-01 satellite launched in July 2022. The fourth member DRO/GTM (aka GECAM-D) was launched in March 2024. With all these instruments, GECAM have been able to provide a full-time monitoring of high energy transients in the full-sky.

Keywords: All-sky Gamma-ray monitor, Gamma-ray burst, Soft Gamma-ray repeater, Gravitational wave electromagnetic counterpart

Contents

1. Introduction

Multi-messenger observation of the first binary neutron star merger (GW170817) both in gravitational wave (GW) and electromagnetic wave (EM) heralded the multi-messenger era of astronomy (Abbott *et al.*, 2017). During the observation campaign of GW170817, the gamma-ray burst (GRB 1701817A) associated with this GW source has played a very important role both in the joint observation and in the physics research of this GW source. Because the GRB 170817A is rather weak in brightness and soft in spectrum, there are only two gamma-ray instruments (i.e., Fermi/GBM and INTEGRAL/SPI-ACS) detected this GRB (Goldstein *et al.*, 2017; Savchenko *et al.*, 2017). On the other hand, the occurrence rate of this kind of GRB associated with binary neutron star merger is estimated to be very low.

Aiming to increase the detection capability of the GRB associated with GW source, we proposed the Gravitational wave high-energy Electromagnetic Counterpart All-sky Monitor (GECAM) mission in 2016, soon after the announcement of the first direct detection of the gravitation wave from compact binary merger (GW150914). The development of the key technology and prototype of GECAM was funded in 2016 by the Chinese Academy of Sciences (CAS). As the first Mission Of Opportunity of CAS, GECAM mission was adopted in 2018 and two GECAM micro-satellites (GECAM-A and

GECAM-B) were launched in 2020. Since these two micro-satellites suffered power supply issues, more instruments (GECAM-C and GECAM-D) have been built and launched to improve and extend the observation capability of the GECAM.

The key science driver of GECAM is to discover new high energy counterpart of GW source and to study if the neutron star-black hole (NSBH) merger and binary black hole (BBH) merger can radiate high energy photons. In addition to the GW-related science topic, the key science cases of GECAM also include Gamma-Ray Burst (GRB), Soft Gamma-ray repeater (SGR) and the high energy counterpart of fast radio burst (FRB).

GRB is the most violent stellar explosive phenomenon in the universe (Zhang *et al.*, 2018). The duration of GRBs vary from a few milliseconds to hours. The long GRB (with duration greater than 2 seconds) are believed to origin from the collapse of massive stars, while the short GRB (with duration less than 2 seconds) is from the merger of binary neutron stars or neutron star and black hole (Abbott *et al.*, 2017; Woosley *et al.*, 2006). Despite of many progress of GRB study in the past decades, the exact physics and mechanisms of GRB are yet to be answered, such as the formation, geometry and composition of the jet, the energy dissipation and radiation mechanisms (Zhang *et al.*, 2004).

SGR is characterized by the intense and brief bursts in soft gamma-rays, typically lasting for tens of milliseconds to a few seconds (Norris *et al.*, 1991). On contrast to the catastrophic nature of the GRB, SGR displays repeating burst from a special type of neutron star with extremely high magnetic field called magnetar (Kouveliotou, 1998; Mereghetti *et al.*, 2015; O'Meara *et al.*, 1999). SGR can be classified as short burst (usually less than 1 second), intermediate burst and giant flare which can lasts for tens of seconds. The observation of an X-ray burst from SGR 1935+2154 associated with the FRB200428 by the Insight-HXMT (Li *et al.*, 2021) and other high energy telescopes indicates that magnetar can emit FRB.

Moreover, GECAM could monitor all kinds of high energy Gamma-ray transient in the space, including the high energy burst from X-ray Binary (XRB), tidal disruption event (TDE), Solar Flare (SFL), Terrestrial Gamma-ray Flash (TGF) and Terrestrial Electron Beam (TEB).

2. GECAM Constellation

The original GECAM mission (Li *et al.*, 2020, 2022) is composed of two identical micro-satellites (GECAM-A and GECAM-B), which was launched to the Low Earth Orbit (LEO) in December 10, 2020. As the subsequent instruments in the GECAM constellation, GECAM-C (aka SATech-01/HEBS)(Zhang *et al.*, 2023a) was launched to LEO onboard the SATech-01 satellite in July 27, 2022, and GECAM-D (aka DRO/GTM) was launched onboard the DRO-A satellite in March 13, 2024. Each GECAM instrument has an all-sky field of view (excluding the sky region blocked by the Earth), wide energy range, trigger and localization of bursts in real-time (Zhao *et al.*, 2021) and prompt down-link of trigger alerts using the Global Short Message Communication of Beidou satellite navigation system (Guo *et al.*, 2023; Huang *et al.*, 2023).

It is worth to note that, a series of novel technologies have been developed and applied in the GECAM mission, including the compact and high performance detector, the new electronics with high time accuracy and high bandwidth of data transmission (Li *et al.*, 2020, 2022; Zhang *et al.*, 2023a). Especially, the near real-time down-link of trigger alerts with Beidou system was firstly proposed and utilized by the GECAM mission (Guo *et al.*, 2023; Huang *et al.*, 2023).

Comprehensive calibrations have been implemented for detectors for all detectors of GECAM instruments (Zheng *et al.*, 2022, 2023, Zhang *et al.*, 2023a, 2023b). The gamma-ray detectors (GRDs or GTPs) are the main detector of all GECAM instrument, which is in charge of monitoring the sky in the hard X-ray and gamma-ray band. However, GRDs are also able to detector charged particles (such as electrons, positrons, protons). Charged particle detectors (CPDs) are equipped to detect charged particles and to monitor the space environment for GECAM-A, GECAM-B and GECAM-C. CPDs are designed to be sensitive to the charged particles and much less sensitive to photons (Xu *et al.*, 2022). Therefore, combination of GRD and CPD could determine whether a burst is caused by gamma-rays or charged particles, which is essential for the burst classification (Zhao *et al.*, 2021).

Burst location in the sky coordinates could be derived from the data of multiple GRD detectors (with different orientation) for each GECAM instrument. Moreover, the time-delay of the burst light

Table 1. The basic information of GECAM Constellation.

Instrument	Launch time	Orbit	Detectors	Energy range	Status
GECAM-A GECAM-B	December 10, 2020	LEO	25 GRDs, 8 CPDs	10 keV–5 MeV	In operation
GECAM-C	July 27, 2022	LEO	12 GRDs, 2 CPDs	8 keV–6 MeV	In operation
GECAM-D	March 13, 2024	DRO	5 GTPs	20 keV–1 MeV	Commissioning

Notes: LEO: Low earth orbit; DRO: Distant retrograde orbit.

curves among multiple GECAM instruments could provide the triangulation localization (Table 1). Especially, the mission orbit of GECAM-D is about 300,000 km to 400,000 km from the Earth, which is beneficial for the time-delay localization of bursts.

To estimate the probability of catching a short transient source randomly occurred all over the sky, we calculate the time-averaged all-sky coverage achieved by the GECAM-B and GECAM-C individually and the joint observation of GECAM-B and GECAM-C. Because GECAM-A has not observing routinely and GECAM-D is in the commissioning phase, here we only consider the GECAM-B and GECAM-C which have been regularly observing. As shown in Figure 1, one can find that the field of view of these two GECAM instruments can cover about 50% of the transients randomly in the sky. As the GECAM-D is joining in, the sky coverage of GECAM constellation is expected to be nearly 100% all-sky all-time.

2.1. *GECAM-A and GECAM-B*

To alleviate the Earth occultation problem for satellite in the LEO, the GECAM mission was designated to consist of two micro-satellites (named as GECAM-A and GECAM-B, Figure 2) (Li *et al.*, 2020, 2022). Each satellite carries 25 GRDs and 8 CPDs, which is capable to monitor the all-sky unblocked by the Earth. These two identical satellites are designed to operate in the same LEO orbit (height of 600 km and inclination angle of 29°) but with opposite orbital phases, providing an instantaneous full-sky coverage.

However, both GECAM-A and GECAM-B satellites suffered power supply issues after the launch, resulted in the limited observation time and sky coverage for these two instruments. For most

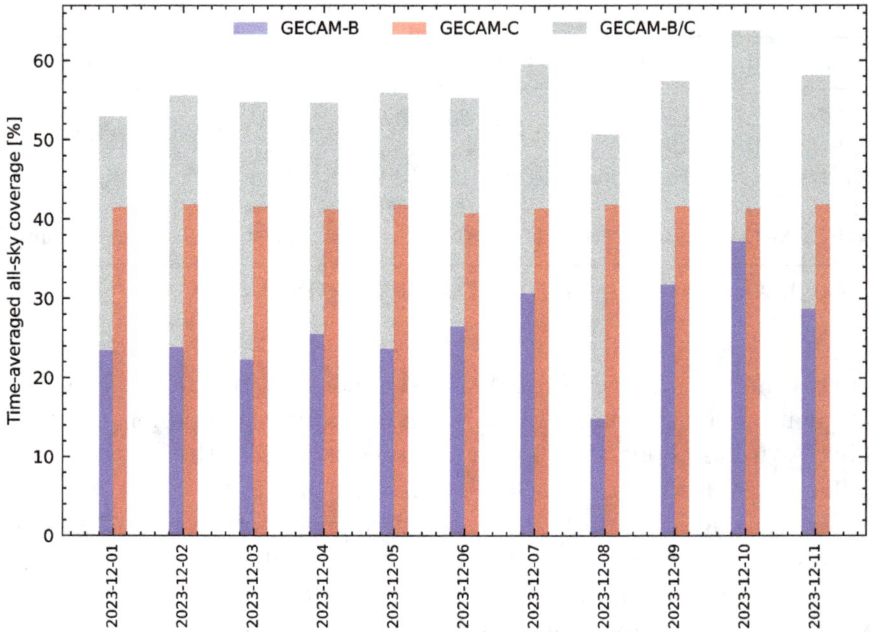

Fig. 1. Time-averaged all-sky coverage for GECAM-B, GECAM-C and the combination of GECAM-B and GECAM-C.

of the operation time, GECAM-B has a half power supply, thus it can only observe about 6 to 10 hours per day. From May 31, 2022 to January 11, 2023, the full power supply of GECAM-B has been recovered, enabling a continuous observation (about 20 hours per day) of GECAM-B. The power supply of GECAM-A is still in the recovering process and a total observation of 300 hours has been made by GECAM-A. After the recovery of the power supply of both micro-satellites, they will be deployed into the opposite orbital phases.

As the main detector of GECAM, each GRD of GECAM-A and GECAM-B operate in two readout channels: high gain (HG) and low gain (LG), which are independent in terms of data processing, transmission, and dead-time (An *et al.*, 2022). On the other hand, there is only one readout channel for each CPD. Comprehensive on-ground and in-flight calibrations have been implemented for GRD and CPD detectors (Zhang *et al.*, 2023b; Zheng *et al.*, 2023).

The relative timing accuracy of GRD and CPD detectors are found to be about 0.1 microsecond by on-ground and in-flight

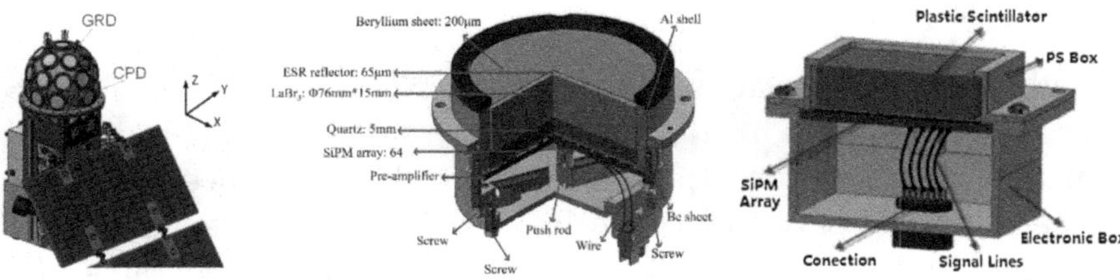

Fig. 2. (a) Illustration of GECAM micro-satellite (GECAM-A and GECAM-B). (b) GRD of the GECAM micro-satellites (An *et al.*, 2022). (c) CPD detector of the GECAM micro-satellites (Xu *et al.*, 2022).

measurements, which is about one order of magnitude better than other gamma-ray monitors of this kind, while the absolute timing accuracy is about several microseconds (Xiao *et al.*, 2023). The lower limit of the detection energy of GRD has been increasing from about 5 keV at the early commissioning phase after the launch to about 20 keV after about three years of operation in the orbit. On the contrast, there is no significant change in the upper limit of the detection energy of GRDs. This degeneration of the lower energy range is primary caused by the radiation effects of the SiPM of GRD detector which is used to read out the scintillation light produced by the crystal (Zhang *et al.*, 2023a).

GECAM-B has triggered many bursts with its real-time in-flight trigger and localization onboard (Zhao *et al.*, 2021), and their trigger alerts have been down-linked in near real-time (time latency is about one minute) through the Beidou satellite navigation system (Guo *et al.*, 2023). The trigger algorithm has been executed for different time scales and energy ranges for each GRD detector. In addition to the real-time in-flight trigger, there is a ground search for bursts in the GECAM data processing pipeline, although the GECAM telemetry data has a time latency about tens of hours.

With the 25 GRDs with different orientation, GECAM-B has proven its localization capability for various bursts (such as GRB, SGR, TGF) by matching the measured counts to the simulated one of all GRD detectors (Zhao *et al.*, 2023). The joint data analysis of GRD and CPD detectors have played an important role in the classification of bursts and especially the observation of the TGF and TEB study.

2.2. *GECAM-C*

To enhance the observation capability of GECAM-A and GECAM-B micro-satellites, GECAM-C (also known as High Energy Burst Searcher, HEBS, Figure 3) (Zhang *et al.*, 2023a) has been launched onboard the SATech-01 (Space Advanced Technology demonstration satellite, (Zhang *et al.*, 2023c)) on July 27, 2022. It operates in a Sun-synchronous orbit at an altitude of 500 km and an inclination angle of 97.4°.

GECAM-C is equipped with 12 GRDs and 2 CPDs, which primarily inherits the design of detectors of GECAM-A and GECAM-B. However, some detailed design of the detector and readout electronics of GECAM-C is different from the GECAM-A and GECAM-B (Zhang *et al.*, 2023a). GECAM-C is also capable of real-time in-flight trigger and localization (Zhao *et al.*, 2021), and it also has the near real-time down-link of trigger alert through the Beidou satellite navigation system (Guo *et al.*, 2023). Moreover, detailed spectral and temporal calibrations have been executed for GECAM-C detectors (Xiao *et al.*, 2023; Zhang *et al.*, 2023b; Zheng *et al.*, 2023).

We note that, owing to the high inclination angle (97.4°) of the orbit, the background variation of GECAM-C is much more complicated than that of GECAM-A and GECAM-B in the orbit of lower inclination (29°), because the flux of charged particle (mostly electron and proton) is very high and vibrant during the high latitude

Fig. 3. Left: overview of the SATech-01 satellite. GECAM-C consists of two detector domes at the top and bottom to monitor the all-sky region unblocked by the Earth (Zhang *et al.*, 2023a). Middle and Right: The detector dome on the top and bottom respectively (Zhang *et al.*, 2023a). GRD and CPD detectors are arranged for different orientation for a large field of view.

regions (i.e., greater than about 30° in the geographic latitude). To make better use of the observation time of GECAM-C, a special working mode is designed for GECAM-C when it pass through the high latitude regions: only turn on a few GRD and CPD detectors to avoid large current of detectors, discard the event-by-event data to save data volume, and disable the in-flight trigger to eliminate the false triggers caused by charged particles (Zhang *et al.*, 2023a).

2.3. *GECAM-D*

To achieve the goal of an instantaneous full-sky monitoring all the time and to further enhance the localization of gamma-ray transients with triangulation method, we proposed the Gamma-ray Transient Monitor (GTM) onboard the DRO-A satellite (Feng *et al.*, 2024) which was launched to deep space on March 13, 2024. Since GTM inherits the technical development and scientific operation of GECAM, it is also dubbed as GECAM-D in the GECAM family.

DRO (distant retrograde orbit) is a highly stable retrograde orbit located within the Earth-Moon system, which is about 310,000 km to 450,000 km from the Earth and 50,000 km to 100,000 km from the Moon. By operating in this deep space orbit, GECAM-D is free of many issues experienced by the instrument in the LEO orbit, such as the turn off of instrument during the SAA passage, significant background variation in the high latitude region, and the Earth blocking of the field of view. On the other hand, the long distance to the Earth can provide a long baseline for the triangulation localization with those GECAM instruments and other telescopes (such as Insight-HXMT, GRID, Fermi/GBM) in the LEO orbit.

GECAM-D consists of 5 Gamma-ray Transient Probe (GTP) detectors, which are installed on the different surface of the DRO-A satellite (as shown in Figure 4). This arrangement ensures the instantaneous all-sky coverage and a moderate localization capability to gamma-ray transients. The detection energy range is from about 10 keV to 1 MeV as there is only one readout channel for GTP detector. Limited by the resource of the satellite, there is no charged particle detector for GECAM-D.

The detailed design of the GTP is different from the GRD of GECAM-A/B/C. For example, each GTP detector has a diameter of 11.5 cm, which results in about two times of the detector area of

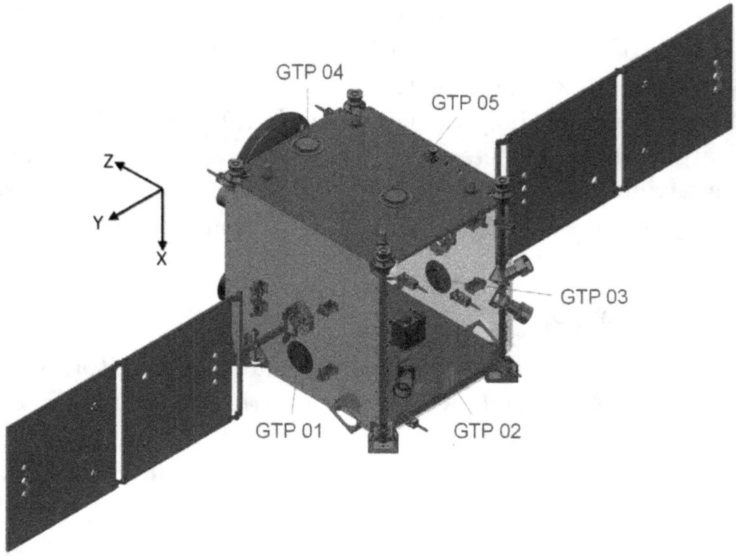

Fig. 4. GECAM-D (aka GTM) onboard the DRO-A satellite. GECAM-D consists of 5 GTP detectors installed on the four sides of the DRO-A.

each GRD. To facilitate the in-orbit calibration, each GTP detector carries a radioactive source of ^{241}Am (Feng *et al.*, 2024).

GECAM-D is also equipped with the real-time in-flight trigger and localization software which is able to find various bursts, to provide an initial classification of the triggered bursts and to produce an enriched trigger alert data for interesting triggers (Zhao *et al.*, 2021). Despite of the deep space orbit, GECAM-D can send down the low-latency trigger alert data to the ground when the telemetry channel of DAO-A satellite is established.

3. Observations

3.1. *GRB*

Since their launch, GECAM-B and GECAM-C have detected many GRBs, including the Brightest-Of-All-Time (BOAT) GRB 221009A (An *et al.*, 2023), the second brightest GRB 230307A (Sun *et al.*, 2023; Xiong *et al.*, 2023; Yi *et al.*, 2023) and many other interesting bursts (such as GRB 210619B, GRB 230812B, GRB 231129C).

Most GRBs were found either by the in-flight trigger software followed by the on-ground search or by the on-ground blind search alone. Those GRBs which are not triggered by the in-flight trigger but found by the on-ground search are usually relatively weak or occurred during a time period with complicated background variation. Some GRBs are even weaker that they can only be (marginally) identified with the targeted search.

Preliminary results of some interesting GRBs detected by GECAM have been reported in the GCN notice and/or GCN circular (Liu *et al.*, 2022). With the in-flight and on-ground processing pipelines of GECAM, the burst time, localization, temporal and spectral properties of the GRBs are analyzed and measured. The detailed information of the GECAM GRBs will be reported in the forthcoming GRB catalogue paper (Zhang *et al.*, In prep.). It is worth to note that GRB 221009A (Figure 5) and GRB 230307A

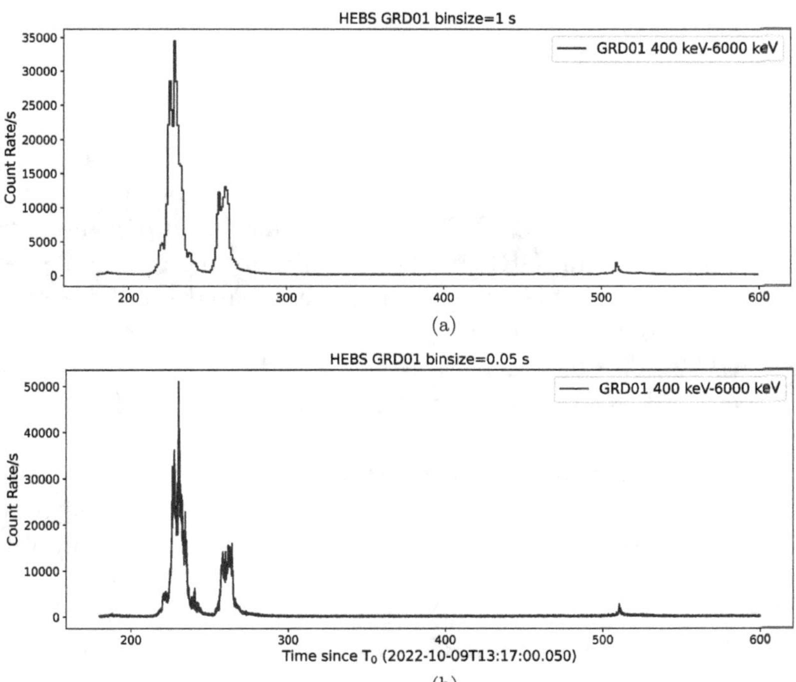

Fig. 5. The GECAM-C light curve of GRB 221009A, the brightest GRB to date (Liu *et al.*, 2022). The time resolution of the light curve is 1 second (a) and 50 ms (b) respectively.

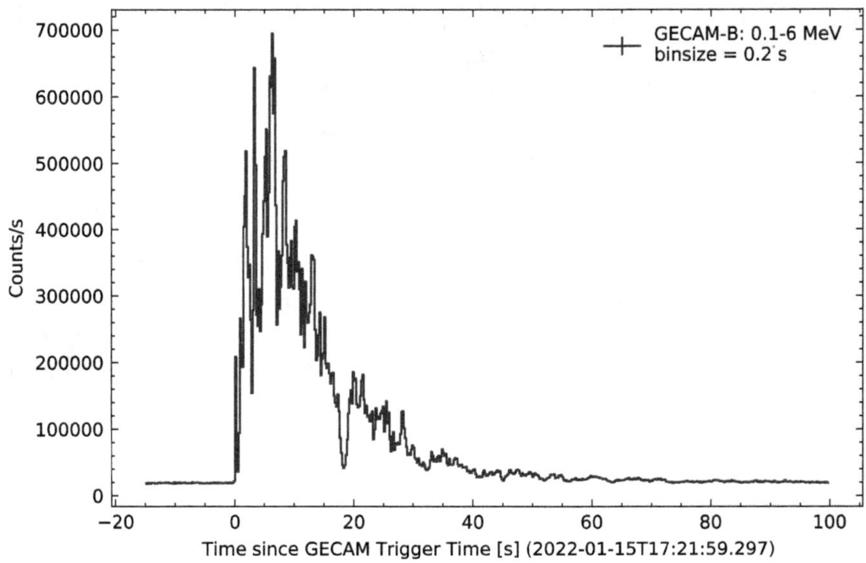

Fig. 6. The GECAM-B light curve of GRB 230307A, the second brightest GRB to date.

(Figure 6) are so bright that many other instruments suffered from data saturation problems (Dalessi, 2023), however, thanks to the dedicated designs to minimize data saturation for bright bursts (Li *et al.*, 2022; Liu *et al.*, 2022), GECAM has accurately measured these extremely bright GRBs without significant instrumental effects (An *et al.*, 2023; Sun *et al.*, 2023). Mostly based on the accurate measurement of GECAM, unprecedented properties of the brightest ever GRB 221009A have been revealed, including the record-breaking brightness and isotropic energy (An *et al.*, 2023), the highest emission line and the power-law decay of the emission line (Zhang *et al.*, 2024a), the close relation between the keV–MeV prompt emission and TeV afterglow emission (Zhang *et al.*, 2024b), as well as the intriguing spectral and temporal properties of the early afterglow and jet break features (Zheng *et al.*, 2024).

For the GRB 230307A, the accurate measurement provided by GECAM show that this GRB displays a self-similarity in its light curves, which posed big challenge in the theoretical interpretations (Yi *et al.*, 2023). Moreover, GECAM firstly observed the cutoff effect in the light curve caused by the jet edge during the prompt emission of a GRB, providing a new approach to measure the geometry of a highly relativistic jet. The temporal and spectral properties measured by GECAM also point towards to a compact merger origin for this burst, even though it has a long duration and extremely brightness (Sun *et al.*, 2023). This finding indicates that this kind of long duration GRB is probably a new type of high energy electromagnetic counterpart of GW sources, which could be detected in the forthcoming multi-messenger observation. It is worth to mention that the extremely brightness of GRB 230307A was firstly found by GECAM-B and reported to the GCN, which initiated a global observation campaign with multi-wavelength telescopes.

3.2. *non-GRB*

GECAM has detected hundreds of X-ray bursts from the SGR J1935+2154, the most prolific Magnetar in the Milky Way. With the observation data from GECAM and other instruments, systemic studies have been carried out to characterize the temporal and spectral properties of these X-ray bursts, especially in the comparison to the GRB and Crab pulsar. It shows that the minimum variation timescale (MVT) peaks at $\sim$2 ms, which corresponds to a light travel time size of about 600 km, lending supports to the magnetospheric origin in pulsar-like models. In addition, there is a linear energy-dependence of the spectral lag for most X-ray bursts from SGR J1935+2154 (Xiao *et al.* 2023a, 2023b). Together with historical observations by Swift/BAT and Fermi/GBM, a period of about 127 days is reported for the x-ray burst activities of SGR J1935+2154 (Xie *et al.*, 2022).

Importantly, GECAM discovered an X-ray burst from SGR J1935+2154 which is associated with FRBs on October 14, 2022 (ATEL #15682). Before this GECAM observation, there is only one X-ray burst found to be matched with FRB 200428 (Li *et al.*, 2021). Therefore, this GECAM observation not only confirmed the association between X-ray bust and radio burst, but also provided crucial data to explore the relation between SGR and FRB and the underling physical mechanisms of these burst phenomenon.

As an all-sky monitor working in the hard X-ray to gamma-ray band with high timing accuracy, GECAM can be used to monitor the high energy emission of pulsars. During the timing calibration of the instrument, the Crab pulse profiles have been clearly detected in

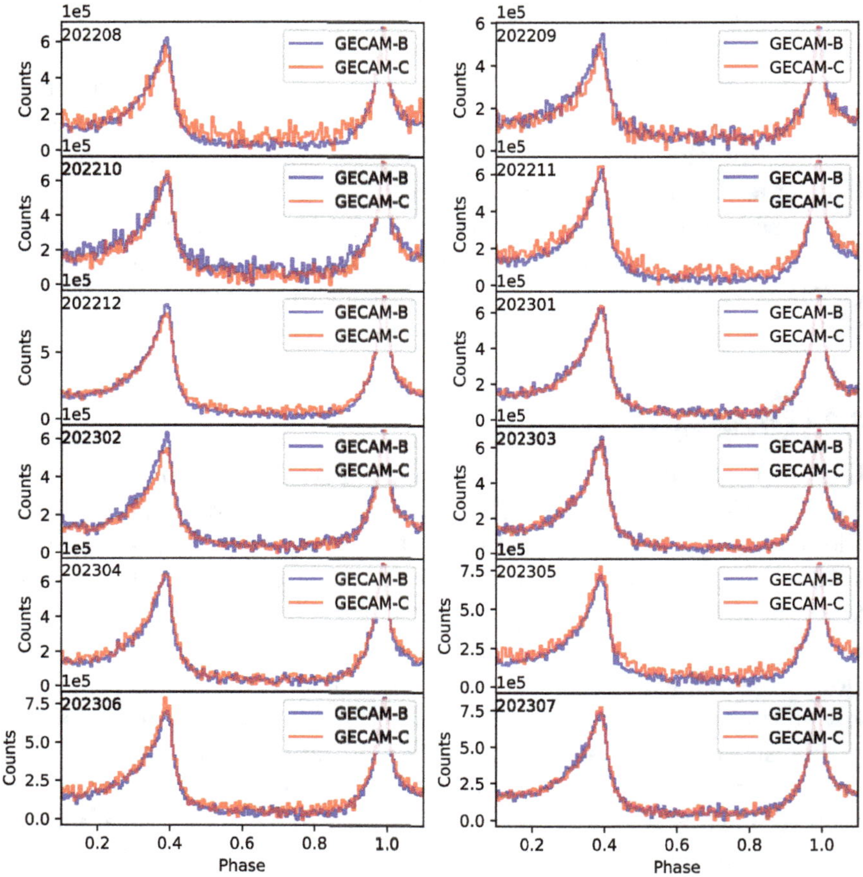

Fig. 7. Crab pulse profiles observed by GECAM-C (red) and GECAM-B (blue).

the GECAM-B and GECAM-C data, as shown in Figure 7. Monitoring of other pulsars with GECAM will be explored.

Last but not least, GECAM can also detect high energy bursts from the Sun-Earth system. Indeed, GECAM discovered the paired quasi-periodic pulsations of hard X-ray emission in a solar flare event on 2022 April 21 at 01:52 UT (Zhao *et al.*, 2023). To explore the high energy radiation from the Earth, a dedicated design has been made for GECAM-A/B/C: both the GRD and CPD detectors are equipped and both types of detector record event-by-event data with high timing accuracy (up to 0.1 microsecond). This compound detector design have helped GECAM to successfully detect both TGFs and TEBs. Importantly, GECAM discovered a new type of event, which has similar features to TEB but has four periodic pulses (Zhao *et al.*, 2023). This new TEB-like event posed challenges on the TEB interpretation thus is subject to further studies.

4. Summary

Initiated by the important science driver of gravitation wave electromagnetic counterpart, we proposed, build, launched and operated a series of GECAM instruments (GECAM-A/B/C/D) to monitor the high energy transients all-sky all-time since early of the year 2016. The first two GECAM-A/B micro-satellites were launched in December 2020, while the latest GECAM-D instrument was launched in March 2024. Novel technologies have been developed and demonstrated for GECAM instruments, including the compact detector, high performance electronics and the prompt communication between satellite and ground based on Beidou navigation system. Thousands of in-flight triggers have been found by the GECAM instruments, among of which hundreds of GRB and SGR bursts were discovered. Based on the unprecedented observations which were made possible by the special design of the instrument, GECAM has refreshed our knowledge of not only the GRB, SGR and FRB from distant universe, but also the energetic activities in the Sun and our Earth.

Acknowledgments

This work is supported by the National Natural Science Foundation of China (Grant Nos. 12273042) and the Strategic Priority Research Program of the Chinese Academy of Sciences (Grant No. XDA15360000, XDA30050000). The GECAM (Huairou-1) mission is funded by the Strategic Priority Research Program on Space Science of the Chinese Academy of Sciences (CAS). S.L.X. appreciate Shuo Xiao for the Crab pulsar plot. We appreciate the GECAM and DRO teams.

Bibliography

Abbott, B. P., Abbott, R., Abbott, T. D. *et al.* (2017). Gravitational waves and gamma-rays from a binary neutron star merger: GW170817 and GRB 170817A. *ApJL* **848**(2), L13.

An, Z. H., Sun, X. L., Zhang, D. L. *et al.* (2022). The design and performance of grd onboard the GECAM satellite. *RDTM* **6**(1), 43–52.

An, Z. H. *et al.* (2023). Insight-HXMT and GECAM-C observations of the brightest-of-all-time GRB 221009A. arXiv:2303.01203.

Feng, P.-Y., An, Z.-H., Zhang, D.-L. *et al.* (2024). Detector performance of the gamma-ray transient monitor onboard DRO-A satellite. arXiv:2401.07513.

Goldstein, A., Veres, P., Burns, E. *et al.* (2017). An ordinary short gamma-ray burst with extraordinary implications: Fermi-GBM detection of GRB 170817A. *ApJL* **848**(2), L14.

Guo, S., Li, G., Zheng, J. *et al.* (2023). Integrated Navigation and Communication Service for LEO Satellites Based on BDS-3 Global Short Message Communication. *IEEE*, 6623–6631.

Huang, Y., Shi, D., Zhang, X. *et al.* (2023). The GECAM Real-Time Burst Alert System. arXiv:2307.04999.

Kouveliotou, C. (1998). Magnetars. In *Abstracts of the 19th Texas Symposium on Relativistic Astrophysics and Cosmology, held in Paris, France, Dec. 14–18, 1998.* J. Paul, T. Montmerle, and E. Aubourg (eds.), (*CEA Saclay*), 3.

Li, X. Q., Wen, X. Y., An, Z. H. *et al.* (2020). The GECAM and its payload. *Sci. Sin. Phys. Mech. & Astrono.* **50**(12), 129508.

Li, C. K. and Lin, L. and Xiong, S. L. and Ge, M. Y. *et al.* (2021). Hxmt identification of a non-thermal x-ray burst from sgr j1935+ 2154 and with frb 200428. *Nat. Astron* **5**(4), 378–384.

Li, X. Q. and Wen, X. Y. and An, Z. H *et al.* (2022). The technology for detection of gamma-ray burst with GECAM satellite. *RDTM* **6**(1), 12–25.

Liu, J.-C., Zhang, Y.-Q., Xiong, S.-L. *et al.* (2022). GRB 221009A: HEBS detection. *GRB Coord. Netw.* **32751**, 1.

Liu, Y.Q., Gong, K., Li, X.Q. *et al.* (2022). The data acquisition algorithm designed for the sipm-based. detectors of GECAM satellite. *RDTM* **6**(1), 70–77.

Mereghetti, S., Pons, J. A., and Melatos, A. (2015). Magnetars: Properties, origin and evolution. *Space Sci. Rev.* **191**, 315-xsxsxs-338.

Norris, J. P., Hertz, P., Wood, K. S., and Kouveliotou, C. (1991). On the nature of soft gamma repeaters. *ApJ* **366**, 240.

O'Meara, K. (1999). Magnetars as soft gamma repeaters. In *APS Northwest Section Meeting Abstracts*, *APS Meeting Abstracts*, B6.14.

Sarah Dalessi (2023). GRB 230307A: Bad time intervals for fermi GBM data. *GRB Coord. Netw.* **33551**, 1.

Savchenko, V., Ferrigno, C., Kuulkers, E. *et al.* (2017). Integral detection of the first prompt gamma-ray signal coincident with the gravitational-wave event GW170817. *ApJL* **848**(2), L15.

Sun, H., Wang, C. W., Yang, J. *et al.* (2023). Magnetar emergence in a peculiar gamma-ray burst from a compact star merger. arXiv:2307.05689.

Woosley, S. E. and Bloom, J. S. (2006). The supernova gamma-ray burst connection. *Annu. Rev. Astron. Astrophy.* **44**(1), 507–556.

Xiao, S., Liu, Y.-Q., Gong, K., An, Z.-H., Xiong, S.-L., Li, X.-Q., Wen, X.-Y., Peng, W.-X., Zhang, D.-L., Tuo, Y.-L. *et al.* (2023). Calibration of the timing performance of GECAM-C. *ApJS* **270**(1), 3.

Xiao, S., Yang, J.-J., Luo, X.-H. *et al.* (2023). The Minimum Variation Timescales of X-Ray Bursts from SGR J1935+2154. *ApJS* **268**(1), 5.

Xiao, S., Tuo, Y.-L., Zhang, S.-N. *et al.* (2023). Discovery of the linear energy dependence of the spectral lag of X-ray bursts from SGR J1935+2154. *MNRAS* **521**(4), 5308–5333.

Xie, S.-L., Cai, C., Xiong, S.-L. *et al.* (2022). Revisit the periodicity of SGR J1935+2154 bursts with updated sample. *MNRAS* **517**(3), 3854–3863.

Xiong, S., Wang, C., Huang, Y., and GECAM Team (2023). GRB 230307A: GECAM detection of an extremely bright burst. *GRB Coord. Netw.*, **33406**, 1.

Yi, S. X., Wang, C. W., and Zhang, B. (2023). Evidence of mini-jet emission in a large emission zone from a magnetically-dominated gamma-ray burst jet. arXiv:2310.07205.

Zhang, B. (2018). *The Physics of Gamma-Ray Bursts*, Cambridge University Press, UK.

Zhang, B. and Meszaros, P. (2004). Gamma-ray bursts: Progress, problems & prospects. *Int. J. Mod. Phys. A* **19**(15), 2385–2472.

Zhang, D. L., Zheng, C., Liu, J. C. *et al.* (2023). The performance of SiPM-based gamma-ray detector (GRD) of GECAM-C. *NIM-A* **1056**, 168586.

Zhang, Y.-Q., Xiong, S.-L., Qiao, R. *et al.* (2023). Cross calibration of gamma-ray detectors (GRD) of GECAM-C. arXiv:2303.00698.

Zhang, X. F., Chen, W., Zhu, X. C. *et al.* (2023). Space advanced technology demonstration satellite. *Sci. China Technol. Sci.*, 1–19.

Zhang, Y.-Q., Xiong, S.-L., Mao, J.-R. *et al.* (2024). Observation of spectral lines in the exceptional GRB 221009A. *Sci. China Phys. Mech. Astron.* **67**, 289511.

Zhang, Y.-Q., Lin, H.-X., Xiong, S.-L. *et al.* (2024). Relation between the keV-MeV and TeV emission of GRB 221009A and its implications. arXiv:2404.03229.

Zhao, X.-Y., Xiong, S.-L., Wen, X.-Y. *et al.* (2021). The in-flight realtime trigger and localization software of GECAM. arXiv:2112.05101.

Zhao, H.-S., Li, D., Xiong, S.-L. *et al.* (2023). Paired quasi-periodic pulsations of hard X-ray emission in a solar flare. *Sci. China Phy. Mech. Astron.* **66**(5), 259611.

Zhao, Y., Liu, J. C., Xiong, S. L. *et al.* (2023). The first GECAM observation results on terrestrial gamma-ray flashes and terrestrial electron beams. *GRL* **50**(14), e2022GL102325.

Zhao, Y., Xue, W.-C., Xiong, S.-L. *et al.* (2023). GECAM localization of high-energy transients and the systematic error. *ApJS* **265**(1), 17.

Zheng, C. and Peng, W. X. and Li, X. B *et al.* (2022). Electron non-linear light yield of labr3 detector aboard GECAM. *NIM-A* **1042**, 167427.

Zheng, C., An, Z.-H., Peng, W.-X. *et al.* (2023). Ground calibration of Gamma-Ray Detectors of GECAM-C. arXiv:2303.00687.

Zheng, C., Zhang, Y.-Q., Xiong, S.-L. *et al.* (2024). Observation of grb 221009a early afterglow in X-ray/gamma-ray energy bands *ApJL* **962**(1), L2.

Part 4

On Pulsars and Compact Stars Science

https://doi.org/10.1142/9789819800643_0019

Chapter 19

Testing Relativistic Gravity with Radio Timing of Neutron Stars

Lijing Shao
Kavli Institute for Astronomy and Astrophysics, Peking University, Beijing 100871, China
National Astronomical Observatories, Chinese Academy of Sciences, Beijing 100012, China
lshao@pku.edu.cn

Neutron stars are the densest material celestial objects. Due to their strong internal gravity, nonperturbative gravitational phenomena might happen in alternative gravity theories. Combined with precision radio timing, neutron stars provide God-given extreme laboratories for strong-field gravity tests. We review the basics of radio pulsar timing, and display a few recent concrete examples where pulsars put unique limits, including anisotropic cubic curvature couplings and scalar-tensor gravity theories. With upcoming new instruments, the future of this field is very bright.

Keywords: Neutron star, Pulsar timing, General relativity, Tests of gravity

Contents

1. Introduction

Physics is a science whose goal is the discovery and formulation of the fundamental laws of the Nature. It deals with the structure of matters and the interactions between the fundamental constituents of our observable Universe. Up to now, the wisdom in basic physics is concisely condensed in, (a) the Standard Model of particle physics that describes matter dynamics and three fundamental interactions (electromagnetic, strong, and weak forces), and (b) the General Relativity (GR) of spacetime that describes the last fundamental interaction, i.e., the gravitational force. However, these two sets of theories conflict at a basic level. Gravity, as it is *apparently* separated from the other three fundamental interactions, might hold a key to a new paradigm in physics (Will, 2014).

Gravity is a long-range force. To understand it thoroughly, one needs to study at various lengthscales. Neutron stars (NSs), especially NSs in binaries, provide unique probes and extreme environments for experimental tests of gravity at length-scales in the order of astronomical units. NSs are the most compact celestial objects after black holes. The strong gravity insides them results in extreme conditions that could be used to reveal strong-field features in certain alternative gravity theories (Damour, 2009).

For a NS in the Milky Way, if its emitting radio beam sweeps towards the direction of the Earth, it could be observed via large radio telescopes. Then it is called a pulsar. A technology called pulsar timing provides exquisite observational precision to such objects. We are specifically interested in binary pulsars. Parameters of the binary system can be determined accurately via pulsar timing, and they reflect the underlying dynamics of the binary, which is in turn controlled by the gravitational interaction between the two binary masses. Consequently, observations of binary pulsars enable precision gravity tests (Shao, 2022; Shao and Wex, 2016; Wex, 2014).

In this chapter, we review some extreme-gravity effects in NSs in Section 2. Basics of pulsar timing are given in Section 3. In Section 4 we present some examples of using pulsar timing to test gravity theories. Finally, some discussions and outlooks are given in Section 5.

2. Neutron Stars with Extreme Gravity

Gravity tests were initially carried out inside the Solar System (Will, 2014). It was already clear at the time when Einstein formulated GR that these tests are weak-field ones. The largest fractional deviation from Newtonian gravity (the famous inverse-square law) one may expect in the Solar System can be estimated by the dimensionless *compactness* of the Sun, denoted as

$$\mathcal{C}_\odot \simeq \frac{GM_\odot}{R_\odot c^2} \sim 10^{-6}\,. \tag{1}$$

For a NS, its compactness is the largest among material objects, reaching $\mathcal{C}_{\rm NS} \sim 0.2$. Even naively speaking, such a large compactness puts NSs among best laboratories for extreme gravity.

Let us take a recent example for illustration. In perturbative Lorentz-violating gravity theories, one could have anisotropic cubic curvature couplings given by an extra term in the Lagrangian (Bailey, 2016),

$$\delta\mathcal{L} = \frac{\sqrt{-g}}{16\pi G}\bar{k}_{\alpha\beta\gamma\delta\kappa\lambda\mu\nu\epsilon\zeta\eta\theta}R^{\alpha\beta\gamma\delta}R^{\kappa\lambda\mu\nu}R^{\epsilon\zeta\eta\theta}, \tag{2}$$

where $R^{\alpha\beta\gamma\delta}$ is the Riemann tensor, and $\bar{k}_{\alpha\beta\gamma\delta\kappa\lambda\mu\nu\epsilon\zeta\eta\theta}$, with physical dimensions of quartic length, are arbitrary constant coefficients for Lorentz violation, satisfying

$$\partial^\rho \bar{k}_{\alpha\beta\gamma\delta\kappa\lambda\mu\nu\epsilon\zeta\eta\theta} = 0, \tag{3}$$

in an asymptotically flat Cartesian coordinate (Kostelecký, 2004). It was shown from field equations that, the only correction to the post-Newtonian metric in GR is in $h_{00} \equiv g_{00} - \eta_{00}$ at an order of $\mathcal{O}(v^4/c^4)$, where v is the characteristic velocity in the system (Bailey, 2016).

Consider a binary pulsar system of a total mass M^{tot} and an orbital size r, the abnormal acceleration introduced by the anisotropic cubic curvature couplings has its symbolic structure (Bailey, 2016),

$$\delta a \sim \frac{GM^{\mathrm{tot}}}{r^2} \times \mathcal{C} \times \frac{K}{\bar{R}^2 r^2}, \tag{4}$$

where GM^{tot}/r^2 is the Newtonian acceleration, K is some linear combination of $\bar{k}_{\alpha\beta\gamma\delta\kappa\lambda\mu\nu\epsilon\zeta\eta\theta}$ components, and $\bar{R}$ is an effective radius of binary bodies,

$$\bar{R} \approx \begin{cases} R_{\mathrm{NS}}, & \text{for a NS-NS binary,} \\ (M^{\mathrm{tot}}/m_{\mathrm{NS}})^{1/3} R_{\mathrm{NS}}, & \text{for a NS with a weak-field companion,} \end{cases} \tag{5}$$

with R_{NS} and m_{NS} respectively the radius and the mass of a NS. The total mass is labeled with a superscript "tot" to be distinguished from the dimensionless $M \equiv M^{\mathrm{tot}}/M_{\odot}$ below. It is evident from Eq. (4) that relativistic small-r binaries containing NSs, whose $\mathcal{C}$ is large and $\bar{R}$ is small, are perfect laboratories to test the perturbative anisotropic cubic curvature couplings in Eq. (2) (Shao and Bailey, 2019).

Gravity with anisotropic cubic curvature couplings is a perturbative example. NSs are even better laboratories when it comes to non-perturbative gravitational effects. About thirty years ago, Damour and Esposito-Farèse (1993) have already shown explicitly that, in a class of scalar-tensor gravity theories where a scalar field φ non-minimally couples to the Ricci scalar R, NSs of certain masses get dressed with scalar fields in a phase-transition-like way. This is true even if the scalar-tensor gravity theory is tuned to agree with GR in the weak field, thus satisfying all gravity tests in the Solar System. In the strong field, the theory still predicts NSs that are very different from their counterparts in GR. This strong-field phenomenon is called *spontaneous scalarization*, and the underlying reason is the activation of a tachyonic-like instability inside NSs (Damour and Esposito-Farèse, 1996; Sennett *et al.*, 2017).

Massive scalar-tensor theories of gravity were also investigated in detail, and similar spontaneous scalarization behaviours were found when the Compton wavelength of the scalar field is much larger than the NS radius (Hu *et al.*, 2021; Ramazanoğlu and Pretorius, 2016; Xu *et al.*, 2020; Yazadjiev *et al.*, 2016). If the Compton wavelength is small, the phenomenon is suppressed in a Yukawa-force fashion and in that case the constraints from binary pulsars to be introduced below can be evaded. Recently, scalar Gauss–Bonnet theories, where a scalar field non-minimally couples to the Gauss–Bonnet term,

$$\mathcal{G} = R_{\alpha\beta\gamma\delta}R^{\alpha\beta\gamma\delta} - 4R_{\alpha\beta}R^{\alpha\beta} + R^2, \tag{6}$$

were also shown to have similar strong-field spontaneous scalarization behaviours (Doneva and Yazadjiev, 2018; Silva *et al.*, 2018; Xu *et al.*, 2022). Therefore, nonperturbative strong-field effects might be common in alternative gravity theories, and NSs provide a valuable observational window.

3. Binary Pulsars Timing

Pulsars are rotating, magnetized NSs, whose radiating beams are observed periodically on the Earth. The most precise measurements of pulsar systems come from radio telescopes with large collecting areas. The times of arrival of pulsar pulses are recorded at telescope sites with the aid of atomic clocks.

In order to probe the perturbative and non-perturbative effects mentioned before, one uses a physical model to connect the observed times of arrival and the emission of these pulses in the proper frame of the pulsar. The model, with a few parameters to be fitted for, is called a *pulsar timing model* (Taylor, 1994). The parameters in general include the sky position, proper motion, spin period and its derivatives, and so on. The pulsar timing model is the key to connect observational data and the theoretical predictions about pulsar systems.

If a pulsar is in a binary system, one also needs to account for its binary motion. The most widely used binary pulsar timing model is the Damour–Deruelle phenomenological timing model (see Refs. (Damour and Deruelle, 1986; Damour and Taylor, 1992)). In this model, besides the usual five Keplerian parameters, $\mathbf{\Theta}^{(\mathrm{K})} = \{P_{\mathrm{b}}, T_0, e, \omega_0, x\}$, there are a set of post-Keplerian parameters,

$\boldsymbol{\Theta}^{(\mathrm{PK})} = \{k, \gamma, r, s, \delta_\theta, \dot{P}_\mathrm{b}, \dot{e}, \dot{x}\}$, to describe relativistic effects of the binary oribt.

In Keplerian parameters $\boldsymbol{\Theta}^{(\mathrm{K})}$, P_b is the orbital period, T_0 is the epoch of periastron passage, e is the orbital eccentricity, ω is the longitude of periastron, and x is the projected semimajor axis of the pulsar orbit. These parameters are quickly determined after a few orbit cycles are observed. After that, with more data, depending on the characteristics of the binary pulsar, post-Keplerian parameters are started to be noticeable. In post-Keplerian parameters $\boldsymbol{\Theta}^{(\mathrm{PK})}$, k is the relativistic periastron advance parameter; γ is the Einstein delay parameter that describes the combined effects from Doppler and gravitational redshifts; r and s are respectively *range* and *shape* parameters in the Shapiro delay that describes the time delay caused by the spacetime curvature due to the companion star; δ_θ is the relativistic orbital deformation parameter; $\dot{P}_\mathrm{b}$, $\dot{e}$, and $\dot{x}$ are respectively time derivatives of P_b, e, and x, which could come from the orbital damping caused by the gravitational radiation.

All the post-Keplerian parameters are zero in a *clean* Newtonian binary with two point masses. However, in Einstein's GR, these parameters are nonzero, and they, in a clean spinless binary, depend only on the Keplerian parameters and the two binary component masses, m_A and m_B (in the Solar unit, $M_\odot$). The expressions of these post-Keplerian parameters for a spinless binary in GR are (Lorimer and Kramer, 2005):

$$k = 3\left(\frac{P_\mathrm{b}}{2\pi}\right)^{-2/3} (T_\odot M)^{2/3} \left(1-e^2\right)^{-1}; \tag{7}$$

$$\gamma = e\left(\frac{P_\mathrm{b}}{2\pi}\right)^{1/3} T_\odot^{2/3} M^{-4/3} m_\mathrm{B}(m_\mathrm{A} + 2m_\mathrm{B}); \tag{8}$$

$$r = T_\odot m_\mathrm{B}; \tag{9}$$

$$s = x\left(\frac{P_\mathrm{b}}{2\pi}\right)^{-2/3} T_\odot^{-1/3} M^{2/3} m_\mathrm{B}^{-1}; \tag{10}$$

$$\delta_\theta = \left(\frac{P_\mathrm{b}}{2\pi}\right)^{-2/3} T_\odot^{2/3} \left(\frac{7}{2}m_\mathrm{A}^2 + 6m_\mathrm{A}m_\mathrm{B} + 2m_\mathrm{B}^2\right) M^{-4/3}; \tag{11}$$

$$\dot{P}_{\rm b} = -\frac{192\pi}{5}\left(\frac{P_{\rm b}}{2\pi}\right)^{-5/3}\left(1+\frac{73}{24}e^2+\frac{37}{96}e^4\right)$$

$$\times\left(1-e^2\right)^{-7/2} T_{\odot}^{5/3} m_{\rm A} m_{\rm B} M^{-1/3}; \tag{12}$$

$$\dot{e} = -\frac{304}{15} e\, T_{\odot}^{5/3}\left(\frac{P_{\rm b}}{2\pi}\right)^{-8/3}\left(1+\frac{121}{304}e^2\right)$$

$$\times (1-e^2)^{-5/2} m_{\rm A} m_{\rm B} M^{-1/3}; \tag{13}$$

$$\dot{x} = \frac{2}{3} x \left(\frac{\dot{P}_{\rm b}}{P_{\rm b}}\right), \tag{14}$$

where $M = m_{\rm A} + m_{\rm B}$ is the total mass in the Solar unit.

In gravity theories other than GR, the functional terms might be different, usually depending on extra parameters of the theory. Therefore, by measuring the post-Keplerian parameters, one can perform gravity tests with pulsar timing. In practice, some of these parameters might be contaminated due to kinematic and dynamic effects caused by, for examples, the presence of a nearby third body or the spins of two binary bodies (Lorimer and Kramer, 2005). After accounting for these extra effects, clean gravity tests are possible. Because of the *a priori* unknown masses, one needs to measure at least three post-Keplerian parameters in order to determine the masses and perform gravity tests.

4. Testing Gravity with Binary Pulsars

Since the first binary pulsar, PSR B1913+16, gravity tests become a routining experiment for pulsar timing of relativistic orbits. The kinds of gravity tests that are possible with pulsar timing are versatile. Readers are referred to Wex (2014) and Shao (2022) for more examples and indepth explanation. Below we use the aforementioned anisotropic cubic curvature couplings and the Damour-Esposito-Farèse scalar-tensor gravity theories as examples. We also mention the unique Double Pulsar system (Kramer *et al.*, 2021) at the end of this section.

With the anisotropic cubic curvature couplings (2), because of the presence of an extra abnormal acceleration (4), the orbital dynamics of binary pulsars are altered. In particular, Shao and Bailey (2019) pointed out that the post-Keplerian parameters k and $\dot{x}$ can be used to test the Lorentz-violating coupling coefficients. These two parameters have extra contributions from the $\bar{k}_{\alpha\beta\gamma\delta\kappa\lambda\mu\nu\epsilon\zeta\eta\theta}$ coefficients, and by verifying that the observed values are consistent with GR predictions given by Eqs. (7) to (14) one can put constraints on the Lorentz-violating coefficients. However, as we noted earlier, the two masses have to be worked out first. It leaves us only a handful of binary pulsar systems that are suitable for the test. We carefully evaluated the case and chose eleven binary pulsars for the test; they are PSRs J0348+0432, J0737−3039A, J0751+1807, J1012+5307, B1534+12, J1738+0333, J1802−2124, J1909−3744, B1913+16, J2043+1711, and B2127+11C. The coefficients that contribute to k and $\dot{x}$ have quite a few independent degrees of freedom, and the use of multiple binary pulsars helped break the degeneracy among them (Shao and Bailey, 2019). This is particularly true for Lorentz-violating theories where rotation invariance is broken. Different binary pulsars are located at different parts of the sky, and the orientations and shapes of orbits are also different. The combination of them offers natural laboratories in testing the anisotropic properties of spacetime. Indeed, stringent limits on some components of $\bar{k}_{\alpha\beta\gamma\delta\kappa\lambda\mu\nu\epsilon\zeta\eta\theta}$ were placed by these binary pulsars for the first time.

For the spontaneously scalarized NSs in binary systems, centers of the inertial masses and the scalar fields differ. Therefore, the strong equivalence principle is violated. Consequently, a dipolar radiation from such a binary system is predicted (Damour and Esposito-Farèse, 1996). In GR, the leading-order gravitational radiation is the quadrupolar radiation, resulting in the $\dot{P}_b$ in Eq. (12). The dipolar radiation introduces an extra term to $\dot{P}_b$,

$$\dot{P}_b^{\rm D} = -2\pi T_\odot \left(\frac{2\pi}{P_b}\right)\left(1+\frac{e^2}{2}\right)\left(1-e^2\right)^{-5/2} m_{\rm A} m_{\rm B} M^{-1} \left(\alpha_{\rm A}-\alpha_{\rm B}\right)^2, \tag{15}$$

where $\alpha_{\rm A}$ and $\alpha_{\rm B}$ are the effective scalar charges of the pulsar and its companion respectively (Damour and Esposito-Farèse, 1993), defined via

$$\alpha_{\rm A} = \frac{\partial \ln m_{\rm A}}{\partial \varphi_0}, \tag{16}$$

with φ_0 the asymptotical value of the scalar field at infinity. The contribution in Eq. (15) can be probed via measuring the post-Keplerian parameter $\dot{P}_b$ and subtracting the GR contribution in Eq. (12) from it. There is an extra layer of complication. The effective scalar charge α_A has dependence on the uncertain equation of state of supranuclear matters that make up of the NS. In order to overcome this, Shao *et al.* (2017) proposed and implemented the usage of multiple binary pulsars with different masses. Even that the spontaneous scalarization happens at different masses for different equations of state, a combination of multiple pulsar systems can probe the whole landscape and evade the uncertainty. Recently, Zhao *et al.* (2022) have updated the analysis with new observations of PSRs J0348+0432, J0737−3039A, J1012+5307, J1738+0333, J1909−3744, J1913+1102, and J2222−0137, and concluded that in the original massless scalar-tensor gravity theories of Damour and Esposito-Farèse (1993), the effective scalar charge for NSs of any mass should be smaller than ~1%, no matter what the real equation of state is.

Finally, it is worth highlighting the unique Double Pulsar system PSR J0737−3039A/B, where both NSs were seen as pulsars, thus providing a double-line observational target. Kramer *et al.* (2021) published the 16.2-yr timing results using combined data from the Parkes 64-m radio telescope in Australia, the Green Bank 100-m radio telescope in the USA, the Westerbork Synthesis Radio Telescope in Netherlands, the Nançay Radio Telescope in France, the Effelsberg 100-m radio telescope in Germany, and the Lovell 76-m radio telescope in the UK. Double Pulsar not only provides best tests of the traditional quadrupolar radiation, Shapiro delay, and so on, but also enables a few new tests, like the light propagation in the strong gravitational field, including retardation and aberrational light bending, etc. Einstein's GR has passed all the tests from the Double Pulsar system with flying colors.

5. Discussion and Outlook

Radio pulsars are superb laboratories for extreme gravity. With the exquisite precision of pulsar timing, combined with the strong internal gravity of NSs, binary pulsars are providing a unique set of strong-field gravity tests in the past half a century.

Nowadays, although we have strong-field gravity tests from gravitational waves, including binary NSs (Abbott *et al.*, 2019) and binary black holes (Abbott *et al.*, 2021), and from black hole shadows (Akiyama *et al.*, 2022), the tests from pulsars have their uniqueness. First, generally a radio pulsar can be observed for decades. With such a long-term observation, tiny changes on binary pulsars' orbits can be captured, in particular for cumulative effects. For example, the precisions in measuring $\dot{P}_{\rm b}$ and k are respectively proportional to $T_{\rm obs}^{-5/2}$ and $T_{\rm obs}^{-3/2}$ (Damour and Taylor, 1992). This is different from the burst nature of gravitational waves from compact binary coalescences. Second, NSs are made of superdense nuclear matters, and some forms of couplings between gravitation and matters, while not available to black holes, can be studied with NSs. Third, the internal strong gravity of NSs and the relatively weak gravity for orbit motions are coupled. As we showed earlier, the internal scalarization of NSs eventually leads to visible effects in the orbital motion, namely the dipolar gravitational radiation in the case of Damour-Esposito-Farèse scalar-tensor gravity. Fourth, the pulsar-timing observation is made with radio waves. Therefore, we are not only investigating effects from the orbital dynamics, but also effects from light propagation in the curved spacetime, as explicitly shown in the Double Pulsar system (Kramer *et al.*, 2021). The study of light propagation shares some similarities with black hole shadow observations. Overall, a combination of complementary strong-field gravity tests from different observations is worth pursuing (Akiyama *et al.*, 2022; Shao *et al.*, 2017).

In the future, a range of new radio instruments will boost the testing gravity experiments with radio pulsars. In China, the Five-hundred-meter Aperture Spherical Telescope (FAST) has been operating successfully for a few years, and the steerable Xinjiang QiTai 110 m radio Telescope (QTT) and the Jingdong 120-m Radio Telescope (JRT) are upcoming soon. At an international collaboration level, the Square Kilometre Array (SKA) is already under construction, and will be operating in a few years to deliver new pulsar sciences in gravity tests (Shao *et al.*, 2015; Weltman *et al.*, 2020). All of these radio telescopes and arrays will provide new sensitivity to radio pulsar observations and result in better timing precision. On the other hand, with better sensitivity and larger field of view, new pulsar systems that are suitable for gravity tests will be discovered, possibly including pulsar — black hole binaries that are crucial for testing black hole spacetimes. Possible synergy between radio pulsar and gravitational wave observations is also promising. For example, combined studies from SKA and the Laser Interferometer Space Antenna (LISA) will provide unprecedented gravity tests from short-orbital-period double NS systems with $P_{\rm b} \sim 10\,{\rm min}$ in the Milky Way (Miao *et al.*, 2021).

Acknowledgments

This work was supported by the National SKA Program of China (2020SKA0120300), the National Natural Science Foundation of China (11975027, 11991053, 11721303), and the Max Planck Partner Group Program funded by the Max Planck Society.

Bibliography

Abbott, B. P. *et al.* (2019). Tests of general relativity with GW170817, *Phys. Rev. Lett.* **123**(1), 011102.

Abbott, R. *et al.* (2021). Tests of general relativity with GWTC-3.

Akiyama, K. *et al.* (2022). First sagittarius A* event horizon telescope results. VI. Testing the black hole metric, *Astrophys. J. Lett.* **930**(2), L17.

Bailey, Q. G. (2016). Anisotropic cubic curvature couplings, *Phys. Rev. D* **94**(6), 065029.

Damour, T. (2009). Binary systems as test-beds of gravity theories, In M. Colpi, P. Casella, V. Gorini, U. Moschella, and A. Possenti (eds.), *Physics of Relativistic Objects in Compact Binaries: From Birth to Coalescence*, **359** (Springer, Dordrecht), 1.

Damour, T. and Deruelle, N. (1986). General relativistic celestial mechanics of binary systems. II. The post-Newtonian timing formula. *Ann. Inst. Henri Poincaré Phys. Théor.* **44**, 263–292.

Damour, T. and Esposito-Farèse, G. (1993). Nonperturbative strong field effects in tensor-scalar theories of gravitation, *Phys. Rev. Lett.* **70**, 2220–2223.

Damour, T. and Esposito-Farèse, G. (1996). Tensor-scalar gravity and binary pulsar experiments, *Phys. Rev. D* **54**, 1474–1491.

Damour, T. and Taylor, J. H. (1992). Strong field tests of relativistic gravity and binary pulsars, *Phys. Rev. D* **45**, 1840–1868.

Doneva, D. D. and Yazadjiev, S. S. (2018). New Gauss-Bonnet black holes with curvature-induced scalarization in extended scalar-tensor theories, *Phys. Rev. Lett.* **120**(13), 131103.

Hu, Z., Gao, Y., Xu, R., and Shao, L. (2021). Scalarized neutron stars in massive scalar-tensor gravity: X-ray pulsars and tidal deformability, *Phys. Rev. D* **104**(10), 104014.

Kostelecký, V. A. (2004). Gravity, Lorentz violation, and the standard model, *Phys. Rev. D* **69**, 105009.

Kramer, M. *et al.* (2021). Strong-field gravity tests with the double pulsar, *Phys. Rev. X* **11**(4), 041050.

Lorimer, D. R. and Kramer, M. (2005). *Handbook of Pulsar Astronomy.* Cambridge University Press, Cambridge, England.

Miao, X. *et al.* (2021). Stringent tests of gravity with highly relativistic binary pulsars in the era of LISA and SKA, *Astrophys. J.* **921**(2), 114.

Ramazanoğlu, F. M. and Pretorius, F. (2016). Spontaneous scalarization with massive fields, *Phys. Rev. D* **93**(6), 064005.

Sennett, N., Shao, L., and Steinhoff, J. (2017). Effective action model of dynamically scalarizing binary neutron stars, *Phys. Rev. D* **96**(8), 084019.

Shao, L. (2022). Radio pulsars as a laboratory for strong-field gravity tests, *Lect. Notes Phys.* **1017**, 385.

Shao, L. and Bailey, Q. G. (2019). Testing the gravitational weak equivalence principle in the standard-model extension with binary pulsars, *Phys. Rev. D* **99**(8), 084017.

Shao, L., Sennett, N., Buonanno, A., Kramer, M., and Wex, N. (2017). Constraining nonperturbative strong-field effects in scalar-tensor gravity by combining pulsar timing and laser-interferometer gravitational-wave detectors, *Phys. Rev. X* **7**(4), 041025.

Shao, L. and Wex, N. (2016). Tests of gravitational symmetries with radio pulsars, *Sci. China Phys. Mech. Astron.* **59**(9), 699501.

Shao, L. *et al.* (2015). Testing gravity with pulsars in the SKA era, *Advancing Astrophysics with the Square Kilometre Array*, **AASKA14**, *Proceedings of Science*, 042.

Silva, H. O., Sakstein, J., Gualtieri, L., Sotiriou, T. P., and Berti, E. (2018). Spontaneous scalarization of black holes and compact stars from a Gauss-Bonnet coupling, *Phys. Rev. Lett.* **120**(13), 131104.

Taylor, J. H. (1994). Binary pulsars and relativistic gravity, *Rev. Mod. Phys.* **66**, 711–719.

Weltman, A. *et al.* (2020). Fundamental physics with the square kilometre array, *Publ. Astron. Soc. Austral.* **37**, e002.

Wex, N. (2014). Testing relativistic gravity with radio pulsars, In S. M. Kopeikin (ed.), *Frontiers in Relativistic Celestial Mechanics: Applications and Experiments*, Vol. 2 (Walter de Gruyter GmbH, Berlin/Boston), 39.

Will, C. M. (2014). The confrontation between general relativity and experiment, *Living Rev. Rel.* **17**, 4.

Xu, R., Gao, Y., and Shao, L. (2020). Strong-field effects in massive scalar-tensor gravity for slowly spinning neutron stars and application to X-ray pulsar pulse profiles, *Phys. Rev. D* **102**(6), 064057.

Xu, R., Gao, Y., and Shao, L. (2022). Neutron stars in massive scalar-Gauss-Bonnet gravity: Spherical structure and time-independent perturbations, *Phys. Rev. D* **105**(2), 024003.

Yazadjiev, S. S., Doneva, D. D., and Popchev, D. (2016). Slowly rotating neutron stars in scalar-tensor theories with a massive scalar field, *Phys. Rev. D* **93**(8), 084038.

Zhao, J. *et al.* (2022). Closing a spontaneous-scalarization window with binary pulsars, *Class. Quant. Grav.* **39**(11), 11LT01.

https://doi.org/10.1142/9789819800643_0020

Chapter 20

The Equation of State of Pulsars

Zhenyu Zhu*,†, Zhiqiang Miao†, and Ang Li†,‡

**Tsung-Dao Lee Institute, Shanghai Jiao Tong University, Shanghai 201210, China*

†Department of Astronomy, Xiamen University, Xiamen, Fujian 361005, China

‡liang@xmu.edu.cn

The equation of state (EoS) of dense nuclear matter plays an important role in the structure of pulsar-like objects and the binary compact-star mergers, among various astrophysical phenomena related to neutron stars. Determining the correct EoS (or pressure-density relationship) that describes the inner structure of neutron stars is a fundamental problem of nuclear physics, particle physics, and astrophysics, helping us understand the fundamental constituents of matter and their interactions under extreme conditions. In this chapter, we will review the recent progress on the EoS from both points of view of nuclear physics and astronomy. The current and upcoming multi-messenger observatories will continue improving the detection of pulsars together with the precise measurements of their global and dynamic properties. Laboratory experiments will provide an emerging understanding of nuclear matter EoS and the transition to exotic matter. The long-standing, open problem of the EoS may be understood in the near future through the confrontation of theoretical calculations with laboratory measurements of nuclear properties & reactions and increasingly accurate observations in astronomy.

Keywords: Equation of state, Dense matter, Neutron star, Quark matter, Hyperons

Contents

1. Introduction

Neutron stars are the densest and smallest stars observed in the Universe, with a mean density exceeding the ordinary nuclear density $\varepsilon_0 \approx 2.8 \times 10^{14}$ g/cm^3 (corresponding to baryon number density of $n_0 \approx 0.16$ fm^{-3}). Most known neutron stars are observed as pulsars and have multi-wavelength emissions. The equation of state (EoS) of dense nuclear matter plays an important role in the structure of pulsar-like objects and in the mergers of compact objects (neutron stars and black holes). "Is it possible to understand the structure of compact stars and matter?" is one of the 21 questions of Astronomy in a recent article jointly published by SJTU and Science (Sanders, 2021).

Once the stellar EoS (pressure p as a function of the density ε, or number density n) is determined, their static or dynamic properties can be calculated from the general-relativistic hydrodynamic equations. Consider an isolated static neutron star with spherical symmetry, the equation could be simplified as the Tolman–Oppenheimer–Volkoff (TOV) equation. Combining the EoS and the boundary conditions, the TOV equation could be solved by integrating i from the stellar center to the surface. When the pressure vanishes, the radial coordinate is identical to the stellar radius R, and the function $m(R)$ could be interpreted as the mass (M) of the star. The magnetic field, the rotation, and the tidal field from the companion will break this spherical symmetry. But an approximated spherical symmetry could still be maintained. The magnetic configuration, moment of inertia, and tidal deformability could be computed by treating their effects as perturbations (Zhu *et al.*, 2020). For a binary neutron star system, the neutron stars in the binary will approach each other and finally merge because of the gravitational wave emission. In this process, strong gravitational waves are radiated and electromagnetic signals (like short gamma-ray bursts and kilonovae) are generated after or during the merger. The EoS also significantly impacts the electromagnetic or gravitational radiations and would imprint in the signal or waveforms.

Therefore, the crucial point of understanding neutron stars is finding out the EoS. Nevertheless, although we have fair good knowledge of the sub-nuclear matter below the neutron drip density ($\varepsilon_{\rm d} \sim 10^{11}{\rm g/cm}^3$), the phase state in the stellar interior is complicated by the possible appearance of a various exotic degree of freedom, such as hyperons (Li *et al.*, 2007), Kaons (Li *et al.*, 2008), Delta isobars or deconfined quarks (Li *et al.*, 2020), even dark matter particles (Li *et al.*, 2012; Miao *et al.*, 2022). "Are there new states of matter at ultrahigh temperatures densities?" is one of the 11 greatest unanswered questions of physics for the New Century selected by U.S. National Academies at the beginning of the current century (Haseltine, 2002).

As seen in Figure 1, the neutron star zoo constitutes (pure) neutron stars (composed mainly of neutrons; possibly with nucleon excitation or meson condensation), hyperon stars (with strangeness-bearing hyperons), hybrid stars (hadronic matter transforms to quark matter after a phase transition or crossover), or quark stars (composed by deconfined nonstrange/strange quarks). These different EoS models or similar EoSs with different compositions may result in differences in observations. At present, utilizing the data

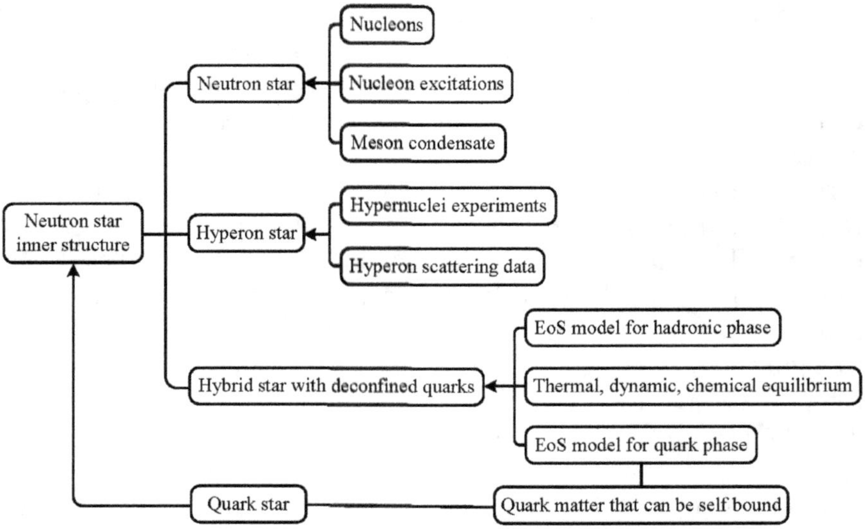

Fig. 1. Four types of EoSs describing the inner structure of neutron stars.

of astronomical observations is a promising way to constrain the EoS and distinguish different types of pulsar-like objects (Li *et al.*, 2016).

The global properties of neutron stars, such as the mass, radius, tidal deformability and moment of inertia, have a one-to-one correspondence to the underlying EoS of neutron-rich matter in beta equilibrium. Promisingly, an increasing number of extremely precise mass measurements have been available, mostly in binary systems with one or two components being pulsars, by taking advantage of the extreme regularity of pulses, such as the presently heaviest PSR J0740+6620 (Cromartie *et al.*, 2020). The maximum mass ($M_{\rm TOV}$) derived from an EoS shall be larger than the masses of neutron stars that we measured. Radius measurement is much more difficult and possible from the pulsed emission caused by hot spots of (millisecond) pulsars, but previous data with large error bars do not help distinguish different EoSs until the recent observations of NICER (Miller *et al.*, 2019, 2021; Riley *et al.*, 2019, 2021). The recent observation of gravitational waves emitted during a coalescing neutron star binary has opened the door to new possibilities for obtaining information on the EoS by means of the measurement of the tidal deformability (Abbott *et al.*, 2017, 2018). Furthermore, measurements are planned for the moment of inertia of PSR J0737-3039 A, the $1.338\,M_\odot$ primary component of the first double pulsar system

PSR J0737-3039, based on the long-term pulsar timing to determine the periastron advance due to relativistic spin-orbit coupling, which will provide complementary constraints on the EoS.

In addition, the Keplerian (mass-shedding) frequency is one of the most studied physical quantities for pulsars. An EoS that predicts Kepler frequencies that are smaller than the observed rotational frequencies is to be rejected, as it is not compatible with observation [see detailed discussions in e.g., Liu *et al.* (2023)]. The calculated highest spin frequencies f_K are all higher than 1000 Hz, while the current observed maximum is $f = 716$ Hz (Hessels *et al.*, 2006) for PSR J1748-2446ad in the globular cluster Terzan 5. A possible reason for this discrepancy is that the star fluid is suffering from r-mode instability (Zhou *et al.*, 2021). And the observed speed-ups (i.e., glitches) in pulsars are widely accepted to be related to the rotating superfluid inside the stars (Packard, 1972). The abundant data from both gravitational wave detection and electromagnetic observations could offer us a large amount of information about an EoS and the microphysics of supra-nuclear matter.

In this chapter, we aim to give an overview of the recent progress on the EoS of neutron stars, using our work as illustrative examples. The theoretical models for EoS are then briefly introduced in the following section. In the subsequent sections, various models of EoSs for neutron stars are discussed and future prospects of EoS and our perspective are presented.

2. Dense Matter and the Equation of State

Since the EoS is not attainable directly by solving quantum chromodynamics (QCD), the underlying theory of the strong interaction, to carry out a study of nuclear many-body system from the quark level, one may first construct a nucleon from confined quarks, by a finite confining region (characterized by a constant energy per unit volume, the bag constant, i.e., the MIT model) or by constituent quarks with a (harmonic oscillator) confining potential. Then nucleons interact with point-like mesons. Since the meson fields modify the internal quark motion, the mesons couple not to point-like nucleons but self-consistently to confined quarks. The effect of the nucleon velocity as well as the effect of antisymmetrisation are usually neglected, and the calculation is done in the mean-field approximation. The many-body nucleonic system is consequently described by the widely used relativistic mean-field approach, following which one obtains the energy density $\varepsilon(n_n, n_p)$, with arbitrary isospin asymmetry $\beta = (n_n - n_p)/n$ (n_n and n_p are the neutron and proton density, respectively) from the energy-momentum tensor. The first model is often called the quark-meson coupling (QMC), and the latter is called the quark mean-field (QMF) model. Though QMC and QMF models were previously distinguished based on their confining mechanisms, recent developments have evolved both models to a similar form, both with large effective quark masses. In this work, we use the name QMF and prefer it for future reference.

The microscopic physics of nuclear matter is modeled within the QMF model and, starting from the same nuclear interaction, a unified description of the crust and core properties has also been developed (see below in the following section for the details of the pulsar inner structure). As seen in Figure 2, the parameter space of the

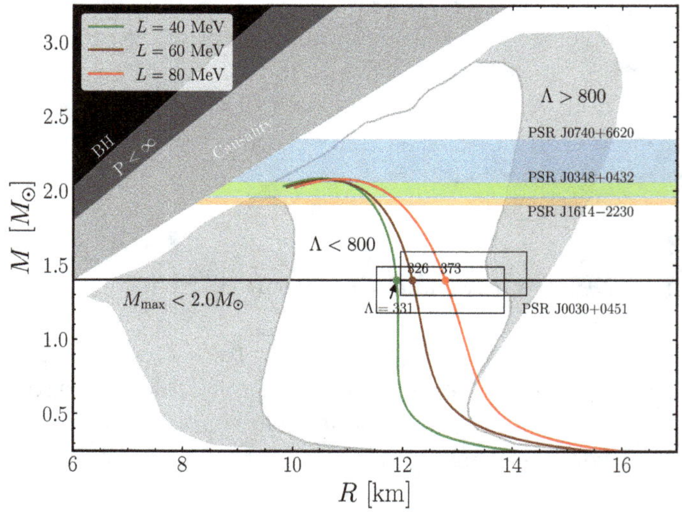

Fig. 2. Mass-radius curves for three EoSs with different values of symmetry energy slope L (40, 60, 80 MeV), together with the mass measurements for three recent massive pulsars. The horizontal black line indicates $M = 1.4\,M_\odot$. Numbers mark the $\Lambda_{1.4}$ values for $1.4\,M_\odot$ stars corresponding to colourful dots. Also shown is the constraining region from binary merger event GW170817, namely, $\Lambda_{1.4} \leq 800$ (Abbott *et al.*, 2017). The tidal deformability describes the magnitude of the induced mass quadrupole moment when reacting to a certain external tidal field. It is zero in the black hole case. Adapted from Zhu *et al.* (2018).

QMF has been constrained for describing neutron stars, following the present measurements of mass, radius, and tidal deformability. The neutron star's maximum mass is approximately $2.1\,M_{\odot}$, with a satisfying reproduction of the nuclear matter properties around the saturation density. The results have a modest dependence on the model parameters, for example the uncertain nuclear symmetry energy: $E_{\text{sym}}(n) \approx [E(n,\beta) - E(n,0)]/\beta^2$, with $E(n,\beta)$ the energy per nucleon of nuclear matter at isospin asymmetry β and density n. Conflicts remain for the symmetry energy [especially its slope $L(n) = dE_{\text{sym}}(n)/dn$] despite significant progress in constraining the symmetry energy around and below the saturation density. It characterizes the density dependence of the symmetry energy and largely dominates the ambiguity and stiffness of the EoS for dense matter at densities approached in neutron star cores, in the case of no strangeness phase transition.

3. The Inner Structure of Pulsars

"What is the nature of neutron stars and dense nuclear matter?" has long been one of the important questions in the field of nuclear astrophysics (Aziz *et al.*, 2021). In the following, we describe a static neutron star for which gravity is balanced by the pressure of the compressed dense matter encountered in its interior. From the star surface to its core, with increasing density the following regions appear sequentially:

Outer crust: non-uniform Coulomb lattice of neutron-rich nuclei embedded in a degenerate electron gas (below the neutron drip density). The properties of the outer crust of neutron stars are well constrained and we use the standard EoS proposed by Baym *et al.* (1971);
Inner crust: non-uniform system composed of more exotic neutron-rich nuclei, degenerate electrons, and superfluid neutrons/superconducting protons (below the saturation density);
Outer core: uniform dense nuclear matter in beta-equilibrium (or weak-interaction equilibrium) with leptons.
Inner core: uniform hyperon matter as described in Section 3.2 or uniform quark matter as described in Section 3.3.

We also elaborate in Section 3.4 on the strong-interaction self-bound quark stars, composed entirely with deconfined quarks, which still serve as viable alternative physical model for pulsar-like objects. Other species (such as Kaons and Delta isobars) might also appear in stellar matter (Li *et al.*, 2020), which the present chapter does not cover. We mention here that, despite the distinct differences among different types of EoS models for neutron stars, there are attempts to test the gravity theory based on (quasi) universal relations between neutron star properties, which is insensitive to the underlying EoS, for example the relations between the moment of inertia, the Love number, and the mass quadrupole (Yagi and Yunes, 2013), the relation between the f-mode frequency and tidal deformability (Chan *et al.*, 2014), etc.

3.1. *Neutron stars and the symmetry energy effect*

To study the structure of neuron stars, we have to calculate the composition and EoS of cold, neutrino-free, catalyzed matter. We require that the neuron star contains charge-neutral matter consisting of neutrons, protons, and leptons (e^-, μ^-) in beta equilibrium. Additionally, since we are looking at neutron stars after neutrinos have escaped, we set the neutrino chemical potentials equal to zero. The energy density of neuron star matter can be written as a function of the different partial densities,

$$\begin{aligned}\varepsilon(n_n, n_p, n_e, n_\mu) &= \varepsilon\left(n_n, n_p\right) + n_\mu m_\mu \\ &\quad + \frac{1}{2m_\mu}\frac{(3\pi^2 n_\mu)^{5/3}}{5\pi^2} + \frac{(3\pi^2 n_e)^{4/3}}{4\pi^2} \end{aligned} \tag{1}$$

where we use ultra-relativistic and non-relativistic approximations for leptons (electrons and muons). Then, the various chemical potentials μ_i of the species ($i = n, p, e, \mu$) can be computed,

$$\mu_i = \partial\varepsilon/\partial n_i, \tag{2}$$

which fulfills beta-equilibrium,

$$\mu_i = b_i\mu_n - q_i\mu_e \tag{3}$$

(b_i and q_i denote the baryon number and charge of species i). Supplemented with the charge neutrality condition,

$$\sum_i n_i q_i = 0 \tag{4}$$

the equilibrium composition $n_i(n)$ can be determined at the given baryon density n, and finally, the EoS is

$$p(n) = n^2 \frac{d(\varepsilon/n)}{dn} = n\frac{d\varepsilon}{dn} - \varepsilon = n\mu_n - \varepsilon \tag{5}$$

for the interior of neutron stars.

The neutron star stable configuration in hydrostatic equilibrium can then be obtained by solving the TOV equation. Experimental nuclear physics provides critical data needed to benchmark theories of nuclear matter EoS contained in neutron stars. Figure 2 shows the mass-radius relations under different symmetry energy slopes L in the range of 40–80 MeV (see above in this section). The TOV mass of the star hardly changes with changing L, while there is a strong positive correlation between the slope parameter and the radius of a $1.4\,M_\odot$ star. As mentioned in the introduction, the EoS governs not only the stable configuration of a single star but also the dynamics of neutron star mergers. During the inspiral phase, the influence of the EoS is evident in the tidal polarizability. The dimensionless tidal deformability Λ is related to the compactness M/R and the Love number k_2 through $\Lambda = \frac{2}{3}k_2(M/R)^{-5}$. In Figure 2, we also include the calculated results of the tidal deformability Λ. We mention that $L \approx 56$ MeV are suggested, following the latest Bayesian analysis (Zhu *et al.*, 2023) on the first multi-messenger event GW170817/AT 2017gfo, as well as the mass-radius simultaneous measurements of PSR J0030+0451 and PSR J0740+6620 from NICER, and the neutron-skin thickness of ^{208}Pb from both PREX-II measurement and *ab initio* calculations.

The above discussions are only for the zero-temperature case, below ~1 MeV for cold neutron stars, lower than the characteristic nuclear Fermi energy, while dense matter is usually hot in heavy-ion collisions and proto-neutron stars, with a temperature as high as ~50 MeV. Although the matter is expected to cool down on timescales of 10^{-22}–10^{-24} seconds and 1–10 seconds, respectively, the thermal effects cannot be ignored, especially in the study of dynamic processes. However, for the equilibrium configurations of cold neutron stars, the EoSs are not affected much by finite temperature. For example, the temperature influence on the maximum mass is very limited, and there is an increase in the neutron star radius for a fixed amount of gravitational mass (Li *et al.*, 2010).

3.2. *Hyperon stars and the hyperon interaction*

As explained above, neutron stars are not made of pure neutrons, since some of them will beta-decay until equilibrium between neutrons, protons and leptons (electrons e and muons μ) is reached, namely $\mu_p = \mu_n - \mu_e$. Such uniform $npe\mu$ matter is the minimal model of the inner neutron-star core. Because the chemical potentials grow with increasing baryon density, additional particle species besides nucleons will be populated when their thresholds are reached. For example, Λ hyperons, the lightest baryon octet $J^P = (1/2)^+$, may appear in high-density neutron-star cores via weak interaction process $p + e^- \longrightarrow \Lambda + \nu_e$, replacing highly energetic neutrons when the Λ chemical potential fulfills the condition $\mu_\Lambda = \mu_n = \mu_p + \mu_e$. Other higher-mass hyperon species (Σ and Ξ hyperons) may also appear. Although unstable under terrestrial conditions, hyperons are stable in dense stellar matter because the Pauli principle blocks their decay into nucleons (Figure 3). Consequently, the core of a massive neutron

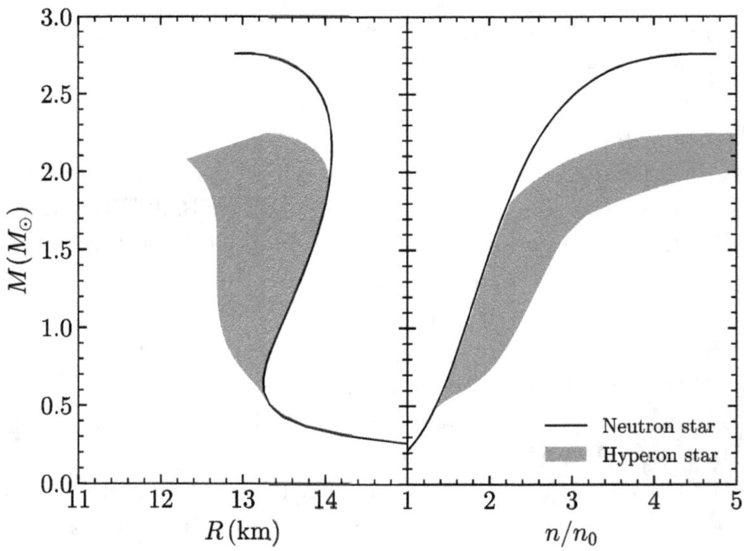

Fig. 3. The most probable mass versus radius & density relations of hyperon stars (at 68% credible level) from the joint analysis of the multimessenger neutron star observations from LIGO/Virgo and NICER, in combination with the study of hypernuclei, compared to those of pure neutrons stars. Taken from Sun *et al.* (2023).

star consists of an inner hyperon core and an outer nucleon shell. These neutron stars are unusually called hyperon stars.

Ever since it is pointed out that hyperons could be present in neutron star cores in Cameron (1959), enormous efforts have been done to learn the composition and EoS of neutron stars including hyperons. However, the parameters of the hyperonic interaction cannot be sufficiently well constrained (Li *et al.*, 2007), and there are still many theoretical and experimental ambiguities regarding baryon interactions in the strangeness sector. Experimentally, hyperonic interaction could be derived from the scattering experiments (several hundred ΛN and ΣN events) and the study of hypernuclei (more than 40 Λ-hypernuclei, a few $\Lambda\Lambda$-hypernuclei and Ξ-hypernuclei), and it is required that hyperonic potentials used should be fitted from the existing nucleon-hyperon scattering data (in microscopic approaches) or be consistent with the properties of hypernuclei (in phenomenological approaches).

Generally, a stiff (soft) nuclear EoS causes an earlier (later) onset and larger (smaller) concentration of hyperons in neutron star interior, which considerably change the star properties. The introduction of hyperons not only softens the neutron star EoS but also results in a more compact star. Recently, the hyperonic interaction is confronted with the recently observed neutron star properties, and one of the first Bayesian inferences of the hyperon-nucleon interaction strengths is performed (Sun *et al.*, 2023) by using the robust multi-messenger neutron star observations, in combination with the study of hypernuclei. For the exemplary EoS case of NL3$\omega\rho$, the hyperon threshold is found to be as low as $\sim 1.5 n_0$, and the hyperon star EoS is moderately stiff with the 68% credible interval of the maximum mass being $M_{\rm TOV} = 2.176^{+0.085}_{-0.202}\, M_\odot$.

3.3. *Hybrid stars and quark-deconfined phase transition*

One of the crucial questions related to the dense matter EoS is whether strange quark matter exists or not. Neutron stars with deconfined quarks in their interior are usually called hybrid stars. The QMF serves as an effective theory for nuclear interaction. However, it is not applicable to deconfined quarks in hybrid stars and quark stars (as discussed further in Section 3.4). In these instances, we employ three different modeling approaches, namely parametrization, the MIT bag model, and effective QCD methods.

By capitalizing on a prominent characteristic observed in a significant class of microscopic quark matter models, where the speed of sound demonstrates a negligible reliance on density, we postulate that the sound speed in quark matter remains unaffected by variations in density [for further elucidation, refer to Miao *et al.* (2020)]. From the first-order transition onset up to the maximum central pressure of a star, the constant-sound-speed (CSS) parametrization (Alford *et al.*, 2013) is applicable to high-density EoSs for which there is a sharp interface (Maxwell construction) between bulk hadronic matter and quark matter, i.e., the quark-hadron surface tension is high enough to disfavor mixed phases (Gibbs construction). One can imagine that a strong first-order phase transition with a sharp interface is the most promising scenario to be tested or distinguished from pure hadronic matter by future observations (Miao *et al.*, 2024).

The CSS corresponds to a general form of parametrization suitable for expressing experimental constraints in a model-independent way. The dimensionless CSS parameters are the squared speed of sound in the high-density phase $c^2_{\rm QM}$ (we work in units where $\hbar = c = 1$), the hadron-quark phase transition density $n_{\rm trans}/n_0$, and the discontinuity in the energy density at the transition $\Delta\varepsilon/\varepsilon_{\rm trans}$, where $n_{\rm trans} \equiv n_{\rm HM}(p_{\rm trans})$ and $\varepsilon_{\rm trans} \equiv \varepsilon_{\rm HM}(p_{\rm trans})$. The value of the sound speed can be varied from $1/\sqrt{3}$ (the perturbative QCD value) to 1 (the casual limit).

There exists a critical value for the energy density jump $\Delta\varepsilon$ below which a stable hybrid star branch connected to the hadronic star branch should be present,

$$\frac{\Delta\varepsilon_{\rm crit}}{\varepsilon_{\rm trans}} = \frac{1}{2} + \frac{3}{2}\frac{p_{\rm trans}}{\varepsilon_{\rm trans}}, \tag{6}$$

which was obtained by performing an expansion in powers of the size of the core of high-density phase, in the presence of a sharp discontinuity in the energy density. For the energy density discontinuities above the critical value, the sequence of stars will become unstable immediately after the central pressure reaches above $p_{\rm trans}$. The twin stars, which represent the scenario that stars with the same mass could have different radii, may exist in the discontinuous mass-radius relation (Miao *et al.*, 2020). One can also formulate EoSs that model quark-hadron interfaces which are mixed (Li *et al.*, 2015, 2008) or feature a smooth crossover, given current uncertainties regarding the nature of the phase transition. Note that the quark matter in neutron star cores is not self-bound and it only appears through a phase transition from hadronic matter at high densities. Figure 4 illustrates the separate parameter space on the $B_{\rm eff}$-a_4 plane (under the bag model description of quark matter; See below in Section 3.4) for hybrid stars and quark stars. Since the strange quark matter in hybrid stars and quark stars are described under different stability conditions, and both of them can't coexist stably (Weber, 2005), a confirmation (or exclusion) of either of them can shed light on the remaining-unsolved QCD at finite baryon density (Li, 2022).

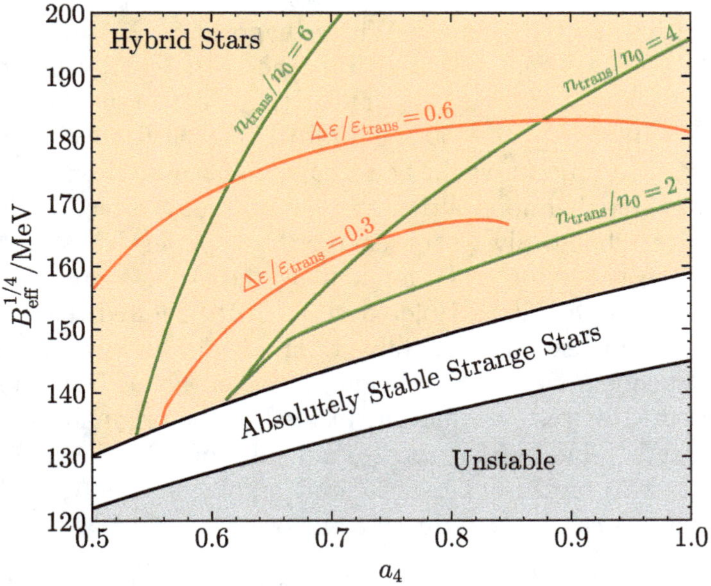

Fig. 4. Separate parameter space for meta-stable quark matter in hybrid star and absolutely stable quark matter constituting the so-called quark stars. The calculations are done in the MIT-type bag model $(B_{\rm eff}, a_4)$. Contour plots for the CSS parameters $n_{\rm trans}/n_0$ and $\Delta\varepsilon/\varepsilon_{\rm trans}$ are also shown. The QMF nuclear EoS is combined for the calculations of hybrid stars, together with either the bag model or the CSS parametrization for the high-density (quark) matter. The grey region is prohibited due to the instability of strange quark matter.

3.4. *Quark stars and the self-bound quark matter*

The possibility that ordinary atomic nuclei could be only a metastable state with respect to the true ground state of baryonic matter at zero pressure was first discussed by Bodmer (1971), and it was pointed out that the possible existence of *uds* matter was not in conflict with the experimental data. It is known as the Bodmer-Witten hypothesis (Bodmer, 1971; Witten, 1984). The hypothetic absolute stability of quark matter, allowed by an ample parameter space in the model calculations of quark matter, not only theoretically supports quark stars as viable alternative physical model for neutron stars (Zhou *et al.*, 2018), but also could have important consequences on various astrophysics and cosmological problems, such as supernova, gamma-ray burst, fast radio burst, pulsar glitch, cosmic rays [see more discussions in Yuan *et al.* (2022) and references therein].

According to the Bodmer–Witten hypothesis, i.e., the absolute stability of quark matter with strangeness, the energy per baryon of *uds* quark matter could be smaller than that of an ^{56}Fe nucleus $E/A(P = 0) = (56m_N - 56 \times 8.8 \text{ MeV})/56 = 930$ MeV. Therefore all pulsar-like objects would be strange quark stars instead of neutron stars. Furthermore, although it was usually regarded that the *uds* matter should always be energetically preferable over the *ud* matter due to the extra Fermi well by the strange quarks, recent studies (Cao *et al.*, 2022; Holdom *et al.*, 2018) showed that *ud* matter could be in general more stable than *uds* matter when taking the flavor-dependent feedback of the quark gas on the QCD vacuum into account. In this case, *ud* matter, instead of the originally proposed *uds* matter, would be the true ground state of cold, dense baryonic matter at zero pressure, i.e., absolutely stable. Generally, the stability conditions can be introduced as follows:

- For self-bound *uds* matter, 1) $(E/A)_{\rm uds} \leq 930$ MeV to ensure the hypothesis of strange matter to be valid; 2) Atomic nuclei should be stable with respect to the formation of droplets of the *ud* matter, that is, nonstrange quark matter in bulk should have an energy per baryon higher than that of the confined phase: $(E/A)_{\rm ud} \geq 930$ MeV; Altogether, $(E/A)_{\rm uds} \leq 930 \text{ MeV} \leq (E/A)_{\rm ud}$.
- For self-bound *ud* matter, $(E/A)_{\rm ud} \leq 930 \text{ MeV} \leq (E/A)_{\rm uds}$.

The internal structure of quark stars is very different from that of neutron stars. Compared with gravity-bound neutron stars, quark stars have a huge surface density of the order of $\sim n_0$, and the central density is only a few times higher than the surface one. It contrasts the 14 orders of magnitude difference of neutron stars. Quantitative studies were carried out a few years after the Bodmer paper (Bodmer, 1971) using the MIT bag model. Since the QCD equation of motion can not be solved on the lattice at finite baryon density, currently the bag model is still the most widely used model. In the model, the effective bag constant B_{eff} accounts for the QCD vacuum's contributions, and a parameter a_4 characterizes the QCD corrections due to gluon-mediated interactions between quarks. The Nambu–Jona–Lasinio (NJL) model is an important and valid effective quark theory, which serves as a suitable approximation to QCD in the low-energy and long-wavelength limit by assuming that gluon degrees of freedom can be frozen into effective pointlike interactions between quarks. Recently, the description of the strong interaction matter is improved at finite baryon number density in the framework of NJL models (Yuan *et al.*, 2022), based on a novel self-consistent mean-field approximation by means of a Fierz transformation. The obtained EoS of interacting quark matter shows a nonlinear (sometimes polytropic) behavior, different from those in the bag model. See more discussions in Yuan *et al.* (2022, 2023) on the recent developments in the quark matter model.

After the release of the gravitational-wave observational data from LIGO/Virgo, in-depth studies have been performed (Li *et al.*, 2020). However, the available mass and radius constraints derived from the two merger events, GW170817 and GW190425, commonly assume neutron star EoSs, with or without quark deconfinement phase transition in their cores. Since the results assume neutron stars cannot be used to study quark star properties which have a sharp surface due to the self-bound feature of their EoS and that the parameterized neutron star EoS cannot catch this behavior, an updated parameter space for quark star EoSs based on the LIGO/Virgo observations are performed in Miao *et al.* (2021) from the beginning. It is found that the quark star EoSs become stiff earlier than the neutron star ones at low densities, indicating the sharp surface of quark stars is of a relatively low density (Li *et al.*, 2017), as low as $\sim 0.22\ \text{fm}^{-3}$ ($\approx 1.4 n_0$), to the 90% confidence level. However, at high densities, quark star EoSs are softer than neutron star ones. The earlier EoS stiffening in quark stars, compared to in neutron stars, has profound influences on the density behavior of the speed of sound, indicating that the pair of quarks starts to play a nontrivial role at such densities [see detailed discussions in Miao *et al.* (2021)].

In addition, different from the neutron star case, where there is a layer of inner crust, and free neutrons are present between the neutron drip density ε_d and the saturation density (i.e., the core part), quark stars may only have a thin nuclear (outer-)crust supported by an electric dipole layer, surrounding the quark matter core (Weber, 2005). Therefore, the maximum density of the quark star crust is limited by the neutron drip density, above which neutrons would gravitate toward the strange-quark matter core. In the NICER estimation of mass and radius, a normal crust is necessary. The crust might add about 1 km to $R_{1.4}$ for neutron stars, which exceeds the design accuracy of the NICER experiment. In Li *et al.* (2021), the analysis of the X-ray emission has assumed that quark stars have a similar crust as neutron stars. Although their effects on the mass and radius are found to be negligible, they may actually be bare or have a different crust regarding, e.g., the thickness, the composition, or even an atmosphere, which would change profoundly the X-ray modeling. An improved study will require a consistent analysis of the NICER observations from the beginning with quark matter assumptions, which might be a considerable effort in the future.

4. Summary and Perspective

It is generally believed that the degree of freedom of dense matter is hadronic around the saturation density, and the color-flavor locked state is expected to be the ground state of three-flavor quark matter at asymptotic densities. The phase states of cold QCD matter at intermediate densities ($\sim$1–10n_0) are unfortunately unknown. The supranuclear matter composition has long been a challenging topic that is directly related to neutron star physics. Thanks to the development of many-body theories of nuclear matter, there is a possibility that the long-standing, open problem of the EoS of dense matter can be understood, by confronting laboratory measurements of nuclear properties & reactions (e.g., RIBLL at HIRFL, CEE at HIAF) and observations in astronomy (e.g., FAST, Insight-HXMT, GECAM, Tianma, LHAASO, eXTP, HUBS, Jingdong, QTT).

4.1. *From the equation of state to the phase state*

Previously, a safe upper limit is found for the neutron star maximum mass at approximately $3.6\,M_\odot$ (Miao *et al.*, 2020) based on the extreme causality EoS. Furthermore, under the assumption that the switching of a hadronic EoS (presently nucleonic EoS) to quark matter EoS is through a first-order phase transition, the maximum mass of neutron stars is predicted to be approximately $2.2\,M_\odot$ from various model calculations (corresponding a moderately stiff neutron-star EoS) (Li *et al.*, 2021; Miao *et al.*, 2022). This conclusion is extremely useful for identifying compact objects' nature with their mass falling into the possible mass gap. The general requirements adopted here (e.g., causality) should also apply to any alternative hadron-quark phase transition scenarios or other types of strangeness phase transitions.

Nevertheless, even if we understand the (effective) stiffness of the EoS, a further challenge is the particle degree of freedom in cold, dense matter. This could help us understand the nonperturbative properties of low-energy QCD (or the parameters of an effective model). However, there are complications to distinguish different types of neutron stars. For example, we discovered that the distinguishing between different states of dense matter (including the onset of phase transition) can hardly be achieved by variations in the sound speed $c_s = \sqrt{dp/d\varepsilon}$ or the adiabatic index $\Gamma = (\varepsilon + p)(dp/d\varepsilon)/p$, since the masquerading of quark matter as hadron-quark mixed phase can be seen in both the c_s and Γ studies (Xia *et al.*, 2021) . Still, some studies have pointed out the difference in the dynamical stability between one-phase stars and multi-phase stars (Pereira *et al.*, 2018). Further efforts regarding the dynamical properties [like neutron star cooling and timing irregularity (Li *et al.*, 2020), binary merger simulation (Zhu and Rezzolla, 2021), etc] of such compact objects, may be necessary for identifying the state of QCD matter at intermediate densities.

4.2. *Hyperon puzzle in neutron stars and possible astrophysical implications*

The presence of hyperons affects the neutron star structure and evolution in various ways, see e.g., Burgio *et al.* (2021); Logoteta (2021) for recent reviews. The formation of hyperons inside neutron stars may lead to a significant softening of the EoS of a neutron-star core, with respect to the $npe\mu$ case, and lower the theoretical maximum mass below the observed pulsar masses, causing the well-known hyperon puzzle problem. The hyperon softening of the EoS may strongly affect the spin evolution of isolated neutron stars; And sufficiently massive stars containing hyperons (or other negatively-charged, strongly-interacting particles) may result in a delayed collapse of the hot newly-born neutron star to a black hole. The presence of hyperons may affect the cooling of neutron stars, the dynamics, and the gravitational wave radiation of neutron star mergers. Also, as soon as hyperons appear, various hyperon reactions contribute to the bulk viscosity of neutron star matter and influence the damping of r-mode instability of neutron stars. The resolution of these problems necessarily requires an accurate description of the hyperonic interaction, which demands joint efforts of physicists and astrophysicists.

4.3. *Constraining phase transition parameters from neutron star observations*

Incorporating the possible first-order phase transition in the neutron star inner core within the QMF+CSS framework (Li *et al.*, 2021), the neutron star EoS can be constrained (Miao *et al.*, 2022) from combined data of LIGO/Virgo and NICER, as well as the mocked data of a future moment of inertia measurement of PSR J0737-3039 A, within the Bayesian statistical approach. The joint analysis is found to prefer the hadron-quark phase transition taking place at not-too-high densities ($n_{\rm trans}/n_0 \approx 2$), and the critical transition pressure is about 16.8 MeV/fm^3, corresponding to a critical baryon chemical potential around 1050 MeV for such zero-temperature beta-stable stellar matter. Also, the sound speed squared $c^2_{\rm QM}$ above the transition should be larger than $\approx$0.4. Nevertheless, the strength of the phase transition can not be well-determined by the neutron star global properties. The exact value of the transition density to quark matter is still a matter of recent debate not only in astrophysics but also within the theory of high energy heavy-ion collisions. For example, mapping the QCD phase diagram is the major scientific goal of the beam energy scan program in heavy-ion collisions (Galatyuk, 2019). Although the dense matter created by nucleus-nucleus collisions in laboratory is instant and the volume is extremely tiny compared with the neutron star matter, it is extremely interesting and helpful if the information on the nuclear EoS from high energy heavy-ion collisions can be related to the physics of neutron star interiors (Huth *et al.*, 2022). See Li *et al.* (2023) for a recent demonstration of the consistency of the phase transition parameters in the beta-stable neutron star matter and in the nearly symmetric nuclear matter formed in heavy-ion collisions.

4.4. *Pulsar glitch and the trigger mechanism*

Glitches, i.e., the sudden spin-up of pulsars with subsequent recovery, were discovered during pulsar timing studies in the Vela pulsar (PSR J0835-4510 or PSR B0833-45) more than 50 yr ago. During glitches, the pulsar spin frequency suddenly increases over a very short time and then usually relaxes to its preglitch rate over a longer time. The glitches in young pulsars, including magnetars, are generally large; Nevertheless, the youngest pulsars, e.g., the Crab pulsar, tend to have more frequent and smaller glitches. They are generally agreed to be the result of the complicated interplay between the superfluid component and the rest of the star.

Many unknown questions concerning both the trigger and the relaxation of a glitch are closely related to the microphysical state of dense matter and hence to the EoS and the structure of the neutron stars (Li *et al.*, 2016). At present, we are still far from a thorough understanding of the general picture of glitches; for example, is there a connection between the stellar interior and the magnetosphere of a star? How can various types of post-glitch behavior be explained? Is there an alternative model besides the superfluid model? The original starquake model suggested that the change in the moment of inertia was due to relaxation in the neutron star (solid) crust to the current equilibrium oblateness but has difficulty explaining the large glitches observed from the Vela pulsar. It is a challenge to quantitatively describe glitch behaviors since the related physical processes are complicated (Shang and Li, 2021). The glitch observation of Vela-like pulsars (especially the details on short-term morphology) may provide a test of the standard superfluid glitch theory. Upcoming higher-time-resolution measurements with post-glitch relaxation should provide more insight into these issues, especially because some real-time automated glitch detection pipelines have been presented (Singha *et al.*, 2021), which can measure the post-glitch recovery phase at a high cadence.

On the other hand, there is increasing evidence of radiative changes accompanying glitch events in magnetars (Kaspi and Beloborodov, 2017) and radio pulsars (Kou *et al.*, 2018; Palfreyman *et al.*, 2018). For example, there are changes in the spin-down state, emission mode changes, and narrowing of the pulse associated with glitch activity in PSR B2035+36 (Kou *et al.*, 2018). Palfreyman *et al.* (2018) detected sudden changes in the pulse shape coincident with the glitch event of the Vela pulsar. In anomalous X-ray pulsar 1E 2259+586, a glitch was observed preceding an X-ray burst (Woods *et al.*, 2004). One of the brightest radio-quiet gamma-ray pulsars, PSR J2021+4026, showed a sudden decrease in gamma-ray emission at the glitch (Zhao *et al.*, 2017). Recently, it was reported that there might be an association between the glitch and polarization change of the Crab pulsar (Feng *et al.*, 2020). They proposed that the sudden spin-up of the crust may cause a change in the configuration of magnetic fields threaded in it and consequently in the corotating magnetosphere.

Studying how glitches affect emission can provide insights into the radiation processes near the pulsars surface, as well as the internal dynamics. For future studies aiming for a thorough understanding of various observed glitch behaviors, a coherent picture involving stellar interiors and their magnetosphere may be necessary. Further information will soon be obtained from high-accuracy facilities with the help of multiband collaborative observations (Liu *et al.*, 2023). We expect this information to give additional constraints to the viability of the superfluid model or to identify the glitch (together with preceding slowdown) that arises by other means, such as magnetospheric activities, hydrodynamic instabilities and turbulence, or starquakes.

Acknowledgments

The XMU neutron star group acknowledges the financial support of National SKA Program of China (No. 2020SKA0120300) and the National Natural Science Foundation of China (Grant Nos. 11873040, 12273028 and 2022M712086). ZZ wishes to acknowledge the support of the China national postdoctoral program for innovation talents (No. BX20220207).

Bibliography

Abbott, B. P. *et al.* (2017). GW170817: Observation of gravitational waves from a binary neutron star inspiral. *Phys. Rev. Lett.* **119**, 161101.

Abbott, B. P. *et al.* (2018). GW170817: Measurements of neutron star radii and equation of state. *Phys. Rev. Lett.* **121**, 161101.

Alford, M. G., Han, S., and Prakash, M. (2013). Generic conditions for stable hybrid stars. *Phys. Rev. D* **88**, 083013.

Aziz, A. A. *et al.* (2021). Progress in nuclear astrophysics of east and southeast asia. *AAPPS Bulletin* **31**, 18.

Baym, G., Pethick, C., and Sutherland, P. (1971). The ground state of matter at high densities: Equation of state and stellar models. *ApJ* **170**, 299.

Bodmer, A. R. (1971). Collapsed nuclei. *Phys. Rev. D* **4**, 1601–1606.

Burgio, G. F., Schulze, H.-J., Vidaña, I., and Wei, J.-B. (2021). Neutron stars and the nuclear equation of state. *Prog. Part. Nucl. Phys.* **120**, 103879.

Cameron, A. G. W. (1959). Pycnonuclear relations and Nova explosions. *ApJ* **130**, 916.

Cao, Z., Chen, L.-W., Chu, P.-C., and Zhou, Y. (2020). GW190814: Circumstantial evidence for up-down Quark Star. *Phys. Rev. D* **106**, 083007.

Chan, T. K., Sham, Y.-H., Leung, P. T., and Lin, L.-M. (2014). Multipolar universal relations between f-mode frequency and tidal deformability of compact stars. *Phys. Rev. D* **90**, 124023.

Cromartie, H. T. *et al.* (2020). Relativistic Shapiro delay measurements of an extremely massive millisecond pulsar. *Nat. Astron.* **4**, 72–76.

Feng, H. *et al.* (2020). Re-detection and a possible time variation of soft X-ray polarization from the Crab. *Nat. Astron.* **4**, 511–516.

Galatyuk, T. (2019). Future facilities for high μ_B physics. *Nucl. Phys. A* **982**, 163–169.

Haseltine, E. (2002). The 11 greatest unanswered questions of physics, *Discover.* **23**, 37–43.

Hessels, J. W. T., Ransom, Scott M., Stairs, I. H., Freire, Paulo C. C., Kaspi, V. M., and Camilo, F. (2006). A radio pulsar spinning at 716 Hz. *Sci.* **311**, 1901–1904.

Holdom, B., Ren, J., and Zhang, C. (2018). Quark matter may not be strange. *Phys. Rev. Lett.* **120**, 222001.

Huth, S. *et al.* (2022). Constraining neutron-star matter with microscopic and macroscopic collisions. *Nature* **606**, 276–280.

Kaspi, V. M., and Beloborodov, A. M. (2017). Magnetars. *Ann. Rev. Astron. Astrophys.* **55**, 261–301.

Kou, F. F., Yuan, J. P., Wang, N., Yan, W. M., and Dang, S. J. (2018). The spin-down state change and mode change associated with glitch activity of PSR B2035+36. *MNRAS* **478**, L24–L28.

Li, A., Zuo, W., and Peng, G. X. (2015). Massive hybrid stars with a first-order phase transition. *Phys. Rev. C* **91**, 035803.

Li, A., Burgio, G. F., Lombardo, U., and Peng, G. X. (2008). Exotic phases in neutron stars. *Inter. J. Mod. Phys. E* **17**, 1635–1647.

Li, A., Dong, J. M., Wang, J. B., and Xu, R. X. (2016). Structures of the Vela Pulsar and the Glitch Crisis from the Brueckner theory. *ApJS* **223**, 16.

Li, A., Zhou, X. R., Burgio, G. F., and Schulze, H.-J. (2010). Protoneutron stars in the Brueckner-Hartree-Fock approach and finite-temperature kaon condensation. *Phys. Rev. C* **81**, 025806.

Li, A., Zhu, Z.-Y., Zhou, E.-P., Dong, J.-M., Hu, J.-N., Xia and C.-J. (2020). Neutron star equation of state: Quark mean-field (QMF) modeling and applications. *J. High Energy Astrophys.* **28**, 19–46.

Li, A., Miao, Z.-Q., Jiang, J.-L., Tang, S.-P., and Xu, R.-X. (2021). Bayesian inference of quark star equation of state using the NICER PSR J0030+0451 data. *MNRAS* **506**, 5916–5922.

Li, A., Huang, F., and Xu, R.-X. (2012). Toomassive neutron stars: The role of dark matter? *Astropart. Phys.* **37**, 70–74.

Li, A., Zuo, W., Schulze J., H., and Lombardo, U. (2008). Hot nuclear matter equation of state and finite temperature Kaon condensation. *Chin. Phys. Lett.* **25**, 4233–4236.

Li, A., Zhang, B., Zhang, N.-B., Gao, H., Qi, B., and Liu, T. (2016). Internal x-ray plateau in short GRBs: Signature of supramassive fast-rotating quark stars?. *Phys. Rev. D* **94**, 083010.

Li, A., Zuo, W., Mi, A.-J., G, B. (2007). Hyperon hyperon interaction in relativistic mean field model. *Chin. Phys.* **16**, 1934–1940.

Li, A., Miao, Z., Han, S., and Zhang, B. (2021). Constraints on the maximum mass of neutron stars with a quark core from GW170817 and NICER PSR J0030+0451 data. *ApJ* **913**, 27.

Li, A. (2022). Can we distinguish quark stars from neutron stars with measurements of global properties? *EPJ web conf.* **260**, 04001.

Li, A., Yong, G.-C., and Zhang, Y.-X. (2023). Testing the phase transition parameters inside neutron stars with the production of protons and lambdas in relativistic heavy-ion collisions. *Phys. Rev. D* **107**, 043005.

Li, A., Zhu, Z.-Y., and Zhou, X. (2017). New equations of state for post-merger supramassive quark stars. *ApJ* **844**, 41.

Liu, X.-J., You, Z.-Q., Chen, Z.-C., Du, S.-S., Li, A., and Zhu, X.-J. (2023). On the spin period distribution of millisecond pulsars. arXiv:2312.12185.

Liu, P., Yuan, J.-P., Ge, M.-Y., Ye, W.-T., Zhou, S.-Q., Dang, S.-J., Zhou, Z.-R., Gügercinoğlu, E., Wang, W.-H., Wang, P., Li, A., Li, D., and Wang, N. (2023). Pulse profile variability associated with the glitch of PSR J1048−5832. arXiv:2312.04305.

Logoteta, D. (2021). Hyperons in neutron stars. *Universe* **7**, 408.

Miao, Z., Li, A., Zhu, Z., and Han, S. (2020). Constraining Hadron-quark phase transition parameters within the quark-mean-field model using multimessenger observations of neutron stars. *ApJ* **904**, 103.

Miao, Z., Li, A., and Dai, Z.-G. (2022). On the moment of inertia of PSR J0737-3039 A from LIGO/Virgo and NICER. *MNRAS* **515**, 5071–5080.

Miao, Z., Jiang, J.-L., Li, A., and Chen, L.-W. (2021). Bayesian inference of strange star equation of state using the GW170817 and GW190425 data. *ApJ* **917**, L22.

Miao, Z., Zhou, E., and Li. A. (2024). Resolving phase transition properties of dense matter through tidal-excited g-mode from inspiralling neutron stars. arXiv:2305.08401

Miao, Z., Zhu, Y., Li, A., and Huang, F. (2022). Dark matter admixed neutron star properties in the light of X-Ray pulse profile observations. *ApJ* **936**, 69.

Miller, M. C. *et al.* (2019). PSR J0030+0451 mass and radius from NICER data and implications for the properties of neutron star matter. *ApJ* **887**, L24.

Miller, M. C. *et al.* (2021). The Radius of PSR J0740+6620 from NICER and XMM-Newton Data. *ApJ* **918**, L28.

Packard, R. E. (1972). Pulsar speedups related to metastability of the superfluid neutron-star core. *Phys. Rev. Lett.* **28**, 1080–1082.

Palfreyman, J., Dickey, J. M., Hotan, A., Ellingsen, S., and van Straten, W. (2018). Alteration of the magnetosphere of the Vela pulsar during a glitch. *Nature* **556**, 219–222.

Pereira, J. P., Flores, C. V., and Lugones, G. (2018). Phase transition effects on the dynamical stability of hybrid neutron stars. *ApJ* **860**, 12.

Riley, T. E. *et al.* (2019). A NICER view of PSR J0030+0451: Millisecond pulsar parameter estimation. *ApJ* **887**, L21.

Riley, T. E. *et al.* (2021). A NICER view of the massive pulsar PSR J0740+6620 informed by radio timing and XMM-newton spectroscopy. *ApJ* **918**, L27.

Sanders, S., (2021). *125 Questions: Exploration and Discovery*, Science/AAAS Custom Publishing Office.

Shang, X., and Li, A. (2021). Revisiting the post-glitch relaxation of the 2000 Vela glitch with the neutron star equation of states in the Brueckner and relativistic brueckner theories. *ApJ* **923**, 108.

Singha, J., Basu, A., Krishnakumar, M. A., Joshi, B. C., and Arumugam, P. (2021). A real-time automated glitch detection pipeline at Ooty Radio Telescope. *MNRAS* **505**, 5488–5496.

Sun, X., Miao, Z., Sun, B., and Li, A. (2022). Astrophysical implications on hyperon couplings and hyperon star properties with relativistic equations of states. *ApJ* **942**, 55.

Weber, F. (2005). Strange quark matter and compact stars. *Prog. Part. Nucl. Phys.* **54**, 193–288.

Witten, E. (1984). Cosmic separation of phases. *Phys. Rev. D* **30**, 272–285.

Woods, P. M. *et al.* (2004). Changes in the X-Ray Emission from the Magnetar Candidate 1E 2259+586 during Its 2002 Outburst. *ApJ* **605**, 378.

Xia, C., Zhu, Z., Zhou, X., and Li, A. (2021). Sound velocity in dense stellar matter with strangeness and compact stars. *C. P. C* **45**, 055104.

Yagi, K., and Yunes, N. (2013). I-Love-Q: Unexpected Universal Relations for Neutron Stars and Quark Stars. *Sci.* **341**, 365–368.

Yuan, W.-L., Li, A., Miao, Z., Zuo, B., and Bai, Z. (2022). Interacting u d and u d s quark matter at finite densities and quark stars. *Phys. Rev. D* **105**, 123004.

Yuan, W.-L., Chao, J., and Li, A,. (2023). Diquark and chiral condensate in a self-consistent NJL-type model, *Phys. Rev. D* **108**, 043008.

Zhao, J. *et al.* (2017). Mode change of a gamma-ray pulsar, PSR J2021+4026. *ApJ* **842**, 53.

Zhou, E.-P., Zhou, X., and Li, A. (2018). Constraints on interquark interaction parameters with GW170817 in a binary strange star scenario. *Phys. Rev. D* **97**, 083015.

Zhou, X., Li, A., and Li, B.-A. (2021). R-mode stability of GW190814's secondary component as a supermassive and superfast pulsar. *ApJ* **910**, 62.

Zhu, Z.-Y., Zhou, E.-P., and Li, A. (2018). Neutron star equation of state from the quark level in light of GW170817. *ApJ* **862**, 98.

Zhu, Z., Li, A., Hu, J., and Shen, H. (2023). Equation of state of nuclear matter and neutron stars: Quark mean-field model versus relativistic mean-field model. *Phys. Rev. C* **108**, 025809.

Zhu, Z., Li, A., and Rezzolla, L. (2020). Tidal deformability and gravitational-wave phase evolution of magnetized compact-star binaries. *Phys. Rev. D* **102**, 084058.

Zhu, Z., and Rezzolla, L. (2021). Fully general-relativistic simulations of isolated and binary strange quark stars. *Phys. Rev. D* **104**, 083004.

https://doi.org/10.1142/9789819800643_0021

Chapter 21

Progress on the Study of Radio Emission Beam of Pulsars

Hong-Guang Wang*, Xiancong Wu and Gaige Deng

Department of Astronomy, Guangzhou University, Guangzhou 510006, China

**hgwang@gzhu.edu.cn*

The structure of radio emission beam is a long standing open question. In this chapter, we briefly reviewed the methods of reconstructing the radio emission beam and the progress on the study of emission beam models. The observational tests for the beam models, especially the test from the precessional pulsar PSR J1906+0746 and its implications are also discussed.

Keywords: Pulsar, Emission mechanism, Radio beam, Magnetosphere

Contents

1. Introduction

The structure of radio emission beam is a matter of long-term debate. Owning to the beaming effect of relativistic charged particles, the emission direction of a charged particle is almost tangent to the magnetic field. This effect, combining with the divergence nature of dipolar magnetic field, leads to a consequence that only a small portion of the radio emission beam is visible as our line of sight sweeps across the beam.

There are two ways to overcome this limitation. The first one resorts to statistics for a large number of mean pulse profiles. Because the mean pulse profile is a product of the emission beam structure and the viewing geometry, the beam structure can be inferred as long as the parameter space of the viewing geometry (the inclination angle and the impact angle) can be fully covered by a large sample of pulsars. The second one is to monitor the pulse profile change for the binary pulsars with strong precession. The precession changes the orientation of pulsar spin axis, therefore changing the viewing angle of our line of sight with respect to the spin axis. The observer will be able to scan the full or a portion of the emission beam during a precessional period. Practically, this method is only suitable for those pulsars with relatively short precessional period, so that a significant portion of emission beam can be scanned by observations within a few decades. Meanwhile, the linear polarization features must enable the measurement of the viewing geometry at every epochs, so that the emission beam can be reconstructed from the observed pulse profiles. At present the most reliable case is PSR J1906+0746 (Desvignes *et al.*, 2019).

Efforts in the first way have been made since the early stage of pulsar astronomy. For most of the pulsars discovered in 1960s and 1970s, the averaged pulse profile is usually of the single-peak or double-peak type. A model of hollow cone coaxial with the magnetic axis was proposed to account for the pulse profiles (Backer, 1976; Komesaroff, 1970; Oster and Sieber, 1976). As more pulsars were discovered, especially those with multi-peak pulse profiles, the emission beam was supposed to consist of nested double cones and a core, so that the model can explain a variety of pulse shapes (Rankin, 1983). Apart from this kind of model, the patchy beam model (Lyne and Manchester, 1988) and the fan beam model (Michel, 1987) were proposed as alternative models in 1980s. In the following 30 years, the conal beam model has been studied more extensively than the other models and became popular as well (e.g., Mitra and Rankin, 2002, 2011; Rankin, 1990, 1993; Rankin *et al.*, 2017; Rankin and Ramachandran, 2003). In 2010s, different types of fan beam model were proposed to explain various properties of radio pulsars (Dyks, 2017; Dyks, *et al.*, 2010; Wang *et al.*, 2014). They have the common feature that the emission beam is elongated radially, which is a natural result of the assumption that particles produce broadband emission when they flow along magnetic flux tubes. However, recently, the 2D radio beam of PSR J1906+0746 (Desvignes *et al.*, 2019) reveals some new features beyond the scenarios of the conal beam and the fan beam models. The emission beam structure is still an open question.

In this chapter, we will briefly review the major emission beam models in Section 2, the interpretation of observational features in Section 3, the observational tests and challenges in Section 4 and the implications to pulsar magnetosphere in Section 5.

2. Three Kinds of Emission Beam Models

The conal beam model proposed two distinct types of component, namely the conal and the core. A typical radio emission beam consists of a core and two nested cones, which are called the inner and the outer cones. Then, the conal single, conal double, conal triple, quadruple and five-component pulse profiles can be accounted for by assuming different impact angles. The emission cone is not necessary to be illuminated evenly. Instead, the intensity distribution in the cones could be patchy (see Figure 1 in Rankin, 1993), so that some kinds of pulse profile, e.g., the core single, triple (with a core and a pair of conal components) and partial cones (only one conal component is visible) can be interpreted.

The patchy beam model proposed that the beam structure is a convolution of a window function that is common to all pulsars and a source function that is unique to a given pulsar. The emission beam consists of randomly distributed patches within a circular or an elliptical boundary (Figure 3 in Manchester, 1995). This model is highly flexible to interpret various kinds of pulse profiles without invoking two distinct emission mechanisms. In this model, different properties of the conal and core emission result from their different locations of the emission region with respect to the magnetic axis, i.e., the core emission is close to the magnetic axis while the conal emission is generated from the outer part of the open field line region.

Despite the essential difference in intensity distribution, the conal and patchy beam models have a consensus that the radio beam has a boundary. In order to form a conal beam or a patchy beam, the emission mechanism needs to be narrow-band, i.e., the particles at a given altitude can only generate narrowband emission. In other words, a single emission frequency is generated within a narrow range of altitude, therefore it forms an emission cone if particles flow along a ring-like flux tube around the magnetic pole, or forms patches if particles are generated in several separated flux tubes. The circular or elliptical boundary is a result of dipole magnetic field, while the patches may originate from local structure of particle distribution related to, e.g., the multipole structure of the magnetic field (Manchester, 1995).

In contrast, the fan beam models are based on the assumption of broadband emission. Assuming that a bunch of particle can generate broadband emission at any altitude, a flux tube will form a radially elongated emission beam, which extends from inner regions near the magnetic pole to outer regions with very large opening angle, accompanying with intensity attenuation outwards. If the emission region consists of a few separated flux tubes, or the distribution of charged particles has complex structure in a large flux tube, the fan beam can account for the diversity of the mean pulse shapes (see Figure 18 in (Dyks, *et al.*, 2010) and Figures 5–12 in (Wang *et al.*, 2014).

To summarize, the beam shape depends on two major factors: the spatial distribution of emission-active flux tubes and the narrowband or broadband nature of radio emission. If the charged particles originate from one or a few nested ring-like flux tubes around the magnetic axis and the emission is narrowband, the emission beam would be conal. Taking the assumption of patchy distribution of particles in the ring-like flux tubes will make the model more flexible to account for the diversity of pulse profiles. A highly separated flux tubes and the narrowband emission produce a patchy beam. However, a combination of separated flux tubes and the broadband emission results in a fan beam.

3. Interpretation of Observations

The most important conjecture of the conal beam model is that the core and conal components are two distinct emission mechanisms. This is based on their distinct observational properties. Comparing with the conal emission, the core emission usually has a steeper spectrum, and is often manifested by moderately large circular polarization and symmetrically alternating sense (Rankin, 1983, 1993). These differences seem to depend on the position, because the excursions of linear polarization position angle indicate that the core emission should come from the central region of the polar cap (close to the magnetic axis), whereas the conal components come from the outer region. This position-dependent features may be a consequence of propagation effects. Some researches found that, owing to propagation effects, the circular polarization of the emission from low altitudes (with small impact angles) could be stronger and the alternating of sign is more often to be seen than the emission from higher altitudes (Andrianov and Beskin, 2010; Beskin and Philippov, 2012; Wang, *et al.*, 2010). Therefore, it is not necessary to assume that the emission mechanisms are different in the inner and outer parts of the open field line region.

In the context of the fan beam models, the position-dependent feature means that the emission properties change from the inner to the outer parts of the beam. The physical mechanisms to account for the position-dependent feature, e.g., the propagation effects, is also valid in the case of the fan beam.

Another prominent feature in many radio pulsars is the anti-correlation between pulse width and emission frequency. The so called radius-to-frequency mapping (RFM), i.e., higher frequency emissions originating from lower altitudes, is a widely-used scenario cooperated with the conal beam model to explain this phenomenon. However, on the basis of an analysis of 150 normal pulsars, Chen and Wang (2014) found that 54% of the sample show noticeable decrease of pulse width with increasing frequency, but the rest 46% pulsars show little variation or even a significant increase with frequency. Among the pulsars with broadening profile at higher frequencies, some of them show rising outermost components at high frequencies, which can be explained as the spectral difference in the core and conal components (Rankin, 1983), but the others can not. In the fan beam model, the diversity is proposed to be caused by different types of spectral variation across the emission beam (Chen and Wang, 2014).

The sub-pulse drifting is noted to be associated with the conal components rather than the core component (Rankin, 1986). It is usually interpreted in term of a carousel pattern of sub-beams circulating around the magnetic pole in the cones (Ruderman and Sutherland, 1975), which may be a consequence of sparks drifting across the magnetic pole in the inner vacuum gap due to the $\mathbf{E} \times \mathbf{B}$ drift. In the conal beam model, different sub-pulse drifting classes are associated with the different locus of the line-of-sight traverse across the emission beam (Basu, *et al.*, 2019; Rankin, 1986). In the case of fan beam, the sub-beams produced by charged particles from the sparks will also circulate around the magnetic pole and show drifting patterns, although the shape of sub-beam is elongated instead of being circular patches in the cone. It is necessary to explore whether the fan beam model can explain different drifting classes.

4. Observational Tests and Challenges

There have been many endeavors to examine the different predictions on some relationships for those emission beam models by statistical analysis of pulsar samples. It was suggested (see Gil and Krawczyk, 1996; Kijak and Gil, 2002) that the frequency dependence for the relative pulse phase between sub-pulses and profile components was highly consistent with the nested conal beam model and inconsistent with the patchy beam model. Early studies on the relationship between pulsar period and beam size (e.g., Gil and Kijak, 1993) favored the conal beam model, but recent analysis based on a much larger sample suggested that it remains difficult to distinguish between the conal and patchy beam models (Johnston and Karastergiou, 2019).

The conal and fan beam models predict the opposite relationships between the impact angle and the pulse width (Wang *et al.*, 2014). As the line-of-sight traverse goes further from the magnetic pole, the pulse width would become narrower for the conal beam, but wider for the fan beam. The positive correlation between the observed mean pulse width and the impact angle for a sample of 64 pulsars favors the fan beam model (Wang *et al.*, 2014). Another way of testing these two kinds of models is the ratio of the peak-to-peak separations between the inner and outer components in the pulse profile. The statistical properties of the ratio for a sample of four- and five-component profiles is more consistent with the fan beams than with the nested-conal beams (Dyks and Pierbattista, 2015).

Apart from the above major methods employing the mean pulse profiles, recently Oswald, *et al.* (2019) developed a method to test the conal and fan beam geometries by comparing the frequency evolution of many single pulses positioned at different phases. The application to PSR J1136+1551 showed that the fan beam model can explain the observations better. This technique is worth being applied to more pulsars.

Comparing with the above methods of testing relationships between some parameters, reconstruction of the 2D beam structure is more straightforward. However, the current statistical method of reconstructing an averaged emission beam based on a sample of pulsar is controversial (Han and Manchester, 2001; Kijak and Gil, 2002), because it is difficult to eliminate the effects of frequency-dependence and period-dependence in pulse profiles.

Unlike the statistical method, monitoring the profile variation of precessional pulsars in binary pulsar systems provides a direct way to reconstruct the emission beam for a single pulsar. Among several pulsars suitable for this method (Philippov and Kramer, 2022; Wang *et al.*, 2014 for the references), PSR J1906+0746 is unique because its linear polarization position angles of the main pulse and inter-pulse follow the standard rotating vector model, therefore the emission geometry can be determined accurately. The emission beams from double poles were successfully reconstructed (Desvignes *et al.*, 2019), which show a general feature of fan beam, but they also present some features challenging the current versions of fan beam model, e.g., an almost constant beam width from the equatorial side to the axial side and a complex 2D intensity distribution in the beam.

5. Implication to the Magnetosphere Structure

In the polar gap theories, secondary particles are generated near the polar cap surface where non-dipolar magnetic fields may play an important role in pair cascades (Ruderman and Sutherland, 1975). Affected by the magnetic field structure, the density distribution of particles is probably nonuniform across the polar cap. For normal pulsars, the particles flow out and emit radio waves at higher altitudes where the dipolar field dominates. Since the low-altitude non-dipolar field is connected to the high-altitude dipolar one, the shape and the intensity structure of the emission beam should be highly related to the geometry of the particle-supplying region and particle distributions therein. In this regards, the radio emission beam could be an indicator to the non-dipolar field. For example, in the case of PSR J1906+0746, the apparent difference in the beam shape and the intensity profiles for the two-pole beams suggest that the magnetic field at the stellar surface differs from the dipolar one (Galishnikova, *et al.*, 2020). The narrow stripe-shaped inter-pulse beam suggests that the relativistic particles may be supplied from a region with a poloidal-shape elongated and toroidal-shape limited cross section in the polar cap, and the difference between the double-pole beams indicates that the particle-supplying regions in the two poles have different geometries. This is inconsistent with the current knowledge of the polar cap shape of either a vacuum rotating dipole (e.g., Cheng, *et al.*, 2000) or a plasma-filled dipolar-field magnetosphere (e.g. Bai and Spitkovsky, 2010; Gralla, Lupsasca, and Philippov, 2017).

In order to explain a variety of pulse profiles and spectral properties, both the conal beam and the fan beam models need to invoke sub-beams,[1] which should be related to local emission-active flux tubes in pulsar magnetosphere. The origin of these separated flux tubes may be related to non-dipolar magnetic field. In the case of millisecond pulsars, this scenario of non-dipolar field seems more necessary because their pulse profiles usually have more components than the normal pulsar do. Comparing with a small polar cap of the normal pulsar, it is natural that a much larger polar cap of millisecond pulsar may contain more non-dipolar field structures, thus, they are more probable to have complex pulse profiles.

Although the non-dipolar magnetic field is a widely-used assumption in many researches when the dipolar field does not work well, and there is evidence of an offset quadrupole magnetic field for PSR

[1]In the conal beam model, the sub-beams refer to the bright patches in the cone.

J0030+0451 through modeling of the NICER X-ray light curve of this pulsar (Kalapotharakos, Wadiasingh, Harding, and Kazanas, 2021), we still lack the knowledge of structure of non-dipolar magnetic field in different pulsar populations. It is worth of exploring the connection between the emission beam model and the non-dipolar field configuration, which may potentially provide a way to obtain more knowledge of non-dipolar field and magnetosphere structure in pulsars.

Acknowledgments

This work is supported by National Natural Science Foundation of China (Grant No. 12133004), National SKA Program of China (Grant No. 2020SKA0120101).

Bibliography

Andrianov, A. S. and Beskin, V. S. (2010). Limiting polarization effect — a key link in investigating the mean profiles of radio pulsars, *Astron. Lett.* **36**, 248–259.

Backer, D. C. (1976). Pulsar average wave forms and hollow-cone beam models., *ApJ* **209**, 895–907.

Bai, X.-N. and Spitkovsky, A. (2010). Modeling of gamma-ray pulsar light curves using the force-free magnetic field, *ApJ* **715**, 1282–1301.

Basu, R., Mitra, D., Melikidze, G. I., and Skrzypczak, A. (2019). Classification of subpulse drifting in pulsars, *MNRAS* **482**, 3757–3788.

Beskin, V. S. and Philippov, A. A. (2012). On the mean profiles of radio pulsars – I. Theory of propagation effects, *MNRAS* **425**, 814–840.

Chen, J. L. and Wang, H. G. (2014). Frequency dependence of pulse width for 150 radio normal pulsars, *ApJS* **215**, 11.

Cheng, K. S., Ruderman, M., and Zhang, L. (2000). A Three-dimensional outer magnetospheric gap model for gamma-ray pulsars: Geometry, pair production, emission morphologies, and phase-resolved spectra, *ApJ* **537**, 964–976.

Desvignes, G., Kramer, M., Lee, K., van Leeuwen, J., Stairs, I., Jessner, A., Cognard, I., Kasian, L., Lyne, A., and Stappers, B. W. (2019). Radio emission from a pulsar's magnetic pole revealed by general relativity, *Sci.* **365**, 1013–1017.

Dyks, J. (2017). The geometry of a radio pulsar beam, *MNRAS* **471**, L131–L134.

Dyks, J. and Pierbattista, M. (2015). The ratio of profile peak separations as a probe of pulsar radio-beam structure, *MNRAS* **454**, 2216–2227.

Dyks, J., Rudak, B., and Demorest, P. (2010). The nature of pulsar radio emission, *MNRAS* **401**, 1781–1795.

Galishnikova, A. K., Philippov, A. A., and Beskin, V. S. (2020). Simulations of the radio polarization of a precessing pulsar PSR J1906+0746, *MNRAS* **497**, 2831–2838.

Gil, J. A. and Kijak, J. (1993). Period dependence of radio emission altitudes in the pulsar magnetosphere, *A&A* **273**, 563–569.

Gil, J. and Krawczyk, A. (1996). Pulsar beams – conal not patchy, *MNRAS* **280**, 143–148.

Gralla, S. E., Lupsasca, A., and Philippov, A. (2017). Inclined pulsar magnetospheres in general relativity: Polar Caps for the Dipole, Quadrudipole, and Beyond, *ApJ* **851**, 137.

Han, J. L. and Manchester, R. N. (2001). The shape of pulsar radio beams, *MNRAS* **320**, L35–L39.

Johnston, S. and Karastergiou, A. (2019). The period-width relationship for radio pulsars revisited, *MNRAS* **485**, 640–647.

Kalapotharakos, C., Wadiasingh, Z., Harding, A. K., and Kazanas, D. (2021). The multipolar magnetic field of the millisecond pulsar PSR J0030+0451, *ApJ* **907**, 63.

Kijak, J. and Gil, J. (2002). Structure of pulsar beams: Conal versus patchy, *A&A* **392**, 189–196.

Komesaroff, M. M. (1970). Possible mechanism for the pulsar radio emission, *Nature* **225**, 612–614.

Lyne, A. G. and Manchester, R. N. (1988). The shape of pulsar radio beams, *MNRAS* **234**, 477–508.

Manchester, R. N. (1995). The shape of pulsar beams, *JA&A* **16**, 107–117.

Michel, F. C. (1987). A pulsar emission model: Observational tests, *ApJ* **322**, 822–830.

Mitra, D. and Rankin, J. M. (2002). Toward an empirical theory of pulsar emission. VII. On the spectral behavior of conal beam radii and emission heights, *ApJ* **577**, 322–336.

Mitra, D. and Rankin, J. M. (2011). Toward an empirical theory of pulsar emission. IX. On the peculiar properties and geometric regularity of Lyne and Manchester's "Partial Cone" Pulsars, *ApJ* **727**, 92.

Oster, L. and Sieber, W. (1976). Pulsar geometries. III. The hollow-cone model, *ApJ* **210**, 220–229.

Oswald, L., Karastergiou, A., and Johnston, S. (2019). Understanding the radio beam of PSR J1136+1551 through its single pulses, *MNRAS* **489**, 310–324.

Philippov, A. and Kramer, M. (2022). Pulsar magnetospheres and their radiation, *ARA&A* **60**, 495–558.

Rankin, J. M. (1983). Toward an empirical theory of pulsar emission. I. Morphological taxonomy, *ApJ* **274**, 333–358.

Rankin, J. M. (1986). Toward an empirical theory of pulsar emission. III. Mode changing, drifting subpulses, and pulse nulling, *ApJ* **301**, 901–922.

Rankin, J. M. (1990). Toward an empirical theory of pulsar emission. IV. Geometry of the core emission region, *ApJ* **352**, 247–257.

Rankin, J. M. (1993). Toward an empirical theory of pulsar emission. VI. The geometry of the conal emission region, *ApJ* **405**, 285–297.

Rankin, J. M., Archibald, A., Hessels, J., van Leeuwen, J., Mitra, D., Ransom, S., Stairs, I., van Straten, W., and Weisberg, J. M. (2017). Toward an empirical theory of pulsar emission. XII. Exploring the physical conditions in millisecond pulsar emission regions, *ApJ* **845**, 23.

Rankin, J. M. and Ramachandran, R. (2003). Toward an empirical theory of pulsar emission. VIII. Subbeam circulation and the polarization-modal structure of conal beams, *ApJ* **590**, 411–423.

Ruderman, M. A. and Sutherland, P. G. (1975). Theory of pulsars: Polar gaps, sparks, and coherent microwave radiation., *ApJ* **196**, 51–72.

Wang, H. G., Pi, F. P., Zheng, X. P., Deng, C. L., Wen, S. Q., Ye, F., Guan, K. Y., Liu, Y., and Xu, L. Q. (2014). A fan beam model for radio pulsars. I. observational evidence, *ApJ* **789**, 73.

Wang, C., Lai, D., and Han, J. (2010). Polarization changes of pulsars due to wave propagation through magnetospheres, *MNRAS* **403**, 569–588.

https://doi.org/10.1142/9789819800643_0022

Chapter 22

The Inner Structure of Pulsars

Xiao-Yu Lai[*,‡] **and Renxin Xu**[†,§]

[*]*Department of Physics and Astronomy,*
Hubei University of Education,
Wuhan 430205, China
[†]*School of Physics and State Key Laboratory of*
Nuclear Physics and Technology, Peking University,
Beijing, People's Republic of China
[‡]*laixy@hue.edu.cn*
[§]*r.x.xu@pku.edu.cn*

The inner structure of pulsars is a question attracting both astronomers and physicists. From the physicists' (theoretical) point of view, it is essentially a question of the fundamental quantum chromo-dynamics (QCD) at the region dominated by non-perturbative effect, the solution of which still remains a challenge. From the astronomers' (observational) point of view, as the remnants of supernova explosions and binary star mergers, the inner structure of pulsars is important for us to understand the related phenomenons, such as the central engine of gamma-ray bursts and sources of gravitational waves. In this chapter, we will give an overall discussion about the matter inside pulsars. We will also introduce the strangeon star model, which is different from both neutron star and quark star models. How to test all these models observationally is an important issue, and we will list some observational properties of pulsars which could be understood by strangeon stars. It is expected that, advanced facilities will identify the existence of strangeon matter, particularly in the new era of multimessenger astronomy.

Keywords: Equation of state, Neutron star, Strange quark star, Strangeons

Contents

1. Introduction

What would happen if the gravity of a star was too strong to avoid contraction, so that atomic nuclei came into close contact? In the year 1932, Landau published a paper in which an idea of gigantic nucleus was presented (Landau, 1932): *We expect that this must occur when the density of matter becomes so great that atomic nuclei come in close contact, forming one gigantic nucleus.* Although the main motivation for Landau to write this paper was to explain the origin of stellar energy (and this explanation is certainly wrong), he recognized for the first time a kind of compact stellar object at about nuclear density, which is surely necessary for us to understand various extreme phenomena through astronomical observations. The gigantic nucleus proposed by Landau was identified in 1934 by Baade and Zwicky (Badde and Zwicky, 1934) to be "neutron stars" which can be the remnants of supernova explosions. The discovery of radio pulsars by Bell and Hewish in 1967 (Hewish *et al.*, 1968) established the concept that radio pulsars are fast-rotating neutron stars. Although the quark model of hadron structure established in the 1960s made people consider that the main ingredient of "neutron stars" might be quarks or other more exotic particles instead of neutrons, the gigantic nucleus proposed by Landau has been developed to be the very elaborated models of conventional neutron stars.

After core-collapsing of an evolved massive star, the supernova-produced rump left behind where normal nuclei are intensely compressed by gravity to form "compressed baryonic matter". From observational point of view, they are strictly called "pulsar-like compact stars" instead of "pulsars" because some of them do not manifest themselves as radio pulsars. Some of the pulsar-like compact stars are manifested as central compact object (CCO), X-ray dim isolated neutron stars (XDINSs), soft gamma-ray repeaters (SGRs) or anomalous X-ray pulsars (AXPs), etc. In most case if it will not cause misunderstandings, they are still called "pulsars" for short.

Although abundant observational data of pulsars has been accumulated and the term "neutron star" is broadly used, astronomers have not reached a consensus about the nature of pulsars. From the typical mass of $1.4M_\odot$ and typical radius of 10 km, we can see that the average density of pulsars $1.4M_\odot/[4\pi(10\ \mathrm{km})^3/3] \sim 2.4\rho_0$, where ρ_0 is the saturated nuclear matter density. This simple estimation implies that the non-perturbative effect of the quantum chromodynamics (QCD) should play an important role in determining the state of matter inside pulsars, and it gives the reason why the interior structure of pulsars is still a controversial topic at present. Some basic problems still remain, such as: Have quarks been de-confined there? Does strangeness play an essential role? Based on different points of view, a variety of structure models for pulsars have been speculated. In Chapter 20, Zhu *et al.* give a review about the equation of state in different models, such us conventional neutron stars, hyperon stars, hybrid stars and quark stars.

In this chapter, we will make an overall introduction about the inner structure of pulsars, showing that there are many difficulties and uncertainties for us to theoretically derive the answer about the nature of pulsars. It essentially depends on the behavior of quantum chromodynamics (QCD) at low-energy scales, which is still a challenge for us to understand. Similar to atomic nuclei, the properties of pulsars are determined by the non-perturbative effects of QCD, and these properties cannot be derived from the first principle. Fortunately, the openings of the window on gravitational wave astronomy bring hope for us to unveil the mystery of pulsars' inner structure.

2. Matter Inside Pulsars: 2- or 3-Flavored?

It is well known that the strong and the electromagnetic interactions are two of the four fundamental interactions that dictate our world. The strong one not only confines quarks inside nucleons, but also holds nucleons together to form atomic nuclei. Atoms that composed of nuclei and electrons are the building blocks of ordinary matter. Microscopic nuclei in atoms, is 2-flavor symmetric (or isospin symmetric) with comparable amounts of protons and neutrons, and this is the reason that an atomic nucleus is electrically positive. In addition, due to the smallness of an atomic nucleus and the relatively weak electromagnetism-force, electrons generally locate outside the nucleus.

2.1. *Neutron matter*

Although macroscopic nuclear matter is hard to be produced in laboratory due to the Coulomb repulsion between positively charged nuclei, gravitational collapse during the supernova explosion can compress the core of the massive star into dense matter with supranuclear densities. The gravitationally compressed baryonic matter will be different from our familiar microscopic nuclei, and electrons should play the key role for us to understand such differences. For a microscopic nucleus, electrons would emerge with approximately half the amount of baryons. If we pile up nucleons (including neutrons and protons) to make a gigantic nucleus, electrons will become almost inside the nucleus and be extremely energetic (up to $\sim$300 MeV), therefore a giant nucleus could hardly keep stable due to those energetic electrons. As the baryon number increases, if its constitutions are still neutrons and protons, the growing nucleus will become more and more neutron matter, via the weak interaction of $e^- + p \to n + \nu_e$. Actually, in 1932, an idea of gigantic nucleus was conceived by Landau (Landau, 1932) for the first time, which was the prototype of standard model of neutron stars.

Neutron stars are mainly composed of neutrons. Most of electrons and protons have been converted into neutrons by inverse beta process. Neutron stars have different compositions from the surface to the center due to different densities (Lattimer and Prakash, 2004). The crust of a neutron star with depth about one kilometer thick is divided into an outer crust and an inner crust. The outer crust starts with a density of about $10^6 \mathrm{g/cm}^3$, and is mainly composed of heavy nuclei and relativistic degenerate electron gas. As the density increases, the Fermi energy of electrons also increases, and the β-equilibrium will gradually make the atomic nuclei become neutron-rich. When the density reaches the neutron droplet density (about $4 \times 10^{11} \mathrm{g/cm}^3$), free neutrons will get out of nuclei. As the density increase further, approaching that of nuclear matter, almost all the protons in the nuclei convert into neutrons. Neutron-rich nuclei, neutrons, electrons and protons constitute the inner crust of the neutron star, where neutrons will be paired into Cooper pairs to form a superfluid state.

Beneath the inner crust, the central area whose density reaches the density of saturated nuclear matter ρ_0 is the core of the neutron star. The core as the main part of the neutron star, which has no ordinary nuclei, is mainly composed of superfluid neutrons and their decayed products (i.e., a few electrons and protons). The density near the center of the star would be more than $2 - 3\rho_0$, which could be high enough to allow the existence of π mesons, K mesons or hyperons.

The above description takes hadrons as fundamental degrees of freedom. It is worth noting that, hadrons could be no longer the fundamental degrees of freedom in the core region, which is two or three denser than nuclear matter. In this region, the density is so high that the hadrons could significantly overlap with each other. Presumptively, when the surface of hadrons disappear, there would be de-confined quarks. Theoretical studies have suggested that in dense core of neutron stars, hadronic matter can undergo phase transition to de-confined quark matter, forming hybrid/mixed stars (Glendenning, 2001). Certainly, hybrid/mixed stars are classified as neutron stars, since they have crusts composed of neutron-rich nuclei and electron gas.

2.2. *Quark matter*

Despite that there are six flavors of quarks, we usually neglect heavy flavors of quarks (charm, top and bottom quarks) in the study of dense matter at a few times of the nuclear saturation density, due to the energy scale, E_{scale}. Estimated with the Heidelberg's relation, the energy scale would be order of $E_{\rm scale} \sim \hbar c/\Delta x \sim 0.5{\rm GeV} < m_{\rm heavy}$, where the separation between quarks is $\Delta x \sim 0.5$ fm and $m_{\rm heavy}$ is the mass scale of heavy flavors of quarks. This energy scale, however, is much higher than the mass difference between strange and up/down quarks ($\Delta m_{\rm uds}$ ~0.1 GeV), $E_{\rm scale} \gg \Delta m_{\rm uds}$. Therefore, we may expect 3-flavored matter at a few times of the nuclear saturation density.

It is a fact that the valence strangeness degree of freedom is absolutely missing in stable nuclei with sizes ~ fm. On one hand, the mass of s quark is larger than that of u/d quarks, so s quarks would convert to u and d quarks via weak interaction. But, on the other hand, the light-flavor symmetry might be restored if the flavor symmetry leads to lowering the energy (as for an example of nuclear analogue, the symmetry energy), especially in case that the strong-interaction matter is very big in size.

Strange quark matter, composed of comparable amounts of up, down and strange quarks, could be more stable than ud quark matter, as proposed by Witten's conjecture and many others (Bodmer, 1971; Witten, 1984). Although it remains uncertified due to the difficulty in non-perturbative quantum chromo-dynamics (QCD), this conjecture seems reasonable, because the additional strange flavor provides an extra degree of freedom to lower the Fermi energy in the free-quark approximation.

A general and extendable way to illustrate the so-called Witten's conjecture is from the concept of symmetry, explicitly the light-flavor (3-flavor) symmetry. The restoration of light-flavor symmetry ensures neutrality without electrons. This is not the case for 2-flavored (u and d quarks) quark matter, because the 2-flavor symmetric strong matter is positively charged and requires electrons to emerge with a number about half the baryon number to ensure neutrality. These electrons in bulk strong matter are relativistic and energetic enough to make it d-rich, i.e., the state with 2-flavor *asymmetry*, which is the case for neutron matter. It is interesting to note that these electrons become insignificant for a micro-nucleus (with 2-flavor symmetry) because it is too small (with baryon number smaller than $A_{\rm c}$) in size so that all the electrons are outside and non-relativistic.

Under some phenomenological models for interacting quarks, the quark matter composed of up and down quarks (udQM) could be more stable than strange quark matter (Holdom *et al.*, 2018), and the astrophysical consequences of compact stars composed of udQM have discussed combined with recent observations (Zhang, 2020). The study of udQM is certainly interesting, showing that the state of dense matter has various possibilities in the theoretical points of view.

2.3. *Strangeon matter*

Certainly, the Witten's conjecture is based on the MIT bag model (Chodos *et al.*, 1974) with negligible interaction between quarks, although it can be illustrated by the "light-flavor maximization" as above. It may be shown that astrophysical compressed baryonic matter should be in a state of strangeon matter by starting from deconfined quark state with the inclusion of strong interaction between quarks.

The state of compressed baryonic matter produced in supernova explosions is essentially the non-perturbative chromodynamics (QCD) problem, which could be seen from an estimation of the strong coupling constant $\alpha_{\rm s}$ (Fischer, 2006),

$$\alpha_{\rm s} = \frac{\alpha_{\rm s}(0)}{\ln(e + a_1 x^{a_2} + b_1 x^{b_2})}, \tag{1}$$

where $\alpha_{\rm s}(0) = 2.972$, $a_1 = 5.292$ GeV$^{-2a_2}$, $a_2 = 2.324$, $b_1 = 0.034$ GeV$^{-2b_2}$, $b_2 = 3.169$, $x = p^2$ with p the typical momentum in GeV. In compact stars with quark number density $n \sim 3n_0$, $p^2 \simeq 0.16$ GeV^2, so the coupling parameter $\alpha_{\rm s}$ would be larger than 2. This means that a weakly coupling treatment could be inadequate for realistic dense matter in compact stars. It is also worth noting that the dimensionless electromagnetic coupling constant is $1/137 < 0.01$, which makes quantum electromagnetic dynamics (QED) tractable. That is to say, a weakly coupling strength comparable with that of QED is possible only if the baryon number density $n_{\rm B} > 10^{123} n_0$, being unbelievable and unrealistically high.

Then what is the exact form of dense matter with three flavors of light quarks? The strong coupling between quarks may naturally render quarks grouped in quark-clusters, and each quark-cluster is composed of several quarks condensating in position space rather than in momentum space. A conjecture of "condensation" in position space with strange quark cluster as the constituent units (Xu, 2003), rather than condensation in momentum space for color superconducting state, was thus made for cold matter at supranuclear density. The strange quark cluster is renamed strangeon, being coined by combining strange nucleon for the sake of simplicity (Xu and Guo, 2017).

Certainly, whether quarks would be grouped in quark-clusters is hard to answer from direct QCD calculations. Some efforts have been made to answer this question by making an analog between quark-clusters and water molecules (Xu, 2009), where the latter are also clusters made of electrons, protons and oxygen nuclei. The much weaker electromagnetic coupling can group particles into clusters, so the strong coupling between quarks may also group them into quark-clusters. The astrophysical compressed baryonic matter, therefore, could be in a state of strangeon matter.

In principle, nucleons can also be viewed as clusters made of three valence quarks. Since the mass of nucleons is small enough with significant quantum effects, nucleons form a liquid phase of nuclear matter. However, at smaller densities, the clustering effects take place and form large nuclei, where various types of crystalline structures are formed. Strangeons, composed of several quarks, are more massive and more likely to form solid structures due to their short quantum wavelength $\sim h/(mc)$. The melting temperature would be much larger than the temperature inside pulsars (Xu, 2003).

2.3.1. *Strangeon stars*

Compact stars composed totally of strangeons are strangeon star. Being similar to strange quark stars, strangeon stars have almost the same composition from the center to the surface. Strangon stars are self-bound by the residual interaction between strangeons, while strange quark stars by bag-like confinement. The in-between interaction of strangeons could be regarded as a natural extension of the nuclear force between nucleons, simply with different flavor numbers of quarks, from 2 to 3. The properties of strangeons, as well as the strangeon matter surface, could then be calculated, to be helpful for understanding different observations of pulsar-like compact stars (see the review (Lai and Xu, 2017) and references therein). Although the state of bulk strong matter is essentially a non-perturbative QCD problem that is difficult to answer from the first principle, astrophysical points of view could provide some hints about the interior structure of pulsars.

Strangeon matter is conjectured to be the compressed baryonic matter of compact stars, where strangeons form due to both the strong and weak interactions and become the dominant components inside those stars. Then what could be a realistic strangeon? We know that Λ particles (with structure *uds*) possess light-flavor symmetry, and one may think that a kind of strangeons would be Λ-like. However, the interaction between Λ's could be attractive, which would render more quarks grouped together. Motivated by recent QCD simulations of the H-dibaryon (with structure *uuddss*), a possible kind of strangeons, H-clusters, is proposed (Lai *et al.*, 2013). Moreover, because of their classical behavior (as a large mass may result in a small quantum wavelength), strangeons that exist in compact stars could locate in periodic lattices (i.e., in a solid state) when temperature becomes sufficiently low. Pulsars are in the low temperature limit, and they could then be solid strangeon star.

In order to better constrain the properties of strangeon stars, a linked bag model (Miao *et al.*, 2022) was proposed to describe the condensed matter by the strong interaction (i.e., strong condensed matter) in both 2-flavored (nucleons) and 3-flavored (hyperons, strangeons, etc.) scenarios. The model parameters are calibrated to reproduce the saturation properties of nuclear matter, which are later applied to hyperon matter and strangeon matter.

3. Observation Tests

It is an important issue for astrophysicists to test the models of pulsars' internal structure, i.e., EOSs, observationally. In principle, various models could be distinguished by multi-band observations, such as radio and X-rays, from which astrophysicists have accumulated abundant observational properties. Significant attention has been paid to the peculiar observational features related to both the global structure and the surface condition of pulsars. The properties of strangeons, as well as the strangeon matter surface, could then be calculated. The results are shown to be helpful for understanding different observations of pulsar-like compact stars (see the review (Lai *et al.*, 2023a; Lai and Xu, 2017) and references therein).

3.1. TOV mass

Combining the EOS and the hydrostatic equilibrium equation for the non-rotational stars in the general relativity, i.e., TOV equation (Tolman, 1939), the mass and radius of the star at a given central density can be calculated. As shown in Figures 1 and 2, each curve represents the mass and radius obtained from different center densities under the same EOS. For a given EOS, the mass generally increases as the central density increases, until it reaches the maximum value corresponding to the highest point on the mass-radius curve. From then on, if the central density increases further, the star will be unstable against gravity and will collapse. This maximum mass for the non-rotating stars is called the TOV mass denoted by M_{TOV}.

Different EOSs will give different mass-radius curves and different M_{TOV}. It is generally believed that the presence of hyperons or strange quark matter inside stars will lead to smaller M_{TOV},

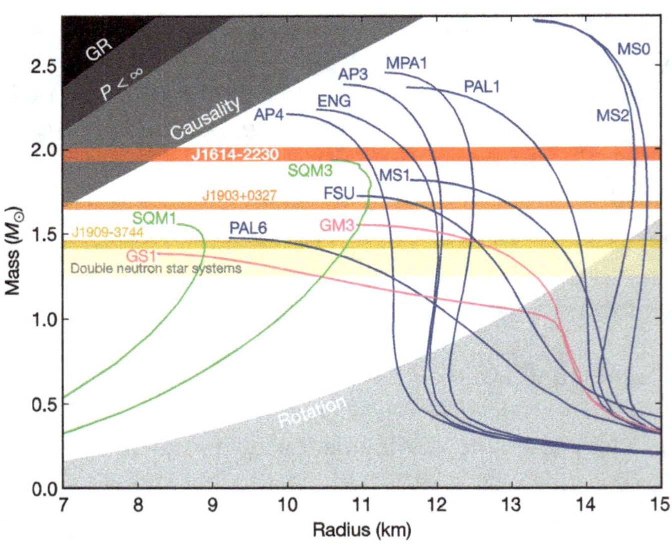

Fig. 1. Mass-radius curves for non-rotating mass versus radius for several typical models of neutron stars (blue and pink curves) and strange quark stars (green curves). The horizontal bands show the observational constraints from PSR J1614-2230 (red rand) and some other pulsars. The grey regions show parameter space that is ruled out by other theoretical or observational constraints (Demorest *et al.*, 2010).

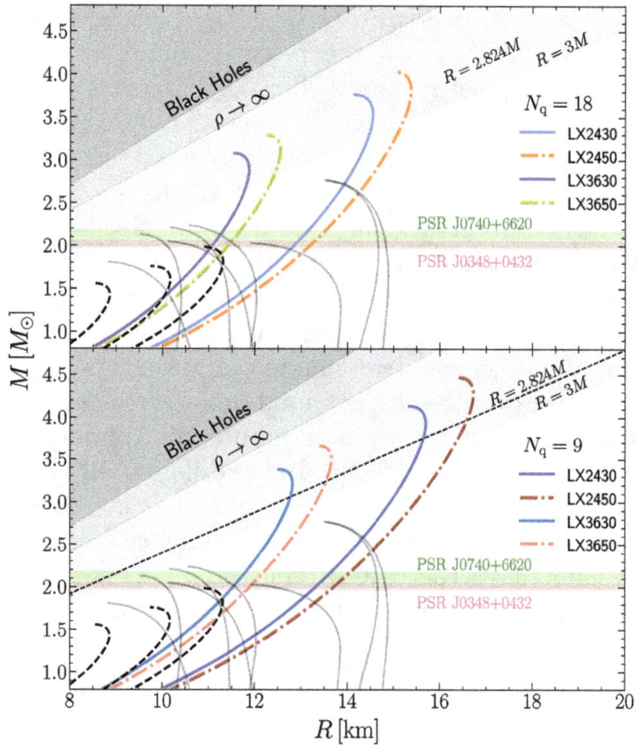

Fig. 2. Mass-radius curves for strangeon stars (Gao *et al.*, 2022). The upper and lower panels are for $N_q = 18$ and $N_q = 9$, respectively (N_q is the number of quarks inside one strangeon), with different combinations of the surface baryonic density and the potential depth between strangeons. The mass-radius curves for neutron stars (grey) and strange quark stars (black) are also shown. The 1σ region of the mass measurements of PSR J0348 + 0432 (Antoniadis *et al.*, 2013) and PSR J0740 + 6620 (Fonseca *et al.*, 2021) are illustrated. Parameter spaces that are ruled out by other theoretical or observational constraints are also shown.

because the additional strange flavor degree of freedom introduced by hyperons and strange quark matter would lower the energy of the system, resulting in a smaller pressure, i.e. softer EOS. If $M_{\rm TOV}$ of one EOS is smaller than one of the known pulsars' masses, this EOS could be excluded. Therefore, we always try to find more massive pulsars to exclude some of the existing EOSs.

The pulsar masses are usually measured by orbital parameters in binary star systems. For example, the first detected massive pulsar is

PSR J1614-2230 with mass 1.97 ± 0.04 $M_\odot$ (Demorest *et al.*, 2010) ($M_\odot$ is the solar mass), and the recently found high-mass pulsar is PSR J0740+6620 with mass 2.08 ± 0.07 $M_\odot$ (at 68.3% credibility) (Fonseca *et al.*, 2021). Under such masses, some quark star models as well as neutron star models with hyperons in the core could be ruled out. However, strangeon stars with stiff EOS (Lai and Xu, 2009) have passed observational tests from massive pulsars.

It is worth mentioning that, the theoretical curves in Figure 2 are obtained for stars without rotation. With rotation, $M_{\rm TOV}$ is not actually the maximum mass of pulsars, because the rotation stabilizes the star and thus increases the maximum mass. The rigid rotation at the largest rotation frequency would increase the maximum mass of pulsars by 20%, but for a pulsar with a spin period ~3 ms, this increase is ≤2% (Berti *et al.*, 2005). In the differentially rotating case, neutron stars can support more mass than that in the case of rigid rotation, and strange quark stars could experience non-axisymmetric instabilities similar to the case of neutron stars, but the degree of differential rotation is quite small (Zhou *et al.*, 2021).

3.2. *Tidal deformability*

The first gravitational wave event GW150914 (Abbott *et al.*, 2016) discovered by LIGO (Laser Interferometer Gravitation Wave Observatory) opened a window of gravitational waves, which marked the beginning of a new era of multi-messenger astronomy. The detections of gravitational event GW170817 (Abbott *et al.*, 2017) and its multiwavelength electromagnetic counterparts (Kasen *et al.*, 2017; Kasliwal *et al.*, 2017) provide us new tools, by which the nature of pulsar-like compact stars could be crucially tested. Perhaps the first important fundamental problem solved in the era of gravitational wave astronomy could be the mystery of pulsar-like compact stars associated with non-perturbative QCD.

When binary stars orbit each other, they will gradually approach each other due to gravitational radiation, which is called the inspiral phase for coalescing binary stars. At the end of inspiral, when each star is deformed in the tidal field of its companion, the finite size of the stars cannot be ignored (Damour *et al.*, 1992). During the late inspiral stage of merging binary compact stars, mass quadrupole moment will be induced by the external tidal field of the companion, which accelerates the coalescence and hence detectable by GW observations (Flanagan and Hinderer, 2008). This property of compact stars can be characterized by the dimensionless tidal deformability (defined as $\Lambda = (2/3)k_2/(GM/c^2R)^5$ where k_2 is the second tidal love number), which depends on EOS (Hinderer, 2008). The smaller Λ of the star, the smaller the tidal deformability under the same external gravitation field, i.e., the harder for the star to deform in the external gravitation field. Conversely, the larger Λ of the star, the easier for the star to deform in the external gravitation field. Different types of EOS predict different values of Λ, and the constraint of Λ obtained by GW 170817 (Abbott *et al.*, 2017) puts robust constraint on EOS of compact stars.

Different EOSs predict different values of Λ. For neutron stars with different EOSs but the same mass, the neutron star with softer EOS will have smaller radius and consequently be more compact, which means that it is harder to deform under the external field, i.e., it has smaller value of Λ. Conversely, for a given mass, a neutron star with stiffer EOS has a larger radius and be therefore less compact, which means that it is more prone to deform under the external field, i.e., it has larger value of Λ. Combining the above conclusions with the discussion in Section 3.1, it can be seen that softer EOSs imply lower $M_{\rm TOV}$ and smaller Λ, while stiffer EOSs imply higher $M_{\rm TOV}$ and larger Λ.

Combining the observations of massive pulsars and the tidal deformability can give strict constraints on EOSs of pulsar-like compact stars. On one hand, the gravitational wave event GW170817 from the binary neutron star inspiral gives an upper limit on Λ, which can exclude many stiff EOSs of neutron stars. On the other hand, the mass measurements of massive pulsars can exclude soft EOSs by giving a lower limit on $M_{\rm TOV}$. In other words, soft EOSs could not support a sufficiently high mass over $2M_\odot$, while stiff EOSs could not give a sufficiently low Λ as implied by GW170817. Therefore, by combining the lower limit of $M_{\rm TOV}$ and the upper limit of Λ, the allowed range of parameter-space for EOSs of pulsar-like compact stars will be narrowed (Annala *et al.*, 2018).

Although strange quark stars and strangeon stars also obey the tendency that $M_{\rm TOV}$ increases with Λ, their quantitative relations between $M_{\rm TOV}$ and Λ are different from that of neutron stars. Neutron stars are gravity-bound, and their surface densities are close to zero. Whereas, both strange quark stars and strangeon stars are self-bound by strong interaction, and their surface densities are still about the density of nuclear matter. Therefore, with the same mass, strange quark stars/strangeon stars are more compact than neutron stars, which means that Λ of the former is smaller than that of the latter. It has been found that (Lai *et al.*, 2019), strangeon stars in a wide range of parameter-space can simultaneously satisfy $M_{\mathrm{TOV}} > 2M_\odot$ and the constraint on Λ given by GW170817.

3.3. *Surface properties*

In this and the following subsections, we will focus on the properties of strangeons, as well as the strangeon matter surface, which can be helpful for understanding different observations of pulsar-like compact stars.

Similar to quark stars, strangeon stars are self-bound on surface. Some problems related to the surface properties of pulsar-like compact stars can be naturally solved in bare strangeon star scenario. (1) A bare strangeon star does not have atomic features in the thermal X-ray emission because no atom exists on its surface. (2) A strangeon matter surface could naturally explain the pulsar magnetospheric activity (Xu *et al.*, 1999; Yu and Xu, 2011) as well as the subpulse-drifting of radio pulsars (Lu *et al.*, 2019). (3) The strong radiation pressure caused by thermal emission from strangeon star might play an important role in promoting core-collapse supernovae (Chen *et al.*, 2007) and even in explaining long-lived plateau of light curves of GRB afterglow (Dai *et al.*, 2011). The neutrino burst observed during supernova 1987A could also be understood in the regime of strangeon star (Yuan *et al.*, 2017). (4) The motion of the electron sea on the magnetized surface was investigated (Xu *et al.*, 2012), and it is found that the absorption feature of 1E 1207.4-5209 could be understood in this hydrocyclotron oscillations model. (5) The strangeness barrier could help us to understand the redshifted O VIII Ly-α emission line and the change in the blackbody radiation area of 4U 1700+24 (Xu, 2014), and the plasma atmosphere of strangeon stars can reproduce the Optical/UV excess observed in X-ray dim isolated neutron stars (Wang *et al.*, 2017).

In addition, the peculiar surface difference between strangeon star and normal neutron star would have profound implication to the magnetospheric activity related to coherent radio emission of compact stars. Besides trying to alleviate the binding energy problem in pulsar radio emission (Xu and Qiao, 1998), one could also find a reliable way to create a bulk of large bunches with energetic $e^{\pm}$-particles just above stellar surface in the strangeon star model, because of the solid and rigidity nature of strangeon matter. The rough solid surface may help triggering a lot of bunches of electron-positron pairs, that would work both for normal radio pulsars radiating regular pulses and for occasional events sending irregular bursts (e.g., repeating fast radio bursts, FRB) (Wang *et al.*, 2022a,b). In the future, studies combining radio pulsar with FRB are welcome in order to understand the electrodynamics and related radiative mechanisms of rotating compact stars with strong magnetic fields.

3.4. *Glitches*

Strangeon stars would be solidified during cooling. The local pressure could be anisotropic in elastic matter, so for solid strangeon stars the radial pressure gradient could be partially balanced by the tangential shear force. A sudden change of tangential force may result in a starquake, with release of both gravitational and elastic energy. The solid stars would be natural to have glitch as the result of starquakes (Lai *et al.*, 2018b; Peng and Xu, 2008; Zhou *et al.*, 2004, 2014).

Dividing the inner motion of the star during starquakes into plastic flow and elastic motion, the behaviors of glitches without significant energy releasing (including the Crab and the Vela pulsars), could be understood in an uniform model (Lai *et al.*, 2018b). The glitch activity, i.e., the statistical feature of glitches, of normal radio pulsars could also be reproduced in the framework of starquake of solid strangeon stars (Wang *et al.*, 2021). Moreover, the glitch recovery with the exponential decay form observed in some glitches can be produced(Lai *et al.*, 2023b).

3.5. *Kilonova*

The multiwavelength electromagnetic counterparts of GW170817, named kilonova, can also give constraints on the nature of compact stars, although not so directly as that given by tidal deformability.

Therefore, it is necessary to explore the kilonova-like event associated with merging strangeon stars.

In the regime of binary neutron star merger, the outer parts of neutron stars are supposed to be made of neutron-rich nuclei as well as neutrons, so merging binary neutron stars would produce neutron-rich ejecta in which r-process nucleosynthesis will happen. The kilonova in this case could be called neutron kilonova. In the regime of binary strangeon star merger, the ejected strangeon nuggets would evaporate strangeons, neutrons and protons, and the ejector would eventually be neutron-rich in the equatorial plan by the decay of strangeons. The binary strangeon star merger would then also give rise to kilonova-like event, called strangeon kilonova (Lai *et al.*, 2018a).

Because of the high maximum mass ($M_{\rm max}$ $\sim 3M_{\odot}$), the merger remnant of binary strangeon stars would usually be very long-lived and even to be stable (if $M < M_{\rm max}$), rather than collapsing quickly to a black hole. The study of the evolution of the ejecta showed that the spin-down process of the long-lived remnant would account for the whole emission of kilonova AT2017gfo associated with GW170817 (Lai *et al.*, 2021). Consequently, the strangeon kilonova would have similar light curves, as well as similar nucleosynthesis, as that of neutron kilonova, passing the test of kilonova observations of merging compact objects.

4. Summary

The study of pulsars opens a unique window that relates fundamental particle physics and astrophysics. It has been more than fifty years since the first neutron star was discovered, while the total number of pulsars discovered is closed to 3,000 and a lot of observational data have been accumulated in the electromagnetic band. On the one hand, due to the lack of knowledge about the non-perturbative properties of strong interaction at low energy scale, it is hard to describe the internal matter of pulsar-like compact stars. On the other hand, pulsars can serve as a kind of extreme astrophysical laboratory to help us explore the non-perturbative QCD problem about EOSs at densities beyond the nuclear matter density.

With the opening of the window of the gravitational waves, people have entered the era of multi-messenger astronomy including electromagnetic waves, gravitational waves and neutrinos, which provides a new opportunity to study the internal structure of pulsar-like compact stars. Certainly, by taking advantage of other facilities, e.g. FAST (Five-hundred-meter Aperture Spherical Telescope), QTT (steerable Xingjiang QiTai 110-m radio Telescope), Insight-HXMT (Hard X-ray Modulation Telescope) and SKA (Square Kilometre Array), it is hopeful to test the EOS-models for pulsar-like compact stars in the near future.

Bibliography

Abbott, B. P. *et al.* (2016). GW150914: The Advanced LIGO detectors in the era of first discoveries, *Phys. Rev. Lett.* **116**(13), 131103.

Abbott, B. P. *et al.* (2017). GW170817: Observation of gravitational waves from a binary neutron star inspiral, *Phys. Rev. Lett.* **119**(16), 161101.

Annala, E., Gorda, T., Kurkela, A., and Vuorinen, A. (2018). Gravitational-wave constraints on the neutron-star-matter equation of state, *Phys. Rev. Lett.* **120**(17), 172703.

Antoniadis, J. *et al.* (2013). A massive pulsar in a compact relativistic binary, *Science* **340**, 6131.

Badde, W. and Zwicky, F. (1934). Supernovae and cosmic rays, *Phys. Rev.* **45**, 138.

Berti, E., White, F., Maniopoulou, A., and Bruni, M. (2005). Rotating neutron stars: An invariant comparison of approximate and numerical space-time models, *MNRAS* **358**(3), 923–938.

Bodmer, A. R. (1971). Collapsed nuclei, *Phys. Rev. D* **4**, 1601.

Chen, A. B., Yu, T. H., and Xu, R. X. (2007). *ApJ* **668**, L55.

Chodos, A., Jaffe, R. L., Johnson, K., Thorn, C. B. and Weisskopf. V. F. (1974) New extended model of hadrons, *Phys. Rev. D* **9**, 3471.

Dai, S., Li, L., and Xu, R. (2011). The Birth of Quark Stars: Photon-driven Supernovae? *Sci. China Phys. Mecha. Astron.* **54**, 1541.

Damour, T., Soffel, M., and Xu, C. (1992). General-relativistic celestial mechanics. II. Translational equations of motion, *Phys. Rev. D* **45**(4), 1017–1044.

Demorest, P., Pennucci, T., Ransom, S., Roberts, M., and Hessels, J. (2010). Shapiro delay measurement of a two solar mass neutron star, *Nature* **467**, 1081–1083.

Fischer, C. S. (2006). Infrared properties of QCD from Dyson-Schwinger equations, *J. Phys. G: Part. Nucl. Phys.* **32**, R253.

Flanagan, É. É. and Hinderer, T. (2008). Constraining neutron star tidal Love numbers with gravitational wave detectors, *Phys. Rev. D* **77**, 021502.

Fonseca, E. *et al.* (2021). Refined mass and geometric measurements of the high-mass PSR J0740+6620, *ApJL* **915**(1), L12.

Gao, Y., Lai, X.-Y., Shao, L., and Xu, R.-X. (2022). Rotation and deformation of strangeon stars in the Lennard-Jones model, *MNRAS* **509**(2) 2758–2779.

Glendenning, N. K. (2001). Phase transitions and crystalline structures in neutron star cores, *Phys. Rep.* **342**(6), 393–447.

Hewish, A., Bell, S. J., Pilkington, J. D., Scott, P. F., and Collins, R. A. (1968). Observation of a rapidly pulsating radio source, *Nature* **217**, 709.

Hinderer, T. (2008). Tidal love numbers of neutron stars, *ApJ* **677**(2), 1216–1220.

Holdom, B., Ren, J., and Zhang, C. (2018). Quark matter may not be strange, *Phys. Rev. Lett.* **120**, 222001.

Kasen, D., Metzger, B., Barnes, J., Quataert, E., and Ramirez-Ruiz, E. (2017). Origin of the heavy elements in binary neutron-star mergers from a gravitational wave event, *Nature* **551**, 80.

Kasliwal, M. *et al.* (2017). Illuminating gravitational waves: A concordant picture of photons from a neutron star merger, *Sci.* **358**, 1559.

Lai, X.-Y., Gao, C.-Y., and Xu, R.-X. (2013). H-cluster stars, *MNRAS* **431**, 3282.

Lai, X.-Y., Wang, W.-H., Yuan, J.-P., *et al.* (2023b). Pulsar glitch in a strangeon star model C III. The recovery, *MNRAS* **523**, 3967.

Lai, X.-Y., Xia, C.-J., and Xu, R.-X. (2023a). Bulk strong matter: The trinity, *Adv. Phys. X* **8**, 2137433.

Lai, X.-Y., Xia, C.-J., Yu, Y.-W., and Xu, R.-X. (2021). Merging strangeon stars II: The ejecta and light curves, *RAA* **21**(10), 250.

Lai, X.-Y. and Xu, R.-X. (2009). Lennard-Jones quark matter and massive quark stars, *MNRAS* **398**, L31.

Lai, X.-Y. and Xu, R.-X. (2017). Strangeon and strangeon star, *J. Phys. Conf. Ser.* **861**(1), 012027.

Lai, X.-Y., Yu, Y.-W., Zhou, E.-P., Li, Y.-Y., and Xu, R.-X. (2018a). Merging strangeon stars, *RAA* **18**, 024.

Lai, X.-Y., Yun, C.-A., Lu, J.-G., *et al.* (2018b). Pulsar glitches in a strangeon star model, *MNRAS* **476**, 3303.

Lai, X.-Y., Zhou, E.-P., and Xu, R.-X. (2019). Strangeons constitute bulk strong matter: Test using GW 170817, *Eur. Phys. J. A* **55**(4), 60.

Landau, L. D. (1932). On the theory of stars, *Phys. Z. Sowjetunion* **1**, 285.
Lattimer, J. M. and Prakash, M. (2004). The physics of neutron stars, *Sci.* **304**, 536.
Lu, J.-G., Peng, B., Xu, R.-X., *et al.* (2019). The radiation structure of PSR B2016+28 observed with FAST, *Sci. China Phys. Mech. Astron.* **62**, 959505.
Miao, Z.-Q., Xia, C.-J., Lai, X.-Y., *et al.* (2022). A bag model of matter condensed by the strong interaction, *Inter. J. Mod. Phys. E* **31**, 2250037.
Peng, C., & Xu, R. X. (2008). Pulsar slow glitches in a solid quark star model, *MNRAS*, **384**, 1034.
Tolman, R. C. (1939). Static solutions of Einstein's field equations for spheres of fluid, *Phys. Rev.* **55**(4), 364–373.
Wang, W. H., Lai, X. Y., Zhou, E. P., *et al.* (2021). Pulsar glitches in a strangeon star model. II. The activity, *MNRAS* **500**, 5336.
Wang, W.-Y., Lu, J.-G., Tong, H., *et al.* (2017). The optical/UV excess of X-ray dim isolated neutron star: I. Bremsstrahlung emission from a strangeon star atmosphere, *Astrophys. J.* **837**, 81.
Wang, W.-Y., Jiang, J.-C., Lu, J., *et al.* (2022a). Repeating fast radio burst: Coherent circular polarization by bunches, *Sci. China Phys. Mech. Astron.* **65**, 289511.
Wang, W.-Y., Yang, Y.-P., Niu, C.-H., Xu, R., and Zhang, B. (2022b). Magnetospheric curvature radiation by bunches as emission mechanism for repeating fast radio bursts, *Astrophys. J.* **927**, 105
Witten, E. (1984). Cosmic separation of phases, *Phys. Rev. D* **30**, 272–285.
Xu, R.-X. (2003). Solid quark matter? *ApJL* **596**, L59–L62.
Xu, R. (2009). The state of cold quark matter: A model-independent view, *arXiv e-prints*, arXiv:0912.0349.
Xu, R.-X. (2014). A solution to the puzzling symbiotic X-ray system 4U 1700+24, *RAA* **14**, 617.
Xu, R.-X., Bastrukov, S.-I., Weber, F., Yu, J.-W., and Molodtsova, I.-V. (2012). Absorption features caused by oscillations of electrons on the surface of a quark star, *Phys. Rev. D* **85**, 023008.
Xu, R. and Guo, Y. (2017). Strange matter: A state before black hole, In C. A. Zen Vasconcellos (ed.), *Centennial of General Relativity: A Celebration*, 119–146.
Xu, R.-X. and Qiao, G.-j. (1998). "Bare" strange stars might be not bare, *Chin. Phys. Lett.* **15**, 934.
Xu, R.-X., Qiao, G.-J., and Zhang, B. (1999). PSR 0943+10: a bare strange star? *Astrophys. J. Lett.* **522**, L109.
Yu, J. W. and Xu, R. X. (2011). The magnetospheric activity of bare strange quark stars, *MNRAS* **414**, 489.

Yuan, M., Lu, J.-G., Yang, Z.-L., Lai, X.-Y., and Xu, R.-X. (2017). Supernova neutrino in a strangeon star model, *RAA* **17**, 092.

Zhang, C. (2020). Probing up-down quark matter via gravitational waves, *Phys. Rev. D* **101**, 043003.

Zhou, A.-Z., Xu, R.-X., Wu, X.-J., and Wang, N. (2004). Quakes in solid quark stars, *Astropart. Phys.* **22**, 73.

Zhou, E.-P., Kiuchi, K., Shibata, M., Tsokaros, A., and Uryu, K. (2021). Evolution of bare quark stars in full general relativity: Single star case, *Phys. Rev. D* **103**(12), 123011.

Zhou, E.-P., Lu, J.-G., Tong, H., and Xu, R.-X. (2014). Two types of glitches in a solid quark star, *MNRAS* **443**, 2705.

https://doi.org/10.1142/9789819800643_0023

Chapter 23

Fast Pulsars, Neutron Stars, and Astrophysical Strange Quark Matter Objects

Delaney Farrell[*,], Fridolin Weber[*,†,††], Milva G. Orsaria[‡,§,††], Ignacio F. Ranea-Sandoval[‡,§,§§], Martín Canullán[‡,§,¶¶], Rodrigo Negreiros[¶,‖‖], Takashi Katayama[‖,***], and Ian Wagaman[‖,†††]**

[*]*Department of Physics, San Diego State University*
San Diego, CA 92182, USA
[†]*Center for Astrophysics and Space Sciences*
University of California at San Diego
La Jolla, CA 92093, USA
[‡]*Grupo de Astrofísica de Remanentes Compactos,*
Facultad de Ciencias Astronómicas y Geofísicas,
Universidad Nacional de La Plata
Paseo del Bosque S/N, La Plata (1900), Argentina
[§]*Consejo Nacional de Investigaciones Científicas y Técnicas (CONICET),*
Buenos Aires (1425), Argentina
[¶]*Catholic Institute of Technology*
Cambridge, MA 02142
Instituto de Física, Universidade Federal Fluminense,
Niterói, Rio de Janeiro, Brazil
[‖]*Department of Physics, San Diego State University*
San Diego, CA 92182, USA

**dfarrell@sdsu.edu
††fweber@sdsu.edu, fweber@ucsd.edu
‡‡morsaria@fcaglp.unlp.edu.ar,
§§iranea@fcaglp.unlp.edu.ar,
¶¶canullanmartin@fcaglp.unlp.edu.ar
‖‖rnegreiros@id.uff.br
***tkatayama0396@sdsu.edu
†††iwagaman6967@sdsu.edu

This book chapter explores key aspects of neutron stars, pulsar glitches, tidal deformability, fast pulsars, the equation of state, and strange quark matter stars. Challenges in directly measuring neutron star radius have led to reliance on spectroscopic and timing techniques, with uncertainties addressed through careful source selection and theoretical modeling. Pulsar glitches reveal insights into the equation of state through angular momentum transfer within the neutron star. Tidal deformability is crucial in gravitational-wave astronomy, exemplified by the GW170817 event. Fast pulsars, instrumental in astrophysical testing, are classified into ordinary pulsars, millisecond pulsars, and magnetars. The EOS is vital for understanding neutron star internal structure, explored through various models. The chapter delves into the theoretical framework for rotating neutron stars, addressing uniform and differential rotation scenarios and their impacts on mass and radius. Additionally, the intriguing concept of quark stars and strange dwarfs is investigated. The various topics discussed in this book chapter contribute to a broader understanding of dense matter physics, astrophysical phenomena, and the potential for transformative discoveries through advanced observational techniques and technologies like gravitational wave detectors, radio telescopes, and X-ray telescopes.

Keywords: Neutron stars, Quark stars, Pulsars, Nuclear equation of state, Strange quark matter, Stellar rotation

Contents

1. Introduction

Neutron stars stand out as the densest observed stellar objects, typically harboring masses within the range of 1 to 2 solar masses ($M_\odot$) and radii spanning from 10 to 15 km. The immense gravitational forces at play within neutron stars compress the matter in their innermost regions to densities exceeding nuclear saturation density (2.5×10^{14} g/cm^3), turning neutron stars into inherent laboratories for investigating ultra-dense matter (Alford, 2001; Alford *et al.*, 2008; Becker, 2009; Blaschke and Chamel, 2018; Haensel *et al.*, 2007; Sedrakian *et al.*, 2023). Notably, neutron star matter is one of the densest forms of matter in the Universe and also exists in its ground state, characterized by temperatures that are less than 10^{10} K (or around 1 MeV), effectively registering as zero temperature on the nuclear scale. The cold, superdense matter encapsulated within neutron stars occupies a unique temperature and density regime that cannot be replicated in terrestrial laboratories, such as the Relativistic Heavy-Ion Collider or the Large Hadron Collider (ATLAS-Collaboration *et al.*, 2008), which are only capable of probing superdense matter in the high-temperature domain. Despite decades of theoretical, experimental, and observational endeavors, understanding the properties of cold, superdense matter remains a forefront area of nuclear and astrophysics.

Determining the equation of state (EOS), which defines the relationship between pressure, density, and temperature of neutron star matter, holds paramount importance (Burgio *et al.*, 2021; Lattimer and Prakash, 2016; Özel and Freire, 2016; Sedrakian, 2007; Sedrakian *et al.*, 2023). This endeavor not only aids in unraveling the nature and characteristics of the strong interaction but also enhances our comprehension of various astrophysical phenomena, including core-collapse supernovae, binary neutron star mergers, and the early Universe.

In addition to the intrinsic significance of neutron stars, a multitude of factors contribute to the evolving landscape of neutron star research. Pulsars, characterized by their intense and precisely timed radiation, offer unique insights into the dynamics of these enigmatic objects. Moreover, the advent of new observational data facilitated by powerful telescopes has significantly advanced our understanding of neutron stars (Kaspi and Gavriil, 2017). Notably, the (1) FAST (Five hundred-meter Aperture Spherical Telescope) stands out as a remarkable instrument capable of detecting extremely faint radio signals from space (Nan *et al.*, 2016). This capability has rapidly expanded the number of pulsar observations, providing invaluable insights into the behavior of neutron stars. In addition to FAST, several other state-of-the-art observatories and instruments lead the forefront of neutron star research. Positioned on the International Space Station, (2) NICER (Neutron star Interior Composition Explorer): NICER serves as an X-ray telescope meticulously designed to scrutinize the interior composition of neutron stars, delving into the intricacies of their dense cores (Gendreau *et al.*, 2016). (3) eROSITA (Extended Roentgen Survey with an Imaging Telescope Array): Functioning as an X-ray instrument on the Spektrum-Roentgen-Gamma mission, eROSITA conducts a comprehensive survey of the X-ray sky (Predehl *et al.*, 2021). Its capabilities significantly contribute to the exploration of neutron stars and other high-energy astrophysical phenomena. (4) Chandra X-ray Observatory: This space telescope, specifically engineered for observing X-rays emanating from high-energy regions of the universe proves invaluable in the study of neutron stars (Weisskopf *et al.*, 2000). (5) XMM-Newton: Serving as an X-ray observatory designed for high-resolution X-ray spectroscopy, XMM-Newton has played a pivotal role in advancing our understanding of neutron stars and their surrounding environments (Jansen *et al.*, 2001). (6) VERITAS (Very Energetic Radiation Imaging Telescope Array System): Contributing to the study of neutron stars emitting high-energy gamma rays, VERITAS enhances our exploration of these celestial objects (Holder *et al.*, 2006). (7) LOFAR (Low-Frequency Array): A vast radio telescope array spanning Europe, LOFAR offers unique capabilities for low-frequency radio observations, providing a valuable avenue for studying pulsars and other radio-emitting neutron star phenomena (van Haarlem *et al.*, 2013). (8) Green Bank Observatory: Instrumental in the examination of pulsars and other radio-emitting neutron star phenomena, the Green Bank Observatory plays a significant role in advancing our understanding of these cosmic entities (Hughes *et al.*, 2006). (9) Jodrell Bank Observatory: The observatory's radio telescopes, including the Lovell Telescope, have been instrumental in observing and monitoring pulsars, aiding in the understanding of these highly magnetized, rotating neutron stars. The precise timing of pulsar signals is crucial for research, including the search for gravitational waves (Jodrell Bank Observatory, Accessed: January 2022). Last but certainly not least, (9) the Parkes telescope is a state-of-the-art facility for radio astronomy, distinguishing itself in the study of neutron stars and pulsars (Johnston *et al.*, 1996). Collectively, these observatories and instruments represent a comprehensive and cutting-edge approach to unraveling the mysteries of neutron stars.

Beyond conventional neutron stars, the intriguing possibility of strange quark stars, composed of strange quark matter, adds an extra layer of complexity to our exploration. Strange quark matter is a speculative and exotic form of matter theorized to exist at exceedingly high densities. According to the hypothesis surrounding strange quark matter, it is envisioned as a composition of up, down, and strange quarks — the fundamental particles constituting protons, neutrons, and hyperons. The proposition suggests that strange quark matter may possess greater stability than conventional nuclear matter, with quarks moving freely rather than being confined within protons and neutrons. Should such matter indeed exist, it could give rise to new categories of compact stellar objects, ranging from compact strange stars, with bulk properties akin to massive neutron stars, to strange quark matter dwarfs, representing the strange counterparts of ordinary white dwarfs. Despite the elusive nature of experimental verification for the existence of strange quark matter, its theoretical implications are pivotal in advancing our understanding of astrophysical phenomena and the fundamental building blocks of matter under ultra-high-density conditions.

In this chapter, we delve into the intricate theoretical modeling of neutron stars, compact strange stars, and strange dwarfs, placing particular emphasis on the influence of rapid stellar rotation.

Our investigation encompasses the examination of both differential and rigid body rotation. The primary objective is to comprehensively explore the properties inherent in these stellar objects and discern the pivotal role played by strange quark matter in interpreting the nature of compact stellar entities.

2. Observation

Observation of neutron stars has been perhaps the most important catalyst to furthering our knowledge on the subject since their theorized existence. In the last two decades, there has been an increasing amount of high-quality, new data from X-ray and gamma-ray telescopes, including precisely timed observations of pulsars (Özel and Freire, 2016). Today, over 3,000 pulsars have been discovered, and their discoveries have provided valuable constraints on various neutron star properties like mass and radius, cooling (Slane *et al.*, 2002), and magnetic fields (Makishima *et al.*, 1999). Additionally, these measurements provide constraints on the cold EOS of superdense matter (Dietrich *et al.*, 2020; Lattimer and Prakash, 2007), a long-standing mystery in nuclear astrophysics. As we move into an era of new-generation X-ray telescopes, these constraints on neutron star properties and EOS are expected to tighten even further; these discoveries will also aid in our understanding of other astrophysical phenomena like core-collapse supernova, r-process nucleosynthesis, and gravitational wave emission (Özel and Freire, 2016).

2.1. *Observed neutron star properties*

2.1.1. *Mass*

The mass of neutron stars is perhaps the most precisely measured neutron star property, especially those coming from radio pulsar timing of neutron star masses in compact binaries (Watts *et al.*, 2016). Observed values of masses of neutron stars (Lattimer, 2016) are extremely valuable measurements as they test theories of nuclear matter and therefore provide constraints on the nuclear EOS. A notable example of such a constraint resulted from the observation of massive neutron stars. For several decades the canonical mass of a neutron star was 1.4 $M_\odot$, but recent observations of heavier neutron stars ($\geq$2 $M_\odot$) have widened the range of masses that must be accounted for with theoretical EOS models. Heavy neutron stars directly challenge theoretical models of dense nuclear matter in quantum chromodynamics (QCD). Observations of massive stars enforced the requirement that EOS models for neutron star matter must produce maximum stable masses of at least 2 $M_\odot$. Such models are referred to as "stiff" meaning they have a large pressure for a given density, so observations of massive neutron stars ruled out many softer theoretical EOS models. An example of such massive observations is MSP J0740+6620 with a measured mass of $2.14^{+0.10}_{-0.09}$ $M_\odot$, making it one of the heaviest millisecond pulsars observed thus far (Cromartie *et al.*, 2020).

Radio pulsars, or highly magnetized, rotating neutron stars, that exist in binary stellar systems generate radio pulses that can be precisely tracked as a method of determining the binary's mass (Özel and Freire, 2016). Radio pulsars have long-term rotational stability, allowing for very precise timing measurements of orbital motion. The binary mass function, $f(M_1, M_2, i)$, can be derived from measurements of five Keplerian parameters: the binary period P_b, the projection of the pulsar's orbital semimajor axis a on the line of sight $x \equiv a_i \sin i$, the orbital eccentricity e, and time T_0 and longitude ω_0 of periastron ω (Weber, 1999). The formula for the mass function is given as

$$f(M_1, M_2, i) \equiv \frac{(M_2 \sin i)^3}{(M_1 + M_2)^2} = \frac{P_b v_1^3}{2\pi G} = \frac{4\pi^2 x^3}{G P_b^2}, \tag{1}$$

where G is Newton's gravitational constant and $v_1 = (2\pi/P_b)x$ is the orbital velocity of star M_1 along the line of sight. Other relativistic effects, like the transverse Doppler effect and gravitational redshift, can also be used to determine the pulsar and companion's masses in a binary system (Özel and Freire, 2016; Weber, 1999).

Alongside other parameters, mass can also be inferred from stellar emission using other observational techniques like waveform modeling, where a Bayesian inference approach is employed to analyze X-ray waveform oscillations emitted by pulsars (Miller *et al.*, 2019; Miller and Lamb, 2015). X-ray oscillations can be observed from accretion-powered pulsars, thermal emission of rotation-powered (or non-accreting) pulsars, or some thermonuclear bursts on accreting neutron stars (Watts *et al.*, 2016). It is believed that the X-ray oscillations are emitted from a "hot spot" on the star's surface, or a region of the stellar surface that is both hotter than its surroundings and offset from the rotational pole of the star. The pulse profile, or waveform, from the hot spot combined with general relativistic effects like light bending and Doppler boosting, can be used to determine the mass and radius of the source. Pulse waveform modeling was pioneered using data from X-ray telescopes like ROSAT and XMM-Newton, but the methodology has become more popular recently with observations from NICER, which has extraordinary absolute timing accuracy when determining the pulse profile of isolated neutron stars and those in binary systems (Miller *et al.*, 2019). For example, NICER data was used to determine estimations for the mass and radius of the isolated millisecond pulsars PSR J0030+0451 (Miller *et al.*, 2019; Riley *et al.*, 2019) and PSR J0740+6620 (together with XMM-Newton data) (Miller *et al.*, 2021; Riley *et al.*, 2021). The latter study showed that although having a mass ~40% larger, the radii of PSR J0740+6620 and PSR J0030+0451 are of the same order (Miller *et al.*, 2021; Riley *et al.*, 2019). This finding is all the more interesting when considering that the determination of neutron star radius has historically been more difficult than mass, which is discussed further below.

2.1.2. *Radius*

Historically, direct measurements of neutron star radius do not exist, and inference of radius using various techniques provides large uncertainties when compared to precise mass measurements. Current methods for determining radius commonly rely on the spectroscopic measurements of surface emission to either measure apparent angular size or look for effects of the emission on neutron star spacetime (Özel and Freire, 2016; Watts *et al.*, 2016). These methods broadly fall into one of two categories: spectroscopic or timing measurements.

Spectroscopic measurements from either the surface emission of quiescent low-mass X-ray binary (LMXB) systems or from thermonuclear bursts can be used to constrain the mass-radius relation. If it is assumed that the surface spectrum is a diluted blackbody and the source of the spectrum is at a known distance D, the mass-radius (M–R) relation of a slowly spinning neutron star can be defined as

$$R^2\left(1-\frac{2GM}{Rc^2}\right)^{-1}=\frac{FD^2f_c^4}{\sigma T_c^4}, \tag{2}$$

where F is the measured flux, σ is the Stefan-Boltzmann constant, and T_c and f_c are the color temperature and color-correction factor, respectively (Watts *et al.*, 2016). The color temperature is assumed to be larger than the star's effective temperature, defined as $T_c = f_c T_{\text{eff}}$. The color-correction factor f_c is dependent on both the assumed atmospheric composition and effective surface gravity, where $f_c \approx 1.3 - 2$ (Watts *et al.*, 2016). Under the same assumed conditions, the observed or apparent radius R_{obs} can be defined as

$$R_{\text{obs}} = R\left(1 - \frac{2GM}{Rc^2}\right)^{-1/2}. \tag{3}$$

The above approximation is reliant on several assumptions and exact measurements and thus may not hold in various instances. In the case of strong magnetic fields, for example, the assumptions of a specific atmospheric composition and a constant surface temperature may be incorrect (Özel and Freire, 2016). For rotating neutron stars that are rotating rapidly (i.e., not spinning slowly), spin-dependent corrections must be applied to the observed angular size as the spacetime can not be described by a Schwarzschild metric due to frame dragging (Özel and Freire, 2016). Additional complications to the approximation are introduced if there is uncertainty when measuring the exact distance D to the star.

Some uncertainties can be alleviated by choosing sources with low magnetic fields or those that exist in globular clusters of known distances, which is why thermal emission from sources like low-mass X-ray binaries (LMXBs) in quiescence are at the forefront of constraining the neutron star mass-radius relation. The observed thermal spectrum can be fit to a well-motivated theoretical model dependent on the assumed atmosphere of the neutron star in a process called spectral fitting. Many of these theoretical models assume quiescent LMXBs have pure hydrogen atmospheres as the companion stars are hydrogen-rich, but other atmospheric compositions like helium and carbon have also been explored in the literature (Steiner *et al.*, 2018).

As mentioned in the section above, the mass-radius relation can also be explored through X-ray timing techniques like waveform modeling, spin measurements, and asteroseismology. While further explanation of these techniques will not be discussed explicitly, the reader can find more information at Miller *et al.* (2019, 2021); Sotani *et al.* (2011); Watts *et al.* (2016).

2.1.3. *Pulsar glitches and moment of inertia*

A pulsar "glitch", or sudden spin jump, is another observable property that can help constrain our knowledge of the EOS of dense neutron star matter. Rotation-powered pulsars will generally experience a slow and gradual spin down from the emission of magnetic dipole radiation, leading to very regular rotational periods (Piekarewicz *et al.*, 2014). However, some pulsars will experience abrupt spin-ups, leading to a deviation from their otherwise regular rotational periods.

The mechanism driving these glitches is an angular momentum transfer between the star's solid outer crust and a portion of the liquid interior (Link *et al.*, 1999). A popular model to explain the angular momentum transfer relies on the theory that a pulsar will form vortex lines in the neutron superfluid formed within the inner crust (at densities of 10^{11} to 10^{14} g/cm^3). A rapidly rotating pulsar can have superfluid vortices with extremely high areal densities, and the spin up or spin down of the superfluid is dependent on increasing or decreasing the vortex density (Alpar *et al.*, 1984; Piekarewicz *et al.*, 2014). The superfluid velocity is thus determined by the spatial arrangement of vortex lines (Weber, 1999). If the vortex lines were assumed to be pinned to a crystal lattice of neutron-rich nuclei, the angular velocity of the superfluid would be largely fixed and appear to be rotating as a rigid body (Piekarewicz *et al.*, 2014). The pulsar will slow down by the emission of magnetic dipole radiation, and as the crust's angular velocity slows down from magnetic torque, the difference between the velocity of the crust and the superfluid regime grows. When this differential rotation between the slower crust and faster superfluid vortices grows too large, some vortex lines become unpinned and suddenly transfer angular momentum to the crust — resulting in a glitch (Weber, 1999).

To date, 671 glitches from 224 pulsars have been recorded using high-precision pulsar timing (an extensive and well-maintained list can be found in the Jodrell bank glitch catalog Basu *et al.*, 2022). While most pulsars have glitches infrequently and irregularly, others like the Vela or Crab pulsar have quasi-periodic glitches (Piekarewicz *et al.*, 2014). An extremely unique case is PSR J0537-6910, the fastest spinning young pulsar recorded, which exhibits large glitches roughly every 100 days (Abbott *et al.*, 2021).

The magnitude of a pulsar glitch is dependent on the star's moment of inertia I, another global neutron star property. A pulsar glitch of magnitude $\Delta\Omega_i$ requires the angular momentum

$$\Delta J_i = I_c \Delta\Omega_i, \tag{4}$$

where I_c is the moment of inertia of both the star's crust and coupled interior core, estimated to make up 90% or more of the star's total moment of inertia (Link *et al.*, 1999). Between each glitch, the rotating interior acts as an angular momentum reservoir at a rate of $\dot{J}_{\text{res}}$. This rate is constrained by the average spin-down rate of the crust $\dot{\Omega}$ and the moment of inertia of the reservoir I_{res} as (Link *et al.*, 1999)

$$\dot{J}_{\text{res}} \leq I_{\text{res}} |\dot{\Omega}|. \tag{5}$$

The ratio of the moment of inertia of the reservoir I_{res} and the moment of inertia of the crust I_c provides the lower limit of

$$\frac{I_{\text{res}}}{I_c} \geq \frac{\Omega}{|\dot{\Omega}|} A = G, \tag{6}$$

where A is the pulsar's glitch activity parameter and G is the coupling parameter that describes the minimum fraction of I that stores and imparts angular momentum to the crust in the glitch (Link *et al.*, 1999; Piekarewicz *et al.*, 2014). The moment of inertia can be expressed in terms of mass and radius, so constraining the upper and lower limits can therefore also provide valuable constraints on the dense matter EOS.

2.1.4. *Tidal deformability and gravitational waves*

The tidal deformability of compact objects is an important physical quantity for gravitational-wave astronomy. The main reason is that it determines the pre-merger GW signal in binary neutron star merger events. Working up to linear order, the tidal deformability λ is given by

$$\lambda = -\frac{\varepsilon_{ij}}{Q_{ij}}, \tag{7}$$

where ε_{ij} denotes the external gravitation field produced by the external source and Q_{ij} is the induced mass-quadrupolar moment of the given object. In addition, λ is related to the second Love number, k_2, and the radius of the compact object, R, by

$$\lambda = \frac{2}{3} k_2 R^5. \tag{8}$$

Finally, the dimensionless tidal deformability, Λ, is defined by

$$\Lambda = \frac{\lambda}{M^5}, \tag{9}$$

where the M denotes the gravitational mass of the compact star.

For details on how theoretical calculations related to this quantity are performed, the interested reader can refer to Refs. Han and Steiner (2019); Hinderer (2008) and references therein.

The detection by LIGO and Virgo observatories of the gravitational waves emitted during event GW170817, a binary neutron star merger (see, for example, (Abbott *et al.*, 2018) and references therein) together with its electromagnetic counterpart (see, for example, (Abbott *et al.*, 2017), and references therein) allow to estimate the possible masses, radius and dimensionless tidal deformability of the merging objects. Additionally, the dimensionless tidal deformability of a canonical neutron star was estimated to be $\Lambda_{1.4} = 190^{+390}_{-120}$ at a 90% level with no strong restrictions to the EOS are imposed (Abbott *et al.*, 2018).

2.2. *Fast pulsars*

Pulsars were first discovered in 1967 by Jocelyn Bell Burnell and Antony Hewish, who detected radio-wave pulses with rapid regularity from an unknown source (Hewish *et al.*, 1979). Just one year later, Thomas Gold proposed that pulsars are rapidly rotating, highly magnetized neutron stars that emit radio waves from their magnetic poles. Since their discovery, pulsars have served as valuable resources for testing theories of relativistic astrophysics and as a tool for the discovery of astrophysical phenomena like the existence of dark matter (De Luca *et al.*, 2021) or the gravitational wave background (Agazie *et al.*, 2023). To date, there have been over 3,000 pulsars observed, and this number will continue to grow as new radio, X-ray, and gamma-ray telescopes are constructed.

Pulsars can broadly be classified into three categories: ordinary pulsars, millisecond pulsars (MSPs), and magnetars. Ordinary or normal pulsars have rotational periods P between 0.1 and 3 s and magnetic fields of roughly 10^{12} G. This class of pulsars will spin rapidly early in their lifetime ($\lesssim 10^4$ years) but slow as they age, eventually disappearing from view (Melrose, 2017). MSPs, also referred to as recycled pulsars, are old neutron stars that have been recycled (spun up) by the accretion of mass and angular momentum from a low-mass companion (Manchester, 2017). MSPs have spin periods on the order of 1.4 to 20 ms and smaller magnetic fields ($B < 10^{11}$ G) when compared to the other two classes of pulsars. The last class, magnetars, are slowly rotating pulsars with extremely high magnetic fields (10^{14}–10^{15} G). The radio emission from all three types of pulsars generally tends to fall into a relatively narrow frequency range of ∼100 MHz to 10 GHz, with a smaller subset of pulsars producing high-energy emission. The three categories of pulsars can be differentiated with two measured parameters: the pulse period P and its rate of change $\dot{P}$, which can be used to determine other important quantities like the star's rotational frequency, spin-down rate, luminosity, characteristic age, and braking index (Melrose, 2017).

As shown by Glendenning *et al.* (1997), a potential signal of quark deconfinement in the core of a pulsar could be the alteration in the timing structure of a pulsar's spin-down. Should quark deconfinement transpire within their cores, the moment of inertia could decrease significantly as the matter in the core undergoes deconfinement to quark matter. This reduction in the moment of inertia leads to an acceleration in rotation, causing an abrupt change in the pulsar's spin-down rate. This transition is identifiable through the examination of the breaking index which would differ drastically from its canonical value of ∼3. Calculations suggest that the braking index may even approach infinity as quark deconfinement is initiated in the stellar core.

As mentioned previously, measuring the stable rotation of pulsars provides valuable constraints on observable properties. The most precise timing measurements come from MSPs, which commonly exist in binary systems with low-density companions like white dwarfs. The small number of isolated MSPs are believed to have lost their companion in some way. For low-mass binaries, this could be from disruptions resulting from the pulsar's strong relativistic winds; for higher-mass double neutron star systems, the orbit could be disrupted by the supernova explosion of the secondary star (Lorimer, 2008).

The fastest pulsar observed thus far is PSR J1748–2446ad, discovered first in 2004 and confirmed in 2005 (Hessels *et al.*, 2006). This radio pulsar is located in the globular cluster Terzan 5, which has the largest known population of globular cluster MSPs. PSR J1748–2446ad spins at an impressive frequency of 716 Hz, with a pulse period P of 1.39595482 ms, and a pulse derivative $\dot{P}$ of $\leq 6 \times 10^{-19}$ (Hessels *et al.*, 2006). Before the discovery of PSR J1748–2446ad, the fastest MSP detected was the first MSP ever detected, PSR 1937+21, spinning at a frequency of 642 Hz (see, for instance, Rawley *et al.*, 1987). A table of various MSPs with spin periods of ≤ 2 ms can be found in Table 1.

3. Equation of State

The essential physics of matter within a neutron star can be succinctly summarized by the EOS, which describes the pressure-density relation within the star. As discussed earlier in the chapter, the matter within the core of a neutron star exists in a low-temperature, high-density regime that cannot be replicated in laboratories or particle colliders. Instead, the EOS of dense neutron star matter must be described by theoretical models that vary widely due to particle composition and mathematical frameworks used. In this section, we discuss the physics of neutron star matter and explore the theoretical implication of the inclusion of quark matter on the EOS.

3.1. *Nuclear matter*

The density of matter within a neutron star varies greatly from the crust to the core, ranging from a few grams per cubic centimeter at the surface to around 10^{15} g/cm^3 at the core. The lower density regime spans from the star's surface to its outer and inner crust regions, while the higher density regime comprises the star's core region. As stated in the Introduction, the density within the innermost part of the core region may experience densities exceeding

Table 1. Sample millisecond pulsars (MSPs) with rotational periods P of $\lesssim 2$ ms not associated with a globular cluster.

PSR name	P (ms)	Year	Telescope
J0034−0534	1.88	1994	PKS
J0955−6150	1.99	2013	PKS
J1023+0038	1.69	2009	GBT
J1036−4353	1.68	2021	MK
J1221−0633	1.93	2017	GBT
J1227−4853	1.69	2014	GMRT
J1301+0833	1.84	2010	GBT
J1555−2908	1.79	2017	GBT
J1723−2837	1.86	2012	PKS
J1747−4036	1.65	2009	PKS
J1804−2858	1.49	2018	PKS
J1810+1744	1.66	2009	GBT
J1831−6503	1.85	2022	MK
J1833−3840	1.87	2015	PKS
J1843−1113	1.85	2004	PKS
J1849+0016	1.81	2020	FAST
J1859+0313	1.61	2019	FAST
J1902−5105	1.74	2009	PKS
J1924+2027	1.95	2020	FAST
J2036−02	1.91	2020	GBT
J2051+50	1.68	2020	GBT

Notes: Included are the rotational periods P, year discovered, and the telescope used in the observation (PKS: Parkes Radio Telescope, GBT: Green Bank Telescope, GMRT: Giant Metre Wave Radio Telescope, MK: MeerKat, FAST: Five Hundred Meter Aperture Spherical Radio Telescope).
Source: Adapted from (Lorimer and Ferrara, 2022).

nuclear saturation density (2.5×10^{14} g/cm^3), giving rise to more exotic types of matter like meson condensates, deconfined quark matter, and strange matter — all of which alter the matter's EOS.

The least dense portions of a neutron star are its surface and outer crust region, where matter is compressed to densities up to 10^6 g/cm^3. The surface region is comprised of ordinary atomic nuclei (such as ^{56}Fe) and non-relativistic electrons and emits the thermal radiation observed by the telescopes discussed in the previous section about observation. In the outer crust, at densities of $7 \times 10^6 - 4.3 \times 10^{11}$ g/cm^3, electrons become relativistic and form a relativistic electron gas. Additionally, the atomic nuclei grow more neutron-rich while forming a solid Coulomb lattice. The final portion of the lower-density regime is the inner crust, existing at densities of $4.3 \times 10^6 - 2 \times 10^{14}$ g/cm^3. In the inner crust, atomic nuclei grow extremely neutron-rich to the point neutrons begin to drip out of the nuclei and populate free states; the neutron-rich heavy metals form clusters in a solid lattice and are then immersed in a gas of relativistic electrons and neutrons (Chamel and Haensel, 2008; Lorenz *et al.*, 1993; Pethick *et al.*, 1995; Pethick and Ravenhall, 1999; Rüster *et al.*, 2006).

The high-density region of a neutron star, its core, is composed of a relativistic Fermi liquid of neutrons and protons in the conventional description. The liquid is extremely neutron-rich while maintaining a certain fraction of protons, electrons, and muons to maintain chemical or β-equilibrium. The condition for β-equilibrium is given as

$$\mu_B = \mu_n + q_B(\mu_e - \mu_{\nu_e}), \tag{10}$$

where μ_B is the chemical potential of baryons B, neutrons (μ_n), electrons (μ_e), and electron neutrinos (μ_{ν_e}), while q_B is the baryon electric charge. The dense, neutron-rich matter in the core may reach densities of up to 10 times nuclear saturation density, and as nucleons at such high densities begin to overlap, the matter is expected to transition into non-nucleonic states of matter. These states can take several forms: the onset of quark degrees of freedom and deconfined quark matter (Baym and Chin, 1976; Chapline and Nauenberg, 1977a,b; Collins and Perry, 1975; Fechner and Joss, 1978; Fritzsch *et al.*, 1973; Itoh, 1970a; Ivanenko and Kurdgelaidze, 1965; Keister and Kisslinger, 1976), boson condensates formed by pions or kaons (Brown and Weise, 1976; Brown *et al.*, 1994; Glendenning and Schaffner-Bielich, 1999; Haensel and Proszynski, 1982; Kaplan and Nelson, 1988; Knorren *et al.*, 1995; Lee *et al.*, 1994; Malik *et al.*, 2021; Mannarelli, 2019; Migdal, 1972; Nelson and Kaplan, 1987; Sawyer and Scalapino, 1973; Schaffner and Mishustin, 1996; Thapa *et al.*, 2021), or the formation of other matter with a high degree of strangeness. Due to the weak interactions in the cold and dense neutron star core, strange states of matter can form hyperons (Ambartsumyan and Saakyan, 1960, 1961; Bethe and Johnson, 1974; Leung and Wang, 1971; Moszkowski, 1974; Pandharipande, 1971) or color superconducting phases (Alford, 2001; Alford *et al.*, 2008; Rajagopal and Wilczek, 2001; Rischke, 2004). Additionally, the existence of self-bound compact stars consisting of strange quark matter has also been discussed in the literature and will be further explained in the next section (Alcock *et al.*, 1986a; Alcock and Olinto, 1988; Madsen, 1999, 2007; Terazawa, 1979, 1991; Weber, 2005; Witten, 1984).

While the matter within the core of a neutron star cannot be replicated in terrestrial laboratories, properties of laboratory neutron-rich nuclei can be used to define the EOS for the crust (up to order 10^{11} g/cm^3). The crust regions are thin, contributing very little to the bulk properties of the star like its mass and radius. Today, two commonly used EOS models for the crust include Baym-Pethick-Sutherland (Baym *et al.*, 1971b) and Baym-Bethe-Pethick (Baym *et al.*, 1971a) (BPS-BBP), or Harrison-Wheler (Harrison *et al.*, 1965) and Negele–Vautherin (Negele and Vautherin, 1973) (HW-NV).

For matter within the core, various theoretical approaches that can broadly be broken into two categories, phenomenological and ab initio, are employed to describe the dense, asymmetric many-body system. Phenomenological models, or relativistic mean-field (RMF) models, are constructed to reproduce properties of finite nuclei and nuclear matter (see ref. Tong *et al.* 2018). For in-depth discussions concerning constraints on the equation of state of high-density matter extracted from nuclear physics and astrophysics, we refer to Lattimer (2019); Lattimer and Steiner (2014); Stone (2021).

Ab initio methods instead start with realistic free-space nucleon-nucleon (NN) interactions applied to the nuclear many-body system (Sammarruca, 2008). NN interactions, based on scattering data from free nucleons and properties of the deuteron, are characterized by a repulsive core at short distances, a strong attraction in an intermediate range, and are dominated by the one-pion exchange at large distances (Machleidt, 1989; Machleidt *et al.*, 1987). Ab initio methods of this kind include the quantum Monte Carlo method (Carlson *et al.*, 2015; Chen and Zhang, 2017; Smith and Brown, 2020), the self-consistent Green's function method (Dickhoff and Barbieri, 2004), and Brueckner-Hartree-Fock (BHF) approaches (Brockmann and Machleidt, 1990; Gross-Boelting *et al.*, 1999; Poschenrieder and Weigel, 1988a,b; Sehn *et al.*, 1997; Ter Haar and Malfliet, 1986; Ter Haar and Malfliet, 1987; Tong *et al.*, 2018; Van Dalen *et al.*, 2004). For the context of this work, we will focus on the RMF approximation (Boguta and Bodmer, 1977; Reinhard, 1989; Serot and Walecka, 1986; Walecka, 1974) but a more complete explanation of the varying approximations can be found in Glendenning (2012a); Spinella and Weber (2020); Weber (1999). In Figure 1, we display the nuclear equations of state computed for three prominent RMF parametrization sets: DD2 (Typel, 2018; Typel *et al.*, 2010), SW4L (Malfatti *et al.*, 2020; Spinella and Weber, 2019), and GM1L (Spinella, 2017; Spinella *et al.*, 2018). The DD2 and GM1L models were recently employed in investigations by Malfatti *et al.* (2019) and Farrell *et al.* (2023), where the properties of hot neutron star matter and proto-neutron stars (Pons *et al.*, 1999) were studied. The Lagrangian is given by (refer also to Glendenning (2012b);

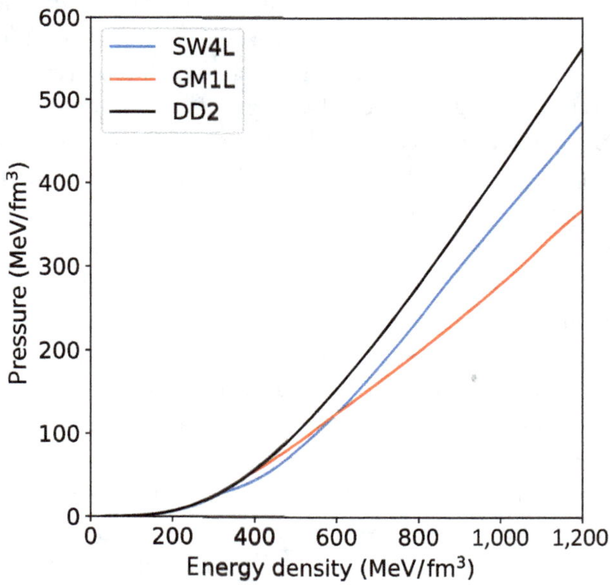

Fig. 1. Pressure as a function of energy density of neutron star matter for three nuclear parameter sets: SW4L (containing the ϕ and $\sigma*$ mesons), GM1L, and DD2.

Sedrakian *et al.* (2022); Spinella and Weber (2020); Weber (1999)):

$$\mathcal{L} = \sum_B \bar{\psi}_B \big[\gamma_\mu (i\partial^\mu - g_{\omega B}\omega^\mu - g_{\rho B}\boldsymbol{\tau}\cdot\boldsymbol{\rho}^\mu) - (m_B - g_{\sigma B}\sigma)\big]\psi_B$$
$$+ \frac{1}{2}(\partial_\mu\sigma\partial^\mu\sigma - m_\sigma^2\sigma^2) - \frac{1}{3}\tilde{b}_\sigma m_N (g_{\sigma N}\sigma)^3 - \frac{1}{4}\tilde{c}_\sigma (g_{\sigma N}\sigma)^4$$
$$- \frac{1}{4}\omega_{\mu\nu}\omega^{\mu\nu} + \frac{1}{2}m_\omega^2\omega_\mu\omega^\mu + \frac{1}{2}m_\rho^2\boldsymbol{\rho}_\mu\cdot\boldsymbol{\rho}^\mu - \frac{1}{4}\boldsymbol{\rho}_{\mu\nu}\cdot\boldsymbol{\rho}^{\mu\nu}, \quad (11)$$

where ψ_B represents the various baryon fields, $g_{\sigma B}$, $g_{\omega B}$, and $g_{\rho B}$ are (density-dependent) meson-baryon coupling constants, and $\tilde{b}_\sigma$ and $\tilde{c}_\sigma$ denote two additional coupling parameters associated with non-linear (cubic and quartic) self-interactions introduced by Boguta and Bodmer (1977). The SW4L models incorporate extra ϕ and σ^* mesons whose Lagrangians are given by (Spinella and Weber, 2019)

$$\mathcal{L}_\phi = -\frac{1}{4}\phi^{\mu\nu}\phi_{\mu\nu} + \frac{1}{2}m_\phi^2\phi_\mu\phi^\mu \quad (12)$$

and

$$\mathcal{L}_{\sigma^*} = \frac{1}{2}(\partial_\mu\sigma^*\partial^\mu\sigma^* - m_{\sigma^*}^2\sigma^*\sigma^*). \quad (13)$$

Note that the incorporation of ϕ and σ^* mesons also modifies the covariant derivative and the baryon B mass terms in the Lagrangian of Equation 11 by adding $-g_{\phi B}\phi^\mu$ and $-g_{\sigma^* B}\sigma^*$ in the respective parentheses. For all parametrizations, the leptons are described by

$$\mathcal{L}_l = \bar{\psi}_l\,(i\gamma_\mu\partial^\mu - m_l)\,\psi_l. \quad (14)$$

The saturation properties of symmetric nuclear matter for these models are shown in Table 2. These properties encompass nuclear saturation density n_0, energy per nucleon E_0, nuclear compressibility K_0, effective nucleon mass m_N^*/m_N, asymmetry energy J_0, asymmetry energy slope L_0, and the value of the nucleon potential U_N. The values of L_0 listed in Table 2 are consistent with the value of the slope of the symmetry energy deduced from nuclear experiments and astrophysical observations (Oertel *et al.*, 2017).

The equations of state shown in Figure 1 include all particles of the spin-$\frac{1}{2}$ baryon octet, comprising the nucleons (n, p) and hyperons $(\Lambda, \Sigma^+, \Sigma^0, \Sigma^-, \Xi^0, \Xi^-)$. Additionally, all states of the spin-$\frac{3}{2}$

Table 2. Properties of symmetric nuclear matter at saturation density for the SW4L, GM1L, and DD2 nuclear parametrizations.

Saturation property	Units	SW4L	GM1L	DD2
n_0	fm^{-3}	0.150	0.153	0.149
E_0	MeV	−16.0	−16.3	−16.02
K_0	MeV	250.0	300.0	242.7
m_N^*/m_N		0.70	0.70	0.56
J_0	MeV	30.3	32.5	32.8
L_0	MeV	46.5	55.0	55.3
U_N	MeV	−64.6	−65.5	−75.2

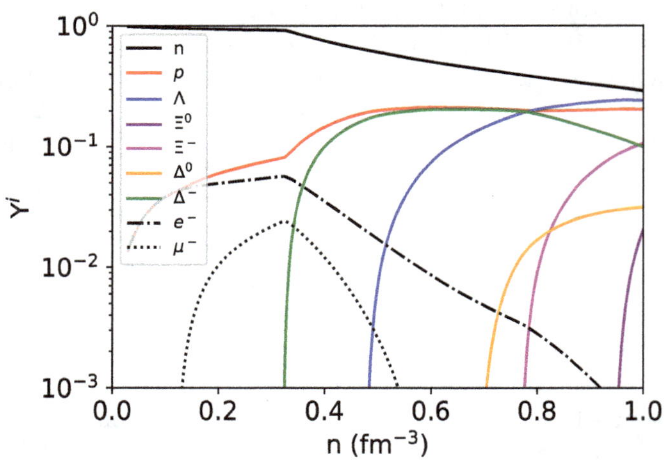

Fig. 2. Relative particle fractions Y^i ($= n_i/n$) in cold ($T = 1$ MeV) neutron star matter determined with the SW4L model.

delta isobar $\Delta(1232)$ ($\Delta^{++}, \Delta^{+}, \Delta^{0}, \Delta^{-}$) are taken into account self-consistently. A detailed discussion of the choices for the meson-baryon coupling constants can be found in Malfatti *et al.* (2019, 2020) and in Spinella and Weber (2020).

Figures 2–5 showcase the relative particle fractions in dense neutron star matter at temperatures ranging from 1 MeV (representing cold matter) to 75 MeV. These compositions adhere to the constraints of electric charge neutrality and chemical equilibrium, which are maintained at every density. The delicate equilibrium between positively and negatively charged particles is crucial to

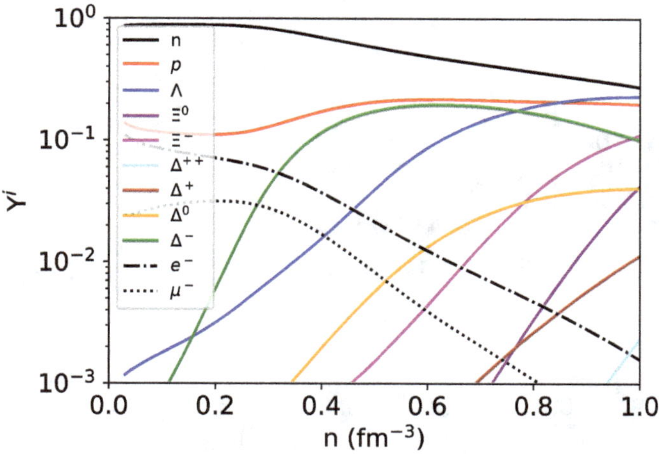

Fig. 3. Same as Figure 2, but for neutron star matter at $T = 25$ MeV.

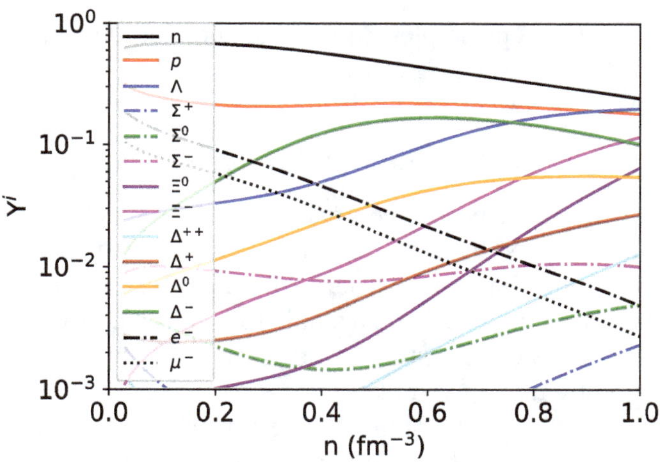

Fig. 4. Same as Figure 2, but for neutron star matter at $T = 50$ MeV.

ensuring the electric charge neutrality of the matter in the stellar core of a neutron star. Finite temperatures induce alterations in the particle composition within the neutron star core due to increased thermal energy. At higher temperatures, particles gain additional kinetic energy, leading to enhanced thermal motion. This process is governed by fundamental principles such as the Pauli exclusion principle, embodied in the Fermi-Dirac distribution, which dictates the occupation of fermions. The thermal population of states is critically

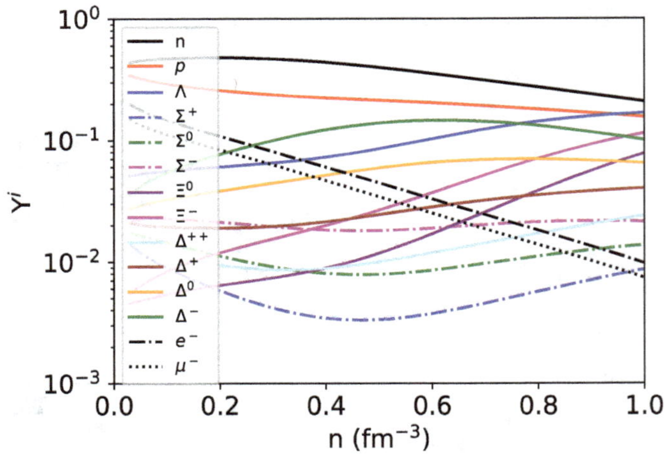

Fig. 5. Same as Figure 2, but for neutron star matter at $T = 75$ MeV.

influenced by these principles, impacting the equilibrium conditions that prevail at lower temperatures.

3.2. *Strange quark stars and strange quark matter*

Quark stars are hypothetical compact objects in which matter would not be held together by gravitational attraction, but rather by the strong interaction between quarks. Notably, quark stars might possess a lower minimum mass and a smaller radius compared to neutron stars. Additionally, quark stars could feature an exposed quark surface capable of emitting radiation at exceptionally high rates (Usov, 2001; Zakharov, 2011) or a thin crust of normal matter (Alcock *et al.*, 1986a; Glendenning *et al.*, 1995; Glendenning and Weber, 1992). The existence of quark stars has not been conclusively confirmed, but certain candidates, including RX J1856.5-37544, PSR J1614-2230, 1E1207, SGR1806, and HESS J1731-347 have been proposed based on observational data (Doroshenko *et al.*, 2022; Horvath *et al.*, 2023; Kurban *et al.*, 2022; Turolla *et al.*, 2004; Weber, 2005; Xu, 2003; Yuan *et al.*, 2022).

The concept of quark stars was initially introduced in the 1970s and 1980s by several researchers (e.g., Itoh 1970b) and gained prominence among theoretical astrophysicists, driven by the hypothesis that strange quark matter, composed of roughly equal numbers of u, d, and s quarks, could be more stable than the most stable atomic nucleus, which is ^{62}Ni.

In 1971, Bodmer (Bodmer, 1971) introduced the concept of potentially collapsed atomic nuclei, a crucial element in the exploration of quark matter. The notion of collapsed atomic nuclei and the emergence of new states of matter with greater stability than conventional atomic nuclei were further articulated by Terazawa in 1979 (Terazawa, 1979) and by Witten in 1984 (Witten, 1984) who proposed that strange quark matter might exhibit greater stability than ^{56}Fe, the most stable element produced in astrophysical contexts, which is known as strange quark matter hypothesis. We note that it is a standard procedure to compare the energy of strange quark matter to ^{56}Fe. However, the energy per particle of ^{56}Fe ranks only third, following ^{62}Ni and ^{58}Fe.

Subsequent theoretical studies suggested that the ground state of deconfined quark matter could exist as a color superconductor with quarks forming diquarks similar to Cooper pairs in ordinary matter (Alford, 2001; Alford *et al.*, 2008; Rajagopal and Wilczek, 2001). One conceivable condensation pattern of color superconducting quark matter is the Color-Flavor-Locked (CFL) phase, yet numerous alternative condensation patterns have been proposed in the literature (Alford, 2001; Weber, 2005). If substantiated, this hypothesis could not only establish the reality of quark stars but also indicate their existence as color superconductors.

Building upon this hypothesis, the existence of strange white dwarfs was also proposed (Glendenning *et al.*, 1995). These would be compact objects distinct from traditional white dwarfs, forming either through the accumulation of normal nuclear matter on their surfaces or by incorporating strangelets found in the galaxy (Glendenning *et al.*, 1995).

Recently, the special case of the neutron star in the supernova remnant HESS J1731-37 has brought quark stars back into the spotlight (Doroshenko *et al.*, 2022). The estimated mass of this neutron star, $M < 0.8\ M_{\odot}$, is much lower than that of any other known neutron star and, in fact, lower than what is believed to be necessary to form a neutron star. Therefore, the question arises of how such a light object could have formed and whether it is indeed a neutron star or a more exotic object that lost part of its mass in a transition from a neutron star to a quark star.

In this context, we dedicate part of this chapter to these exotic and hypothetical compact objects. Numerous studies investigating the properties of quark stars have been conducted over the past several decades. For a recent examination, we direct readers to the work of Zhou *et al.* (2021) and the references contained therein.

3.2.1. *Equation of state of quark matter and stability windows*

To construct self-bound quark stars, we will use two descriptions for the EOS of the matter that composes them: quark matter with corrections due to the strong coupling constant, and quark matter in color superconducting CFL state.

Corrections in bulk quark matter due to strong coupling constant, α_s, were included by (Farhi and Jaffe, 1984) in 1984. In the case where the quark gas interacts strongly, $O(\alpha_s)$ corrections (being α_s the strong coupling constant among quarks) must be made through the perturbative expansion of QCD, developing an improved renormalization group. To do this, a renormalization point ρ where $m(\rho)$ and $\alpha_s(\rho)$ are defined must be chosen. Specifically, $\rho = M_N/3 = 313$ MeV was chosen, where M_N is the mass of the neutron. Thus, the pressures for light quarks are as follows:

$$P_u = \frac{\mu_u^4}{4\pi^2}\left(1 - \frac{2\alpha_s}{\pi}\right), \tag{15}$$

$$P_d = \frac{\mu_d^4}{4\pi^2}\left(1 - \frac{2\alpha_s}{\pi}\right), \tag{16}$$

$$P_s = \frac{1}{4\pi^2}\left\{\mu_s(\mu_s^2 - m_s^2)^{1/2}\left(\mu_s^2 - \frac{5}{2}m_s^2\right)\right.$$
$$+ \frac{3}{2}m_s^4 \ln\frac{\mu_s + (\mu_s^2 - m_s^2)^{1/2}}{m_s}$$
$$- \frac{2\alpha_s}{\pi}\left[3\left(\mu_s(\mu_s^2 - m_s^2)^{1/2} - m_s^2 \ln\frac{\mu_s + (\mu_s^2 - m_s^2)^{1/2}}{\mu_s}\right)^2\right.$$

$$- 2(\mu_s^2 - m_s^2)^2 + 3m_s^4 \ln\left(\frac{m_s}{\mu_s}\right)^2 + 6 \ln\left(\frac{\rho}{\mu_s}\right)$$

$$\times\left(\mu_s m_s^2 (\mu_s^2 - m_s^2)^{1/2} - m_s^4 \ln \frac{\mu_s + (\mu_s^2 - m_s^2)^{1/2}}{m_s}\right)\Bigg]\Bigg\}, \quad (17)$$

where we have considered massless u and d quarks, the strange quark mass, $m_s = 96$ MeV; $\mu_i = \mu + q_i\mu_e$ is the chemical potential of the quark i, with electric charge q_i in β-equilibrium and μ_e is the chemical potential of the electron. The contribution of the electrons to the pressure is given by

$$P_e = \frac{\mu_e^4}{12\pi^2}. \quad (18)$$

For simplicity, muons are not included because their contribution to quark matter at zero temperature is negligible. The equation of state for quarks with corrections due to the strong coupling constant, within the MIT Bag model, arises from a mechanical-statistical or thermodynamic analysis in which the energy density is given by

$$\varepsilon = \sum_{i=u,d,s,e} (\mu_i n_i - P_i) + B, \quad (19)$$

where i ranges over the three quark flavors, and electron, arising from the condition of chemical equilibrium among the system's particles. The bag constant, B, acts as a vacuum energy per unit volume (or pressure) and takes into account the long-range interactions of QCD. The pressure due to the contribution of the three quarks in the CFL phase is given by (Smith *et al.*, 2009)

$$P_{\mathrm{CFL}} = \frac{3\mu^4}{4\pi^2}(1-c) - \frac{3\mu^2}{4\pi^2}(m_s^2 - 4\Delta^2) - B, \quad (20)$$

where $c = 2\alpha_s/\pi$ and Δ is the color superconducting gap, treated here as a free parameter. We will consider c=0 for the CFL phase. The energy density in this case is given by

$$\varepsilon_{\mathrm{CFL}} = -P_{\mathrm{CFL}} + \sum_{i=u,d,s} \mu_i n_i, \quad (21)$$

where $n_i = \partial P_{\mathrm{CFL}}/\partial \mu_i$.

It is worth noting that matter in a CFL phase has properties similar to those of a superfluid (Alford *et al.*, 2008). In our phenomenological treatment of this phase, we have assumed electric and color charge neutrality. We have mentioned that quarks pair up to form Cooper pairs, implying that all quarks of different spins and flavors have the same Fermi momentum and, therefore, the same number density. This leads to the CFL phase being electrically neutral on average, allowing us to consider the presence of electrons negligible. While each Cooper pair contains one quark of each of the three fundamental colors, we can also assume that the color combination in each Cooper pair is such that the total color charge of the CFL phase is zero.

As mentioned earlier, the hypothesis of absolute stability of quark matter suggests that this matter is the ground state of hadronic interactions, i.e., it is more stable than nuclear matter. In this definition, the critical pressure is zero: the true ground state of matter is always quark matter. Figure 6 shows the stability window of quark matter considered in this chapter. The black line in both figures corresponds to the mass of the most stable known element in nature that can be produced in astrophysical contexts, iron ^{56}Fe. Any combination of parameters below the black line in panels (a) and (b) of Figure 6 gives us an appropriate equation of state for the construction of self-bound stars.

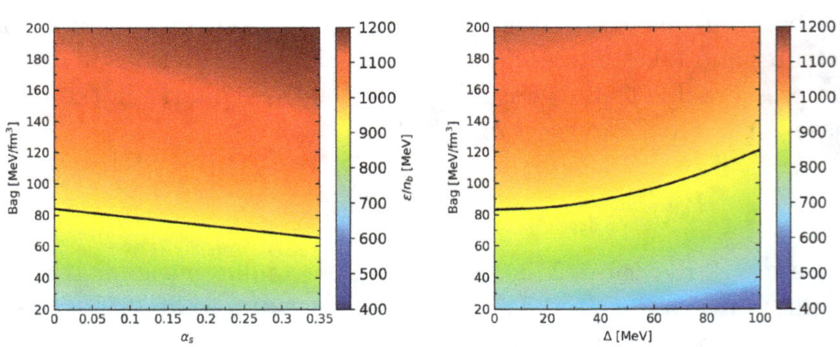

(a) Color map for strange quark matter with strong coupling correction.

(b) Color map for CFL quark matter.

Fig. 6. Absolute stability of strange quark matter with respect to an ^{56}Fe nucleus. Energy per baryon is presented as a color map in the B-α_s plane, panel (a), (B-Δ plane, panel (b), for strange (CFL) quark matter within the MIT Bag model (see text for details). The black curve represents the constraint corresponding to the ^{56}Fe mass. Only B-α_s and B-Δ pairs below the black curves result in combinations for EOS that lead to stable astrophysical configurations.

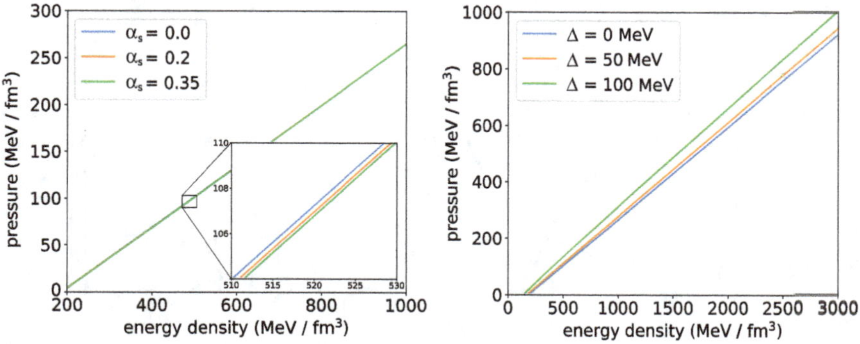

Fig. 7. Pressure as a function of energy density for quark matter modeled with the MIT Bag Model, where $B = 45$ MeV/fm^3. The right panel demonstrates the impact of modifying the strong coupling constant α_s, while the left panel shows variations in the superconducting gap Δ within the Color-Flavor-Locked (CFL) phase. (See text for details).

The EOS for the two models of quark matter considered in this chapter are shown in Figure 7. Increasing the strong coupling constant α_s or the superfluid gap Δ leads to a stiffened quark EOS: for the same energy density, the pressure is higher when these two parameters increase.

4. Properties of Strange Dwarfs

Continuing our exploration, we now direct our focus toward the unique characteristics of hypothetical strange dwarfs. These intriguing objects bear a resemblance to white dwarf stars but distinguish themselves by possessing cores composed of strange quark matter, which are enclosed within a conventional matter crust. Figure 8 illustrates the comprehensive mass-radius relationship of strange quark matter stars, encompassing a wide range from the compact members ('A') to the enigmatic strange dwarf objects ('C' to 'E'). The minimum masses of these sequences are confined within the rectangular area labeled 'B.' Notably, at 'D,' we find the maximum-mass white dwarfs. However, it's essential to note that stars between 'D' and 'E' are prone to unstable radial oscillations. At point 'E,' the strange matter core at the center of white dwarfs has effectively shrunk to zero, resulting in the transformation into an ordinary white dwarf situated within the oscillatory unstable region. Additionally, mass–radius relations for strange quark matter EOS models are compared

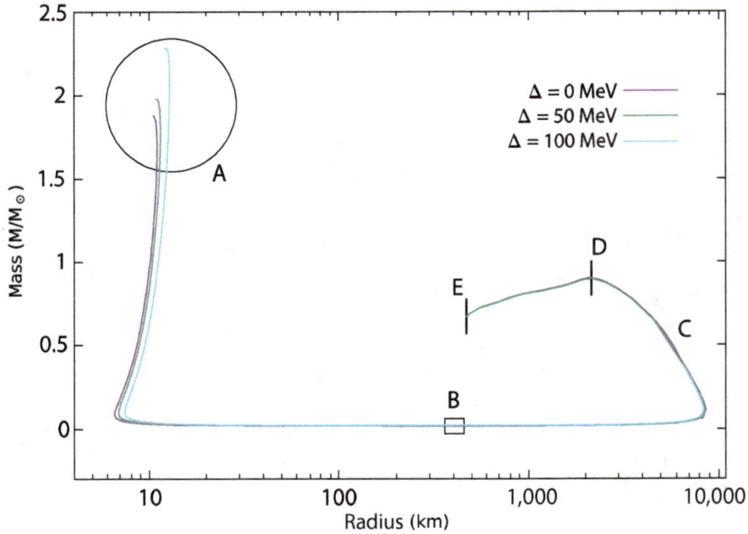

Fig. 8. Mass-radius relations for quark stars constructed with the MIT Bag Model varying the CFL superfluid gap Δ. The nuclear crust is described by the Baym-Pethick-Sutherland EOS. (See text for details.)

to modern astronomical observations of lower mass and higher mass stars in Figure 9. The theoretical mass-radius curves are shown to agree with these observations, particularly for lower-mass stars.

The stability of strange dwarfs against radial perturbations was called into question by Alford and their collaborators in 2017, casting uncertainty on the potential existence of such celestial bodies (Alford *et al.*, 2017). Subsequent investigations by various research groups (Di Clemente *et al.*, 2023; Goncalves *et al.*, 2023) have shed light on this matter, demonstrating that strange dwarfs can indeed achieve stability under certain circumstances, provided that the transitions between quarks and hadrons occur at a slower rate than the radial perturbations themselves. Here, we proceed with the presumption that these conditions hold.

While it is not anticipated that strange matter endured the extreme conditions prevalent in the early Universe (Alcock and Farhi, 1985), there exist alternative scenarios for its generation. One such possibility is the formation of strange matter through the gravitational collapse of stellar cores during supernova explosions, with subsequent dissemination into the interstellar medium occurring

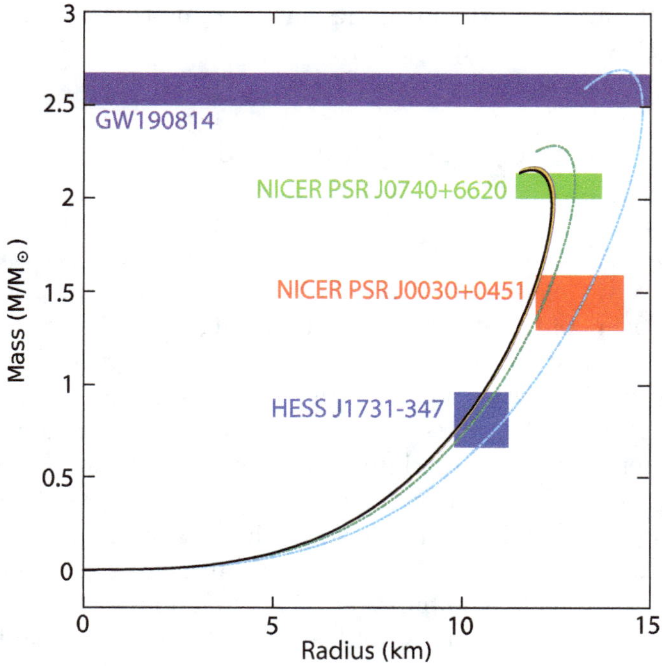

Fig. 9. Mass-radius relations for quark stars constructed within the MIT Bag Model, computed for different coupling constants α_s (shown with solid lines) and superfluid gaps Δ (shown with dashed lines), with $B = 45$ MeV/fm^3, for the quark EOS models considered in this chapter. The results are compared with recent compact-star observations.

during the merger of neutron stars with other compact objects (Bucciantini *et al.*, 2022). Despite these potential sources, strange stars and strange dwarfs have, thus far, eluded conclusive detection due to their striking resemblance to neutron stars and white dwarfs, respectively. Recently, reports have emerged suggesting possible candidates for strange dwarfs (Kurban *et al.*, 2022). Nevertheless, this situation may witness a transformative shift in the forthcoming years, chiefly owing to the advancements in gravitational wave astronomy. In particular, Perot *et al.* (Perot *et al.*, 2023), have demonstrated the feasibility of identifying hidden strange dwarfs within white dwarf binary systems. This can be achieved through precise measurements of their tidal deformability utilizing future space-based gravitational-wave detectors, such as the Laser Interferometer Space Antenna (LISA). These groundbreaking developments hold the promise of unraveling the mysteries surrounding these novel celestial objects (Perot and Chamel, 2023).

5. Properties of Rotating Neutron Stars

Neutron stars can exhibit rigid body rotation, wherein the entire star rotates as a singular entity, maintaining a consistent rotation rate. Alternatively, neutron stars may display differential rotation, with different layers of the star rotating at varying rates. As neutron stars cool over time, diverse mechanisms come into play, leading to the damping of differential rotation and the establishment of uniform rotation. A crucial process in this transition is Eckman pumping, a phenomenon wherein the interaction between the star's magnetic field and its interior matter creates a torque that aligns the rotation axis of the star with its magnetic axis. This alignment process aids in reducing differential rotation by redistributing angular momentum within the star. The combination of angular momentum conservation during the collapse phase and subsequent processes like Eckman pumping guides the evolution of neutron stars from initially hot and differentially rotating objects to cooler ones with more uniform rotation. Understanding these processes is essential for comprehending the internal dynamics, magnetic field evolution, and the overall life-cycle of neutron stars.

In this section, we provide a brief overview of the theoretical framework that elucidates the behavior of rotating neutron stars. We present results detailing the properties of neutron stars, focusing on both differentially rotating and uniformly rotating scenarios.

5.1. *Uniform rotation*

To calculate neutron star properties, the stellar structure equations must be solved in the framework of Einstein's theory of general relativity. These equations are derived from Einstein's field equation and depend on the stellar matter's equation of state, or the underlying relationship between pressure P and energy density ϵ.

The simplest form of the stellar structure equations describes non-rotating, spherically symmetric objects in hydrostatic equilibrium. In this case, the stellar structure simplifies to an ordinary differential equation describing the pressure gradient within the star, known as the Tolman-Oppenheimer-Volkoff (TOV) equation (given in geometrical units where $G = c = 1$):

$$\frac{dP}{dr} = -\frac{(\epsilon + P)(m + 4\pi r^3 P)}{r^2 \left(1 - \frac{2m}{r}\right)}, \tag{22}$$

where m is the gravitational mass enclosed within a sphere of radius r, P is the pressure, and ϵ is the energy density. When integrating the TOV equation, one first would specify a value for the central density, ϵ_c, which in turn is used to determine the pressure value at the center of the star, P_c, using the specified EOS; at the center of the star ($r = 0$), the mass m and pressure differential P' disappear. For an infinitesimal increase in r, the mass of the star can be solved for as

$$\frac{dm}{dr} = 4\pi r^2 \epsilon. \tag{23}$$

The values ϵ_c, P_c, and m are used to solve for P', thus allowing for the determination of P at the next radial step, which thus determines ϵ from the EOS. The computation of these parameters continues successively, increasing r until the pressure vanishes at the edge of the star. Then, the total gravitational mass M of a star can be deduced from the full radius R of the star as $M \equiv m(R) = 4\pi \int_0^R dr\ r^2 \epsilon$.

Rotating neutron star calculations are performed in the framework of general relativity and depend on the matter's nuclear equation of state. Modeling rotating neutron stars is more complicated than spherically symmetric, non-rotating stars for a few reasons: rotation deforms neutron stars, stabilizes them against collapse, and drags along the local inertial frames inside and outside of them so that they co-rotate with the stars (Mellinger Jr *et al.*, 2017; Weber, 1999). Rapid rotation can cause deformation where the star flattens at the pole and radially expands in the equatorial direction; this deformation results in the stellar structure equations not only being dependent on the radial coordinate r but also on the polar coordinate θ. Rotation is a physical mechanism that stabilizes massive stars against collapse, resulting in rotating neutron stars being able to sustain up to 20% more mass than their non-rotating counterparts (Kalogera and Baym, 1996). The increased mass of rotating stars alters the geometry of spacetime by introducing a dependence on the star's rotational frequency to the line element and self-consistency condition to the stellar structure equations to account for the dragging of local inertial frames (Weber, 1999).

To solve for properties of rotating stars, we begin with the metric, which has the form

$$ds^2 = -e^{\gamma-\rho}dt^2 + e^{2\alpha}(dr^2 + r^2 d\theta) + e^{\gamma-\rho}r^2 \sin^2\theta(d\phi - \omega dt)^2, \tag{24}$$

where γ, ρ, α, and ω are metric functions dependent on the radial coordinate r and polar angle θ, and ω describes frame dragging due to rotation. These functions also implicitly depend on the star's angular velocity Ω. The metric functions are computed from Einstein's field equation

$$R^{\kappa\sigma} - \frac{1}{2}Rg^{\kappa\sigma} = 8\pi T^{\kappa\sigma}, \tag{25}$$

where $R^{\kappa\sigma}$ is the Ricci tensor, R is the curvature scalar, and $g^{\kappa\sigma}$ is the metric tensor. $T^{\kappa\sigma}$ is the energy momentum tensor given by

$$T^{\kappa\sigma} = (\epsilon + P)\, u^{\kappa}u^{\sigma} + g^{\kappa\sigma}P, \tag{26}$$

where ϵ and P are supplied by the equation of state. The partial differential equations for the metric functions γ, ρ, α, and ω, which follow from Eq. (25), are to be solved self-consistently for a given rotational stellar frequency (Paschalidis and Stergioulas, 2017). The absolute constraint on rapid rotation is defined by mass shedding, occurring at the Kepler frequency, Ω_K. However, this limit might not be reached due to instabilities driven by gravitational radiation-reaction-driven instabilities known as r and f-modes (Andersson and Kokkotas, 2001; Bratton *et al.*, 2022; Paschalidis and Stergioulas, 2017). Certain classes of instabilities may be associated with the emission of gravitational waves from neutron stars (Orsaria *et al.*, 2019).

The Kepler frequency is determined by the centrifugal force and gravity. In classical Newtonian mechanics, the Kepler frequency can be defined as

$$\Omega_K = \sqrt{\frac{MG}{R^3}}, \tag{27}$$

but this formula is drastically modified in general relativity. Following the metric function shown in Eq. (24), the general relativistic Kepler frequency is defined in terms of the metric functions as

$$\Omega_K = \frac{e^{\rho}}{r} V + \omega \tag{28}$$

where the quantity V denotes the equatorial velocity given by

$$V = \frac{\omega'}{(8 + \gamma' - \rho')} r e^{-\rho} + \sqrt{\frac{\gamma' + \rho'}{(8 + \gamma' - \rho')} + \left(\frac{\omega'}{(8 + \gamma' - \rho')} r e^{-\rho} \right)^2}. \tag{29}$$

Eqs. (28) and (29) are to be evaluated self-consistently at the star's equatorial radius.

Once a set of self-consistent solutions are found for the metric potentials, bulk properties of neutron stars like mass, radius, and Ω can be determined.

5.1.1. *Results for uniformly rotating compact objects*

This section presents comprehensive insights into the bulk properties of uniformly rotating neutron stars and quark stars. The calculations for uniformly rotating neutron stars utilize the nuclear DD2 model, as detailed in Section 3.1. Figure 10 provides results for both

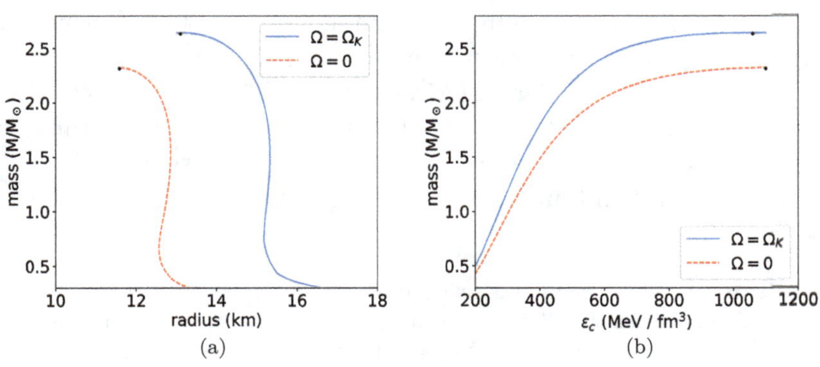

Fig. 10. Gravitational mass versus radius (a) and gravitational mass versus central stellar density (b) for both non-rotating ($\Omega = 0$) neutron stars and neutron stars rotating at the Kepler frequency ($\Omega = \Omega_K$), modeled with the DD2 nuclear EOS. Mass peaks are highlighted with black dots.

a sequence of non-rotating stars, computed using the TOV equation, as expressed in Eq. (22), and a sequence of uniformly rotating stars at their mass-shedding limits ($\Omega = \Omega_K$).

In Figure 10, the left pane illustrates the mass-radius relationships for both non-rotating and uniformly rotating stars. Correspondingly, the right pane of Figure 10 presents the gravitational mass as a function of central energy density, with each sequence's maximum mass denoted by a solid black dot. The mass peak values for both the non-rotating and uniformly rotating sequences derived from the DD2 model are provided in Table 3, along with their respective radii and central energy density values. Notably, uniform rotation at the Kepler limit enhances the mass peak for the DD2 model by around 14%.

To calculate the properties of uniformly rotating quark stars, the EOS for quark matter is modeled within the MIT Bag Model with a bag constant $B = 45$ MeV/fm^3, as explained in Section 3.2.1. As for the DD2 model, a sequence of non-rotating stars and a sequence of uniformly rotating stars at their mass-shedding limit ($\Omega = \Omega_K$) for quark matter are shown in Figure 11.

The strange quark star sequences depicted in Figure 11 are calculated for bare quark stars, consisting entirely of strange quark matter from the core to the surface. As discussed by Alcock *et al.* (1986a,b), strange quark stars however do not necessarily have to be bare; they can be surrounded by nuclear crusts, with densities at the crust's base lower than neutron drip density. Examples of strange stars with nuclear crusts are presented in Figure 8. The thickness of the nuclear crust of the compact members of the stars shown there is typically just a few hundred meters or less (see Figure 12) and the mass is below $\sim 10^{-5} M_\odot$ (Alcock *et al.*, 1986a). However, rapid rotation can moderately increase these values, as demonstrated by Glendenning and Weber (1992). The value of the mass peaks for both the non-rotating and uniformly rotating sequences for quark matter

Table 3. Maximum gravitational mass with corresponding radius and central density of non-rotating and uniformly rotating neutron stars modeled with the DD2 parameter set.

Rotation	Mass ($M_\odot$)	Radius (km)	ϵ_c (MeV/fm^3)
Non-rotating (TOV)	2.316	11.589	1,100
Uniform (Ω_K)	2.637	13.111	1,060

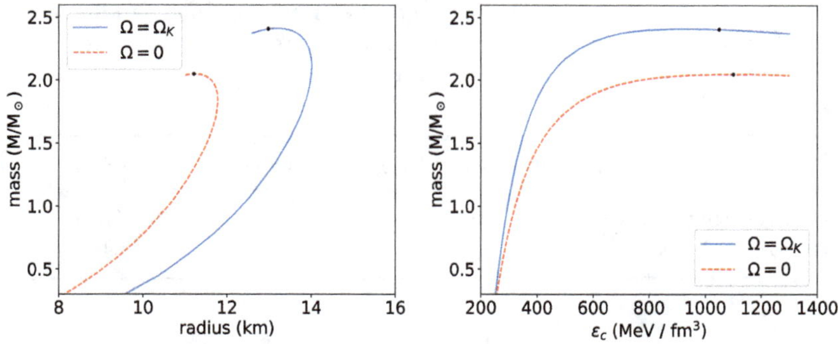

Fig. 11. Gravitational mass versus radius (a) and gravitational mass versus central density (b) of non-rotating ($\Omega = 0$) quarks stars and quark stars rotating at the Kepler frequency ($\Omega = \Omega_K$), for quark matter modeled within the MIT Bag Model with $B = 45$ MeV/fm^3. The mass peaks are indicated with black dots.

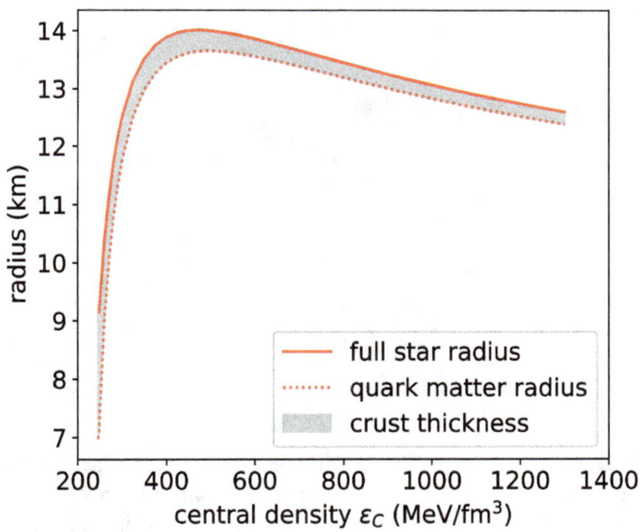

Fig. 12. The thickness of the nuclear crust of quark stars, with quark matter modeled using the MIT Bag Model, featuring a bag constant of $B = 45$ MeV/fm^3.

are given in units of $M_\odot$ in Table 4 along with their corresponding radii and central energy density values. Uniform rotation at the Kepler limit for quark matter modeled within the MIT Bag Model increases the mass peak by 18% when compared to the non-rotating counterpart, a larger increase than that of the DD2 model.

Table 4. Maximum gravitational mass with corresponding radius and central density of non-rotating and uniformly rotating quark stars for quark matter modeled within the MIT Bag Model with a bag constant $B = 45$ MeV/fm^3.

Rotation	Max. Mass ($M_{\odot}$)	Radius (km)	ϵ_c (MeV/fm^3)
Non-rotating (TOV)	2.047	11.217	1,100
Uniform (Ω_K)	2.413	13.250	900

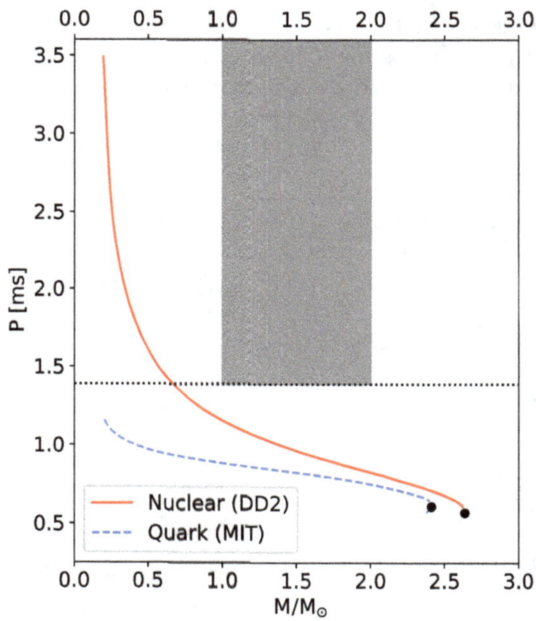

Fig. 13. Rotational periods P (in milliseconds) versus gravitational mass of both neutron stars and strange quark stars. Neutron stars are calculated using the DD2 nuclear EOS, while strange stars are modeled within the MIT Bag Model with the coupling constants $\alpha_s = 0$, the superfluid gap $\Delta = 0$ MeV, and a bag constant of $B = 45$ MeV/fm^3. The black dotted line represents the period of the fastest observed MSP ($P = 1.3889$ ms), and the grey-shaded region denotes the range encompassing all observed pulsars to date.

The rotational period, P, of the uniformly rotating sequences from both the DD2 model and the MIT Bag Model can also be seen in Figure 13. At lower mass values, P is much higher for the nuclear EOS model DD2, but at their mass peaks, both the nuclear and quark EOS models see a similar value for P. In Figure 13, the horizontal black dotted line indicates the rotational period of the fastest observed MSP, PSR J1748–2446ad, with a rotational period of 1.3889 ms. The grey-shaded region schematically shows the period-mass range of all observed pulsars to date. The position of this region will shift downwards to shorter rotational periods if future pulsar discoveries unveil objects rotating below 1.3899 ms. Particularly, super-rapid rotation below approximately 1 ms, contingent on the mass, could suggest the presence of strange quark stars.

5.2. *Differential rotation*

While the majority of pulsars are believed to be rotating uniformly, some neutron stars form with an appreciable amount of differential rotation following extreme astrophysical events like core-collapse supernovae or binary neutron star mergers.

A newly formed proto-neutron star may experience some degree of differential rotation following a core-collapse supernova, but the phenomena would be short-lived. Hydrodynamical forces in the form of Ekman pumping and turbulent mixing will damp out any differential rotation in the star within a few days (Weber, 1999). A more likely avenue for a star to experience differential rotation is following a binary neutron star merger event. When a neutron star binary coalesces, the two stars are most likely irrotational just prior to merging. The collision would result in a substantial velocity discontinuity at the surface of contact, giving rise to differentially rotating hypermassive or supramassive remnant stars. The remnant stars' fate depends largely on the dynamics and mass of the binary prior to merging. Many previous studies (Baumgarte *et al.*, 1999; Gondek-Rosińska *et al.*, 2017; Morrison *et al.*, 2004; Shapiro, 2000) have explored the implications of massive, differentially rotating neutron stars, but we extend this line of research to also include differentially rotating strange quark stars.

Differential rotation has physical implications that allow stars to remain stable in otherwise unstable configurations. Most notably, differentially rotating stars can support more mass than their uniformly rotating or non-rotating counterparts (Morrison *et al.*, 2004). In the case of massive remnant stars of binary merger events, differential rotation is one of the mechanisms that provide extra centrifugal support to stabilize the star well above the TOV and uniform rotation mass limits.

The modeling of differential rotation begins with the same metric as uniform rotation. Once solved for, the metric functions are used to solve the equation of hydrostatic equilibrium for a barotropic fluid:

$$h(P) - h_p = \frac{1}{2}[\gamma_p + \rho_p - \gamma - \rho - \ln(1 - v^2) + F(\Omega)] \tag{30}$$

where $h(P)$ is the enthalpy as a function of pressure, and γ_p and ρ_p are the values of the metric potentials at the pole. The quantity v is defined as

$$v = (\Omega - \omega)\, r \, \sin\, \theta e^{-\rho}, \tag{31}$$

and the last term, $F(\Omega)$, defines the rotational law for the matter. Following the form in (Cook *et al.*, 1992, 1994; Komatsu *et al.*, 1989), the rotational law is set to

$$F(\Omega) = A^2(\Omega_c - \Omega), \tag{32}$$

where Ω_c is the central value for the angular velocity. In this equation, the parameter A is a constant scaling factor of the degree of differential rotation and is used to determine the length scale over which the frequency changes (Morrison *et al.*, 2004).

Simplifying Eq. (30) and including the rotation law gives

$$(\Omega_c - \Omega) = \frac{1}{\hat{A}^2}\left[\frac{(\Omega - \omega)\, s^2(1 - \mu^2)\, e^{-2\rho}}{(1 - s)^2 - (\Omega - \omega)^2\, s^2\, (1 - \mu^2)\, e^{-2\rho}}\right], \tag{33}$$

where the rotational frequency Ω can be isolated and solved numerically using a root-finding algorithm. The other non-defined parameters, s and μ, represent unitless radial and polar coordinates. Equation (33) uses a modified version of the rotation parameter, labeled $\hat{A}$, which is scaled by the equatorial radius r_e: $\hat{A} \equiv A/r_e$. Uniform rotation is achieved in the limit $\hat{A}^{-1} \to 0$, and has an upper bound of 1. The gravitational and matter fields are solved in a self-consistent fashion; once a set of self-consistent solutions to these equations is obtained, physical quantities like mass, radius, angular momentum, and rotational kinetic energy are outputted.

5.2.1. *Results for differentially rotating compact objects*

In this section, M-R relations for differentially rotating neutron stars and quark stars are shown. The M-R relations are calculated for different levels of differential rotation, varying the rotation parameter A in Eq. (32). Because the rotation parameter A repeatedly appears as $\hat{A}^{-1}$ in the equations described in Section 5.2, we follow the lead of previous work which parameterized sequences by values of $\hat{A}^{-1} =$ 0.3, 0.5, 0.7, and 1.0 (Cook *et al.*, 1992; Galeazzi *et al.*, 2012; Morrison *et al.*, 2004). It's important to note that using the numerical scheme described in Section 5.2, the most extreme case of differential rotation ($\hat{A}^{-1} = 1.0$) may fail to find solutions at high central densities depending on the underlying EOS. In Figure 14, the relationship between mass and central density ϵ_c is illustrated for the DD2 parameter set. This comparison involves three levels of differential rotation characterized by the rotation parameter $\hat{A}^{-1}$, contrasting them with the uniformly rotating (Kepler) and non-rotating (TOV) scenarios discussed in Section 5.1.1. The figure underscores the influence of varying degrees of differential rotation on the mass-central density relationship. Table 5 provides detailed information on the maximum gravitational mass, corresponding radius, and central

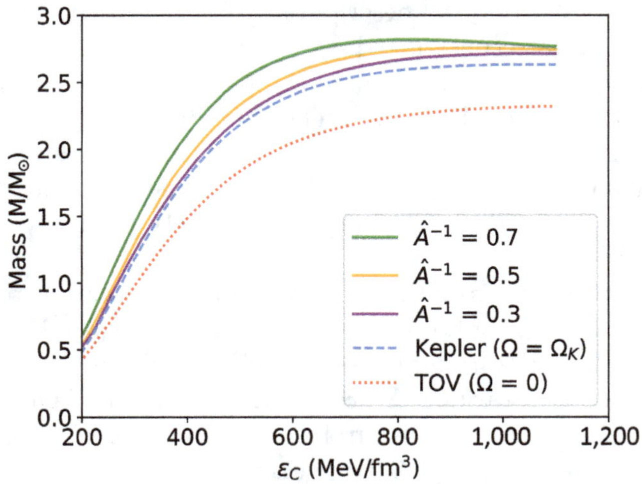

Fig. 14. Mass as a function of central density ϵ_c for the DD2 parameter set, comparing three degrees of differential rotation characterized by the rotation parameter $\hat{A}^{-1}$ to the uniformly rotating (Kepler) and non-rotating (TOV) cases.

Table 5. Maximum gravitational mass with corresponding radius and central density of differentially rotating neutron stars, parametrized by three degrees $\hat{A}^{-1}$ of differential rotation, modeled with the DD2 parameter set.

$\hat{A}^{-1}$	Mass ($M_\odot$)	Radius (km)	ϵ_c (MeV/fm^3)
0.3	2.66	13.05	1,100
0.5	2.73	13.35	920
0.7	2.75	13.59	760

density for differentially rotating neutron stars. These models are parametrized by three levels of the rotation parameter $\hat{A}^{-1}$: 0.3, 0.5, and 0.7, as modeled with the DD2 parameter set. The results demonstrate how the introduction of differential rotation affects the stellar properties, with decreasing values of $\hat{A}^{-1}$ gradually converging toward the uniform rotation scenario. This analysis highlights the nuanced impact of differential rotation on the stellar structure and provides a comprehensive understanding of the mass-central density relationship for neutron stars with varying degrees of rotation.

We next turn to investigating the impact of differential rotation on quark stars. Following the description in Section 3.2.1, six equation of state models are constructed: three models of quark matter with corrections due to the strong coupling constant, and three models of quark matter in the CFL phase. The three models with corrections due to the strong coupling constant, α_s, vary in the value of α_s (0.0, 0.2, and 0.35). The other three CFL models vary the parameter Δ, setting equal to 0, 50, or 100. For all six models, the bag constant B is set to 45 MeV/fm^3.

M-R relations for four different values of the rotation parameter $\hat{A}^{-1}$ (0.3, 0.5, 0.7, and 1.0) are constructed for each EOS model. The M-R relations varying α_s are shown visually in Figure 15, and the M-R relations varying Δ are shown visually in Figure 16. The maximum gravitational mass from each curve for the four values of $\hat{A}^{-1}$ are given in Table 6. The general trend for maximum mass expected when varying the value of $\hat{A}^{-1}$ is to see higher mass values as $A^{-1} \to 1.0$, which is reflected in Table 6. Additionally, stellar models with higher degrees of differential rotation struggle to find

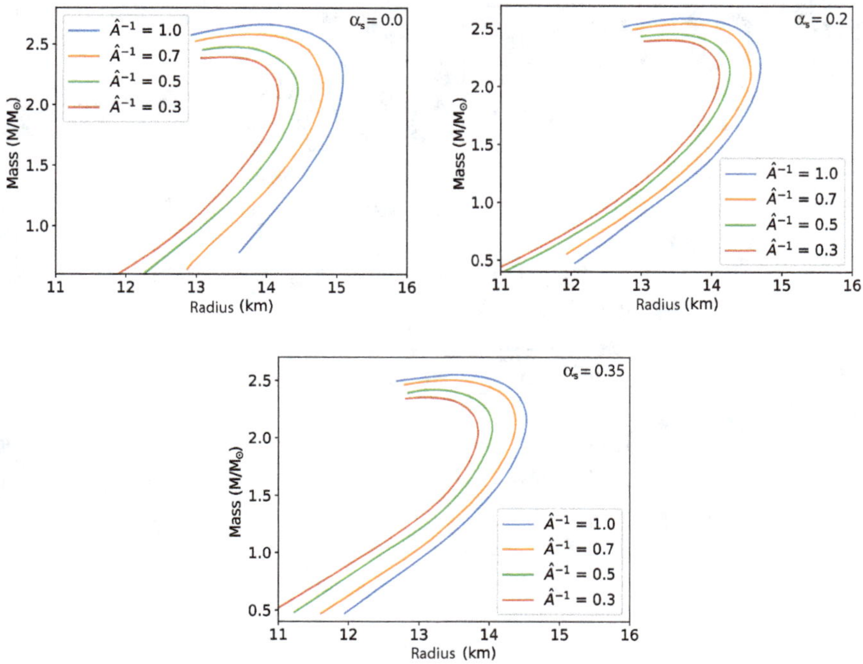

Fig. 15. Mass-radius relations of quark stars for the MIT Bag Model EOS with three values of the strong coupling constant α_s (0.0 on top, 0.2 on the bottom left, and 0.35 on the bottom right). Each figure depicts four degrees of differential rotation characterized by the scaled rotation parameter $\hat{A}^{-1}$.

physical results at lower masses, as reflected in Figures 15 and 16 for both sets of quark matter EOS models.

The mass-radius relations in Figures 15 and 16 are constructed for a range of constant central densities ϵ_c. Two additional parameters are specified for each curve: the ratio of polar to equatorial ratio for the stellar models r_{ratio} and the given rotation parameter $\hat{A}^{-1}$. As the magnitude of differential rotation increases in a star, signified by $\hat{A}^{-1} \to 1$, the star grows more deformed, expanding at the equator, and thus decreases the r_{ratio}. For extremely high levels of differential rotation in massive stars, the star may even form the shape of a torus as $r_{\text{ratio}} \to 0$. The numerical scheme described in Section 5.2 uses the central density as the maximum density within the star which limits the lower threshold for the parameter r_{ratio}, so only quasi-toroidal configurations are possible for extreme differential rotation. The reader can find more information on modifying the specified

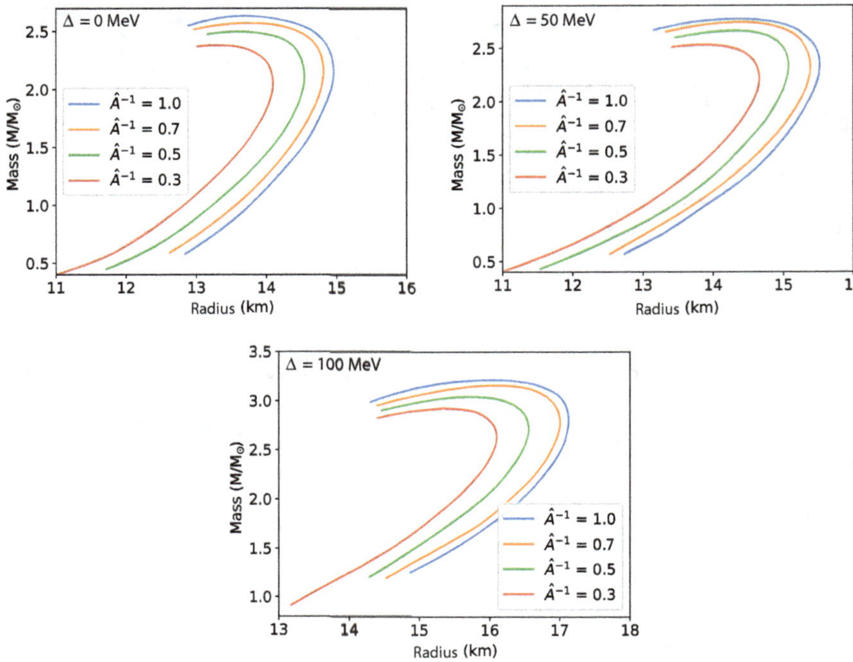

Fig. 16. Mass-radius relations of quark stars computed for the MIT Bag Model EOS with three values of Δ in the CFL phase (0 MeV on top, 50 MeV on the bottom left, and 100 MeV on the bottom right). Each figure depicts four degrees of differential rotation characterized by the scaled rotation parameter $\hat{A}^{-1}$.

Table 6. Maximum gravitational mass of quark stars (in units of $M_\odot$) for four degrees of differential rotation characterized by the rotation parameter $\hat{A}^{-1}$.

Parametrization	$\hat{A}^{-1} = 0.3$	$\hat{A}^{-1} = 0.5$	$\hat{A}^{-1} = 0.7$	$\hat{A}^{-1} = 1.0$
$\alpha_s = 0.0$	2.384	2.476	2.579	2.664
$\alpha_s = 0.2$	2.391	2.443	2.546	2.597
$\alpha_s = 0.35$	2.353	2.411	2.506	2.549
$\Delta = 0$ MeV	2.379	2.492	2.576	2.651
$\Delta = 50$ MeV	2.525	2.647	2.732	2.767
$\Delta = 100$ MeV	2.917	3.037	3.162	3.193

Note: The first three rows vary the strong coupling constant α_s, and the last three vary the superfluid gap Δ in the CFL phase.

numerical scheme to include fully toroidal configurations in Kaplan *et al.* (2014) and Morrison *et al.* (2004).

6. Summary and Concluding Remarks

This chapter covers various topics related to neutron stars, addressing historical challenges in measuring their radius and the consequent uncertainties in size determination. Methods for radius estimation involve spectroscopic and timing techniques, with considerations for factors like magnetic fields and rapid rotation. Pulsar glitches or sudden spin jumps in neutron stars, provide insights into dense matter's EOS, revealing details about angular momentum transfer and interior dynamics. Tidal deformability, crucial in gravitational-wave astronomy, is explored in the context of binary neutron star mergers, with emphasis on the GW170817 event's contributions to mass, radius, and deformability estimates. The discussion extends to fast pulsars, categorizing them based on rotation, magnetic fields, and the role they play in astrophysical discoveries. The EOS of neutron stars is detailed, encompassing lower-density crust models and high-density core approaches. The chapter concludes by delving into quark stars, strange dwarfs, and the transformative possibilities of understanding these exotic objects through future observations and gravitational wave detectors like LISA. Additionally, properties of rotating neutron stars are thoroughly examined, covering uniform and differential rotation scenarios, with insights into the impact of rotation on mass and radius, particularly after binary neutron star mergers. The comparison of maximum gravitational masses offers valuable information about the stability and deformations of rotating compact objects.

Key insights derived from the examination of fast pulsars, neutron stars, and astrophysical objects featuring strange quark matter include the following:

(1) Historical challenges related to direct measurements of neutron star radius have resulted in uncertainties, necessitating the reliance on spectroscopic and timing techniques. Spectroscopic methods, particularly in low-mass X-ray binary systems, involve making assumptions about emitted spectra to infer mass-radius

relations. However, challenges arise in cases of strong magnetic fields or rapid rotation.

(2) Pulsar glitches offer a unique perspective into the equation of state of dense neutron star matter, as the proposed models involve angular momentum transfer between the solid crust and the extremely dense interior, specifically the neutron superfluid.

(3) Tidal deformability, crucial in gravitational-wave astronomy, is elaborated in the context of binary neutron star mergers. Theoretical calculations, along with the groundbreaking GW170817 event, contribute to estimates of masses, radii, and dimensionless tidal deformability.

(4) Fast pulsars have played a pivotal role in testing theories of relativistic astrophysics. With over 3,000 observed pulsars to date, their numbers continue to grow rapidly with advancements in telescopes. Measuring pulsar rotation provides essential constraints on their bulk properties and the nature of the matter in their central cores. Superfast rotation, ideally with rotational periods below 1 millisecond, is expected to provide an additional observational window into the inner workings of neutron stars.

(5) The equation of state (EOS), governing the pressure-density relationship within neutron stars, is crucial for understanding their internal structure. Models for the lower-density crust and high-density core are briefly discussed, emphasizing their role in simulations and observational analyses.

(6) The intriguing concept of strange quark stars is explored. If they exist, such stars could be composed of quark matter in the color-flavor-locked phase, among other condensation patterns. The properties of strange dwarfs, characterized by conventional nuclear crusts and strange quark matter cores, are reviewed. Recently, potential implications for gravitational wave astronomy have been highlighted by Perot and Chamel (2023).

The discovery of stellar objects composed of absolutely stable strange quark matter remains an intriguing and challenging prospect in astrophysics. Detecting such objects presents a significant challenge due to the numerous observational signatures that overlap with those of ordinary neutron stars and white dwarfs. Distinguishing features may arise from ultra-rapid rotation (well below 1 ms), small radii (less than 10 km), and photon emission from the surface of a bare strange star. Proposed candidates, like the compact star RX J1856.5-3754, PSR J1614-2230, PSR J0348+0432, 4U 1820-30, and HESS J1731-347 have been considered, but definitive evidence is yet to be established. Advanced observational techniques and future space-based observatories, coupled with refined theoretical models, offer hope in the ongoing quest to unravel the mysteries of strange quark matter and identify potential stellar objects made of such matter.

(7) Theoretical frameworks for calculating properties of rotating neutron stars are presented, covering uniform and differentially rotating scenarios. The impact of rotation on mass and radius is discussed using the DD2 model for the nuclear equation of state. Insights into stability and deformations are gained through comparisons of maximum gravitational masses.

Acknowledgments

This research was supported by the National Science Foundation (USA) under Grant No. PHY-2012152. MGO, IFR-S, and MC thank CONICET, UNLP and MinCyT (Argentina) for financial support under grants PIP-0169, 11/G187 and PICT 2019-3662. MC thanks Dr. Mauro Mariani and Marcos Celi for their assistance in creating the 3D figures for this chapter.

Bibliography

Abbott, B. P., Abbott, R., Abbott, T. D., *et al.* (2018). Gw170817: Measurements of neutron star radii and equation of state, *Phys. Rev. Lett.* **121**, 161101.

Abbott, B. P. *et al.* (2017). Multi-messenger observations of a binary neutron star merger, *ApJ* **848**, 2, L12.

Abbott, R., Abbott, T., Abraham, S., Acernese, F., Ackley, K., Adams, A., Adams, C., Adhikari, R. X., Adya, V., Affeldt, C., *et al.* (2021). Constraints from ligo o3 data on gravitational-wave emission due to r-modes in the glitching pulsar PSR j0537–6910, *ApJ* **922**, 1, 71.

Agazie, G., Anumarlapudi, A., Archibald, A. M., Arzoumanian, Z., Baker, P. T., Bécsy, B., Blecha, L., Brazier, A., Brook, P. R., Burke-Spolaor, S., *et al.* (2023). The nanograv 15 yr data set: Evidence for a gravitational-wave background, *ApJL* **951**, 1, L8.

Alcock, C. and Farhi, E. (1985). Evaporation of strange matter in the early universe, *Phys. Rev. D* **32**, 1273–1279.

Alcock, C., Farhi, E., and Olinto, A. (1986a). Strange stars, *ApJ* **310**, 261–272.

Alcock, C., Farhi, E., Olinto, A., and Gleiser, M. (1986b). Quark stars and strange stars, *ApJ* **310**, 772–778.

Alcock, C. and Olinto, A. V. (1988). Exotic phases and hadronic matter and their astrophysical application, *Ann. Rev. Nucl. Part. Sci.* **38**, 161.

Alford, M. (2001). Color-superconducting quark matter, *Ann. Rev. Nucl. Part. Sci.* **51**, 131.

Alford, M. G., Harris, S. P., and Sachdeva, P. S. (2017). On the stability of strange dwarf hybrid stars, *ApJ* **847**, 2, 109.

Alford, M. G., Schmitt, A., Rajagopal, K., and Schäfer, T. (2008). Color superconductivity in dense quark matter, *Revi. Mod. Phys.* **80**, 4, 1455–1515.

Alpar, M., Langer, S. A., and Sauls, J. (1984). Rapid postglitch spin-up of the superfluid core in pulsars, *Astrophysical Journal* **282**, 533–541.

Ambartsumyan, V. A. and Saakyan, G. S. (1960). The degenerate superdense gas of elementary particles, *Soviet Ast.* **4**, 187.

Ambartsumyan, V. A. and Saakyan, G. S. (1961). On equilibrium configurations of superdense degenerate gas masses, *AZh* **38**, 785.

Andersson, N. and Kokkotas, K. D. (2001). The r-mode instability in rotating neutron stars, *Int. J. Mod. Phys. D* **10**, 381–441.

ATLAS-Collaboration, G. A., Abat, E., Abdallah, J., Abdelalim, A. A., *et al.* (2008). The atlas experiment at the CERN large hadron collider, *JINST* **3**, 08, S08003.

Basu, A., Shaw, B., Antonopoulou, D., Keith, M. J., Lyne, A. G., Mickaliger, M. B., Stappers, B. W., Weltevrede, P., and Jordan, C. A. (2022). The Jodrell bank glitch catalogue: 106 new rotational glitches in 70 pulsars, *MNRAS* **510**, 3, 4049–4062.

Baumgarte, T. W., Shapiro, S. L., and Shibata, M. (1999). On the maximum mass of differentially rotating neutron stars, *ApJ* **528**, 1, L29.

Baym, G., Bethe, H. A., and Pethick, C. J. (1971a). Neutron star matter, *Nucl. Phys. A* **175**, 2, 225–271.

Baym, G. and Chin, S. (1976). Can a neutron star be a giant MIT bag? *Phys. Lett. B* **62(b)**, 241.

Baym, G., Pethick, C., and Sutherland, P. (1971b). The ground state of matter at high densities: equation of state and stellar models, *ApJ* **170**, 299.

Becker, W. (ed.) (2009). *Neutron Stars and Pulsars*, Astrophysics and Space Science Library, Springer, Berlin, Heidelberg.

Bethe, H. A. and Johnson, M. B. (1974). Dense baryon matter calculations with realistic potentials, *Nucl. Phys. A* **230**, 1, 1–58.

Blaschke, D. and Chamel, N. (2018). *Phases of Dense Matter in Compact Stars*, chap. 7 (Springer International Publishing), 337–400.

Bodmer, A. R. (1971). Collapsed nuclei, *Phys. Rev. D* **4**, 1601–1606.

Boguta, J. and Bodmer, A. R. (1977). Relativistic calculation of nuclear matter and the nuclear surface, *NPA* **292**, 413.

Bratton, E. L., Lin, Z., Weber, F., Orsaria, M. G., Ranea-Sandoval, I. F., and Saavedra, N. (2022). Gravitational-wave instabilities in rotating compact stars, *Galaxies* **10**, 5.

Brockmann, R. and Machleidt, R. (1990). Relativistic nuclear structure. i. nuclear matter, *Phys. Rev. C* **42**, 5, 1965.

Brown, G. and Weise, W. (1976). Pion condensates, *Phys. Rep.* **27**, 1, 1–34.

Brown, G. E., Lee, C.-H., Rho, M., and Thorsson, V. (1994). From kaon-nuclear interactions to kaon condensation, *Nucl. Phys. A* **567**, 4, 937–956.

Bucciantini, N., Drago, A., Pagliara, G., Traversi, S., and Bauswein, A. (2022). Formation and evaporation of strangelets during the merger of two compact stars, *Phys. Rev. D* **106**, 103032.

Burgio, G., Schulze, H.-J., Vidaña, I., and Wei, J.-B. (2021). Neutron stars and the nuclear equation of state, *Prog. Part. Nucl. Phys.* **120**, 103879.

Carlson, J., Gandolfi, S., Pederiva, F., Pieper, S. C., Schiavilla, R., Schmidt, K., and Wiringa, R. B. (2015). Quantum monte carlo methods for nuclear physics, *Rev. Mod. Phys.* **87**, 3, 1067.

Chamel, N. and Haensel, P. (2008). Physics of neutron star crusts, *Living Rev. Relativ.* **11**, 1, 10.

Chapline, G. and Nauenberg, M. (1977a). On the possible existence of quark stars, *Ann. N. Y. Acad. Sci.* **302**, 191.

Chapline, G. and Nauenberg, M. (1977b). Asymptotic freedom and the Baryon-quark phase transition, *Phys. Rev. D* **16**, 450.

Chen, X. and Zhang, Y. (2017). Ab initio nuclear structure calculations with quantum monte carlo methods, *J. Comput. Phys.* **92**, 4, 567–578.

Collins, J. C. and Perry, M. J. (1975). Superdense matter: Neutrons or asymptotically free quarks? *Phys. Rev. Lett.* **34**, 1353–1356.

Cook, G. B., Shapiro, S. L., and Teukolsky, S. A. (1992). Spin-up of a rapidly rotating star by angular momentum loss-effects of general relativity, *ApJ* **398**, 203–223.

Cook, G. B., Shapiro, S. L., and Teukolsky, S. A. (1994). Rapidly rotating neutron stars in general relativity: Realistic equations of state, *ApJ* **424**, 823–845.

Cromartie, H. T., Fonseca, E., Ransom, S. M., Demorest, P. B., Arzoumanian, Z., Blumer, H., Brook, P. R., DeCesar, M. E., Dolch, T., Ellis, J. A., *et al.* (2020). Relativistic Shapiro delay measurements of an extremely massive millisecond pulsar, *Nat. Astron.* **4**, 1, 72–76.

De Luca, V., Franciolini, G., and Riotto, A. (2021). Nanograv data hints at primordial black holes as dark matter, *Phys. Rev. Lett.* **126**, 4, 041303.

Di Clemente, F., Drago, A., Char, P., and Pagliara, G. (2023). Stability and instability of strange dwarfs, *A&A* **678**, L1.

Dickhoff, W. and Barbieri, C. (2004). Self-consistent green's function method for nuclei and nuclear matter, *Prog. Part. Nucl. Phys.* **52**, 2, 377–496.

Dietrich, T., Coughlin, M. W., Pang, P. T., Bulla, M., Heinzel, J., Issa, L., Tews, I., and Antier, S. (2020). Multimessenger constraints on the neutron-star equation of state and the Hubble constant, *Sci.* **370**, 6523, 1450–1453.

Doroshenko, V., Suleimanov, V., Pühlhofer, G., and Santangelo, A. (2022). A strangely light neutron star within a supernova remnant, *Nat. Astron.* **6**, 12, 1444–1451.

Farhi, E. and Jaffe, R. L. (1984). Strange matter, *Phys. Rev. D* **30**, 11, 2379–2390.

Farrell, D., Alp, A., Weber, F., Spinella, W., Malfatti, G., Orsaria, M. G., and Ranea-Sandoval, I. F. (2023). *Hot Neutron Star Matter and Protoneutron Stars*, Chapter 5 (World Scientific), 199–259.

Fechner, W. B. and Joss, P. C. (1978). Quark stars with 'realistic' equations of state, *Nature* **274**, 347.

Fritzsch, H., Gell-Mann, M., and Leutwyler, H. (1973). Advantages of the color octet gluon picture, *Phys. Lett. B* **47**, 4, 365–368.

Galeazzi, F., Yoshida, S., and Eriguchi, Y. (2012). Differentially-rotating neutron star models with a parametrized rotation profile, *A & A* **541**, A156.

Gendreau, K. C., Arzoumanian, Z., *et al.* (2016). The neutron star interior composition explorer (nicer): Mission description, *ApJ* **832**, 2, 189.

Glendenning, N. (2012a). *Compact Stars: Nuclear Physics, Particle Physics and General Relativity*, Astronomy and Astrophysics Library, Springer New York.

Glendenning, N. (2012b). *Compact Stars: Nuclear Physics, Particle Physics and General Relativity*, Astronomy and Astrophysics Library, Springer New York.

Glendenning, N., Kettner, C., and Weber, F. (1995). Possible new class of dense white dwarfs, *Phys. Rev. Lett.* **74**, 18, 3519.

Glendenning, N. K., Pei, S., and Weber, F. (1997). Signal of quark deconfinement in the timing structure of pulsar spin-down, *Phys. Rev. Lett.* **79**, 9, 1603–1606.

Glendenning, N. K. and Schaffner-Bielich, J. (1999). First order kaon condensate, *Phys. Rev. C* **60**, 2, 025803.

Glendenning, N. K. and Weber, F. (1992). Nuclear solid crust on rotating strange quark stars, *ApJ* **400**, 647.

Goncalves, V. P., Jimenez, J. C., and Lazzari, L. (2023). Revisiting the stability of strange-dwarf stars and strange planets, *The European Physical Journal A,* **56**, p. 251.

Gondek-Rosińska, D., Kowalska, I., Villain, L., Ansorg, M., and Kucaba, M. (2017). A new view on the maximum mass of differentially rotating neutron stars, *ApJ* **837**, 1, 58.

Gross-Boelting, T., Fuchs, C., and Faessler, A. (1999). Covariant representations of the relativistic brueckner t-matrix and the nuclear matter problem, *Nucl. Phys. A* **648**, 1-2, 105–137.

Haensel, P., Potekhin, A. Y., and Yakovlev, D. G. (2007). *Neutron Stars 1* (Springer).

Haensel, P. and Proszynski, M. (1982). Pion condensation in cold dense matter and neutron stars, *ApJ* **258**, 306–320.

Han, S. and Steiner, A. W. (2019). Tidal deformability with sharp phase transitions in binary neutron stars, *Phys. Rev. D* **99**, 083014.

Harrison, B. K., Thorne, K. S., Wakano, M., and Wheeler, J. A. (1965). Gravitation theory and gravitational collapse, *Gravitation Theory and Gravitational Collapse.*

Hessels, J. W., Ransom, S. M., Stairs, I. H., Freire, P. C., Kaspi, V. M., and Camilo, F. (2006). A radio pulsar spinning at 716 HZ, *Sci.* **311**, 5769, 1901–1904.

Hewish, A., Bell, S. J., Pilkington, J. D., Frederick Scott, P., and Collins, R. A. (1979). Observation of a rapidly pulsating radio source, in *A Source Book in Astronomy and Astrophysics, 1900–1975* (Harvard University Press), 498–504.

Hinderer, T. (2008). Tidal love numbers of neutron stars, *ApJ* **677**, 2, 1216.

Holder, J., Acciari, V. A., Aliu, E., *et al.* (2006). The first veritas telescope, *Astropart. Phys.* **25**, 6, 391–401.

Horvath, J., Rocha, L., de Sá, L., Moraes, P., Barão, L., de Avellar, M., Bernardo, A., and Bachega, R. (2023). A light strange star in the remnant hess j1731-347: Minimal consistency checks, *A&A* **672**, L11.

Hughes, D. H., Briggs, F. H., Vogel, S. N., *et al.* (2006). The green bank telescope, *ApJ* **131**, 450.

Itoh, N. (1970a). Hydrostatic equilibrium of hypothetical quark stars, *Prog. Theor. Phys.* **44**, 291–292.

Itoh, N. (1970b). Hydrostatic equilibrium of hypothetical quark stars, *Prog. Theor. Phys.* **44**, 1, 291–292.

Ivanenko, D. D. and Kurdgelaidze, D. F. (1965). Hypothesis concerning quark stars, *Astrophysics* **1**, p. 251.

Jansen, F., Lumb, D., Altieri, B., *et al.* (2001). Xmm-newton observatory. i. the spacecraft and operations, *AA* **365**, 1, L1.

Jodrell Bank Observatory (Accessed: January 2022). Official website: Jodrell bank observatory.

Johnston, S., Ball, L., Bonnardeau, M., *et al.* (1996). The parkes multibeam pulsar survey, *MNRAS* **279**, 1026–1046.

Kalogera, V. and Baym, G. (1996). The maximum mass of a neutron star, *ApJ* **470**, 1, L61.

Kaplan, D. B. and Nelson, A. E. (1988). Kaon condensation in dense matter, *Nucl. Phys. A* **479**, 273–284.

Kaplan, J., Ott, C., O'Connor, E., Kiuchi, K., Roberts, L., and Duez, M. (2014). The influence of thermal pressure on equilibrium models of hypermassive neutron star merger remnants, *ApJ* **790**, 1, 19.

Kaspi, V. M. and Gavriil, F. P. (2017). Magnetars, *ApJ* **845**, 2, 148.

Keister, B. D. and Kisslinger, L. S. (1976). *Phys. Lett. B* **64**, 117.

Knorren, R., Prakash, M., and Ellis, P. J. (1995). Strangeness in hadronic stellar matter, *Phys. Rev. C* **52**, 6, 3470–3482.

Komatsu, H., Eriguchi, Y., and Hachisu, I. (1989). Rapidly rotating general relativistic stars–i. numerical method and its application to uniformly rotating polytropes, *MNRAS* **237**, 2, 355–379.

Kurban, A., Huang, Y.-F., Geng, J.-J., and Zong, H.-S. (2022). Searching for strange quark matter objects among white dwarfs, *Phys. Lett. B* **832**, 137204.

Lattimer, J. (2016). Observed neutron star masses, https://stellarcollapse.org/index.php/nsmasses.html.

Lattimer, J. M. (2019). Neutron star mass and radius measurements, *Universe* **5**, 7.

Lattimer, J. M. and Prakash, M. (2007). Neutron star observations: Prognosis for equation of state constraints, *Phys. Rep.* **442**, 1-6, 109–165.

Lattimer, J. M. and Prakash, M. (2016). The physics of neutron stars, *Phys. Rep.* **621**, 127–164.

Lattimer, J. M. and Steiner, A. W. (2014). Constraints on the symmetry energy using the mass-radius relation of neutron stars, *EPJ A* **50**, 2, 40.

Lee, C.-H., Jung, H., Min, D.-P., and Rho, M. (1994). Kaon-nucleon scattering from chiral Lagrangians, *Phys. Lett. B* **326**, 1-2, 14–20.

Leung, Y. C. and Wang, C. G. (1971). Properties of hadron matter. II. dense baryon matter and neutron stars, *ApJ.* **170**, 499.

Link, B., Epstein, R. I., and Lattimer, J. M. (1999). Pulsar constraints on neutron star structure and equation of state, *Phys. Rev. Lett.* **83**, 17, 3362.

Lorenz, C. P., Ravenhall, D. G., and Pethick, C. J. (1993). Neutron star crusts, *Phys. Rev. Lett.* **70**, 379–382.

Lorimer, D. and Ferrara, E. (2022). Galactic millisecond pulsars, https://www.astro.umd.edu/~eferrara/GalacticMSPs.html.

Lorimer, D. R. (2008). Binary and millisecond pulsars, *Living Rev. Relativ.* **11**, 1, 1–90.

Machleidt, R. (1989). Advances in nuclear physics, *Adv. Nucl. Phys.* **19**, 188.

Machleidt, R., Holinde, K., and Elster, C. (1987). The nucleon-nucleon interaction, *Phys. Rep.* **149**, 1, 1.

Madsen, J. (1999). Physics and astrophysics of strange quark matter, *Lect. Notes Phys.* **516**, 162.

Madsen, J. (2007). Strange quark matter in astrophysics, *ASR* **40**, 9, 1389–1402.

Makishima, K., Mihara, T., Nagase, F., and Tanaka, Y. (1999). Cyclotron resonance effects in two binary x-ray pulsars and the evolution of neutron star magnetic fields, *ApJ* **525**, 2, 978.

Malfatti, G., Orsaria, M. G., Contrera, G. A., Weber, F., and Ranea-Sandoval, I. F. (2019). Hot quark matter and (proto-) neutron stars, *Phys. Rev. C* **100**, 015803.

Malfatti, G., Orsaria, M. G., Ranea-Sandoval, I. F., Contrera, G. A., and Weber, F. (2020). Delta baryons and diquark formation in the cores of neutron stars, *Phys. Rev. D* **102**, 063008.

Malik, T., Banik, S., and Bandyopadhyay, D. (2021). New equation of state involving Bose-Einstein condensate of antikaon for supernova and neutron star merger simulations, *EPJST* **230**, 2, 561–566.

Manchester, R. (2017). Millisecond pulsars, their evolution and applications, *JA & A* **38**, 3, 42.

Mannarelli, M. (2019). Meson Condensation, *Particles* **2**, 3, 411–443.

Mellinger Jr, R. D., Weber, F., Spinella, W., Contrera, G. A., and Orsaria, M. G. (2017). Quark deconfinement in rotating neutron stars, *Universe* **3**, 1, 5.

Melrose, D. (2017). Coherent emission mechanisms in astrophysical plasmas, *Rev. Mod. Plasma Phys.* **1**, 1–81.

Migdal, A. B. (1972). Vacuum stability and limiting fields, *Sov. Phys. Uspe.* **14**, 6, 813–813.

Miller, M., Lamb, F. K., Dittmann, A., Bogdanov, S., Arzoumanian, Z., Gendreau, K. C., Guillot, S., Harding, A., Ho, W., Lattimer, J., *et al.* (2019). Psr j0030+0451 mass and radius from nicer data and implications for the properties of neutron star matter, *ApJL* **887**, 1, L24.

Miller, M. C., Lamb, F., Dittmann, A., Bogdanov, S., Arzoumanian, Z., Gendreau, K., Guillot, S., Ho, W., Lattimer, J., Loewenstein, M., *et al.* (2021). The radius of PSR j0740+6620 from nicer and xmm-newton data, *ApJL* **918**, 2, L28.

Miller, M. C. and Lamb, F. K. (2015). Determining neutron star properties by fitting oblate-star waveform models to x-ray burst oscillations, *ApJ* **808**, 1, 31.

Morrison, I. A., Baumgarte, T. W., and Shapiro, S. L. (2004). Effect of differential rotation on the maximum mass of neutron stars: realistic nuclear equations of state, *ApJ* **610**, 2, 941.

Moszkowski, S. A. (1974). Energy of neutron-star matter, *Phys. Rev. D* **9**, 1613–1625.

Nan, R., Li, D., Zhang, B., Fan, C., Qian, H., Cheng, J., Jin, Z., Lu, D., Li, X., Shao, Y., Jing, Z., Wu, X., Zhang, X., Cao, H., Yan, J., Yu, J., Wang, D., Wang, Q., Liu, D., Zheng, X., Yue, Y., Wang, G., Jiang, P., Zhang, K., Yang, J., Wang, Y., Yan, Y., Jin, Y., Jin, Y., Zhu, X., Fan, Y., Gao, Y., Dai, S., Han, Y., Zhang, H., Xu, Y., Wang, Y., Chen, Z., Zhang, S., You, X., Zhu, L., Qian, L., Wen, L., Jin, C., Zheng, W., Kong, L., Zhang, Y., Li, L., Qiu, K., Zhu, G., Xiao, L., Zhou, L., Gong, J., Lou, Y., Yang, S., Hu, K., Zhao, J., Chen, X., Wang, J., Zhang, Z., Deng, W., Zhang, K., Wang, J., Wang, H., Wang, J., Zhang, Z., Zhang, Z., Zhang, F., Xu, J., Li, H., Zhu, W., Yang, Y., Zhang, L., Lu, H., Zhang, X., Wang, X., Zhang, X., Yao, G., Lin, J., Li, J., Qi, Y., Zhang, Y., Yang, G., Zhu, M., Wu, X., Zhang, L., Chen, X., Cao, D., Cheng, J., Wang, Y., and Zheng, X. (2016). The five-hundred-meter aperture spherical radio telescope (fast) project, *Inter. J. Mod. Phys. D* **25**, 08, 1641003.

Negele, J. W. and Vautherin, D. (1973). Neutron star matter at sub-nuclear densities, *Nucl. Phys. A* **207**, 2, 298–320.

Nelson, A. E. and Kaplan, D. B. (1987). Strange condensate realignment in relativistic heavy ion collisions, *Phys. Lett. B* **192**, 1-2, 193–197.

Oertel, M., Hempel, M., Klähn, T., and Typel, S. (2017). Equations of state for supernovae and compact stars, *Rev. Mod. Phys.* **89**, 015007.

Orsaria, M. G., Malfatti, G., Mariani, M., Ranea-Sandoval, I. F., García, F., Spinella, W. M., Contrera, G. A., Lugones, G., and Weber, F. (2019). Phase transitions in neutron stars and their links to gravitational waves, *J. Phys. G: Nucl. Part. Phys.* **46**, 7, 073002.

Özel, F. and Freire, P. (2016). Masses, radii, and the equation of state of neutron stars, *Ann. Rev. Astron. Astrophys.* **54**, 401–440.

Pandharipande, V. R. (1971). Hyperonic matter, *Nucl. Phys. A* **178**, 123–144.

Paschalidis, V. and Stergioulas, N. (2017). Rotating stars in relativity, *Living Rev. Relati.* **20**, 1, 7.

Perot, L. and Chamel, N. (2023). Role of quark matter and color superconductivity in the structure and tidal deformability of strange dwarfs, *Universe* **9**, 9.

Perot, L., Chamel, N., and Vallet, P. (2023). Unmasking strange dwarfs with gravitational-wave observations, *Phys. Rev. D* **107**, 103004.

Pethick, C., Ravenhall, D., and Lorenz, C. (1995). The inner boundary of a neutron-star crust, *Nucl. Phys. A* **584**, 4, 675–703.

Pethick, C. J. and Ravenhall, D. G. (1999). The Physics of Neutron Star Crusts, in Z. Arzoumanian, F. Van der Hooft, and E. P. J. van den Heuvel (eds.), *Pulsar Timing, General Relativity and the Internal Structure of Neutron Stars*, 177.

Piekarewicz, J., Fattoyev, F., and Horowitz, C. (2014). Pulsar glitches: The crust may be enough, *Phys. Rev. C* **90**, 1, 015803.

Pons, J. A., Reddy, S., Prakash, M., Lattimer, J. M., and Miralles, J. A. (1999). Evolution of proto-neutron stars, *ApJ* **513**, 780–804.

Poschenrieder, P. and Weigel, M. (1988a). Nuclear matter problem in the relativistic green's function approach, *Phys. Rev. C* **38**, 1, 471.

Poschenrieder, P. and Weigel, M. (1988b). Nuclear matter properties in the relativistic λ-approximations, *Phys. Lett. B* **200**, 3, 231–234.

Predehl, P., Andritschke, R., Becker, W., *et al.* (2021). The EROSITA x-ray telescope on SRG, *A & A* **647**, A1.

Rajagopal, K. and Wilczek, F. (2001). *The Condensed Matter Physics of QCD*, in M. Shifman (ed.), *At the Frontier of Particle Physics/Handbook of QCD* (World Scientific), 1.

Rawley, L., Taylor, J., Davis, M., and Allan, D. (1987). Millisecond pulsar PSR 1937+21: A highly stable clock, *Sci.* **238**, 4828, 761–765.

Reinhard, P.-G. (1989). Relativistic mean field theory - foundations and applications, *Rep. Prog. Phys.* **52**, 4, 439–513.

Riley, T. E., Watts, A. L., Bogdanov, S., Ray, P. S., Ludlam, R. M., Guillot, S., Arzoumanian, Z., Baker, C. L., Bilous, A. V., Chakrabarty, D., *et al.* (2019). A nicer view of PSR j0030+0451: millisecond pulsar parameter estimation, *ApJ* **887**, 1, L21.

Riley, T. E., Watts, A. L., Ray, P. S., Bogdanov, S., Guillot, S., Morsink, S. M., Bilous, A. V., Arzoumanian, Z., Choudhury, D., Deneva, J. S., *et al.* (2021). A nicer view of the massive pulsar PSR j0740+6620 informed by radio timing and XMM-Newton spectroscopy, *ApJL* **918**, 2, L27.

Rischke, D. H. (2004). Color superconductivity in cold dense quark matter, *Prog. Part. Nucl. Phys.* **52**, 197–299.

Rüster, S. B., Hempel, M., and Schaffner-Bielich, J. (2006). Outer crust of nonaccreting cold neutron stars, *Phys. Rev. C* **73**, 035804.

Sammarruca, F. (2008). The "ab initio" approach to the nuclear equation of state: review and discussion, *arXiv preprint arXiv:0807.0263*.

Sawyer, R. F. and Scalapino, D. J. (1973). Pion condensation in superdense nuclear matter, *Phys. Rev. D* **7**, 953–964.

Schaffner, J. and Mishustin, I. N. (1996). Hyperon-rich matter in neutron stars, *Phys. Rev. C* **53**, 3, 1416–1429.

Sedrakian, A. (2007). The physics of dense hadronic matter and compact stars, *Prog. Par. Nucl. Phys.* **58**, 1, 168–246.

Sedrakian, A., Li, J.-J., and Weber, F. (2022). *Hyperonization in Compact Stars* (WORLD SCIENTIFIC Publishing Company Incorporated).

Sedrakian, A., Li, J. J., and Weber, F. (2023). Heavy baryons in compact stars, *Prog. Part. Nucl. Phys.* **475**, 104041.

Sehn, L., Fuchs, C., and Faessler, A. (1997). Nucleon self-energy in the relativistic Brueckner approach, *Phys. Rev. C* **56**, 1, 216.

Serot, B. D. and Walecka, J. D. (1986). Recent progress in quantum hadrodynamics, *Adv. Nucl. Phys.* **16**, 1–327.

Shapiro, S. L. (2000). Differential rotation in neutron stars: Magnetic braking and viscous damping, *ApJ* **544**, 1, 397.

Slane, P. O., Helfand, D. J., and Murray, S. S. (2002). New constraints on neutron star cooling from Chandra observations of 3c 58, *ApJ* **571**, 1, L45.

Smith, A. and Brown, D. (2020). Quantum Monte Carlo techniques in nuclear structure calculations, in R. Jones and S. White (eds.), *Adv. Nucl. Phys.* (Springer), 123–145.

Smith, A. M., McFarquhar, G. M., Rauber, R. M., Grim, J. A., Timlin, M. S., Jewett, B. F., and Jorgensen, D. P. (2009). Microphysical and thermodynamic structure and evolution of the trailing stratiform regions of mesoscale convective systems during BAMOEX. Part i: Observations, *Mon. Weather Rev.* **137**, 4, 1165–1185.

Sotani, H., Yasutake, N., Maruyama, T., and Tatsumi, T. (2011). Signatures of hadron-quark mixed phase in gravitational waves, *Phys. Rev. D* **83**, 024014.

Spinella, W. M. (2017). *A Systematic Investigation of Exotic Matter in Neutron Stars*, Ph.D. thesis, Claremont Graduate University & San Diego State University.

Spinella, W. M. and Weber, F. (2019). Hyperonic neutron star matter in light of gw170817, *Astron. Nachr.* **340**, 1-3, 145–150.

Spinella, W. M. and Weber, F. (2020). *Dense Baryonic Matter in the Cores of Neutron Stars*, Chapter 4 (World Scientific), 85–152.

Spinella, W. M., Weber, F., Orsaria, M. G., and Contrera, G. A. (2018). Neutrino emissivity in the Quark-Hadron mixed phase, *Universe* **4**, 5, 64.

Steiner, A. W., Heinke, C. O., Bogdanov, S., Li, C. K., Ho, W. C., Bahramian, A., and Han, S. (2018). Constraining the mass and radius of neutron stars in globular clusters, *MNRAS* **476**, 1, 421–435.

Stone, J. R. (2021). Nuclear physics and astrophysics constraints on the high density matter equation of state, *Universe* **7**, 8, 257.

Ter Haar, B. and Malfliet, R. (1986). Equation of state of nuclear matter in the relativistic Dirac-Brueckner approach, *Phys. Rev. Lett.* **56**, 12, 1237.

Ter Haar, B. and Malfliet, R. (1987). Nucleons, mesons and deltas in nuclear matter a relativistic Dirac-Brueckner approach, *Phys. Rep.* **149**, 4, 207–286.

Terazawa, H. (1979). Dynamical subquark model of pregauge and pregeometric interactions, Tech. rep., Institute for Nuclear Study, University of Tokyo.

Terazawa, H. (1991). Are super-hypernuclei found in cosmic rays? *JPSJ* **60**, 6, 1848–1851.

Thapa, V. B., Sinha, M., Li, J. J., and Sedrakian, A. (2021). Massive Δ-resonance admixed hypernuclear stars with antikaon condensations, *Phys. Rev. D* **103**, 063004.

Tong, H., Ren, X.-L., Ring, P., Shen, S.-H., Wang, S.-B., Meng, J., *et al.* (2018). Relativistic Brueckner-Hartree-Fock theory in nuclear matter without the average momentum approximation, *Phys. Rev. C* **98**, 5, 054302.

Turolla, R., Zane, S., and Drake, J. J. (2004). Bare quark stars or naked neutron stars? The case of rx j1856. 5–3754, *ApJ* **603**, 1, 265.

Typel, S. (2018). Relativistic mean-field models with different parametrizations of density dependent couplings, *Particles* **1**, 1, 3–22.

Typel, S., Röpke, G., Klähn, T., Blaschke, D., and Wolter, H. H. (2010). Composition and thermodynamics of nuclear matter with light clusters, *Phys. Rev. C* **81**, 015803.

Usov, V. V. (2001). Thermal emission from bare quark matter surfaces of hot strange stars, *ApJL* **550**, L179–L182.

Van Dalen, E., Fuchs, C., and Faessler, A. (2004). The relativistic Dirac–Brueckner approach to asymmetric nuclear matter, *Nucl. Phys. A* **744**, 227–248.

van Haarlem, M. P., Wise, M. W., Gunst, A. W., *et al.* (2013). Lofar: The low-frequency array, *A&A* **556**, A2.

Walecka, J. D. (1974). A theory of highly condensed matter, *Ann. Phys.* **83**, 491–529.

Watts, A. L., Andersson, N., Chakrabarty, D., Feroci, M., Hebeler, K., Israel, G., Lamb, F. K., Miller, M. C., Morsink, S., Özel, F., Patruno, A., Poutanen, J., Psaltis, D., Schwenk, A., Steiner, A. W., Stella, L., Tolos, L., and van der Klis, M. (2016). Colloquium: Measuring the neutron star equation of state using x-ray timing, *Rev. Mod. Phys.* **88**, 021001.

Weber, F. (1999). *Pulsars as Astrophysical Laboratories for Nuclear and Particle Physics (Series in High Energy Physics, Cosmology and Gravitation)* (CRC Press).

Weber, F. (2005). Strange quark matter and compact stars, *Progress in Particle and Nucl. Phys.* **54**, 1, 193–288.

Weisskopf, M. C., O'Dell, S. L., van Speybroeck, L. P., *et al.* (2000). The chandra x-ray observatory, *ApJ* **536**, 2, L81.

Witten, E. (1984). Cosmic separation of phases, *Phys. Rev. D* **30**, 272–285.

Xu, R. X. (2003). Strange Quark Stars - A Review, in X. D. Li, V. Trimble, and Z. R. Wang (eds.), *High Energy Processes and Phenomena in Astrophysics*, Vol. 214 (Astronomical Society of the Pacific), 191.

Yuan, W.-L., Li, A., Miao, Z., Zuo, B., and Bai, Z. (2022). Interacting *ud* and *uds* quark matter at finite densities and quark stars, *Phys. Rev. D* **105**, 123004.

Zakharov, B. G. (2011). Photon emission from bare quark stars, *JETP* **112**, 1, 63–76.

Zhou, E., Kiuchi, K., Shibata, M., Tsokaros, A., and Uryū, K. b. o. (2021). Evolution of bare quark stars in full general relativity: Single star case, *Phys. Rev. D* **103**, 123011.

https://doi.org/10.1142/9789819800643_0024

Chapter 24

Axisymmetric Oscillations of Rotating Neutron Stars: Gravitational Wave Emission

Stéphane Vincent* **and José A. de Freitas Pacheco**†

Université de la Côte d'Azur, Observatoire de la Côte d'Azur
06304 Nice Cedex, France
**vincent@oca.eu*
†pacheco@oca.eu

In this chapter quasi-radial oscillations of rotating neutron stars are reviewed. Then, new calculations of axisymmetric modes are presented based on linearized relativistic field equations that take into account both fluid and metric perturbations. Our numerical procedure was checked by comparing results derived from the present code with relativistic hydrodynamic calculations and adopting a common polytropic equation of state. Such a comparison indicates that mode frequencies differ by less than 2% excepting for rotation velocities near the mass shedding limit, where differences increase up to 10%. Rapidly rotating neutron star models were computed for different "realistic" equations of state characterized by nuclear compressibilities ranging from 140 MeV up to 1200 MeV. The fundamental and the first overtone frequencies of quasi-radial and quadrupole modes were calculated for a sequence of neutron star models having the same baryonic mass ($M_b = 1.5\ M_\odot$) but different rotation velocities. The present results confirm that quadrupole frequencies are lower than those of quasi-radial modes, contrary to results derived from models built with polytropic equations of state. Excitation of quasi-radial modes and gravitational wave emission were considered in a scenario in which a pure hadronic star suffers a mini-collapse due to a hadron-quark phase transition in the core. The released gravitational energy, which depends on the adopted equation of state for the nuclear matter, is of the order of 1.98×10^{52} erg for our considered neutron star model. However, only 8.9% is channeled into gravitational waves. Nevertheless, the planned European gravitational Einstein Telescope would be able to detect such a signal up to distances of about 114 Mpc.

Keywords: Neutron stars, Axisymmetric modes, Gravitational waves

Contents

1. Introduction

Oscillations of relativistic stars have been extensively studied in the past fifty years, aiming to investigate stability conditions of these compact objects and the detection possibility of such pulsations in the domain either of the electromagnetic spectrum or gravitational waves. In the same way as the new asteroseismology field provides information about the stellar interiors, the eventual detection of neutron star oscillations might impose severe constraints on the equation of state of nuclear matter (Andersson and Kokkotas, 1998). The first studies constituting the basis of the stellar stability theory in a relativistic framework were developed in the early seventies by Chandrasekhar and Friedman (1972a,b) and Schutz (1972) among others. These investigations generalized the early conclusions derived from approaches based on Newtonian dynamics and established conditions for the onset of instabilities along equilibrium model sequences (Friedman and Schutz, 1975).

Different mechanisms able to excite neutron star pulsations have been discussed in the literature as, for instance, the core collapse of a massive star (Moenchmeyer *et al.*, 1991; Ott *et al.*, 2006; Zwerger and Mueller, 1997), crustal cracking when solid crusts are present (Ruderman 1991a,b,c, Burdyuzha *et al.* 1996, de Freitas Pacheco 1998a) and core phase transitions, which are able to trigger a "mini" collapse as a consequence of the softening of the equation of state (Haensel *et al.*, 1989; Lin *et al.*, 2006; Marranghello *et al.*, 2002; Miniutti *et al.*, 2003).

In the one hand, pure radial modes by their simplicity have been the first to be investigated. Radial oscillations of non-rotating neutron stars have been analyzed in the pioneering studies by Chandrasekhar (1964), Bardeen, Thorne, and Meltzer (1966) and, more recently by Kokkotas and Ruoff (2001). On the other hand, non-radial modes play an important role in the physics of neutron stars because they are able to generate gravitational waves that carry information about the internal structure of these stars, since the mode frequencies and damping timescales are functions of the gravitational mass and radius of the configuration. The very first studies on non-radial oscillations of non-rotating neutron stars and their associated gravitational radiation were performed in the late sixties (Price and Thorne, 1969; Thorne and Campolattaro, 1967; Thorne, 1969). Subsequent investigations have shown that if the released elastic energy during a starquake is of the order of 10^{45} erg (de Freitas Pacheco, 1998a), then the gravitational radiation from quadrupole modes excited by shear waves could be detected up to distances of several kpc by advanced laser interferometers like the LIGO-VIRGO collaboration.

Pure radial modes do not emit gravitational waves and, in this case oscillations are damped essentially by internal viscous forces and neutrino emission. However, the situation is quite different if the star rotates since the degeneracy in the index m is removed: prograde ($m < 0$) modes are now distinct from retrograde ($m > 0$) modes and a non-rotating l-mode is split into $2l + 1$ modes (Stergioulas, 2003). Moreover, rotation couples a polar l-term to an axial $l \pm 1$ term and the coupling to the $l + 1$ mode is strongly favored over the $l - 1$

mode, similar to what occurs for the axial modes (Chandrasekhar and Ferrari, 1991; Kojima, 1992; Stergioulas and Friedman, 1998). As a consequence of all these rotational couplings, even the lowest $l = 0$ quasi-radial mode may radiate gravitational waves.

In the one hand, the first investigations on axisymmetric modes of relativistic rotating stars were performed by using the so-called "slow-rotation approximation", which neglects, at first order, deviations of the stellar shape from spherical symmetry (Chandrasekhar and Friedman, 1971; Datta *et al.*, 1998; Hartle and Friedman, 1975; Hartle *et al*, 1972). On the other hand, axisymmetric oscillations of rapidly rotating relativistic stars were investigated only in the past decades (Font *et al.*, 2001; Yoshida and Eriguchi, 2001) but within the Cowling approximation or, in other words, fluid perturbations are taken into account but not those of the background metric (McDermott *et al.*, 1983). More recently, numerical relativistic hydrodynamic codes have been developed to follow the long-term dynamical evolution of relativistic stars, in particular of rapidly rotating neutron stars undergoing quasi-radial and quadrupole oscillations (Dimmelmeier *et al.*, 2006; Font *et al.*, 2002).

All these past investigations have provided an important insight on the expected oscillation frequencies and how they are affected by an increasing rotation, despite the fact that most of these studies was based on simple polytropic equations of state, assumed to describe all the stellar interior. In this article, the fundamental and the first overtone frequencies of $l = 0$ and $l = 2$ modes of rotating neutron stars are calculated by using a linear perturbation approach. Moreover, in the present investigation both fluid and metric perturbations are taken into account and unperturbed neutron star models are computed by using a metric adequate to describe a fast rotating star. Stellar models were calculated for several "realistic" equations of state characterized by different nuclear compressibilities K_N. A grid of models of constant baryonic mass but with different rotation velocities was computed in order to investigate the effects of rotation and nuclear matter properties on the resulting eigenfrequencies. The detection possibility of gravitational wave emission from these oscillations was also considered within a scenario in which these axisymmetric modes are excited as a consequence of a core mini-collapse resulting from a deconfinement phase transition. This article is organized as follows: in Section 2, relativistic rotating models are discussed as well as the linearized equations and the quasi-radial modes; in Section 3, the main numerical results are presented; in Section 4 the gravitational wave emission and its detection by laser interferometers are discussed and finally, in Section 5 the main conclusions are given.

2. Rotating Neutron Star Models

As mentioned above, the present investigation will be focused on axisymmetric oscillation modes of rotating neutron stars, since they can be potential sources of gravitational waves that can be detected by the future generation of laser interferometers. In our approach, despite of using a system of linearized equations, we will show that the computed eigenfrequencies have an accuracy comparable to results derived from non-linear simulations.

The unperturbed configuration is represented by an axisymmetric star rotating uniformly with an angular velocity Ω as seen by a distant observer at rest. In this case, the geometry is defined by the following line element (Carter, 1969; Cook *et al.*, 1994), (the usual convention $G = c = 1$ was adopted but physical constants are recovered always when necessary)

$$ds^2 = -e^{2\nu}dt^2 + e^{2\psi}(d\phi - \omega dt)^2 + e^{2\mu}(dr^2 + r^2 d^2\theta), \qquad (1)$$

where the metric functions ν, μ, ψ and ω depend on the coordinates r and θ only. Rotating stars "drag" local inertial frames, leading to the recognition of local or comoving observers and "distant" ones. Eulerian observers have world-lines normal to hypersurfaces of constant t but the description of the system is simplified by introducing local "zero angular momentum observers" (ZAMO) (Bardeen, 1970).

Matter is considered as an ideal fluid with an energy-momentum tensor defined by

$$T^{ab} = (\varepsilon + p)u^a u^b + pg^{ab}, \qquad (2)$$

where ε and p are respectively the energy density and the pressure of the fluid measured by a comoving observer. The fluid 4-velocity components are

$$u^a = \frac{e^{-\nu}}{\sqrt{1 - v^2}}(t^a + \Omega\phi^a), \qquad (3)$$

where v is the fluid velocity measured by an Eulerian observer defined by

$$v = (\Omega - \omega)e^{(\psi-\nu)}, \tag{4}$$

and where t^a, ϕ^a are commuting Killing vectors associated to time and to rotational symmetries of the metric. In the above equation ω measures the dragging of the inertial frames in the sense that a ZAMO observer moves along trajectories whose angular velocity relative to infinity is Ω.

2.1. *The linearized equations*

We consider variations due to displacements of world-lines in a 4-dimensional spacetime manifold M, corresponding to coordinate variations $x^a \rightarrow x^a + \Delta x^a$ and variations of the metric tensor $g_{ab} \rightarrow g_{ab} + \delta g_{ab}$ at fixed points in M. Since we are interested in the time evolution of similar stellar configurations having the same baryon number and entropy, we introduce a family of time-dependent solutions characterized by the parameter λ, e.g.,

$$A(\lambda) \equiv [g_{ab}(\lambda), u^a(\lambda), \varepsilon(\lambda), p(\lambda)] \,. \tag{5}$$

Then, to the first order in λ, the perturbed functions $A(\lambda)$ are compared to their equilibrium values. First order departures from equilibrium can be described in two ways: either as Eulerian perturbations δA, which denotes variations in A as measured with respect to a fixed observer in the coordinate grid or as Lagrangian perturbations ΔA, which can be interpreted as variations of A as measured by observers moving along fluid lines. Lagrangian and Eulerian variations are related by

$$\Delta A = \delta A + \pounds_\xi A, \tag{6}$$

where $\pounds_\xi$ denotes the Lie derivative along the vector ξ^a connecting fluid elements in the equilibrium configuration to the corresponding elements in the perturbed spacetime, namely, $\xi^a \equiv \Delta x^a$. In this way perturbed quantities can be expressed in terms of Eulerian changes in the metric and a Lagrangian displacement vector.

For the derivation of the linearized equations, one assumes an Eulerian metric perturbation h_{ab}, supposed to be small in comparison with the background metric g_{ab}. Hence, the perturbed spacetime is described by the metric

$$\tilde{g}_{ab} = g_{ab} + h_{ab}, \tag{7}$$

and the lowering (rising) of indices is performed by using the background metric. The perturbed Einstein equations are

$$\delta R_{ab} = 8\pi \left(\delta T_{ab} - \frac{1}{2} g_{ab} T - \frac{1}{2} h_{ab} T \right). \tag{8}$$

The perturbed field equations should be considered together with the perturbed energy-momentum conservation, e.g.,

$$\delta(\nabla_b T^{ab}) = 0. \tag{9}$$

Under these conditions, to the first order in h_{ab} the perturbed Ricci tensor is

$$\delta R_{ab} = \nabla_c \nabla_a h^c_b - \frac{1}{2}(\nabla_a \nabla_b h^c_c + \nabla^c \nabla_c h_{ab}). \tag{10}$$

According to Priou (1992), the components of δR_{ab} were computed for a stationary, axisymmetric background in its most general form, prior to any choice of gauge and these results are used here. Now, perturbing the energy-momentum tensor, one obtains

$$\delta T_{ab} = (\delta\varepsilon + \delta p) u_a u_b + (\varepsilon + p)(u_a \delta u_b + u_b \delta u_a) + g_{ab} \delta p + p h_{ab}, \tag{11}$$

while the perturbed contracted tensor is

$$\delta T = 3\delta p - \delta\varepsilon. \tag{12}$$

From the normalization condition $u^a u_a = -1$ and using the fact that the Lagrangian variation does not change the velocity direction, one obtains

$$\Delta(u^a u_a) = 0 \Rightarrow \Delta u^a = \frac{1}{2} u^a u^b u^c \Delta g_{bc}. \tag{13}$$

Moreover, using the well known relation

$$\pounds_\xi g_{ab} = \nabla_a \xi_b + \nabla_b \xi_a, \tag{14}$$

the Lagrangian variation of the metric can be recast as

$$\Delta g_{ab} = h_{ab} + \nabla_a \xi_b + \nabla_b \xi_a. \tag{15}$$

Therefore, combining Eqs. 13 and 15, the fluid velocity perturbations can be written as

$$\delta u^a = q^a_b \pounds_u \xi^b + \frac{1}{2} u^a u^b u^c h_{bc}, \tag{16}$$

where the tensor q^a_b is the projection operator orthogonal to u^a, e.g.,

$$q^{ab} = g^{ab} + u^a u^b. \tag{17}$$

Under these conditions, the Lagrangian perturbation of the energy density and pressure can be expressed in terms of h_{ab} and ξ^a as

$$\frac{\Delta p}{p\Gamma} = \frac{\Delta \varepsilon}{(\varepsilon + p)} = -\frac{1}{2} q^{ab} \Delta g_{ab}, \tag{18}$$

where the adiabatic index is defined as

$$\Gamma = \frac{(\varepsilon + p)}{p} \frac{\partial p}{\partial \varepsilon}. \tag{19}$$

Because the equilibrium configuration is stationary and axisymmetric, linear perturbations can be decomposed into modes proportional to $e^{\imath\sigma t + m\phi}$. Moreover, due to the axial symmetry, the metric potentials and the fluid state parameters depend only on the coordinates r and θ. Thus, any unperturbed function $f(r, \theta)$ can be expanded in terms of Legendre polynomials.

Perturbed quantities are expressed as functions of the metric perturbation tensor h_{ab} and of the displacement vector ξ^a, which are commonly expanded in terms of scalar, vector and tensor spherical harmonics. In this case, the Lagrangian components of the displacement vector can be written as

$$\xi^r \propto \sum_{l=0}^{\infty} \sum_{m=-l}^{l} a_{lm} Y_l^m, \tag{20}$$

$$\xi^\theta \propto \sum_{l=0}^{\infty} \sum_{m=-l}^{l} \left(a_{lm} \partial_\theta Y_l^m + b_{lm} \frac{1}{\sin\theta} \partial_\phi Y_l^m \right), \tag{21}$$

$$\xi^\phi \propto \sum_{l=0}^{\infty} \sum_{m=-l}^{l} \left(a_{lm} \partial_\phi Y_l^m + b_{lm} \sin\theta \partial_\theta Y_l^m \right). \tag{22}$$

In the above equations, a_{ij} and b_{ij} are complex coefficients distinct for each component but not independent. We adopted the gauge freedom inherent to this formalism to set $\xi^t = 0$. Following the Regge-Wheeler notation for spherical harmonics, the Eulerian perturbations of the metric tensor take the form

$$htt, h_{rr}, h_{tr} \propto \sum_{l=0}^{\infty} \sum_{m=-l}^{l} a_{lm} Y_l^m, \tag{23}$$

$$h_{t\theta}, h_{t\phi}, h_{r\theta}, h_{r\phi} \propto \sum_{l=0}^{\infty} \sum_{m=-l}^{l} \left(a_{lm}\Phi_{l\theta}^m + b_{lm}\Psi_{l\phi}^m\right), \tag{24}$$

$$h_{\theta\phi} \propto \sum_{l=0}^{\infty} \sum_{m=-l}^{l} \left(a_{lm}\Phi_{l\theta\phi}^m + b_{lm}\Phi_{l\theta\phi}^m + c_{lm}\chi_{l\theta\phi}^m\right). \tag{25}$$

The coefficients appearing in the last three equations are complex and distinct for each metric component.

2.2. *Equations for quasi-radial modes*

Quasi-radial oscillations are modes that would be purely radial in the absence of rotation. The coupling between polar and axial modes induced by slow rotation was recognized more than 30 years ago (Chandrasekhar and Ferrari, 1991; Kojima, 1992), while the generalization of the selection rules was established by Stergioulas and Friedman (1998), who have also found that all harmonics contribute to the mode.

The next step is to choose an adequate gauge transformation able to reduce and simplify the system of equations. Consider the vector transformation $x^a \to \tilde{x}^a + \eta^a(x)$, where η^a has an infinitesimal amplitude and transforms like a vector. In this case, the metric coefficients transform as

$$\tilde{h}_{ab} = h_{ab} + \eta_{a;b} + \eta_{b;a}. \tag{26}$$

This gauge transformation provides four degrees of freedom associated to the choice of the vector η^a. In Stergioulas and Friedman (1998) adopted a gauge satisfying the conditions $h_{r\theta} = h_{r\phi} = 0$, $h_{t\phi} = \omega h_{\phi\phi}$ and $h_{\phi\phi} = h_{\theta\theta}e^{2(\psi-\phi)}/r^2$. Under these conditions, the six remaining equations describing the perturbed metric and the displacement vector can be respectively written as (using the notation by Priou, 1992)

$$h_{ab} = \sum_l e^{\imath\sigma t} \begin{pmatrix} H_{0l}(r)Y_l^0(r) & H_{1l}(r)Y_l^0(r) & m_l(r)\partial_\theta Y_l^0 & 0 \\ H_{1l}(r)Y_l^0(r) & H_{2l}(r)Y_l^0(r) & q_l(r)\partial_\theta Y_l^0 & 0 \\ H_{2l}(r)Y_l^0(r) & q_l(r)\partial_\theta Y_l^0 & r^2\left(a_l(r) + b_l(r)\partial_{\theta\theta}\right)Y_l^0(r) & 0 \\ 0 & 0 & 0 & 0 \end{pmatrix}, \tag{27}$$

and

$$\xi^a = \sum_l \left(0, X_1(r), X_2(r)\partial_\theta, X_3(r)\sin\theta\partial_\theta\right) Y_l^0 e^{\imath\sigma t}. \tag{28}$$

The resulting system of equations involving the functions $X_1(r)$, $X_2(r)$, $X_3(r)$, $H_0(r)$, $H_1(r)$, $H_2(r)$, $m(r)$, $q(r)$, $a(r)$, $b(r)$ and their derivatives is rather cumbersome. Notice that when we focus only on axisymmetric perturbations, we have 10 equations and 10 unknowns. However, for $m \neq 0$ the adopted gauge is not adequate since we have now three extra unknowns.

The solution of the system of linearized equations must satisfy appropriate boundary conditions. The perturbed functions must be regular everywhere and, in particular, they must be finite at the stellar center ($r = 0$) and at the surface ($r = R$). The metric perturbations exterior to the star are described differently. In the case of a non-rotating star, the external perturbations are described by the Regge–Wheeler–Zerilli (RWZ) equation with a suitable defined potential for polar and axial modes. The external solution must be matched to the internal solution at the stellar surface. Solutions of the RWZ equation are ingoing and outgoing waves. The former excite the star while the later damp the oscillations, constituting quasi-normal modes with complex frequencies. Excepting for pure spacetime modes or $w-modes$, for non-rotating stars, polar quasi-normal modes have imaginary parts smaller than the real part. The behavior of the fields near $r = 0$ can be obtained by expanding the corresponding functions in Taylor series whose coefficients can be estimated from the field equations. In practice we have a linear algebraic system of equations that can be represented in a matrix form as

$$A\vec{X} = B, \tag{29}$$

where A is a matrix including the expansion coefficients, $\vec{X}$ is a vector whose components include eigenfunction parameters and B is a matrix constituted by expansion coefficients of the perturbed metric. An additional boundary condition imposes that pressure variations vanish at the perturbed stellar surface. This condition is equivalent to the requirement that the Lagrangean variation of the pressure is zero at the unperturbed surface of the star or, in other words, the condition $\Delta p(R) = 0$ becomes

$$\delta p + \xi^r \partial_r p = 0. \tag{30}$$

3. Numerical Results

The computation of the unperturbed 2-D stellar models was performed by using a Runge-Kutta scheme. In order to check the accuracy of our code, neutron star models and eigenfrequencies were compute initially by using different grids (r, θ) with the following number of points: (601, 129), (301, 251) and (601, 409). The last one was finally adopted in all subsequent calculations, since it gives comparable errors either for slow or rapid rotating configurations. The perturbed equations have coupling terms involving infinite series including Legendre polynomials. In practice, the required accuracy is rapidly achieved by the inclusion of only the first three terms in agreement with other similar studies (Stavridis, Passamonti, and Kokkotas, 2007; Stergioulas and Friedman, 1995).

In order to check our solutions, neutron star models based on a polytropic equation of state $p = K\rho^{(N+1)/N}$ were initially computed in order to be compared with non-linear calculations (Dimmelmeier *et al.*, 2006; Font *et al.*, 2002), expected to estimate the eigenfrequencies with higher accuracy. In fact, the models described in those references correspond to relativistic dynamical calculations in which the mode frequencies were extracted from a Fourier analysis of non-linear perturbations. We have adopted the same values defining the equation of state, that is $K = 100$ (in units where $G = c = 1$, $M = 1\ M_\odot$) and $N = 1$. Similarly to the aforementioned authors, a sequence of models with different rotation velocities but with a constant central energy density equal to $\varepsilon_c = 8.93 \times 10^{14}\ erg.cm^{-3}$ were computed as well as the resulting quasi-radial frequencies.

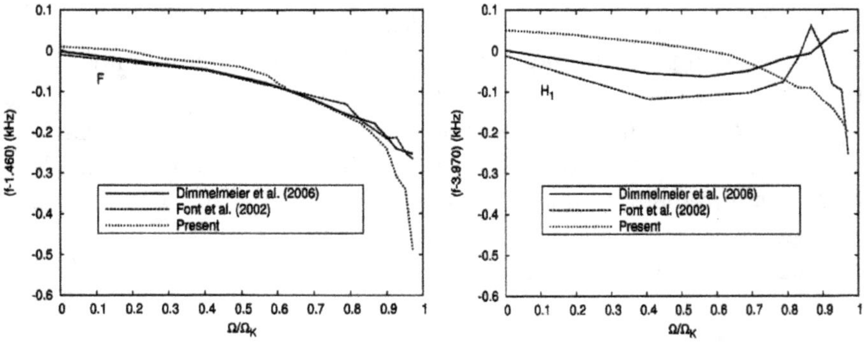

Fig. 1. In the horizontal axis is given the ratio between the rotation frequency Ω and the critical rotational frequency Ω_K corresponding to the mass-shedding limit. In the vertical axis is shown (in percent) the difference between the quasi-radial mode frequency at a given Ω and the corresponding value for a non-rotating star. The left panel corresponds to the fundamental mode F and the right panel corresponds to the first harmonic H_1. Present results are represented by dotted lines as indicated by the labels inserted in figures.

In Figure 1 are plotted the resulting quasi-radial frequencies for the fundamental mode F (left panel) and the first harmonic H_1 (right panel) as a function of the rotational frequency given in terms of the critical Keplerian frequency Ω_K or the so-called mass-shedding limit, defined as the angular velocity of a particle in a circular Keplerian orbit at the stellar equator, whose expression is given by

$$\Omega_K = \frac{\omega'}{2\psi'} + e^{(\nu-\psi)}\left[\frac{\nu'}{\psi'} + \left(\frac{\omega'}{2\psi'}e^{(\psi-\nu)}\right)^2\right]^{1/2} + \omega \qquad (31)$$

where the metric coefficients correspond to those in Eq. 1 and the notation $y' = dy/dr$ was adopted.

In order to compare the present results with those by Dimmelmeier *et al.* (2006), Font *et al.* (2002), we have plotted in ordinates the mode frequency difference (in percent) between the value at a given Ω and the corresponding value for a non-rotating star. Up to rotation velocities of $\Omega/\Omega_K \simeq 0.85$, the differences between our results and those derived from a non-linear approach are of the order of 1–2% for the F-mode and 2–3% for the first harmonic. Differences increase as the mass-shedding limit is approached, attaining values up to 10%, indicating that our method is less accurate near the Keplerian limit. Similar results are obtained when quadrupole modes are considered.

Once our code was checked and its limitations near the mass-shedding limit established, the next step was to build neutron star models with different "realistic" equations of state. In our code, the equations representing the nuclear matter in the core of the star were matched to those describing the outer crust: for densities in the range 7.8–1200 $g.cm^{-3}$ the results by Feynman *et al.* (1949) for a pure iron crystal were used; for higher densities up to values around $(3-4) \times 10^{12}$ $g.cm^{-3}$, where the neutron drip occurs, the equations of state computed by Baym *et al.* (1971) and Baym and Pethick (1975) were adopted.

Here are reported three sequences of models having distinct nuclear compressibilities defined as $K_N = 9(\partial p/\partial n)$ and computed at at the so-called saturation density, that is $n_0 = 0.16 fm^{-3}$. All the three sequences have the same baryonic mass $M_b = 1.5\ M_\odot$ but different rotation velocities, which affect the flattening and the central density of the star. The first sequence of models ($eosA$) was based on a equation of state relatively soft in which two-body interactions were modelled by a Reid potential (Pandharipande, 1971). The resulting compressibility is $K_N \simeq 140$ MeV, leading to quite compact neutron star configurations. The characteristics of this sequence of models are listed in Table 1. Columns give respectively: (1) the ratio between the polar and the equatorial radius; (2) the central energy density

Table 1. Properties of models "eosA" — $K_N \simeq 140$ MeV and $\Omega_K = 8324\ s^{-1}$.

R_p/R_{eq}	ε_c	M_{gr}	R_{eq}	Ω/Ω_K	F	H_1	^2f	$^2p_1$
0.998	1.696	1.334	9.680	0.072	3.26	8.60	2.40	5.30
0.987	1.684	1.335	9.733	0.222	3.25	8.59	2.39	5.29
0.969	1.665	1.336	9.822	0.346	3.24	8.57	2.38	5.27
0.942	1.637	1.339	9.963	0.470	3.22	8.56	2.36	5.26
0.905	1.596	1.341	10.479	0.594	3.19	8.53	2.34	5.24
0.858	1.545	1.345	10.465	0.710	3.16	8.51	2.32	5.21
0.824	1.509	1.348	10.700	0.777	3.15	8.49	2.31	5.20
0.787	1.469	1.351	10.983	0.837	3.13	8.45	2.29	5.19
0.737	1.417	1.355	11.418	0.901	3.11	8.47	2.26	5.16
0.674	1.358	1.360	12.079	0.958	3.08	8.48	2.23	5.13
0.619	1.318	1.363	12.806	0.988	3.00	8.50	2.19	5.10

(ε_c/c^2) in units of 10^{15} $g.cm^{-3}$; (3) the gravitational mass in solar units; (4) the equatorial radius in km; (5) the angular velocity in terms of the Keplerian velocity. The last four columns give respectively the fundamental (F) and the first overtone (H_1) frequencies for quasi-radial modes and the quadrupole mode frequencies ^2f, $^2p_1$, all in kHz.

The second sequence was built with a rather stiff equation of state ("$eosL$") computed by Pandharipande *et al.* (1976), which has a nuclear compressibility $K_N \simeq 1200$ MeV. This equation derived from a relativistic N-body interaction model always satisfies the causality principle, that is the sound velocity is always less than the light velocity. Neutron star models computed with this equation of state have larger dimensions when compared to the $eosA$ model sequence having the same baryonic mass ($M_b = 1.5\ M_\odot$) and rotation velocity. Results are shown in Table 2 and columns have the same meaning and units as in Table 1.

Finally, the third sequence of models was computed with an equation of state ("$eos - SLy$") characterized by an intermediate compressibility ($K_N \simeq 360$ MeV) describing both the crust and a "liquid" core. This equation of state is based on an effective nuclear interaction of the Skyrme type and is suitable to describe a neutron-rich nuclear matter (Duchin and Haensel, 2001). Results are listed in Table 3 and the meaning of columns is the same as in the previous tables.

Table 2. Properties of models "eosL" — $K_N \simeq 1200$ MeV and $\Omega_K = 4753\ s^{-1}$.

R_p/R_{eq}	ε_c	M_{gr}	R_{eq}	Ω/Ω_K	F	H_1	^2f	$^2p_1$
0.994	0.430	1.394	14.863	0.136	2.65	5.53	1.32	4.80
0.987	0.429	1.394	14.919	0.205	2.64	5.52	1.31	4.78
0.972	0.427	1.395	15.014	0.302	2.63	5.51	1.30	4.76
0.949	0.424	1.396	15.182	0.406	2.60	5.49	1.28	4.72
0.913	0.419	1.398	15.462	0.527	2.57	5.46	1.26	4.69
0.869	0.413	1.401	15.843	0.636	2.54	5.47	1.25	4.67
0.822	0.406	1.403	16.296	0.731	2.51	5.48	1.23	4.65
0.746	0.395	1.407	17.183	0.843	2.47	5.50	1.21	4.60
0.671	0.385	1.412	18.333	0.919	2.42	5.54	1.18	4.57
0.625	0.380	1.413	19.252	0.948	2.36	5.58	1.15	4.50

Table 3. Properties of models "eos-SLy" — $K_N \simeq 360$ MeV and $\Omega_K = 6481\ s^{-1}$.

R_p/R_{eq}	ε_c	M_{gr}	R_{eq}	Ω/Ω_K	F	H_1	${}^2f$	${}^2p_1$
0.994	0.954	1.362	11.746	0.112	3.23	6.95	2.20	4.88
0.975	0.946	1.364	11.868	0.300	3.22	6.94	2.18	4.86
0.941	0.931	1.366	12.078	0.461	3.20	6.93	2.16	4.83
0.918	0.921	1.368	12.228	0.538	3.18	6.91	2.13	4.79
0.891	0.909	1.369	12.419	0.614	3.16	6.88	2.11	4.76
0.858	0.894	1.371	12.669	0.690	3.14	6.86	2.08	4.72
0.814	0.874	1.374	13.040	0.773	3.11	6.82	2.05	4.68
0.760	0.850	1.378	13.568	0.853	3.08	6.79	2.03	4.70
0.693	0.822	1.382	14.372	0.923	3.05	6.75	2.00	4.71
0.627	0.800	1.385	15.405	0.965	3.01	6.70	1.98	4.73
0.599	0.794	1.386	15.952	0.975	2.98	6.68	1.95	4.75

From simple inspection of these tables, we notice that: 1) as expected, the flattening "e" of the star, defined as $e = (R_{eq} - R_p)/R_{eq}$ increases as the rotation velocity increases, a well-known result. However, the nuclear compressibility does not affect considerably the flattening but a small effect is seen, since at comparable values of Ω/Ω_K, models with softer equations of state are less flat than models built with higher nuclear compressibility. This is understandable since for the same baryonic mass hard equations of state produce extended models with lower central density, which feel more easily the rotation effects than compact stars resulting from soft equations of state. For instance, at $\Omega/\Omega_K \simeq 0.8$, the model computed with $K_N = 140$ MeV has $e \simeq 0.19$ while the model characterized by $K_N \simeq 1200$ MeV has $e \simeq 0.23$; 2) the central density of the models decreases slightly as the rotation velocity increases as well as the "softness" of the equation of state increases, which are also expected results; 3) concerning the mode frequencies, the general behavior is a small decrease with an increasing rotation. The effect of the "hardness" of the equation of state is noticed mainly in the first harmonic H_1 whose values at $\Omega/\Omega_K \simeq 0.8$ decreases from 8.5 kHz for the *eosA* model to 5.5 kHz for the *eosL* sequence. For the other modes, the decrease is less noticeable.

Effects of the nuclear compressibility on the mode frequency are better seen in a diagram in which frequencies of a given model sequence are plotted as a function of the compactness of the star defined as $GM_{gr}/R_{eq}c^2$ (see Figure 2). For the fundamental mode, the frequency grows as the compactness of the star increases as it can be seen on the left panel of Figure 2. The sequence of models having an intermediate compressibility value (K_N = 360 MeV) have frequencies higher that the other sequences while the "hardest" sequence (K_N = 1,200 MeV) has the lowest frequencies. The behavior of the first harmonic is more complex. Firstly, there is a clear dependence on the nuclear compressibility since the frequencies are smaller as the value of K_N increases for each sequence of models. Secondly, the frequency decrease with the compactness observed for the fundamental mode is not observed here: initially there is a slight decrease of the mode frequency as the compactness increases and then, beyond a critical compactness value that depends on K_N, the mode frequency increases with the star compactness. This behavior is more accentuated for the sequence corresponding to the hardest equation of state.

It is worth mentioning that significant differences exist between neutron star models built with polytropic and "realistic" equations of state. In the latter, there are surfaces defining changes in the stiffness of matter, in particular at the neutron drip density: the interior region of this surface constitutes the high density core while the exterior constitutes the low density envelope composed by neutron-rich nuclei. These distinct regions respond for the existence

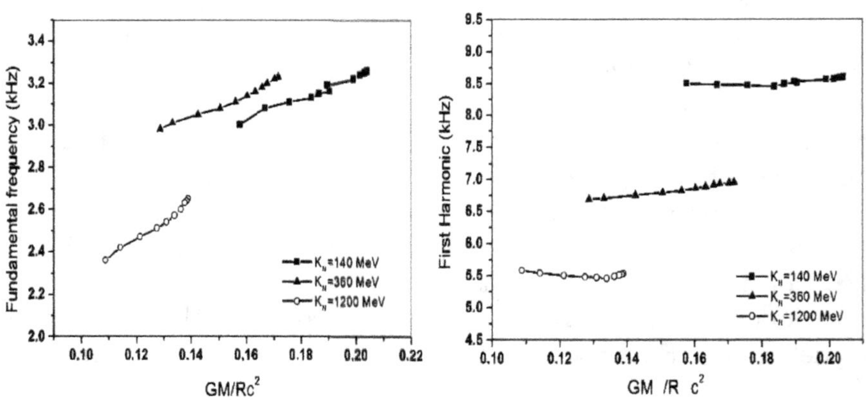

Fig. 2. The fundamental mode (a) and the first harmonic (b) frequencies as a function of the compactness of the star for model sequences characterized by a given nuclear compressibility.

of independent families of modes (Gondek and Zdunik, 1999). Radial-mode frequencies derived from non-rotating models computed with "realistic" equations of state and forming a sequence of increasing central densities (Kokkotas and Ruoff, 2001), indicate the following behavior of different modes: as the central density increases, the mode frequency increases until a critical density, whose value depends on the adopted equation of state and on the considered mode. For densities higher than the critical value, the frequencies decrease and, in the particular case of the fundamental mode, go to zero at a certain density corresponding to the onset of the radial instability. Results of our model computations indicate that frequencies of the fundamental quadrupole mode are smaller than frequencies of the fundamental F-mode, contrary to what occurs for polytropic models of comparable mass.

In Figure 3 are shown the frequency variation of the fundamental radial and quadrupole modes as a function of the central density for non-rotating neutron star models. In the left panel are displayed models built with a soft equation of state ($K_N = 140$ MeV) while on the right panel model sequences were computed with a stiff equation of state. F-modes data were taken from (Kokkotas and Ruoff, 2001) whereas quadrupole frequencies are those computed by

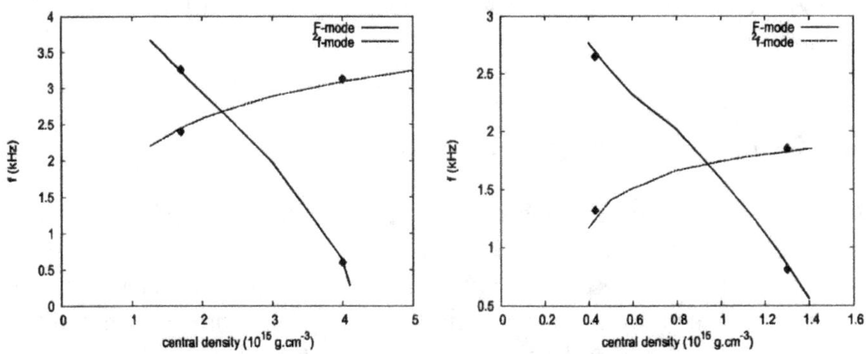

Fig. 3. Variation of the fundamental radial (F) and quadrupole (^2f) frequencies as a function of the central density. On the left panel models computed with a soft ($K_N = 140$ MeV) equation of state and on the right panel sequences computed with a stiff equation of state ($K_N = 1{,}200$ MeV). Curves correspond to non-rotating models computed by Kokkotas and Ruoff (2001); Lindblom and Deitweiler (1983), while diamonds correspond to low-rotation models computed in the present work.

Lindblom and Deitweiler (1983). Notice that in both cases there is a crossing between radial and quadrupole modes, indicating that at high densities the quadrupole frequency is higher than that of the fundamental mode. Although the present investigation concerns rapid rotating stars, we have plotted in Figure 3 frequency values derived from our lowest rotation models, which are indicated by filled diamonds. Notice that our low-rotation results agree with those previous calculations, confirming the reliability of our results. For polytropic models the situation is more complex since for l=2 modes (fundamental and first overtone) there is a slight increase of the frequency with rotation until $\Omega/\Omega_K \sim 0.6$ and then a decrease as the Keplerian limit is approached. Such a decrease with an increasing rotation is probably due to the decreasing central density and to the fact that quasi-radial modes are comparable to pure radial modes in stars with the same central density. However, is worth mentioning that: a) in this case, for polytropic models with the same central density, the F-mode frequency should remain practically constant and decrease up to values of the rotation velocity $\sim 0.8\Omega_K$; b) in the case of models built with "realistic" equations of state, the central densities are above the critical value defined previously, implying that for a slight decrease in the density, the mode frequency is expected to increase (see Figure 3). However, according to Hartle and Friedman (1975) such a frequency variation is probably related to the star compactness.

4. Gravitational Wave Emission

As discussed in the previous sections, quasi-radial oscillations present in rotating neutron stars can emit gravitational waves but an excitation mechanism is necessary. Once the external solid crust is formed, stresses induced by strong magnetic fields and rotation may trigger a tectonic activity. In this case, the stored elastic energy can be released as a consequence of star-quakes that may excite radial and non-radial oscillating modes. However, the maximum elastic energy channeled into oscillating modes is likely to be of the order of 10^{45} erg (de Freitas Pacheco, 1998a), restricting the detection only to galactic objects. Other more promising mechanisms able to excite neutron star oscillations have been proposed in the literature as the delayed collapse of the object resulting from the merger of two neutron stars (Shibata and Uryu, 2000) or quakes induced by a phase transition in the core (Cheng and Dai, 1998; Miniutti *et al.*, 2003). Examples of these phase transitions are the formation of a kaon condensate or the deconfinement of quarks (Glendenning, 2000). The later possibility is the scenario that will be explored in the present investigation.

In our picture, we have initially a low-mass binary system constituted by a red dwarf and a pure hadronic neutron star as sketched in Figure 4 (see, for instance, a recent review on these systems in Bahramian and Degenaar (2022)). In the first phase or (*phase I*), the red dwarf evolves, engulfs its Roche lobe and transfer mass to the companion. Since the gas flowing through the inner Lagrange point has a net angular momentum with respect to the neutron star, an accretion disk is formed, source of X-rays. In this process, the disk transfers mass and angular momentum to the compact object, producing an increase of the rotation velocity that may lead to the

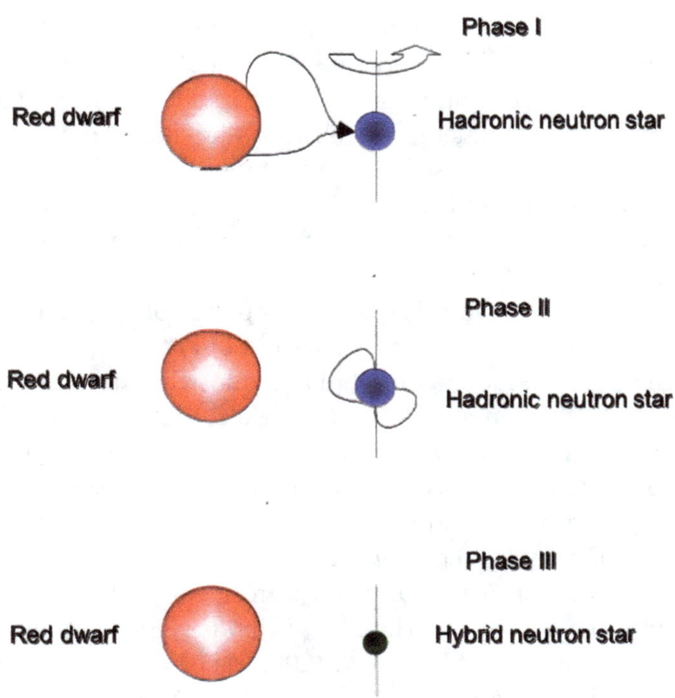

Fig. 4. Sketch of the three main evolutionary phases of a low-mass X-ray binary.

birth of a millisecond X-ray pulsar. At the beginning of phase I, the mass of the neutron star is not high enough and the central density is below the critical value at which quarks become deconfined. Moreover, during the lifetime of the accretion disk (about 10^8 yr) the neutron star may accrete an amount of mass of about 0.5 $M_\odot$. Such a mass increase is enough for the central density to attain the critical value at which the deconfinement would occur if the star had a low rotation. However, as we have seen, the central density decreases as the rotation increases. Hence, at the end of phase I the neutron star has a mass high enough to trigger the deconfinement in the core but its rotation velocity is sufficiently elevated to prevent the phase transition.

In phase II, the red dwarf has lost most of its external envelope and the mass transfer to the neutron star is very small. In this situation, the magnetic dipole emission mechanism becomes dominant in the balance of "gain and loss" of angular momentum and the rotation of the star is decelerated. As the rotation velocity decreases, the central density increases and reaches the critical density. When this occurs, a phase transition describing the quark deconfinement takes place in the core. The sudden softening of the equation of state leads to a new equilibrium configuration, which is more compact. In this mini-collapse, energy is released and non-radial oscillating modes are excited. In phase III, the binary system is constituted by an evolved red dwarf and a hybrid neutron star that is, a star with a small core constituted by a mix of quarks and an outer hadronic envelope.

The gravitational radiation from a rapidly rotating neutron star ($1.4\,ms < P < 1.0\,ms$) was investigated in Lin *et al.* (2006), who considered the collapse induced by a phase transition from nuclear to a mixed quark-hadron matter in the core. According to their hydrodynamic simulations, quasi-radial and quadrupole modes were excited, having comparable amplitudes. It should be emphasized that the analysis of the gravitational wave power spectrum derived from other simulations (Dimmelmeier *et al.*, 2006) indicates that the ^2f-mode emits an amount of energy under the form of gravitational waves higher by a factor of $\sim$20 than de F-mode. The damping of the oscillations estimated by Lin *et al.* (2006) is of the order of few milliseconds and is due essentially to the growth of the differential rotation and dissipation bu strong shock waves. This conclusion is not to much affected by the Newtonian character of their approach. Notice that Font *et al.* (2002) investigated the migration of a neutron star from an initial unstable state to a stable configuration concluding that oscillations are damped via shock heating.

4.1. *Phase transition parameters*

Above a certain critical nuclear density, quarks are expected to be deconfined from their hadronic "bags". The process is usually treated as a phase transition whose physical parameters are estimated by imposing the equality of pressure and chemical potentials describing both phases (in fact, for the chemical potential, the condition is $\mu_H = 3\mu_q$). Clearly, under these conditions, the parameters defining the phase transition depend on the adopted equations of state both for the hadronic and deconfined phases.

In the present calculations, the hadronic matter present in the core of the star is a mixture of electrons, protons, neutrons and Λ-hyperons. Interactions between nucleons are described by a two-body potential UV14 plus a three-body contribution UVII (Wiringa, Fiks, and Fabrocini, 1998). The "cold" quark matter is suppose to be composed by quarks u, d and s. The first two are assumed to be massless while the strange quark has a mass of 280 MeV. The phase transition parameters are listed in Table 4 and details of calculations can be found in de Freitas Pacheco (1998b). The critical hadron density at which the transition occurs is $n_t = 1.08\ fm^{-3}$, a value about 6.7 times the saturation density. Notice that there is a jump in the energy density, which is higher in the quark phase. The difference between both values corresponds to the latent heat of the transition, that is $L_t = (\varepsilon_q - \varepsilon_H) = 358\ MeV.fm^{-3}$. Hence, energy must be given to star in order that the phase transition could take place. In fact, since just after the phase transition the star contracts and a fraction of the released gravitational energy is used to provide the required latent heat as we shall see latter. The discontinuity in the energy density distribution may generate instabilities and in order that the star remains stable against radial perturbations, the Seidov criterion (Seidov, 1971) must be satisfied that is

$$\frac{\varepsilon_q}{\varepsilon_H} < \frac{3}{2}\left(1 + \frac{P_t}{\varepsilon_H}\right). \tag{32}$$

Table 4. Physical parameters of the deconfinement phase transition.

Phase	Pressure $(MeV.fm^{-3})$	Energy density $(MeV.fm^{-3})$	Chemical potencial (MeV)
Hadron	422.4	1270.0	1558.5
Quark	422.4	1628.8	519.5

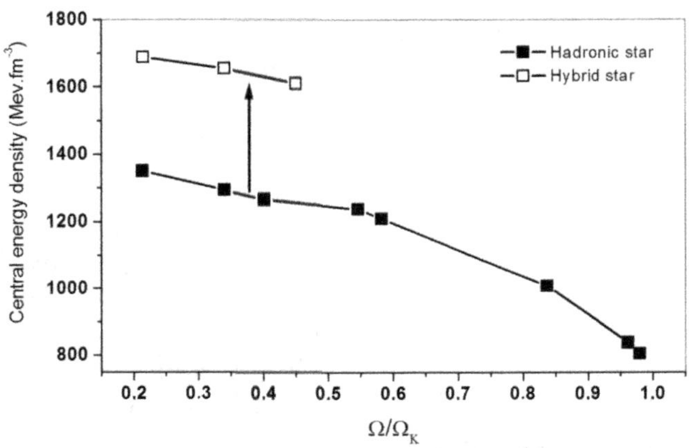

Fig. 5. Evolution of the central density as a function of the rotation velocity normalized at $\Omega_K = 8537\ s^{-1}$. The upper arrow indicates the point at which the phase transition occurs.

The Seidov criterion demands that the energy jump, measured by the ratio between the energy density of both phases at the transition point be smaller than the right side of Eq. 32. Using the values given in Table 4, it is trivial to shown that stars with those parameters at the critical surface are stable.

Using our code, a sequence of hadronic models with a baryonic mass $M_b = 1.965\ M_\odot$ were computed using the equation of state described above and with different rotation velocities. The sequence starts with models near the mass-shedding limit that have a rotation period of about 0.75 ms increasing up to 3.45 ms. A similar sequence of hybrid stars having a small deconfined core was also computed. The evolution of the central density as a function of the rotation velocity for both sequences is shown in Figure 5 (notice that the evolution goes from right to left, in the sense of an increasing rotation period). As the rotation period increases due to the magnetic dipole breaking, the central density also increases and when $P \simeq 2.16$ ms the phase transition takes place. If the magnetic field of the reaccelerated neutron star is about 10^9 G, then the magnetic breaking timescale is about 50 Myr and the time interval for the period to increase from 0.75 to 2.16 ms is about 360 Myr.

During the phase transition, the star contracts due to the formation of a small quark core whose mass is about 7.8×10^{-4} $M_\odot$, passing from an equatorial radius of 10.29 km to 10.18 km (a contraction of about 110 m). The gravitational mass decreases from 1.815 $M_\odot$ to 1.804 $M_\odot$, representing an energy release of 1.98×10^{52} erg. A fraction of this energy is used for the latent heat of the transition $E_t \simeq LM_c c^2/\varepsilon_q \simeq 3.0 \times 10^{50}$ erg, representing about 1.5% of the gravitational energy released during the collapse.

4.2. *Detection of the gravitational signal*

In the previous section, we have seen that taking into account the energy used for the transition, the remaining amount, i.e., about 1.95×10^{52} erg are dissipated either in the form of heat or gravitational waves. Adopting the same approach as in Marranghello *et al.* (2002), the fraction of the available energy channeled into gravitational waves (fundamental mode) is

$$f_g = \frac{1}{(1 + \frac{\tau_g}{\tau_h})}, \tag{33}$$

where τ_g is the gravitational damping timescale of the quasi-radial fundamental mode and τ_h is the timescale of heat dissipation by shock waves, which is about 20 ms according to simulations by Lin *et al.* (2006). For the gravitational damping, the same expression given by Marranghello *et al.* (2002) was adopted that permits to estimate $\tau_g \simeq 0.21s$. Under these conditions, from eq.33 one obtains $f_g \simeq 0.09$ or $E_g \simeq 1.75 \times 10^{51}$ erg.

Assume that the gravitational signal represented by the quasi-radial fundamental mode is simply modelled by

$$h(t) = h_0 e^{-t/\tau_g} \cos(\sigma t), \tag{34}$$

where h_0 is the signal amplitude and $\sigma = 2\pi f_0$ is related to the mode frequency. The signal amplitude depends on the energy E_g

channeled into the mode, the distance to the source D and on the damping timescale by emission of gravitational waves τ_g, namely

$$h_0 = \frac{2}{\pi D f_0}\left(\frac{GE_g}{\tau_g c^3}\right)^{1/2}. \tag{35}$$

After filtering the signal, the expected signal-to-noise (S/N) ratio is given by

$$\left(\frac{S}{N}\right)^2 = 4\int_0^\infty \frac{|\tilde{h}(f)|^2}{S_n(f)}df, \tag{36}$$

where $\tilde{h}(f)$ is the Fourier transform of the signal (see Eq. 34) and $S_n(f)$ is the noise power spectrum. of the detector, which is shown in Figure 6 both for the advanced LIGO and the planned European Einstein Telescope (Horta and Gair, 2010). Performing the calculations, the (S/N) ratio can be written as

$$\left(\frac{S}{N}\right)^2 = \frac{1}{5}h_0^2\left[\frac{\tau_g}{S_n(f_0)}\right]F, \tag{37}$$

where

$$F = \frac{4Q^2}{1+4Q^2}, \tag{38}$$

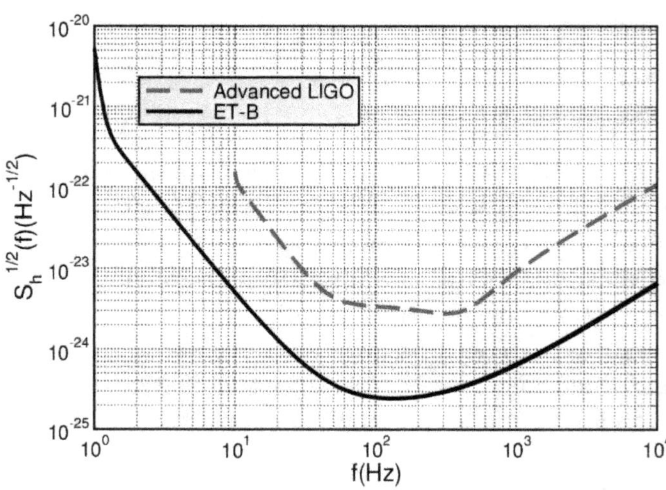

Fig. 6. Sensibility of the advanced LIGO laser interferometer and that of the planned European Einstein Telescope (version B).

and $Q = \pi f_0 \tau_g$ is the mode quality factor. The frequency of the fundamental mode f_0 computed from our code for the resulting hybrid star is 670 Hz, implying that $Q = 442$ and $F \simeq 1$. Combining equations 35 and 37, the S/N ratio can be rewritten as

$$\left(\frac{S}{N}\right)^2 = \frac{4}{5\pi^2 D^2} \frac{GE_g}{c^3 f_0^2 S_n(f_0)}. \tag{39}$$

Fixing a signal-to-noise ratio $S/N = 5$ for a trustful detection, the equation above permits to estimate the maximum distance at which the gravitational signal can be detected: for the ad-LIGO, D=11 Mpc while the ET will allow detection up to 114 Mpc.

The expected event rate is more difficult to be estimated since the rate of a mini-collapse in the Galaxy is quite uncertain. Estimates of accretion-induced collapse of white dwarfs into neutron stars (Fryer, Benz, Heraut, and Colgate, 1999) or population synthesis of low or intermediate-mass X-ray binaries (Pfhal, Rappaport, and Podsiadlowski, 2003) suggest galactic formation rates in the range $(1.0-4.5) \times 10^{-5}\ yr^{-1}$. The rate of mini-collapse of neutron stars forming hybrid stars is probably comparable to rates estimated for LMXB evolution or for accretion-induced collapse of white dwarfs. In this case, taking a rate of one event each 10^5 years per galaxy, the expected event rate within a volume of 114 Mpc radius that can be probed by the ET laser interferometer will be around three events each two years.

5. Conclusions

In the present chapter past investigations on radial and quasi-radial oscillation modes of neutron stars were reviewed, in particular the coupling of modes when the star rotates since in this case gravitational waves can be emitted. New calculations of axisymmetric modes of fully relativistic rotating stars are reported. The frequencies of the fundamental and of the first overtone of l=0 and l=2 modes of rotating neutron stars were computed by using a linear perturbation theory. Both fluid and metric perturbations were taken into account and the background stellar models were built by using a metric adequate to describe rapidly rotating stars. The different stellar models were computed with several "realistic" equations of state for the nuclear matter that are characterized by the nuclear compressibility. The present numerical code was checked by comparing mode frequencies resulting from neutron star models built with polytropic equations of state with mode frequencies issued from non-linear evolutionary simulations. Maximum deviations of about (9–10)% occur only for rotation velocities near the mass-shedding limit.

The present results indicate that the quasi-radial fundamental mode frequency increases with the compactness of star but the variation is less pronounced for the first harmonic. The effect of the nuclear compressibility is complex. Neutron star model sequences built with the **same baryonic mass** and with the "hardest" nuclear equation of state ($K_N = 1200\ MeV$) have fundamental mode frequency around ~2.6 kHz) while models built with softer equations of state ($K_N \sim 140{-}360 MeV$) have F ~ 3.2 kHz. A similar behavior is seen for quadrupole modes. The present investigation confirms that neutron star models built with "realistic" equations of state have quadrupole frequencies lower than those of quasi-radial modes, contrary to the results derived from models built with polytropic equations of state.

In LMXB systems the accreting neutron star may gain mass and angular momentum. In this situation the central density is close but below the critical value at which a hadron-quark phase transition occurs in the core. As the rotation period increases due to the magnetic dipole breaking, the central density also increases and attains the critical density, triggering a mini-collapse, consequence of the sudden softening of the nuclear equation of state. The pure hadronic star becomes a hybrid star constituted by a small quark core and a hadronic mantle. The modeling of such a scenario indicates that a hadronic star of mass 1.815 $M_\odot$ becomes a hybrid star of mass 1.804 $M_\odot$ but with same original baryonic mass. During the mini-collapse an amount of energy of about 1.98×10^{52} erg is released. However, as we have seen, only 8.9% of the released energy is channeled into gravitational waves. The expected gravitational signal could be detected up to distances of about 114 Mpc by the planned ET interferometer with an (uncertain) rate of about three events each two years.

Bibliography

Andersson, N. and Kokkotas, K. D. (1998). Pulsation modes for increasingly relativistic polytropes, *MNRAS* **297**, 493.

Bahramian, A. and Degenaar, N. (2022). Low-mass x-ray binaries in handbook of x-ray and γ-ray astronomy, arXiv:2206.10053.

Bardeen, J. M. (1970). A variational principle for rotating stars in general relativity, *ApJ* **162**, 71.

Bardeen, J. M., Thorne, K. S., and Meltzer, D. W. (1966). A catalogue of methods for studying the normal modes of radial pulsation of general-relativistic stellar models, *ApJ* **145**, 505.

Baym, G., Pethick, C., and Sutherland, P. (1971). The ground state of matter at high densities: Equation of state and stellar models, *ApJ* **170**, 299.

Baym, G. and Pethick, C. (1975). Neutron stars, *Ann. Rev. Nucl. Sci.* **25**, 27.

Burdyuzha, V., Kautz, V. L., and de Freitas Pacheco, J. A. (1996). Appearance and disappearance of the 511 keV line as a consequence of changes in the annihilation regime, *A&A* **313**, 485.

Carter, B. (1969). Killing horizons and orthogonally transitive groups in space-time, *J. Math. Phys.* **10**, 70.

Chandrasekhar, S. (1964). The dynamical instability of gaseous masses approaching the schwarzschild limit in general relativity, *ApJ* **40**, 417.

Chandrasekhar, S. and Friedman, J. L. (1972). On the stability of axisymmetric systems to axisymmetric perturbations in general relativity. II. A criterion for the onset of instability in uniformly rotating configurations and the frequency of the fundamental mode in case of slow rotation, *ApJ* **176**, 745.

Chandrasekhar, S. and Friedman, J. L. (1972). On the stability of axisymmetric systems to axisymmetric perturbations in general relativity. III. Vacuum metrics and carter's theorem, *ApJ* **177**, 745.

Chandrasekhar, S. and Ferrari, V. (1991). On the non-radial oscillations of slowly rotating stars induced by the lense-thirring effect, *Proc. Math. Phys. Sci.* **433**, 423.

Chandrasekhar, S. and Friedman, J. L. (1971). Criterion for the instability of a uniformly rotating configuration in general relativity, *Phys. Rev. Lett.* **26**, 1047.

Cheng, K. S. and Dai, Z. G. (1998). Gravitational waves from phase transitions of accreting neutron stars, *ApJ* **492**, 281.

Cook, G. B., Shapiro, S. L., and Teukolsky, S. A. (1994). Rapidly rotating neutron stars in general relativity: Realistic equations of state, *ApJ* **424**, 823.

Datta, B., Hasan, S. S., Sahu, P. K., and Prasanna, A. R. (1998). Radial modes of rotating neutron stars in the Chandrasekhar-Friedman formalism, *Int. J. Mod. Phys. D* **7**, 49.

de Freitas Pacheco, J. A. (1998a). Do soft gamma repeaters emit gravitational waves? *A&A* **336**, 397.

de Freitas Pacheco, J. A. (1998b). Equations of state for dense matter: Application to collapsed objects in structure and stability of nucleon and nuclear systems, A.A. Raduta, S. Stoica and I.I. Ursu, (eds.), World Scientific, Singapore. p. 547.

Dimmelmeier, H., Stergioulas, N., and Font, J. A. (2006). Non-linear axisymmetric pulsations of rotating relativistic stars in the conformal flatness approximation, *MNRAS* **368**, 1609.

Duchin, F. and Haensel, P. (2001). A unified equation of state of dense matter and neutron star structure, *A&A* **380**, 151.

Feynman, R. P., Metropolis, N., and Teller, E. (1949). Equations of state of elements based on the generalized fermi-thomas theory, *Phys. Rev.* **75**, 1561.

Font, J. A., Dimmelmeier, H., Gupta, A., and Stergioulas, N. (2001). Axisymmetric modes of rotating relativistic stars in the Cowling approximation, *MNRAS* **325**, 1463.

Font, J. A., Goodale, T., and Iyer, S. (2002). *et al.*, Three-dimensional numerical general relativistic hydrodynamics. II. Long-term dynamics of single relativistic stars, *Phys. Rev. D* **65**, 084024.

Friedman, J. L. and Schutz, B. F. (1975). On the stability of relativistic systems, *ApJ* **200**, 204.

Fryer, C., Benz, W., Heraut, M., and Colgate, S. A. (1999). What can the accretion-induced collapse of white dwarfs really explain? *ApJ* **516**, 892.

Glendenning, N. K. (2000). Compact stars: Nuclear physics, particle physics, and general relativity, Springer, New York.

Gondek, D. and Zdunik, J. L. (1999). Avoided crossings in radial pulsations of neutron and strange stars, *A&A* **344**, 117.

Haensel, P., Zdunik, J. L., and Schaeffer, R. (1989). Phase transitions in dense matter and radial pulsations of neutron stars, *A&A* **217**, 137.

Hartle, J. B. and Friedman, J. L. (1975). Slowly rotating relativistic stars. VIII. Frequencies of the quasi-radial modes of an $N = 3/2$ polytrope, *ApJ* **196**, 653.

Hartle, J. B., Thorne, K. S., and Chitre, S. M. (1972). Slowly rotating relativistic stars. VI. Stability of the quasiradial modes, *ApJ* **176**, 177.

Horta, E. A. and Gair, J. R. (2010). Intermediate-mass-ratio inspirals in the Einstein Telescope: signal-to-noise ratio calculations, *Phys. Rev. D* **83**, 044020.

Kojima, Y. (1992). Equations governing the nonradial oscillations of a slowly rotating relativistic star, *Phys. Rev. D* **46**, 4298.

Kojima, Y. (1992). Equations governing the nonradial oscillations of a slowly rotating relativistic star, *Phys. Rev. D* **46**, 4289.

Kokkotas, K. D. and Ruoff, J. (2001). Radial oscillations of relativistic stars, *A&A* **366**, 565.

Lin, L. -M., Cheng, K. S., Chu, M. -C., and Suen, W. -M. (2006). Gravitational waves from phase-transition-induced collapse of neutron stars, *ApJ* **639**, 382.

Lindblom, L. and Deitweiler, S. L. (1983). The quadrupole oscillations of neutron stars, *ApJS* **53**, 73.

Marranghello, G. F., Vasconcellos, C. A. Z., and de Freitas Pacheco, J. A. (2002). Phase transitions in neutron stars and gravitational wave emission, *Phys. Rev. D* **66**, 064027.

McDermott, P. M., van Horn, H. M., and Scholl, J. F. (1983). Nonradial g-mode oscillations of warm neutron stars, *ApJ* **268**, 837.

Miniutti, G., Pons, J. A., Berti, E., Gualtieri, L., and Ferrari, V. (2003). Non-radial oscillation modes as a probe of density discontinuities in neutron stars, *MNRAS* **338**, 389.

Moenchmeyer, R., Schaeffer, G., Mueller, G., and Kates, R. E. (1991). Gravitational waves from the collapse of rotating stellar cores, *A&A* **246**, 417.

Ott, C. D., Burrows, A., Dessart, L., and Livne, E. (2006). A new mechanism for gravitational-wave emission in core-collapse supernovae, *Phys. Rev. Lett.* **96**, 201102.

Pandharipande, V. R. (1971). *Hyperonic matter*, *Nucl. Phys. A* **178**, 123.

Pandharipande, V. R., Pines, D., and Smith, R. A. (1976). Neutron star structure: Theory, observation, and speculation, *ApJ* **208**, 550.

Pfhal, E., Rappaport, S., and Podsiadlowski, Ph. (2003). The galactic population of low- and intermediate-mass X-ray binaries, *ApJ* **597**, 1036.

Price, R. and Thorne, K. S. (1969). Non-radial pulsation of general-relativistic stellar models. II. Properties of the gravitational waves, *ApJ* **155**, 163.

Priou, D. (1992). The perturbations of a fully general relativistic and rapidly rotating neutron star. I — Equations of motion for the solid crust, *MNRAS* **254**, 435.

Ruderman, M. (1991). Neutron star crustal plate tectonics. I. Magnetic dipole evolution in millisecond pulsars and low-mass x-ray binaries, *ApJ* **366**, 261.

Ruderman, M. (1991). Neutron star crustal plate tectonics. II. Evolution of radio pulsar magnetic fields, *ApJ* **382**, 576.

Ruderman, M. (1991). Neutron star crustal plate tectonics. III. Cracking, glitches, and gamma-ray bursts, *ApJ* **382**, 587.

Schutz, B. F. (1972). Linear pulsations and starility of differentially rotating stellar models. II. General-relativistic analysis, *ApJ* **24**, 343.

Seidov, Z. F. (1971). The stability of a star with a phase change in general relativity theory, *Astro. Zhur.* **48**, 443.

Shibata, M. and Uryu, K. (2000). Simulation of merging binary neutron stars in full general relativity: $\Gamma = 2$ case, *Phys. Rev. D* **61**, 064001.

Stavridis, A., Passamonti, A., and Kokkotas, K. (2007). Nonradial oscillations of slowly and differentially rotating compact stars, *Phys. Rev. D* **75**, 064019.

Stergioulas, N. and Friedman, J. L. (1995). Comparing models of rapidly rotating relativistic stars constructed by two numerical methods, *ApJ* **444**, 306.

Stergioulas, N. and Friedman, J. L. (1998). Nonaxisymmetric neutral modes in rotating relativistic stars, *ApJ* **492**, 301.

Stergioulas, N. (2003). Rotating stars in relativity, *Liv. Rev. in Relat.* **6**, 3.

Thorne, K. S. and Campolattaro, A. (1967). Non-radial pulsation of general-relativistic stellar models. I. Analytic analysis for $L \geq 2$, *ApJ* **149**, 591.

Thorne, K. S. (1969). Nonradial pulsation of general-relativistic stellar models. III. Analytic and numerical results for neutron stars, *ApJ* **158**, 1.

Yoshida, S. and Eriguchi, Y. (2001). Quasi-radial modes of rotating stars in general relativity, *MNRAS* **322**, 389.

Zwerger, T. and Mueller, E. (1997). Dynamics and gravitational wave signature of axisymmetric rotational core collapse, *A&A* **320**, 209.

Wiringa, R. Fiks, V., and Fabrocini, A. (1988). Equation of state for dense nucleon matter, *Phys. Rev. C* **38**, 1010.

Part 5
Summary and Outlook

https://doi.org/10.1142/9789819800643_0025

Chapter 25

Summary and Outlook

Zhifu Gao[*,¶], **Renxin Xu**[†,‖], **Jorge Horvath**[‡,**], **and César Augusto Zen Vasconcellos**[§,††]

[*]*Xinjiang Astronomical Observatory, Chinese Academy of Sciences, Urumqi, 830011, Xinjiang, Peoples Republic of China*
[†]*School of Physics and State Key Laboratory of Nuclear Physics and Technology, Peking University, Beijing 100871, China*
Kavli Institute for Astronomy and Astrophysics, Peking University, Beijing 100871, China
[‡]*Departamento de Astronomia, Instituto de Astronomia, Geofísica e Ciências Atmosféricas (IAG), Universidade de São Paulo, R. do Matão 1226, C. Universitária, São Paulo 05508-090, Brazil*
[§]*Instituto de Fsica, Universidade Federal do Rio Grande do Sul (UFRGS), Porto Alegre, Brazil*
International Center for Relativistic Astrophysics Network (ICRANet), Pescara, Roma, Italy
[¶]*zhifugao@xao.ac.cn*
[‖]*r.x.xu@pku.edu.cn*
[**]*foton@iag.usp.br*
[††]*cesaraugustozenvasconcellos@gmail.com*

The state-of-the-art of China's astronomical instrumentation relevant to pulsars and recent scientific achievements are focused in this book. After a general review of pulsar astronomy, the chapters are categorized into facilities of either "existing" or "in progress". We summarize briefly the highlights obtained by China's astronomical mega-science projects, as well as prospects for the future development of pulsar astronomy in China.

Keywords: Pulsars, Neutron stars, Gravitational waves, Chinese astronomy, FAST, *Insight*-HXMT, TianMa, LHAASO, QTT, JRT, eXTP, GECAM, Hubbs

Contents

Introduction

China, as one of great civilizations in human history, has contributed significantly to historical astronomy, from planets to stars, especially the astonishing event dating back to 1054 AD when Chinese astronomers recorded a supernova explosion (the Crab pulsar as a rump left behind). Modern astronomy follows the establishment of quantum theory and relativity, targeting on extremely relativistic objects like neutron stars and black holes, as well as the whole Universe, with multi-messengers from full-band electromagnetic waves to neutrinos, cosmic rays and even gravitational waves. It is worth noting that, in addition to non-negligible effects of the general relativity, the equation of state of supra-nuclear matter inside neutron stars is still challenging, and pulsar astronomy is thus the focus of this book.

Pulsars have provided us with the opportunity to explore physics under extreme conditions since its discovery in 1967. With the rapid economic development, the Chinese government has organized to strengthen technological innovation and to accelerate the construction of astronomical facilities, including several large radio telescopes and a series of deep space exploration satellites in the next 10 years. These efforts would certainly be beneficial to pulsar-related sciences, understanding fundamental forces (gravity and strong interaction) and detecting nano-Hz gravitational waves, for instance.

We have spent a few years to prepare and edit this book. In September 2020, the world-renowned publisher, World Scientific Publishing Company, contacted Professor César Augusto Zen Vasconcellos from the International Center for Relativistic Astrophysics (ICRA, Italy) to plan the publication of a book dedicated to introducing China's major astronomical equipment and achievements in pulsar astronomy. Subsequently, César invited Professor Renxin Xu from Peking University, Professor Zhifu Gao from Xinjiang Astronomical Observatory of the Chinese Academy of Sciences, and Professor Jorge Ernesto Horvath from the Universidade de São Paulo in Brazil to form an editorial team. We began planning for the content and organizational structure of the book, focusing on China's existing facilities and scientific achievements, key facilities under construction in China, and some theoretical researches on pulsars, with a plan to complete all manuscripts and publishing within two years. More than twenty distinguished scientists and outstanding young scholars from the field of pulsar sciences accepted our editorial team's invitation and devoted themselves wholeheartedly to the writing and editing of the book. They come from the following universities and research institutions: Peking Universit; Xiamen University; Shanghai Jiao Tong University; Guangzhou University; Hubei University of Education; Xiangtan University; National Astronomical Observatories of the Chinese Academy of Sciences; Yunnan Observatories; Key Laboratory for Particle Astrophysics of the Institute of High Energy Physics, Beijing; Kavli Institute for Astronomy and Astrophysics, Peking University; Xinjiang Astronomical Observatory; Yunnan Observatories; Institute of High Energy Physics of the Chinese Academy of Sciences; Xinjiang Key Laboratory of Radio Astrophysics; School of Astronomy and Space Science from the University of Chinese Academy of Sciences, Beijing; Tsung-Dao Lee Institute from Shanghai Jiao Tong University; Shanghai Astronomical Observatory; University of Tübingen in Germany; Leibniz Institute for Astrophysics, Potsdam, Germany; Universidade de São Paulo, Brazil; Italian National Institute for Astrophysics (INAF), San Diego State University, USA; Center for Astrophysics and Space Sciences, University of California at San Diego, USA; Facultad de Ciencias Astronómicas y Geofísicas, Universidad Nacional de La Plata, Argentina; Consejo Nacional de Investigaciones Científicas y Técnicas (CONICET), Buenos Aires, Argentina; Catholic Institute of Technology, Cambridge, USA; Instituto de Física, Universidade Federal Fluminense, Niterói, Rio de Janeiro, Brazil; Núcleo de Astrofísica e Cosmologia (Cosmo-Ufes), Departamento de Física, Universidade Federal do Espírito Santo, ES, Brazil; Divisão de Astrofísica, Instituto Nacional de Pesquisas Espaciais, Brazil; Instituto Tecnológico de Aeronáutica, São José dos Campos, SP, Brazil; Instituto de Física, Universidade Federal do Rio Grande do Sul, Porto Alegre, Brazil; International Center for Relativistic Astrophysics Network (ICRANet), Rome, Pescara, Italy.

In January 2020, our editorial team sent out invitation letters to all authors of the book. Our initial plan was to complete all drafts within a year and a half and aim to officially publish the book by the end of 2022. However, due to the impact of the global COVID-19 pandemic, we had to postpone the publishing plan by one year and a few months. It is gratifying that, over the past three years, China has made some achievements in pulsar research as well as in the construction of large astronomical facilities. For example, due to its highest sensitivity in comparison to other facilities in the world, the Five hundred meter Aperture Spherical Telescope (FAST) has continuously discovered new pulsar-like celestial bodies, increasing the number of publicly released pulsars from 150 three years ago to more than 800 now. The Large High-Altitude Air Shower Observatory project (LHAASO), which was under construction three years ago but runs regularly now, has opened the window of the highest energy γ-ray astronomy at TeV energy. The Xinjiang Qitai 110 m Radio Telescope (QTT) project was conceived as an unproven scientific plan three years ago but has now become a real major scientific project under construction at the national level.

The book is divided into five parts, comprising 25 chapters. In the first part, we provide a review of pulsar astrophysics based on magnetars, isolated neutron stars, and radio pulsars. The second part of the book introduces four Chinese facilities that have been constructed and successfully operated: from FAST to *Insight*-HXMT, Tianma, and LHAASO. We believe that, through reading the chapters, readers will gain an understanding of the structure, performance, advantages, scientific goals, and latest scientific achievements of these devices. In the third part, we provide comprehensive introductions to five facilities currently under construction or planning: the QTT, JRT, the enhanced X-ray Timing and Polarimetry Mission (eXTP), the Gravitational wave high-energy Electromagnetic Counterpart All-sky Monitor (GECAM) and the Hot Universe Baryon Surveyor (HUBS). The fourth part provides a diversive approach to pulsars and compact stars science, covering their internal structure, equation of state, radiation mechanisms, and tests of gravity theories. In the fifth part, we summarize the entire book and provide a future outlook for pulsar research.

Below we give an overview of all the chapters for this book.

Review of Pulsar Astrophysics: Magnetars, Isolated Neutron Stars, and Radio Pulsars

1. Magnetars

Magnetars are believed to be neutron stars mainly powered by their huge magnetic fields, which can reach up to 10^{16} G. They are rare but very important for studying matter and radiation under extreme conditions, as well as the origin and evolution of various energetic events in the Universe, such as gamma-ray bursts, supernovae, fast radio bursts, and gravitational waves. The author of Chapter 1, Sandro Maregetti, reviews some of the recent results on magnetars, many of which have been achieved with the Chinese facilities described in this book. These include the identification of magnetar giant flares in other galaxies, the detection of polarization in their soft X-ray emission, and the confirmation of the link between magnetars and fast radio bursts. As stated by Sandro Maregetti, in addition to the significant contributions already obtained in this field of knowledge from Chinese facilities, more achievements are expected in the near future. Among the most promising fields, it is important to mention the observations in the radio band exploiting the great sensitivity of FAST, and further advances in X-ray polarimetry studies with the eXTP satellite.

2. High-Energy Emission in SGRs/AXPs

The authors of Chapter 2, Ronaldo V. Lobato, Jaziel G. Coelho, and Manuel M. Malheiro, address the high-energy emission in SGRs/AXPs, neutron stars with extreme magnetic fields, identified as magnetars. On basis on the outer gap accelerator model, the authors describe the generation of high-energy emission far away from the surface of the star in a combination of curvature radiation and inverse Compton scattering. According to the authors, the high-energy emission may be constrained by employing death-lines, the condition for pair-production and the generation of hard X and soft gamma rays, topics that deserve further investigation.

3. The Demographics of Isolated Neutron Stars from X-Ray Survey Data

The ROSAT mission discovered seven radio-quiet isolated neutron stars (INSs, M7) not very far from the Sun. Their X-ray emission is steady and thermal, but hard to detect since they have no counterparts at other wavelength bands. They are ideal laboratory for testing atmosphere models and thermal evolution of matter under extreme gravity and magnetic field. The M7 spin slower than standard radio pulsars. In Chapter 3, Adriana Mancini Pires reviews the X-ray properties of INSs, suggesting crust heating by field decay and a link with magnetars. They may be part of a large, hidden population of neutron stars, as only one candidate was known beyond 1 kpc. The eROSITA instrument on the Spectrum-RG mission will provide the largest catalogue of X-ray sources ever, with unprecedented sensitivity and resolution. This will enable a better sampling of the elusive INSs that are missed by other surveys. Searching for new thermally emitting INSs requires the use of the high X-ray-to-optical flux ratios and the absence of counterparts of the M7. Follow-up identification campaigns require state-of-the-art facilities. The new INSs will be younger and hotter than the M7 on average, and will be located near massive star associations, open clusters, and supernova remnants. New thermally emitting INSs are found: three candidates have no optical counterparts and a soft thermal spectrum like the M7. Follow-up observations using advanced facilities will reveal more on the INS candidates. The M7 shows spectral complexity and long-term variations, challenging the theory and modeling of magnetized neutron stars.

4. Evolution of Long-Period Compact Radio Sources Driven by Winds

In Chapter 4, Jorge E. Horvath, Lucas M. de Sá, Lívia S. Rocha, Gustavo Y. Chinen, Lucas G. Barão and Márcio G.B. de Avellar address the nature and evolution of the long-period compact star radio sources, which has recently added several unexpected members. The central hypothesis is that particle winds drive their evolution, being an important factor for these relatively old sources. The authors show the consistency of this picture and remark some unsolved problems and caveats within it.

Fig. 1. The Five hundred meter Aperture Spherical Telescope (FAST).

FAST: The Five Hundred Meter Aperture Spherical Telescope; HXMT: The Hard X-Ray Modulation Telescope; Tianma: The Tian Ma 65 m Radio Telescope; LHAASO: Large High Altitude Air Shower Observatory

5. The Fundamental Performance of FAST with 19-Beam Receiver at L Band

FAST is the world's largest single-dish radio telescope, which was opened in 2016 and the first new pulsar discovered by FAST was made in 2017. It has achieved breakthroughs in pulsar, spectral line, and fast radio burst researches. In Chapter 5, Peng Jiang provides a comprehensive account of FAST's design, site selection, construction, and maintenance. This chapter consists of five sections, each presenting a different aspect of FAST. The first section introduces FAST and its main features. The second section explains how FAST was designed and commissioned. The third section details the 19-beam L-band receiver, the primary instrument for observations with FAST. The fourth section describes the backends and the standing waves that affect the data quality. FAST has unparalleled sensitivity, pursues five major scientific goals and new frontiers, such as fast radio bursts and exoplanets. FAST has already produced remarkable results and discoveries. Its ultimate goal is to promote our understanding of the cosmos.

6. Single-pulse study of radio pulsars using FAST

Pulsars' single pulses are essential for understanding their radio emission. Scientists have been studying pulsars for over 50 years, but their emission process remains a mystery. The author of Chapter 6, Ji-Guang Lu, explains that this is because different pulsars have different emission properties, and the integrated pulse data are not enough to capture the full details of the emission. Certainly, the emission changes due to the pulsar's rotation, which adds another dimension to the problem. The single pulses are relatively random, showing the details of the particles in the magnetosphere. FAST is the world's best telescope for observing single pulses, as it has the highest signal-to-noise ratio. Ji-Guang Lu says that telescope sensitivity, not integration time, determines the quality of single pulse data. FAST is the most sensitive telescope and can get the best data. Pulsars' single pulses are both correlated and random. By analyzing these features from FAST data, we hope to solve the puzzle of pulsar emission.

7. FAST's Impact on Fast Radio Bursts

Fast Radio Bursts (FRBs) are short and intense radio flashes that have been a hot topic in astronomy since its discovery in 2007. In Chapter 7, Weiyang Wang and Kejia Lee review the current knowledge of FRBs and the CRAFTS survey conducted by FAST with high sensitivity. CRAFTS discovered five new FRBs, of which four were non-repeating and one was repeating. The non-repeating FRBs have large dispersion measures (DM) and are likely located at cosmological distances. The repeating FRB is very active and nearby, but had a complex environment. The polarization properties of FRBs have also been investigated with FAST. FRB 20180301A and FRB 20201124A show circular polarization and evidence of a magnetized medium along the line of sight. FRB 20201124A exhibits diverse morphologies and a 30-ms burst period. FRB 20200428A originates from a magnetar that transitioned to a radio pulsar. Its phase is different from the regular radio pulses, implying different emission regions for FRBs.

8. The Hard X-Ray Modulation Telescope (HXMT)

Chapter 8, written by Min-Yu Ge and Fang-Jun Lu, introduces *Insight*-HXMT, China's first space astronomy satellite. It is a versatile X-ray telescope that explores high-energy objects such as black holes and neutron stars. The authors provide an overview of its development, launch, parameters, objectives, proposals, plan, performance, and data. They also emphasize its contributions to pulsar astronomy. *Insight*-HXMT, which was proposed in 1994 and launched in 2017, has three X-ray telescopes that cover the energy range of 1–250 keV. The engineering team of *Insight*-HXMT has successfully accomplished all the predefined engineering tasks, such as

Fig. 2. The Hard X-ray Modulation Telescope (HXMT) and Electromagnetic Counterpart All-sky Monitor Mission (GECAM).

design, construction, ground calibration, and in-orbit testing. The in-orbit test results have shown that *Insight*-HXMT has met its design specifications in terms of effective area, energy resolution, and time resolution, laying a solid foundation for the research of high-energy celestial phenomena such as black holes and neutron stars, as well as the observation of intense short-duration events.

9. Advances in Neutron Stars from *Insight*-HXMT

In Chapter 9, Min-Yu Ge summarizes the achievements of *Insight*-HXMT in studying neutron stars. *Insight*-HXMT has discovered new cyclotron lines, including the highest-energy one at 146 keV in Swift J0243.6+6124, the first ultra-luminous X-ray source in the Hanoi catalog. It has also detected signs of a radiation pressure-dominated disk in this source. Additionally, *Insight*-HXMT has observed the Type I X-ray burst 4U1636-536 and found no hard X-rays between 40 and 70 keV. Furthermore, *Insight*-HXMT has identified the X-ray counterpart of FRB 200428 from a magnetar and analyzed the wide-band spectra of the Crab pulsar. The wide-band phase-resolved spectra of the Crab pulsar have also been investigated using *Insight*-HXMT's data. *Insight*-HXMT is currently in its extended mission phase, focusing on observations of neutron star-like objects, and more discoveries are expected.

10. HXMT Views on the Accreting Millisecond X-Ray Pulsars

In Chapter 10, Zhao-Sheng Li reviews how *Insight*-HXMT observes accreting X-ray pulsars. These pulsars are important for China's astronomical projects and for understanding binary evolution, burst oscillations, and recycling theory. Only 23 of them have been found, and their emission sites and timing noise are debated. *Insight*-HXMT measures their spin frequencies, magnetic fields, and nuclear burning effects. The observations by *Insight*-HXMT have significant implications for understanding the evolutionary mechanisms of pulsar binaries, and they will contribute to the preparatory research for China-led eXTP high-energy telescope system, which aims to measure the mass and radius of neutron stars.

11. Pulsars at Shanghai Tian Ma 65 m Radio Telescope

The Tianma, a 65-m radio telescope in Shanghai that operates from 1.3 to 50 GHz, is the topic of Chapter 11. In this chapter, the authors, Zhen Yan, Zhi-Qiang Shen, and Yajun Wu, Rongbing Zhao, and Qinghui Liu, address that the Tianma Radio Telescope has defined pulsar astronomy as one of its important scientific goals, and has carried out relevant observation technology and astrophysics research, considering the important physical significance and potential application value of pulsars. As reported in Chapter 11, a digital terminal system was introduced abroad and localized to meet the requirements of pulsar searching and online integration mode observations. A fast radio burst real-time search system and VLBI terminal were developed too. This chapter also reports on the Tianma's achievements in observing pulsars in the Galactic Center, monitoring magnetars powered with extremely large magnetic fields, discovering pulsar glitches (a sudden increase in spin-up), and improving the timing precision. The first high-frequency observations of multiple pulsars internationally, nearly doubling the integrated profiles of pulsars at 8.60 GHz are presented. Using the Tianma, the scientists are conducting research projects on high-frequency deep searches around the Galactic Center black hole and pulsar interstellar scintillation, aiming to achieve a series of breakthroughs in these fields.

Fig. 3. The Tian Ma 65 m Radio Telescope (TianMa).

12. The Large High Altitude Air Shower Observatory

The Large High-Altitude Air Shower Observatory (LHAASO), the new generation of cosmic ray observatory, is a national major scientific and technological infrastructure in China, with cosmic ray observation research as its core science. The authors of Chapter 12, Xin-Hua Ma and Zhen Cao, review the LHAASO experiments. In this chapter, the authors reveal that the LHAASO Mission has three arrays and other systems that cover 1.3 square kilometers, adopts various detection techniques, which can measure the reactions of gamma rays and cosmic rays in the atmosphere and reconstruct their

Fig. 4. Large High Altitude Air Shower Observatory (LHAASO).

basic information, and its unique multiple detection methods, cross-validation capabilities, and powerful gamma-ray and cosmic ray component discrimination abilities ensure the accuracy and reliability of the measurement results. They report that LHAASO can discover a large number of gamma-ray sources and achieve internationally leading sensitivity in the research of ultra-high-energy gamma-ray sources and cosmic ray component spectra and will provide an important experimental platform for conducting frontier interdisciplinary scientific research in areas such as atmosphere, environment, and space weather. LHAASO becomes a scientific base for multilateral international cooperation and high-level research.

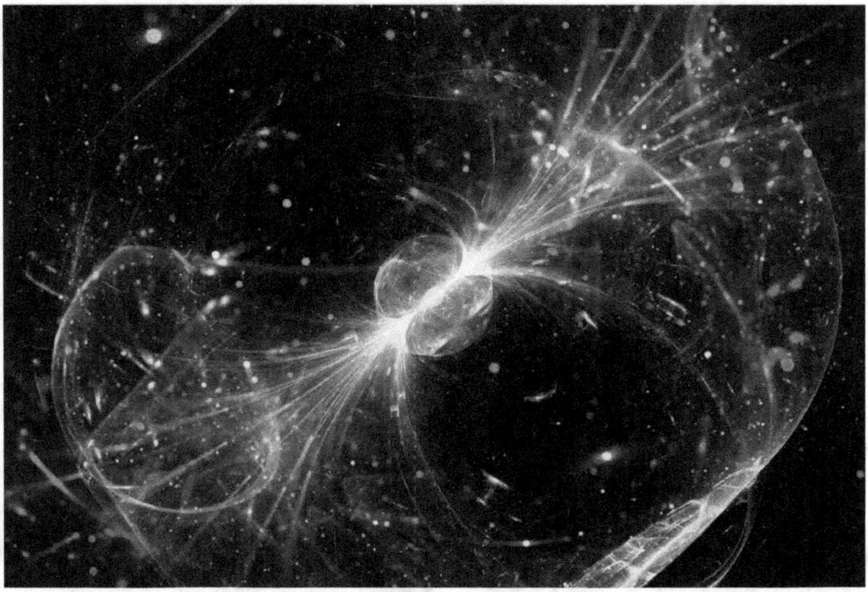

Fig. 5. Recently, scientists at the Large High Altitude Air Shower Observatory (LHAASO) have detected the highest-energy gamma rays from pulsars and PeVatrons sources in the Milky Way, such as the Crab Nebula. The image shows an artistic representation of gamma rays produced from the two steams of jets formed when supermassive stars collapse into themselves at the end of their lives. Image from https://newatlas.com.

13. The Highest Energy Gamma Ray from Pulsars Detected by LHAASO

A century after the discovery of cosmic rays, their origin remains a mystery. One of the main physical goals of LHAASO is to search for gamma-ray sources and discover the origin of cosmic rays from this research perspective. The authors of Chapter 13, Xin-Hua Ma and Zhen Cao summarizes the findings of LHAASO, a project that studies gamma-ray sources and cosmic rays. LHAASO has detected the highest-energy gamma rays from pulsars and other sources in the Milky Way, such as the Crab Nebula. These sources are called PeVatrons and can accelerate particles to very high energies. In this chapter, Xin-Hua Ma and Zhen Cao describe some of the PeVatrons discovered by LHAASO that may be related to pulsars, PWNs and TeV halos, including the Crab Nebula, LHAASO J0621+3755,

LHAASO J0621+3755, LHAASO J0341+5258. LHAASO entered the full array scientific operation stage in October 2021. It can be expected that LHAASO will discover more pulsars and other types of gamma-ray sources, and discover and observe super PeVatrons by extending the spectrum to several PeV or above, thus providing strong experimental evidence for the solution of the mystery of the origin of cosmic rays.

Fig. 6. Qitai Radio telescope (QTT).

QTT: QiTai Radio Telescope; JRT: Jingdong 120 m Radio Telescope; eXTP: X-Ray Timing and Polarimetry Mission; GECAM: Electromagnetic Counterpart All-sky Monitor Mission; Hubbs: Hot Universe Baryon Surveyor Space-Based X-Ray Telescope

14. The QiTai Radio Telescope (QTT)

In Chapter 14, Na Wang reviews the overall design scheme of the QiTai Radio Telescope (QTT), which is a fully steerable Gregorian type telescope with a main reflector diameter of 110 meters. The observing frequency range of QTT is 150 MHz to 115 GHz, and it is equipped with ultra-wideband, multi-beam and phased array feed receiver. Na Wang points out that QTT can observe more than three quarters of the sky, covering the 12° sky area to the south of the Galaxy center, and will become a world-class observation platform for Nanahertz gravitational wave detection through pulsar timing array, as well as the discovery of binary black hole systems, the exploration of dark matter and the origin of life in the universe. The chapter emphasizes that QTT will also improve the performance of Chinese and international VLBI networks, perform highly sensitive and high-resolution observations of galaxies and gravitational lensing systems, and help improve the accuracy of celestial reference systems for deep astrometric observations. In addition, QTT will provide strong technical support for future space activities such as exploration of planets and remote solar system, and become a scientific experiment platform for searching for extraterrestrial civilizations. Construction of QTT is scheduled for completion in 2028, and we look forward to many more exciting discoveries and results from QTT in the future.

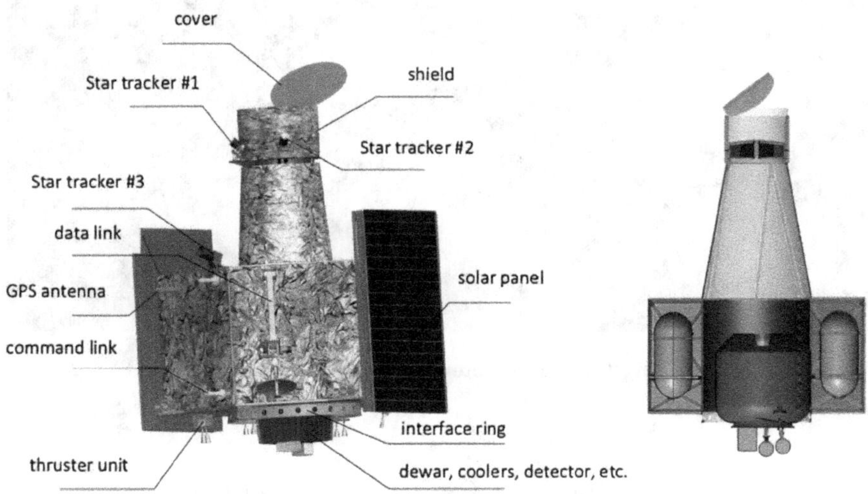

Fig. 7. Hot Universe Baryon Surveyor (HUBS).

15. Hot Universe Baryon Survey (HUBS)

Chapter 15 reviews the Hot Universe Baryon Survey (HUBS) project, which will search for the "missing" baryons in the universe with a high-resolution X-ray imaging satellite. It will also study the properties of neutron stars and other compact objects by observing their surface spectral lines. Zhao-Sheng Li claims that HUBS is a major scientific endeavor that will enhance our understanding of galaxy formation and evolution, and bridge the technological gaps in China. This chapter further explains how HUBS will determine the mass function, gravitational redshift, and dense matter state of neutron stars by observing the Doppler shifts and surface emissions of various atomic lines. These spectral lines will enable the estimation of the mass function of binary systems, the gravitational redshift at the surface of neutron stars, and the limits on the dense matter state.

Fig. 8. Jing Dong 120-meter Pulsar Radio Telescope (JRT).

16. The 120-m Jingdong Radio Telescope (JRT)

In Chapter 16, Longfei Hao, Min Wang, and Yonghua Xu review the overall design scheme of the Jingding Radio Telescope (JRT), which will be the world's largest single-dish decimeter-wave radio telescope for pulsar research. The JRT will be a key window for Chinese astronomy and provide more accurate pulsar data. In this chapter, the authors describes the JRT's construction, scientific goals, and related technologies, such as pulsar detection, radio telescope location, design schemes, and digital backends. This chapter also discusses other related research, such as the role of FAST in advancing pulsar science research, and the NANOGrav project, etc. In the future, radio astronomy research will achieve more breakthrough progress, such as more accurate observations and parameter measurements of pulsars, which will help to better understand the pulsar-related phenomenon in the universe. The completion of JRT will also help to deepen the understanding of the pulsar phenomenon and related astrophysical problems in the universe. At the same time, JRT will also become an important window for the field of Chinese astronomy, attracting more scientists and research institutions to participate in radio astronomy research. More exciting discoveries and achievements are expected in the future.

Fig. 9. Enhanced X-ray Timing and Polarimetry Mission (eXTP).

17. The Enhanced X-Ray Timing and Polarimetry Mission (eXTP)

In Chapter 17, Yu-Peng Xu, Y.P. Xu, A. Santangelo, and Shuang-Nan Zhang review the project of Enhanced X-ray Timing and Polarimetry Mission, (eXTP), which is China's next flagship X-ray satellite after *Insight*-HXMT. It aims to study extreme gravity, magnetic fields, and density of compact objects like black holes, neutron stars, or quark stars. Yu-Peng Xu, Y.P. Xu, A. Santangelo, and Shuang-Nan Zhang emphasize that eXTP will measure the mass and spin of many black holes, test general relativity and quantum theories in strong fields, determine the equation of state of compact objects, explore neutron matter or quark matter, and also monitor and follow up various explosive events. They describe that, with the support of the Strategic Priority Program on Space Science of the Chinese Academy of Sciences, eXTP has completed the key technology research and scheme design work by the end of 2022. The European cooperation team of eXTP has obtained the support from their countries, and carried out the Phase A and Phase B research work simultaneously. At present, eXTP has been included in the medium and long-term development plan of China's national space science, and is undergoing project evaluation, planning to be launched around 2029.

18. GECAM: A Gamma-Ray Monitor Constellation

The authors of Chapter 18, Shaolin Xiong, Yanqiu Zhang and Chenwei Wang, present the project called Gravitational Wave High-energy Electromagnetic Counterpart All-sky Monitor (GECAM). This project was proposed in 2016 and approved in 2018 by the Chinese Academy of Sciences, with the aim of exploring gravitational-wave gamma-ray bursts and other high-energy events. The GECAM-A and -B satellites were launched in 2020, and GECAM-C in 2022. Each satellite is equipped with both gamma-ray and electrically charged particle detectors. GECAM's energy range is about 6 keV to 5 MeV. The authors of this chapter highlight the scientific and technological achievements made by GECAM since its launch. The GECAM team found and measured the brightest gamma-ray burst ever observed (GRB 221009A), as well as the second fastest high-energy radio burst (FRB-XRB 221014) and a new high-energy radiation phenomenon from Earth. The project observed more than 1,000 explosions in orbit, including more than 100 gamma-ray bursts and magnetic explosions, issued more than 600 warnings, and conducted monitoring and joint observations with several different observation equipment. GECAM also used a new space-based high-energy radiation detector, a lanthanum bromide and SiPM detector, and realized real-time detection and warning of several high-energy explosion events.

On Pulsars and Compact Stars Science

19. Testing Relativistic Gravity with Radio Timing of Neutron Stars

Pulsars, as high-precision clocks, are in the strong gravitational field environment of neutron stars, providing an ideal laboratory for testing strong gravitational fields. The author of Chapter 19, Lijing Shao, introduces the non-perturbative gravitational behavior that may occur in neutron stars under strong gravitational fields from the perspective of strong-field gravity. If neutron stars undergo phase transitions such as strong-field scalarization, they will be reflected in the orbital dynamics of binary systems, and thus have the hope of being measured by high-precision pulsar timing. Lijing Shao explains that, if there is a cubic correction of the Riemann tensor in the gravitational interaction, it will also cause the violation of the strong equivalence principle in the strong gravitational field, changing the orbit of the binary motion. Under the precise timing observation of the new generation of radio telescopes, the strong-field gravity test will usher in a new era.

20. The Equation of State of Pulsars

In Chapter 20, Zhenyu Zhu, Zhiqiang Miao, and Ang Li review the equation of state (EOS) of pulsars. Neutron stars are the densest and smallest stars, with unclear interiors due to QCD's complexity. They are mostly nuclear matter with some electrons and muons, balanced by gravity and pressure. Hyperons or quark matter may exist in their center, forming hybrid stars. Quark stars are hypothetical objects with only deconfined quarks. Zhenyu Zhu, Zhiqiang Miao, and Ang Li explain that the QMF (quantum mean field) parameter space can be well constrained by mass, radius, and tidal deformability measurements by using QMF to model dense matter, and the hyperon puzzle may be solved, with hyperon stars' maximum mass of 2.2 solar masses using hypernuclei data and hyperon-nucleon interaction, They also introduce that the earlier EOS stiffening in quark stars affects the sound speed, showing the importance of quark pairs. However, the phase transition strength is not well-determined by global properties. The particle degree of freedom in cold, dense matter is also uncertain.

This could help us understand low-energy QCD or effective model parameters. Thus, it is crucial to combine efforts from different fields and discuss common interests and problems. Experimental nuclear physics provides data to test dense matter EOS theories in pulsars. The current and future multimessenger observatories (e.g., FAST, *Insight*-HXMT, GECAM, Tianma, LHAASO, eXTP, HUBS, Jingdong, QTT) will improve pulsar detection and measurement, as well as their evolution. Hopefully, the high-density EOS behavior can be determined soon, solving many open problems in nuclear physics and high-energy astrophysics.

21. Progress on the Study of Radio Emission Beam of Pulsars

The authors of Chapter 21, Hong-Guang Wang, Xiancong Wu, and Gaige Deng review the research progress on the radiation beam and radiation zone structures of radio pulsars, which emit radio beams from their poles and appear as pulses when they sweep across observers. Since observers only see part of the beam, it is difficult to determine the intensity, spectrum, polarization and structure of the radiation zone in the magnetosphere. The authors introduce two methods to study the beam structure: one is to use statistical data from many pulsars to construct a model, and the other is to use the precession effect of binary pulsars to scan the beam directly. The first method has various models, such as core hollow-cone, core conal-double, plaque, fan beam, etc. The second method is more recent and limited to a few pulsars. To infer the structure of the radiation zone from the beam structure, it is essential to understand the radiation geometry, magnetic field structure, relativistic effects, acceleration process and radiation mechanism in the magnetosphere.

22. The Inner Structure of Pulsars

In Chapter 22, Xiao-Yu Lai explores the challenging inner structure and EOS of pulsar-like compact objects and introduces a strangeon star model, which differs from neutron star and quark star models.

Xiao-Yu Lai explains that both perturbative and non-perturbative QCD effects may cause quark flavour maximization and quark localization inside strangeons, as well as a hard core repulsion between strangeons, and the strangeon star EOS should be very stiff with the lowest energy per baryon if quarks are localized and a hard core repulsion exists. This chapter also summarizes that various manifestations of pulsar-like compact stars, from surface and global properties to the kilonova radiation of merging binaries, will be explained by strangeon stars, and advanced facilities may confirm the existence of strangeon matter, especially in the multimessenger astronomy era. Research progress about radiation beam and radiation zone structures of radio pulsars.

23. Fast Pulsars, Neutron Stars, and Astrophysical Strange Quark Matter Objects

In Chapter 23, Delaney Farrell, Fridolin Weber, Milva G. Orsaria, Ignacio F. Ranea-Sandoval, Martín Canullán, Rodrigo Negreiros, Takashi Katayama, and Ian Wagaman explore key aspects of neutron stars, pulsar glitches, tidal deformability, fast pulsars, the equation of state, and strange quark matter stars. Challenges in directly measuring neutron star radius have led to reliance on spectroscopic and timing techniques, with uncertainties addressed by the authors through careful source selection and theoretical modeling. The authors emphasize that Pulsar glitches reveal insights into the equation of state through angular momentum transfer within the neutron star and that Tidal deformability is crucial in gravitational-wave astronomy, exemplified by the GW170817 event. Fast pulsars, instrumental in astrophysical testing, are classified by the authors into ordinary pulsars, millisecond pulsars, and magnetars. The EOS, a vital ingredient for understanding neutron star internal structure, are explored through various models. The chapter delves into the theoretical framework for rotating neutron stars, addressing uniform and differential rotation scenarios and their impacts on mass and radius. Additionally, the intriguing concept of quark stars and strange dwarfs is investigated. The various topics discussed in this book chapter contribute to a broader understanding of dense matter physics, astrophysical phenomena, and the potential for transformative discoveries through advanced observational techniques and technologies like gravitational wave detectors, radio telescopes, and X-ray telescopes.

24. Axisymmetric Oscillations of Rotating Neutron Stars: Gravitational Wave Emission

Stéphane Vincent and José A. de Freitas Pacheco review, in Chapter 24, review quasi-radial oscillations of rotating neutron stars and present new calculations of axisymmetric modes based on linearized relativistic field equations that take into account both fluid and metric perturbations. Their numerical procedure was checked by comparing results derived from the present code with relativistic hydrodynamic calculations and adopting a common polytropic equation of state. Such a comparison indicates that mode frequencies differ by less than 2% excepting for rotation velocities near the mass shedding limit, where differences increase up to 10%. In this chapter, rapidly rotating neutron star models were computed for different "realistic" equations of state characterized by nuclear compressibilities ranging from 140 up to 1200 MeV. The fundamental and the first overtone frequencies of quasi-radial and quadrupole modes were calculated, by Stéphane Vincent and José A. de Freitas Pacheco, for a sequence of neutron star models having the same baryonic mass ($M_b = 1.5\ M_\odot$) but different rotation velocities. The present results confirm that quadrupole frequencies are lower than those of quasi-radial modes, contrary to results derived from models built with polytropic equations of state. Excitation of quasi-radial modes and gravitational wave emission were considered in a scenario in which a pure hadronic star suffers a mini-collapse due to a hadron-quark phase transition in the core. The released gravitational energy, which depends on the adopted equation of state for the nuclear matter, is of the order of $1.98 \times 10^{52}\ erg$ for the considered neutron star model. However, only 8.9% is channeled into gravitational waves. Nevertheless, the planned European gravitational Einstein Telescope would be able to detect such a signal up to distances of about 114 Mpc.

Conclusion

In a word, we have summarized the current and future important astronomical facilities in China, including FAST, *Insight*-HXMT, Tianma, QTT, JRT, eXTP and LHAASO, as well as their scientific objectives, achievements and significance. Chinese astronomers and engineers have already made great efforts to build a number of world-class astronomical facilities, and to contribute in revealing the mysteries of the Universe.

In recent years, the process of detecting pulsars and our understanding of these objects subject to extreme conditions has evolved substantially. The study of pulsars implies knowledge in several areas of fundamental physics, such as general relativity, particle physics, nuclear physics, condensed matter physics, processes that involve radiation of electric and magnetic fields under extreme physical conditions. The pulsar observation facilities built by China are among the most advanced in the world, revolutionizing the search for these mysterious objects thus ushering in a new era of discovery for the field of compact stars, astrophysics and cosmology. Therefore, we may anticipate an exciting period of pulsar astronomy to come.

Acknowledgments

Zhifu Gao is grateful to the National Key Research and Development Program of China (No. 2022YFC2205202), and the Natural Science Foundation of Xinjiang Uygur Autonomous Region (2022D01A155).

https://doi.org/10.1142/9789819800643_bmatter

Index

S

Z

www.ingramcontent.com/pod-product-compliance
Lightning Source LLC
LaVergne TN
LVHW010323220926
848484LV00004BB/834
* 9 7 8 9 8 1 9 8 0 0 6 3 6 *